Fourth Edition

COMPUTATIONAL FLUID DYNAMICS VOLUME I

KLAUS A. HOFFMANN

STEVE T. CHIANG

A Publication of Engineering Education System™, Wichita, Kansas, 67208-1078, USA

www.EESbooks.com

ISBN 0-9623731-0-9
Second Print : January 2004

This book is typeset by Jeanie Duvall dba SciTech Computer Typesetting of Austin, Texas.

To obtain information on purchasing this or other texts published by EES, please write to:

Engineering Education System™
P.O. Box 20078
Wichita, KS 67208-1078
USA

Or visit:

www.EESbooks.com

CONTENTS

Chapter Four: Stability Analysis 113

Chapter Five: Elliptic Equations 152

Chapter Six: Hyperbolic Equations 185

Appendices

PREFACE

This three-volume text is designed for use in introductory, intermediate, and advanced courses in computational fluid dynamics (CFD) and computational fluid turbulence (CFT). The fundamentals of computational schemes are established in the first volume, presented in nine chapters. The first seven chapters include basic concepts and introductory topics, whereas Chapters 8 and 9 cover advanced topics. In the second volume, the fundamental concepts are extended for the solution of the Euler, Parabolized Navier-Stokes, and Navier-Stokes equations. Finally, unstructured grid generation schemes, finite volume techniques, and finite element method are explored in the second volume. In the third volume, turbulent flows and several computational procedures for the solution of turbulent flows are addressed.

The first two volumes are designed such that they can be easily adapted to two sequential courses in CFD. Students with an interest in fluid mechanics and heat transfer should have sufficient background to undertake these courses. In addition, fundamental knowledge of programming and graphics is essential for the applications of methods presented throughout the text. Typically, the first course is offered at the undergraduate level, whereas the second course can be offered at the graduate level. The third volume of the text is designed for a course with the major emphasis on turbulent flows.

The general approach and presentation of the material is intended to be brief, with emphasis on applications. A fundamental background is established in the first seven chapters, where various model equations are presented, and the procedures used for the numerical solutions are illustrated. For purposes of analysis, the numerical solutions of the sample problems are presented in tables. In many instances, the behavior of a solution can be easily analyzed by considering graphical presentations of the results; therefore, they are included in the text as well. Before attempting to solve the problems proposed at the end of each chapter, the student should try to generate numerical solutions of the sample problems, using codes developed individually or available codes modified for the particular application. The results should be verified by comparing them with the solutions presented in the text. If an analytical solution for the proposed problem is available, the numerical solution should be compared to the analytical solution.

The emphasis in the first volume is on finite difference methods. Chapter 1 classifies the various partial differential equations, and presents some fundamental concepts and definitions. Chapter 2 describes how to achieve approximate representation of partial derivatives with finite difference equations. Chapter 3 discusses procedures for solving parabolic equations. Stability analysis is presented in Chapter 4. The order for Chapters 3 and 4 can be reversed. In fact, the results of stability analysis are required for the solution of parabolic equations in Chapter 3. The reason that the solution procedure of parabolic equations is developed first in Chapter 3 is to spread the computer code developments, since they require a substantial amount of time compared to other assignments. This will prevent the concentration of code development in the latter part of the course. Procedures for solving elliptic and hyperbolic partial differential equations are presented in Chapters 5 and 6, respectively. Chapter 7 presents a scalar model equation equivalent of the Navier-Stokes equations. In this chapter numerical algorithms are investigated to solve a scalar model equation which includes unsteady, convective, and diffusive terms.

The solution schemes established in the first seven chapters are extended to the solution of a system of partial differential equations in Chapter 8. In particular, the Navier-Stokes equations for incompressible flows in primitive variables, as well as vorticity-stream function formulations, are reviewed. Subsequently, the numerical schemes and specification of appropriate boundary conditions are introduced. Finally, Chapter 9 is designed to introduce the structured grid generation techniques. Various schemes, along with applications, are illustrated in this chapter.

While every attempt has been made to produce an error-free text, it is inevitable that some errors still exist. The authors would greatly appreciate the reader's input on any corrections, so that they may be incorporated into future printings. Furthermore, we would appreciate any comments and/or suggestions from the readers on the improvement of the text. Please forward your comments by mail to:

Klaus Hoffmann
P.O. Box 20078
Wichita, KS 67208-1078
or e-mail to: **Klaus.Hoffmann@Wichita.edu**

In addition to this three-volume text, *Computational Fluid Dynamics,* a three-volume text, *Student Guide to CFD,* has been developed. The text, *Student Guide to CFD,* includes computer codes, description of input/output, and additional example problems. However, it is important to emphasize that computer code development is an important aspect of CFD, and that, in fact, one learns a great deal about the numerical schemes and their behaviour as one develops, debugs, and validates his or her own computer code. Therefore, it is important to state here that the computer codes provided in the text *Student Guide to CFD* should not be used as an avenue to replace that aspect of CFD and that code development must be an important objective of the learning process. However, these codes can be used as a basis upon which one may develop other codes, or the codes can be modified for other applications.

The authors greatly appreciates the support and help of many friends and colleagues — in particular, Dr. John Bertin and Dr. James Forsythe of the U.S. Air Force Academy, Dr. Walter Rutledge of Sandia National Laboratories; Dr. Dennis Wilson and Dr. Douglas Cline of The University of Texas at Austin; Dr. Shamoun Siddiqui of the Ministry of Defense, Pakistan; Mr. John Buratti of IBM; Mr. Shigeki Harada of Hewlett-Packard, Japan; Dr. Yildirim B. Suzen of University of Kentucky; Mr. Apichart Devahastin; Mr. Jean-Francois Dietiker; and Mr. Henri-Marie Damevin of Wichita State University. Furthermore, we are indebted to many of our students at The University of Texas, The Wichita State University, and those who have participated in various CFD correspondence and short courses offered by AIAA, EES, and ASME.

Finally, we greatly appreciate the efforts of Mrs. Karen Rutledge for editorial work, Mr. Tim Valdez for art work, and Ms. Jeanie Duvall for her skillful typing of the manuscript.

Klaus A. Hoffmann
Steve T. Chiang

INTRODUCTION

The task of obtaining solutions to the governing equations of fluid mechanics represents one of the most challenging problems in science and engineering. In most instances, the mathematical formulations of the fundamental laws of fluid mechanics are expressed as partial differential equations (PDE). Second-order partial differential equations appear frequently and, therefore, are of particular interest in fluid mechanics and heat transfer. Generally, the governing equations of fluid mechanics form a set of coupled, nonlinear PDEs which must be solved within an irregular domain subject to various initial and boundary conditions.

In many instances, analytical solutions of the equations of fluid mechanics are limited. This is further restricted due to the imposed boundary conditions. For example, a PDE subject to a Dirichlet boundary condition (i.e., values of the dependent variable on the boundary are specified) may have an analytical solution. However, the same PDE subject to a Neumann boundary condition (where normal gradients of the dependent variable on the boundary are specified) may not have an analytical solution.

Experimental fluid mechanics can provide some information regarding a particular flowfield. However, the limitation on the hardware, such as the model and tunnel size and the difficulty in adequately simulating the prototype flowfield, makes it an impractical means of obtaining flowfields for many problems. Nevertheless, the flowfield information from experiments is valuable in validating mathematical solutions of the governing equations. Thus, experimental data is used along with computational solutions of the equations for design purposes.

A technique that has gained popularity in recent years is computational (numerical) fluid dynamics. Of course, numerical analysis has been around for many years. However, improvements in computer hardware, resulting in increased memory and efficiency, have made it possible to solve equations in fluid mechanics using a variety of numerical techniques. These advancements have stimulated the introduction of newer numerical techniques which are being proposed almost on a daily basis. Unlike experimental fluid mechanics, the geometry and flow conditions can be easily varied to obtain various design goals. The solution that any such numerical program generates should be validated by comparing it to a set of experimental data; but once its validity has been established, the program can be used for various design purposes, within the limits imposed by the assumptions on which it was based.

The fundamental concept of numerical schemes is based on the approximation of partial derivatives by algebraic expressions. Once the partial differential equation has been approximated by an algebraic equation, it can be solved numerically with the aid of a computer. The schemes by which the approximations to partial differential equations can be develoepd may be categorized into three groups. They are: finite difference (FD) methods, finite volume (FV) methods, and finite element (FE) methods. The finite difference methods are used in conjunction with structured grids, whereas the finite volume or finite element methods are typically used in conjunction with unstructured grids. The emphasis in this text is on finite difference methods, even though both finite volume and finite element methods are introduced in Volume II.

In the first seven chapters, we will explore the fundamental concepts of numerical methods used to solve PDEs, investigate how various methods are to be applied to the proposed model equations, and analyze the resulting solutions. In Chapter 8, the concepts of the computational schemes for the solution of a system of PDEs are explored.

The numerical procedures introduced in the first volume are extended to the solutions of Euler, Parabolized Navier-Stokes, and Navier-Stokes equations in Volume II. Furthermore, unstructured grid generation schemes, finite volume schemes, and finite element method are explored in the second volume of the text.

Turbulent flows and numerical considerations for the solution of turbulent flows are provided in Volume III. Fundamental concepts and definitions are established in Chapter 20. Subsequently, the modification of the Navier-Stokes equation to include the effect of turbulence and turbulence models is introduced in Chapter 21. Compact finite difference formulation is developed in Chapter 22. Finally, Large Eddy Simulation and Direct Numerical Simulation are discussed in Chapter 23.

Chapter 1

Classification of Partial Differential Equations

1.1 Introductory Remarks

Since the solution procedure of a partial differential equation (PDE) depends on the type of the equation, it is important to study various classifications of PDEs. Imposition of initial and/or boundary conditions also depends on the type of PDE. Most of the governing equations of fluid mechanics and heat transfer are expressed as second-order PDEs and therefore classification of such equations is considered in this chapter. In addition, a system of first-order PDEs and a system of second-order PDEs are considered as well.

1.2 Linear and Nonlinear PDEs

Partial differential equations can be classified as linear or nonlinear. In a linear PDE, the dependent variable and its derivatives enter the equation linearly, i.e., there is no product of the dependent variable or its derivatives. Individual solutions of this type of PDE can be superimposed, e.g., two solutions to the governing equation can be added together to give a third solution to the original equation. An example of a linear PDE is the one-dimensional wave equation

$$\frac{\partial u}{\partial t} = -a\frac{\partial u}{\partial x}$$

where a is the speed of sound which is assumed constant.

On the other hand, a nonlinear PDE contains a product of the dependent variable and/or a product of its derivatives. Two solutions to a nonlinear equation cannot be added to produce a third solution that also satisfies the original equation. An example of a nonlinear PDE is the inviscid Burgers equation:

$$\frac{\partial u}{\partial t} = -u\frac{\partial u}{\partial x}$$

If a PDE is linear in its highest order derivatives, it is called a quasi-linear PDE.

1.3 Second-Order PDEs

To classify the second-order PDE, consider the following equation

$$A\frac{\partial^2 \phi}{\partial x^2} + B\frac{\partial^2 \phi}{\partial x \partial y} + C\frac{\partial^2 \phi}{\partial y^2} + D\frac{\partial \phi}{\partial x} + E\frac{\partial \phi}{\partial y} + F\phi + G = 0 \tag{1-1}$$

where, in general, the coefficients A, B, C, D, E, F, and G are functions of the independent variables x and y and of the dependent variable ϕ. Assume that $\phi = \phi(x, y)$ is a solution of the differential equation. This solution describes a surface in space, on which space curves may be drawn. These curves patch various solutions of the differential equation and are known as the characteristic curves. Some fundamental concepts of characteristics are provided in Appendix A.

By definition, the second-order derivatives along the characteristic curves are indeterminate and, indeed, they may be discontinous across the characteristics. However, no discontinuity of the first derivatives is allowed, i.e., they are continuous functions of x and y. Thus, the differentials of ϕ_x and ϕ_y, which represent changes from location (x, y) to $(x + dx,\ y + dy)$ across the characteristics, may be expressed as

$$d\phi_x = \frac{\partial \phi_x}{\partial x}dx + \frac{\partial \phi_x}{\partial y}dy = \frac{\partial^2 \phi}{\partial x^2}dx + \frac{\partial^2 \phi}{\partial x \partial y}dy \tag{1-2}$$

and

$$d\phi_y = \frac{\partial \phi_y}{\partial x}dx + \frac{\partial \phi_y}{\partial y}dy = \frac{\partial^2 \phi}{\partial x \partial y}dx + \frac{\partial^2 \phi}{\partial y^2}dy \tag{1-3}$$

The original equation, i.e., Equation (1-1), may be expressed as follows

$$A\frac{\partial^2 \phi}{\partial x^2} + B\frac{\partial^2 \phi}{\partial x \partial y} + C\frac{\partial^2 \phi}{\partial y^2} = H \tag{1-4}$$

where

$$H = -\left(D\frac{\partial \phi}{\partial x} + E\frac{\partial \phi}{\partial y} + F\phi + G\right)$$

Now Equation (1-4), along with Equations (1-2) and (1-3), can be solved for the second-order derivatives of ϕ. For example, using Cramer's rule,

$$\frac{\partial^2 \phi}{\partial x \partial y} = \frac{\begin{vmatrix} A & H & C \\ dx & d\phi_x & 0 \\ 0 & d\phi_y & dy \end{vmatrix}}{\begin{vmatrix} A & B & C \\ dx & dy & 0 \\ 0 & dx & dy \end{vmatrix}} \tag{1-5}$$

Since it is possible to have discontinuities in the second-order derivatives of the dependent variable across the characteristics, these derivatives are indeterminate. Thus, setting the denominator equal to zero,

$$\begin{vmatrix} A & B & C \\ dx & dy & 0 \\ 0 & dx & dy \end{vmatrix} = 0 \tag{1-6}$$

yields the equation

$$A\left(\frac{dy}{dx}\right)^2 - B\left(\frac{dy}{dx}\right) + C = 0 \tag{1-7}$$

Solving this quadratic equation yields the equations of the characteristics in physical space:

$$\left(\frac{dy}{dx}\right)_{\alpha,\beta} = \frac{B \pm \sqrt{B^2 - 4AC}}{2A} \tag{1-8}$$

Setting the numerator of (1-5) equal to zero provides a set of characteristic curves in the ϕ_x, ϕ_y plane. These are known as hodograph characteristics. Depending on the value of $(B^2 - 4AC)$, characteristic curves can be real or imaginary. For problems in which real characteristics exist, a disturbance can propagate only over a finite region, as shown in Figure 1-1. The downstream region affected by a disturbance at point A is called the zone of influence (indicated by horizontal shading). A signal at point A will be felt only if it originated from a finite region called the zone of dependence of point A (vertical shading).

The second-order PDE previously expressed as Equation (1-1) is classified according to the sign of the expression $(B^2 - 4AC)$. It will be

(a) elliptic if $B^2 - 4AC < 0$

(b) parabolic if $B^2 - 4AC = 0$ or

(c) hyperbolic if $B^2 - 4AC > 0$

Note that the classification depends only on the coefficients of the highest order derivatives.

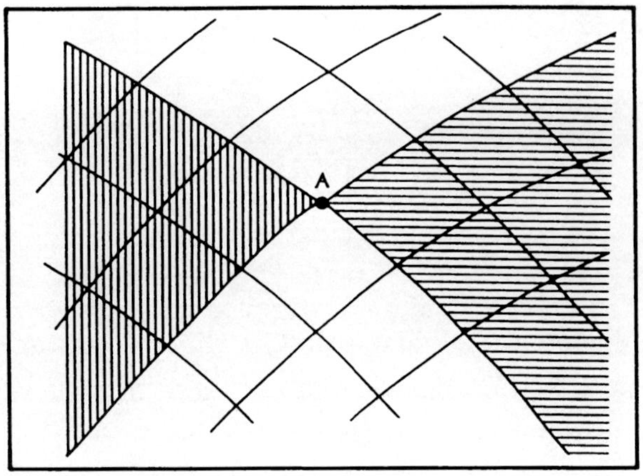

Figure 1-1. Zone of influence (horizontal shading) and zone of dependence (vertical shading) of point A.

1.4 Elliptic Equations

A partial differential equation is elliptic in a region if $(B^2 - 4AC) < 0$ at all points of the region. An elliptic PDE has no real characteristic curves. A disturbance is propagated instantly in all directions within the region. Examples of elliptic equations are Laplace's equation

$$\frac{\partial^2 \phi}{\partial x^2} + \frac{\partial^2 \phi}{\partial y^2} = 0 \tag{1-9}$$

and Poisson's equation

$$\frac{\partial^2 \phi}{\partial x^2} + \frac{\partial^2 \phi}{\partial y^2} = f(x, y) \tag{1-10}$$

The domain of solution for an elliptic PDE is a closed region, R, shown in Figure 1-2. On the closed boundary of R, either the value of the dependent variable, its normal gradient, or a linear combination of the two is prescribed. Providing the boundary conditions uniquely yields the solution within the domain.

1.5 Parabolic Equations

A partial differential equation is classified as parabolic if $(B^2 - 4AC) = 0$ at all points of the region. The solution domain for a parabolic PDE is an open region, as shown in Figure 1-3. For a parabolic partial differential equation there exists one characteristic line. Unsteady heat conduction in one dimension

$$\frac{\partial T}{\partial t} = \alpha \frac{\partial^2 T}{\partial x^2} \tag{1-11}$$

and diffusion of viscosity, expressed as

$$\frac{\partial u}{\partial t} = \nu \frac{\partial^2 u}{\partial y^2} \tag{1-12}$$

are examples of parabolic PDEs. An initial distribution of the dependent variable and two sets of boundary conditions are required for a complete description of the problem. The boundary conditions are prescribed as the value of the dependent variable or its normal derivative or a linear combination of the two. The solution of the parabolic equation marches downstream within the domain from the initial plane of data satisfying the specified boundary conditions. The parabolic partial differential equation is the counterpart to an initial value problem in an ordinary differential equation (ODE).

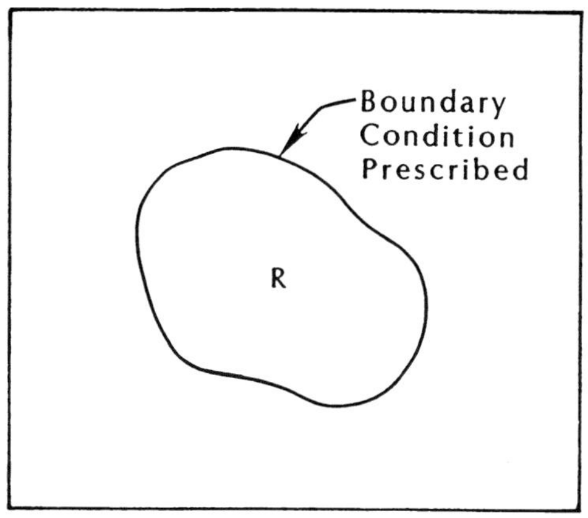

Figure 1-2. The domain of solution for an elliptic PDE.

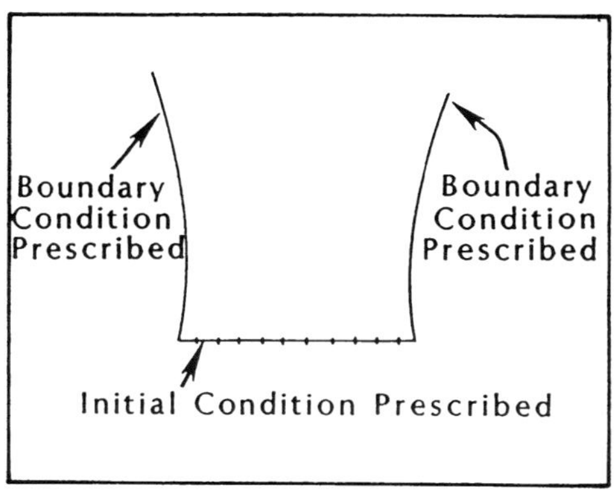

Figure 1-3. The domain of solution for a parabolic PDE.

1.6 Hyperbolic Equations

A partial differential equation is called hyperbolic if $(B^2 - 4AC) > 0$ at all points of the region. A hyperbolic PDE has two real characteristics. An example of a hyperbolic equation is the second-order wave equation:

$$\frac{\partial^2 \phi}{\partial t^2} = a^2 \frac{\partial^2 \phi}{\partial x^2} \tag{1-13}$$

A complete description of the flow governed by a second-order hyperbolic PDE requires two sets of initial conditions and two sets of boundary conditions. The initial conditions at $t = 0$ may be expressed as

$$\phi(x, 0) = f(x)$$

and

$$\phi_t(x, 0) = g(x)$$

where the functions f and g are specified for a particular problem.

For a first-order hyperbolic equation, such as

$$\frac{\partial \phi}{\partial t} = -a \frac{\partial \phi}{\partial x}$$

only one initial condition needs to be specified. Note that the initial condition cannot be specified along a characteristic line.

A classical method of solving a hyperbolic PDE with two independent variables is the method of characteristics (MOC). Along the characteristic lines, the PDE reduces to an ODE, which can be easily integrated to obtain the desired solution. Details of MOC and the appropriate solution schemes will not be discussed here. However, some essential elements of characteristics are provided in Appendix A. Additional materials on MOC may be found in References [1-1] or [1-2].

To illustrate classification of a second-order PDE, an example is proposed as follows:

Example 1.1: Classify the steady two-dimensional velocity potential equation.

$$\left(1 - M^2\right) \phi_{xx} + \phi_{yy} = 0$$

Solution: According to notations used in Equation (1-1),

$$A = (1 - M^2), \quad B = 0, \quad \text{and} \quad C = 1$$

Thus, $(B^2 - 4AC) = -4(1 - M^2)$. If $M < 1$ (subsonic flow), then $(B^2 - 4AC) < 0$ and the equation is elliptic. For $M = 1$ (sonic flow), $(B^2 - 4AC) = 0$ and the equation is parabolic. For $M > 1$ (supersonic flow), $(B^2 - 4AC) > 0$ and the equation is hyperbolic.

Now consider the physical interpretation of various classifications. Assume that a body moving with a velocity u in an inviscid fluid is creating disturbances which propagate with the speed of sound, a. If the velocity u is smaller than a, that is, if the flow is subsonic, then the disturbance is felt everywhere in the flowfield (Figure 1-4a). Note that this is what happens for an elliptic PDE.

As the speed of the body u increases and approaches the speed of sound, a front is developed, with a region ahead of it which does not feel the presence of the disturbance (Figure 1-4b). This region is known as the zone of silence. Thus the disturbance is felt only behind the front. This region is known as the zone of action. When the speed u is further increased, to the extent that it exceeds the speed of sound, a conical front (in three-dimensional analysis) is formed (Figure 1-4c). The effect of the disturbance is felt only within this cone.

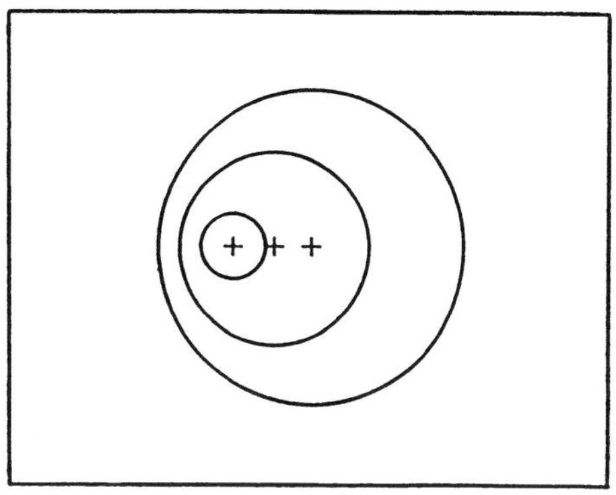

Figure 1-4a. Propagation of disturbance in subsonic flow.

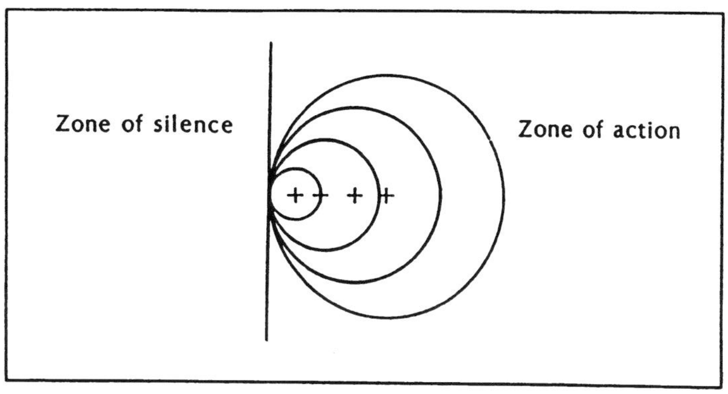

Figure 1-4b. Propagation of disturbance in sonic flow.

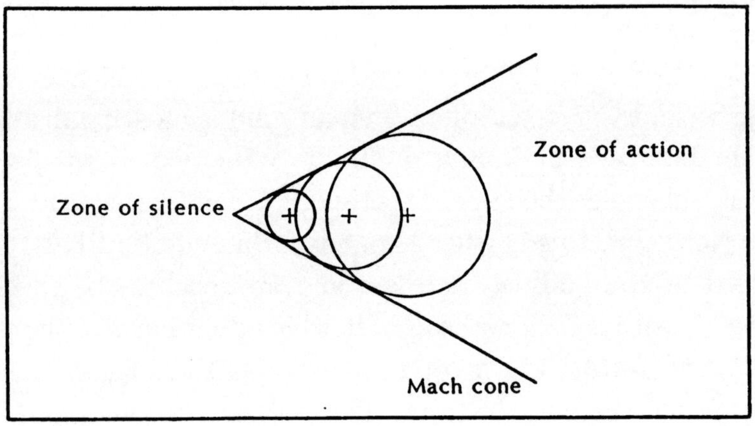

Figure 1-4c. Propagation of disturbance in supersonic flow.

This conical front is known as the Mach cone in three-dimensional space or as Mach lines in two-dimensional space. Mach lines patch two different solutions of the PDE and thus represent the characteristics of the PDE.

1.7 Model Equations

Several partial differential equations will be used as model equations in the following chapters. These equations will be used to illustrate the application of various finite differencing techniques and stability analyses. By observing and analyzing the behavior of the numerical methods when applied to simple model equations, an understanding should be developed which will be useful in studying more complex problems. The selected equations are primarily derived from principles of fluid mechanics and heat transfer. However, this selection should not limit our discussion to problems in fluid mechanics. Many PDEs in science and engineering may be represented by the selected model equations investigated here.

The selected model PDEs which will be used in the next chapters are as follows:

1. Laplace's equation:

$$\frac{\partial^2 \phi}{\partial x^2} + \frac{\partial^2 \phi}{\partial y^2} = 0 \tag{1-14}$$

2. Poisson's equation:

$$\frac{\partial^2 \phi}{\partial x^2} + \frac{\partial^2 \phi}{\partial y^2} = f(x, y) \tag{1-15}$$

3. The equation for unsteady heat conduction:

$$\frac{\partial T}{\partial t} = \alpha \left(\frac{\partial^2 T}{\partial x^2} + \frac{\partial^2 T}{\partial y^2} \right) \tag{1-16}$$

4. The y-component of the Navier-Stokes equation reduced to Stokes' first problem:

$$\frac{\partial u}{\partial t} = \nu \frac{\partial^2 u}{\partial y^2} \tag{1-17}$$

5. The wave equation:

$$\frac{\partial^2 u}{\partial t^2} = a^2 \left(\frac{\partial^2 u}{\partial x^2} \right) \tag{1-18}$$

6. The Burgers equation:

$$\frac{\partial u}{\partial t} = -u \frac{\partial u}{\partial x} \tag{1-19}$$

These equations are expressed in one- or two-space dimensions in the Cartesian coordinate system. Some of the model equations in two-space dimensions will be reduced to one-space dimension in the upcoming discussions.

In most cases, the selected model equation subject to imposed initial and boundary conditions has an analytical solution. In such instances, the analytical solution is used as a basis for comparison with various numerical solutions. These comparisons are very useful in determining the accuracy of the various numerical algorithms employed.

1.8 System of First-Order PDEs

The equations of fluid motion are composed of conservation of mass, conservation of momentum, and conservation of energy. The governing equations may be expressed by partial differential equations, thus forming a system of second-order PDEs. For certain classes of problems, the governing equations are reduced to a system of first-order PDEs. For example, the equations of fluid motion for inviscid flowfields, known as the Euler equations, belong to this category. Furthermore, in some applications a higher-order PDE may be reduced to a system of first-order PDEs by introducing new viariables. In this section, the conditions under which a system of first-order PDEs is classified will be explored. Consider a set of first-order PDEs expressed in the following form

$$\frac{\partial \Phi}{\partial t} + [A]\frac{\partial \Phi}{\partial x} + [B]\frac{\partial \Phi}{\partial y} + \Psi = 0 \tag{1-20}$$

where Φ represents a vector (or column matrix) containing the unknown variables. The elements of the coefficient matrices $[A]$ and $[B]$ are functions of x, y, and t; and the vector Ψ is a function of Φ, x, and y. For example, a set of two first-order PDEs could be represented by the following equations:

$$\frac{\partial u}{\partial t} + a_1 \frac{\partial u}{\partial x} + a_2 \frac{\partial v}{\partial x} + a_3 \frac{\partial u}{\partial y} + a_4 \frac{\partial v}{\partial y} + \Psi_1 = 0 \tag{1-21}$$

and

$$\frac{\partial v}{\partial t} + b_1\frac{\partial u}{\partial x} + b_2\frac{\partial v}{\partial x} + b_3\frac{\partial u}{\partial y} + b_4\frac{\partial v}{\partial y} + \Psi_2 = 0 \tag{1-22}$$

where

$$\Phi = \begin{vmatrix} u \\ v \end{vmatrix}, \quad [A] = \begin{bmatrix} a_1 & a_2 \\ b_1 & b_2 \end{bmatrix}, \quad [B] = \begin{bmatrix} a_3 & a_4 \\ b_3 & b_4 \end{bmatrix}, \text{ and } \Psi = \begin{vmatrix} \Psi_1 \\ \Psi_2 \end{vmatrix}$$

If the eigenvalues of the matrix $[A]$ are all real and distinct, the set of equations is classified as hyperbolic in t and x. For complex eigenvalues of $[A]$, the system of equations is elliptic in t and x. Similarly, the set of equations is hyperbolic in t and y if all the eigenvalues of matrix $[B]$ are real and distinct; otherwise, for complex eigenvalues, the set of equations is classified as elliptic. If the system of equations has the following form (the steady-state form of Equation (1-20)),

$$[A]\frac{\partial \Phi}{\partial x} + [B]\frac{\partial \Phi}{\partial y} + \Psi = 0, \tag{1-23}$$

then the set of equations is classified according to the sign of

$$H = R^2 - 4PQ \tag{1-24}$$

where

$$P = |A| \text{ (determinant of } A\text{)}, \quad Q = |B|$$

and

$$R = \begin{vmatrix} a_1 & a_4 \\ b_1 & b_4 \end{vmatrix} + \begin{vmatrix} a_3 & a_2 \\ b_3 & b_2 \end{vmatrix} \tag{1-25}$$

The set of first-order PDEs is hyperbolic for $H > 0$, parabolic for $H = 0$, and elliptic for $H < 0$. This classification is presented in Ref. [1-3].

At this point, consider a general form of a system of first-order PDEs and its classification. This consideration should also clarify the origin of relation $H = R^2 - 4PQ$, i.e., Equation (1-24). In the arguments to follow, the mathematical details are omitted; instead, applications are emphasized. Recall that characteristics represent a family of lines across which the properties are continuous, whereas there may be discontinuities in their derivatives. Now, define S to represent characteristic surfaces and normal to these surfaces denoted by $\vec{n}$. In the Cartesian system, we may write (for 2-D problems)

$$\vec{n} = n_x\vec{i} + n_y\vec{j}$$

At this point, we seek a relation whereby the number of possible characteristics may be determined. If the characteristic normals are all real, then the system is classified as *hyperbolic*. If they are complex, then the system is *elliptic*. For mixed real and complex values, the system is *mixed elliptic/hyperbolic*. The system is classified as *parabolic* if there is less than K real characteristics, where K is the number of PDEs in the system. To introduce the required relation, consider the following model equation:

$$A\frac{\partial \Phi}{\partial x} + B\frac{\partial \Phi}{\partial y} = 0 \tag{1-26}$$

A wave-like solution (characteristics direction) for the system may be obtained if

$$|T| = 0 \tag{1-27}$$

where

$$[T] = [A]\,n_x + [B]\,n_y \tag{1-28}$$

For mathematical details, see Reference [1-4].

The matrices $[A]$ and $[B]$ were previously defined from Equations (1-21) and (1-22). Now, matrix $[T]$ is formed as

$$[T] = \begin{bmatrix} a_1 n_x & a_2 n_x \\ b_1 n_x & b_2 n_x \end{bmatrix} + \begin{bmatrix} a_3 n_y & a_4 n_y \\ b_3 n_y & b_4 n_y \end{bmatrix} = \begin{bmatrix} a_1 n_x + a_3 n_y & a_2 n_x + a_4 n_y \\ b_1 n_x + b_3 n_y & b_2 n_x + b_4 n_y \end{bmatrix}$$

from which the determinant is computed as

$$|T| = (a_3 b_4 - b_3 a_4) n_y^2 + (a_1 b_2 - a_2 b_1) n_x^2 + (a_1 b_4 + a_3 b_2 - a_2 b_3 - b_1 a_4) n_x n_y = 0$$

Divide by n_x^2 to obtain

$$(a_3 b_4 - b_3 a_4)\left(\frac{n_y}{n_x}\right)^2 + (a_1 b_4 + a_3 b_2 - a_2 b_3 - b_1 a_4)\left(\frac{n_y}{n_x}\right) + (a_1 b_2 - a_2 b_1) = 0$$

This equation may be written as

$$Q\left(\frac{n_y}{n_x}\right)^2 + R\left(\frac{n_y}{n_x}\right) + P = 0$$

from which

$$\left(\frac{n_y}{n_x}\right) = \frac{-R \pm \sqrt{R^2 - 4PQ}}{2Q} = \frac{-R \pm \sqrt{H}}{2Q}$$

Note that the notation previously defined for Equation (1-23) is used in the equation above. Therefore it is seen that, if $H > 0$, the system is hyperbolic; if $H < 0$, the system is elliptic; and if $H = 0$, the system is parabolic.

To illustrate the procedure described above, the following applications are proposed.

Example 1.2 Classify the following system of partial differential equations.

$$\frac{\partial u}{\partial x} + \frac{\partial v}{\partial y} = 0$$

$$\frac{\partial v}{\partial x} - \frac{\partial u}{\partial y} = 0$$

Solution: The system may be expressed in a vector form as

$$A\frac{\partial \vec{q}}{\partial x} + B\frac{\partial \vec{q}}{\partial y} = 0$$

where

$$\vec{q} = \begin{bmatrix} u \\ v \end{bmatrix} , \quad A = \begin{bmatrix} 1 & 0 \\ 0 & 1 \end{bmatrix} \quad \text{and} \quad B = \begin{bmatrix} 0 & 1 \\ -1 & 0 \end{bmatrix}$$

Therefore,

$$P = 1 \, , \quad Q = 1 \quad \text{and}$$

$$R = \begin{vmatrix} 1 & 1 \\ 0 & 0 \end{vmatrix} + \begin{vmatrix} 0 & 0 \\ -1 & 1 \end{vmatrix} = 0$$

Now, from Equation (1-24), H is determined as $H = R^2 - 4PQ = -4$. Since H is negative, the system is classified as elliptic.

As a second approach, Equation (1-27) may be used. For this purpose, $[T]$ is determined as

$$[T] = [A]\, n_x + [B]\, n_y = \begin{bmatrix} n_x & 0 \\ 0 & n_x \end{bmatrix} + \begin{bmatrix} 0 & n_y \\ -n_y & 0 \end{bmatrix} = \begin{bmatrix} n_x & n_y \\ -n_y & n_x \end{bmatrix}$$

The determinant $|T|$ is

$$|T| = n_x^2 + n_y^2$$

According to the requirement (1-27),

$$\begin{aligned} n_x^2 + n_y^2 &= 0 \quad \text{or} \\ \left(\frac{n_y}{n_x}\right)^2 + 1 &= 0 \end{aligned}$$

from which both values of (n_y/n_x) are imaginary and, therefore, the system is elliptic.

The elliptic nature of the system can be further verified by combining the two equations. That is accomplished by eliminating either u or v from the system. The result is

$$\frac{\partial^2 u}{\partial x^2} + \frac{\partial^2 u}{\partial y^2} = \frac{\partial^2 v}{\partial x^2} + \frac{\partial^2 v}{\partial y^2} = 0$$

which was previously classified as elliptic.

When the system of equations exceeds two, the general relation given by (1-27) must be used for classification of the system. To illustrate this point, consider the following example.

Example 1.3 The governing nondimensional equations of fluid motion for steady, inviscid and incompressible flow in two dimensions are given by:

$$\frac{\partial u}{\partial x} + \frac{\partial v}{\partial y} = 0$$

$$u\frac{\partial u}{\partial x} + v\frac{\partial u}{\partial y} + \frac{\partial p}{\partial x} = 0$$

$$u\frac{\partial v}{\partial x} + v\frac{\partial v}{\partial y} + \frac{\partial p}{\partial y} = 0$$

Classify the system of equations.

Solution: Select the unknown vector, Q, as

$$Q = \begin{bmatrix} u \\ v \\ p \end{bmatrix}$$

Therefore, the vector formulation is written as

$$A\frac{\partial Q}{\partial x} + B\frac{\partial Q}{\partial y} = 0$$

where

$$A = \begin{bmatrix} 1 & 0 & 0 \\ u & 0 & 1 \\ 0 & u & 0 \end{bmatrix} \quad \text{and} \quad B = \begin{bmatrix} 0 & 1 & 0 \\ v & 0 & 0 \\ 0 & v & 1 \end{bmatrix}$$

Following (1-28), the matrix $[T]$ is

$$[T] = [A]\,n_x + [B]\,n_y = \begin{bmatrix} n_x & 0 & 0 \\ un_x & 0 & n_x \\ 0 & un_x & 0 \end{bmatrix} + \begin{bmatrix} 0 & n_y & 0 \\ vn_y & 0 & 0 \\ 0 & vn_y & n_y \end{bmatrix}$$

or

$$[T] = \begin{bmatrix} n_x & n_y & 0 \\ un_x + vn_y & 0 & n_x \\ 0 & un_x + vn_y & n_y \end{bmatrix}$$

from which

$$\begin{aligned} |T| &= n_x\left[-n_x(un_x + vn_y)\right] - n_y\left[n_y\,(un_x + vn_y)\right] \\ &= -\,(un_x + vn_y)\left(n_x^2 + n_y^2\right) \end{aligned}$$

Therefore, according to (1-27),

$$|T| = -(un_x + vn_y)\,(n_x^2 + n_y^2) = 0$$

Dividing the above equation by n_x^3 yields

$$\left(\frac{n_y^2}{n_x^2} + 1\right)\left(\frac{n_y}{n_x}v + u\right) = 0$$

from which

$$\frac{n_y}{n_x} = -\frac{u}{v}\ , \quad \frac{n_y}{n_x} = \pm\sqrt{-1}$$

Since mixed real and complex values result, the system is a mixed hyperbolic/elliptic system.

1.9 System of Second-Order PDEs

On many occasions, the system of PDEs will include second-order derivatives. In fluid mechanics, for example, the viscous terms in the momentum equation and the heat conduction term in the energy equation are second order. The classification of such a system is facilitated if one reduces the second-order system to a first-order system and, subsequently, applies the procedure described previously for classification of a first-order system. To illustrate the procedure, consider the steady, incompressible equations of motion in nondimensional form given as

$$\frac{\partial u}{\partial x} + \frac{\partial v}{\partial y} = 0 \tag{1-29}$$

$$u\frac{\partial u}{\partial x} + v\frac{\partial u}{\partial y} = -\frac{\partial p}{\partial x} + \frac{1}{Re}\left(\frac{\partial^2 u}{\partial x^2} + \frac{\partial^2 u}{\partial y^2}\right) \tag{1-30}$$

$$u\frac{\partial v}{\partial x} + v\frac{\partial v}{\partial y} = -\frac{\partial p}{\partial y} + \frac{1}{Re}\left(\frac{\partial^2 v}{\partial x^2} + \frac{\partial^2 v}{\partial y^2}\right) \tag{1-31}$$

In order to reduce the system to a first-order system, introduce the following auxiliary variables:

$$a = \frac{\partial v}{\partial x}\ , \quad b = \frac{\partial v}{\partial y}\ , \quad \text{and } c = \frac{\partial u}{\partial y}$$

Note that, from the continuity Equation (1-29),

$$\frac{\partial u}{\partial x} = -\frac{\partial v}{\partial y} = -b$$

The new variables may be related by cross-differentiation. For example,

$$\frac{\partial}{\partial x}\left(\frac{\partial v}{\partial y}\right) = \frac{\partial^2 v}{\partial x \partial y} = \frac{\partial b}{\partial x}$$

and

$$\frac{\partial}{\partial y}\left(\frac{\partial v}{\partial x}\right) = \frac{\partial^2 v}{\partial x \partial y} = \frac{\partial a}{\partial y}$$

Equating the two expressions,

$$\frac{\partial b}{\partial x} - \frac{\partial a}{\partial y} = 0$$

Similarly,

$$\frac{\partial c}{\partial x} + \frac{\partial b}{\partial y} = 0$$

Now, the system of first-order equations is written as

$$\frac{\partial u}{\partial y} = c$$

$$\frac{\partial u}{\partial x} + \frac{\partial v}{\partial y} = 0$$

$$\frac{\partial b}{\partial x} - \frac{\partial a}{\partial y} = 0$$

$$\frac{\partial c}{\partial x} + \frac{\partial b}{\partial y} = 0$$

$$\frac{1}{Re}\left(-\frac{\partial b}{\partial x} + \frac{\partial c}{\partial y}\right) - \frac{\partial p}{\partial x} = -ub + vc$$

$$\frac{1}{Re}\left(\frac{\partial a}{\partial x} + \frac{\partial b}{\partial y}\right) - \frac{\partial p}{\partial y} = ua + vb$$

This system is written in the vector form as

$$A\frac{\partial Q}{\partial x} + B\frac{\partial Q}{\partial y} = C$$

where

$$Q = \begin{bmatrix} u \\ v \\ a \\ b \\ c \\ p \end{bmatrix} \qquad A = \begin{bmatrix} 0 & 0 & 0 & 0 & 0 & 0 \\ 1 & 0 & 0 & 0 & 0 & 0 \\ 0 & 0 & 0 & 1 & 0 & 0 \\ 0 & 0 & 0 & 0 & 1 & 0 \\ 0 & 0 & 0 & -\frac{1}{Re} & 0 & -1 \\ 0 & 0 & \frac{1}{Re} & 0 & 0 & 0 \end{bmatrix}$$

$$B = \begin{bmatrix} 1 & 0 & 0 & 0 & 0 & 0 \\ 0 & 1 & 0 & 0 & 0 & 0 \\ 0 & 0 & -1 & 0 & 0 & 0 \\ 0 & 0 & 0 & 1 & 0 & 0 \\ 0 & 0 & 0 & 0 & \frac{1}{Re} & 0 \\ 0 & 0 & 0 & \frac{1}{Re} & 0 & -1 \end{bmatrix} \qquad C = \begin{bmatrix} c \\ 0 \\ 0 \\ 0 \\ 0 \\ -ub + vc \\ ua + vb \end{bmatrix}$$

Now the matrix $[T]$ is formed as,

$$[T] = \begin{bmatrix} n_y & 0 & 0 & 0 & 0 & 0 \\ n_x & n_y & 0 & 0 & 0 & 0 \\ 0 & 0 & -n_y & n_x & 0 & 0 \\ 0 & 0 & 0 & n_y & n_x & 0 \\ 0 & 0 & 0 & -\frac{n_x}{Re} & \frac{n_y}{Re} & -n_x \\ 0 & 0 & \frac{n_x}{Re} & \frac{n_y}{Re} & 0 & -n_y \end{bmatrix}$$

and, following (1-27),

$$|T| = \frac{1}{Re} n_y^2 \left(n_x^2 + n_y^2\right)^2 = 0$$

Thus,

$$n_x^2 + n_y^2 = 0$$

or

$$\left(\frac{n_y}{n_x}\right)^2 + 1 = 0$$

since (n_y/n_x) is shown to be imaginary, the system is classified as elliptic.

Now, an example is proposed where the system is time dependent and, therefore, falls in the category of the system given by (1-21) and (1-22).

Example 1.4 The governing equations of motion for one-dimensional, inviscid flows are given by the Euler equations. If the assumption of perfect gas is imposed, the system is written as

$$\frac{\partial \rho}{\partial t} + u\frac{\partial \rho}{\partial x} + \rho\frac{\partial u}{\partial x} = 0$$

$$\frac{\partial u}{\partial t} + u\frac{\partial u}{\partial x} + \frac{1}{\rho}\frac{\partial p}{\partial x} = 0$$

$$\frac{\partial p}{\partial t} + u\frac{\partial p}{\partial x} + \rho a^2\frac{\partial u}{\partial x} = 0$$

It is required that this system be classified.

Solution. Define the variable vector Q as

$$Q = \begin{bmatrix} \rho \\ u \\ p \end{bmatrix}$$

The system is written in the vector form as

$$\frac{\partial Q}{\partial t} + A\frac{\partial Q}{\partial x} = 0$$

where

$$A = \begin{bmatrix} u & \rho & 0 \\ 0 & u & \dfrac{1}{\rho} \\ 0 & \rho a^2 & u \end{bmatrix}$$

The eigenvalues of the system are obtained from

$$\begin{vmatrix} u-\lambda & \rho & 0 \\ 0 & u-\lambda & \dfrac{1}{\rho} \\ 0 & \rho a^2 & u-\lambda \end{vmatrix} = 0$$

or

$$(u-\lambda)\left[(u-\lambda)(u-\lambda)-\left(\frac{1}{\rho}\right)\left(\rho a^2\right)\right]=0$$

or

$$(u-\lambda)\left[(u-\lambda)^2-a^2\right]=0$$

from which

$$\begin{aligned}\lambda_1 &= u\\ \lambda_2 &= u-a\\ \lambda_3 &= u+a\end{aligned}$$

Since all the eigenvalues are real, the system is classified as hyperbolic.

1.10 Initial and Boundary Conditions

In order to obtain a unique solution of a PDE, a set of supplementary conditions must be provided to determine the arbitrary functions which result from the integration of the PDE (compared to arbitrary constants in ODE). The supplementary conditions are classified as boundary or initial conditions.

An initial condition is a requirement for which the dependent variable is specified at some initial state.

A boundary condition is a requirement that the dependent variable or its derivative must satisfy on the boundary of the domain of the PDE.

Various types of boundary conditions which will be encountered are:

1. *The Dirichlet boundary condition.* If the dependent variable along the boundary is prescribed, it is known as the Dirichlet type.

2. *The Neumann boundary condition.* If the normal gradient of the dependent variable along the boundary is specified, it is called the Neumann type.

3. *The Robin boundary condition.* If the imposed boundary condition is a linear combination of the Dirichlet and Neumann types, it is known as the Robin type.

4. *The Mixed boundary condition.* Frequently the boundary condition along a certain portion of the boundary is the Dirichlet type and, on another portion of the boundary, a Neumann type. This type is known as a mixed boundary condition.

As an example, consider transient conduction in two-space dimensions. Assume that a long rectangular bar has been heated to a temperature distribution of $T = f(x, y)$. An initial condition would then be prescribed such that

$$\text{for } t = 0, \quad T = f(x, y)$$

Now, place the bar in an environment in which the lower and right sides are in contact with a convecting fluid of temperature T_f and a constant film coefficient of h, while the left side is insulated (adiabatic) and the upper side is kept at a constant temperature.

The corresponding boundary conditions are:

$$t \geq 0 \quad x = 0 \quad , \quad \frac{\partial T}{\partial x} = 0$$

$$x = L \quad , \quad \frac{\partial T}{\partial x} = -\frac{h}{k}(T - T_f)$$

$$y = 0 \quad , \quad \frac{\partial T}{\partial y} = \frac{h}{k}(T - T_f)$$

and

$$y = H \quad , \quad T = T_c$$

These are shown in Figure 1-5.

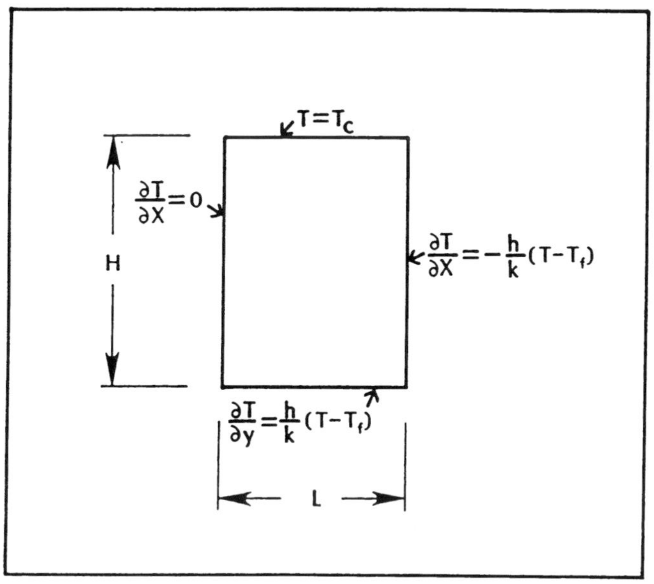

Figure 1-5. Sketch illustrating the imposed boundary conditions on the rectangular bar.

1.11 Remarks and Definitions

In order to solve a given PDE by numerical methods, the partial derivatives in the equation are approximated by finite difference relations. These representations of the partial derivatives are obtained from Taylor series expansions, as will be shown in the next chapter. The resulting approximate equation, which represents the original PDE, is called a finite difference equation (FDE).

To illustrate the objectives and the procedures to be developed, consider a two-dimensional rectangular domain. We wish to solve a PDE within this domain subject to imposed initial and boundary conditions. The rectangular domain is divided into equal increments in the x and y directions. Denote the increments in the x direction by Δx and the increments in the y direction by Δy. Note that the increments in the x direction do not need to be equal to the increments in the y direction. These increments may be defined as mesh size, step size, or grid size. The location of mesh points, grid points, or nodes is designated by i in the x direction and by j in the y direction. The maximum number of grid points in the x and y directions are denoted by IM and JM, respectively. These nomenclatures are shown in Figure 1-6.

The finite difference equation that approximates the PDE is an algebraic equation. This algebraic equation is written for each grid point within the domain. The solution of the finite difference equations provides the values of the dependent variable at each grid point. The objectives are to study the various schemes to approximate the PDEs by finite difference equations and to explore numerical techniques for solving the resulting approximate equations.

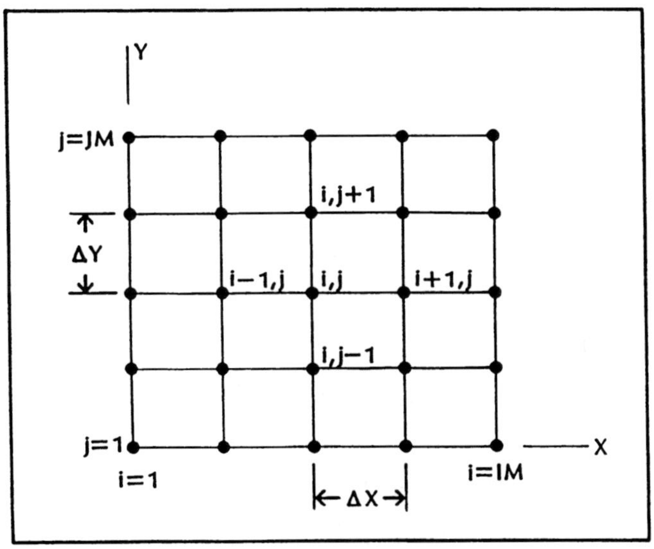

Figure 1-6. Sketch illustrating the nomenclature of computational space.

Before proceeding with an analysis of numerical techniques, it is necessary to define additional terminology for concepts which will be investigated in the upcoming chapters.

1. *Consistency:* A finite difference approximation of a PDE is consistent if the finite difference equation approaches the PDE as the grid size approaches zero.

2. *Stability:* A numerical scheme is said to be stable if any error introduced in the finite difference equation does not grow with the solution of the finite difference equation.

3. *Convergence:* A finite difference scheme is convergent if the solution of the finite difference equation approaches that of the PDE as the grid size approaches zero.

4. *Lax's equivalence theorem:* For a FDE which approximates a well-posed, linear initial value problem, the necessary and sufficient condition for convergence is that the FDE must be stable and consistent.

5. *The conservative (divergent) form of a PDE:* In this formulation of a physical law, the coefficients of the derivatives are either constant or, if variable, their derivatives do not appear anywhere in the equation. For example, the conservation of mass for steady two-dimensional flow is written in conservative form as

$$\nabla \cdot (\rho \vec{V}) = 0$$

   or in Cartesian coordinate system as

$$\frac{\partial}{\partial x}(\rho u) + \frac{\partial}{\partial y}(\rho v) = 0$$

   If this equation is written in expanded form as

$$\rho \frac{\partial u}{\partial x} + \rho \frac{\partial v}{\partial y} + u \frac{\partial \rho}{\partial x} + v \frac{\partial \rho}{\partial y} = 0$$

   it is known as the nonconservative form of the equation.

6. *Conservative property of a FDE:* If the finite difference approximation of a PDE maintains the integral property of the conservation law over an arbitrary region containing any number of grid points, it is said to possess a conservative property.

In closing, a few points should be emphasized.

(a) First-order PDEs occur only occasionally in engineering problems. Almost all first-order equations have real characteristics and thus behave much like hyperbolic equations of the second order.

(b) Equations with more than two independent variables may not fit neatly into the classification of PDEs described in this chapter. However, the concepts of elliptic, parabolic, and hyperbolic can be extended to such PDEs.

(c) Not all problems are expressed as purely elliptic, parabolic, or hyperbolic problems. On many occasions, a problem is expressed as a mixture of elliptic and parabolic equations, i.e., the governing equations are elliptic in one region and parabolic in another region of the domain.

1.12 Summary Objectives

After completing this chapter, you should be able to do the following:

1. Define and give examples of:

    a. Linear and nonlinear PDEs

    b. Elliptic, parabolic, and hyperbolic PDEs

    c. Initial and boundary conditions

    d. The conservative form of a PDE

2. Define:

    a. Zone of influence and zone of dependence

    b. Convergence

    c. Consistency

    d. Stability

3. Solve the problems for Chapter One.

1.13 Problems

Classify the following second-order partial differential equations.

1.1 $$3\frac{\partial^2\phi}{\partial x^2} + \frac{\partial^2\phi}{\partial x\partial y} + 2\frac{\partial^2\phi}{\partial y^2} = 0$$

1.2 $$\frac{\partial^2\phi}{\partial z^2} + \frac{\partial\phi}{\partial z} + \frac{\partial^2\phi}{\partial r^2} + \frac{\partial\phi}{r\partial\theta} + K = 0$$

1.3 $$\frac{\partial\phi}{\partial t} + \beta\frac{\partial\phi}{\partial x} + \alpha\frac{\partial^2\phi}{\partial x^2} = 0$$

1.4 $$4\frac{\partial^2\phi}{\partial x^2} + y^2\frac{\partial\phi}{\partial x} + x\frac{\partial\phi}{\partial x} + \frac{\partial^2\phi}{\partial y^2} + 4\frac{\partial^2\phi}{\partial x\partial y} - 4xy = 0$$

Determine the values of x and y to make the following partial differential equations elliptic, parabolic, or hyperbolic.

1.5 $$x\frac{\partial^2\phi}{\partial x^2} + x\frac{\partial^2\phi}{\partial x\partial y} + y\frac{\partial^2\phi}{\partial y^2} = 0$$

1.6 $$x\frac{\partial^2\phi}{\partial x^2} + y\frac{\partial^2\phi}{\partial x\partial y} + K = 0$$

1.7 $$x^2y\frac{\partial^2\phi}{\partial x^2} + xy\frac{\partial^2\phi}{\partial x\partial y} - y^2\frac{\partial^2\phi}{\partial y^2} = 0$$

1.8 $$\sin x\frac{\partial^2\phi}{\partial x^2} + 2\cos x\frac{\partial^2\phi}{\partial x\partial y} + \sin x\frac{\partial^2\phi}{\partial y^2} = 0$$

1.9 Classify the following system of equations:

$$\frac{\partial u}{\partial t} + a\frac{\partial u}{\partial x} + b\frac{\partial v}{\partial x} = 0$$

$$\frac{\partial v}{\partial t} + b\frac{\partial u}{\partial x} + a\frac{\partial v}{\partial x} = 0$$

1.10 The x-component of the momentum equation for an incompressible flow with zero pressure gradient is given by

$$u\frac{\partial u}{\partial x} + v\frac{\partial u}{\partial y} = \nu\frac{\partial^2 u}{\partial y^2}$$

Assume ν to be a constant. Reformulate the equation into the conservative form.

1.11 The governing equations for stationary, shallow water are expressed as

$$u\frac{\partial u}{\partial x} + v\frac{\partial u}{\partial y} + g\frac{\partial h}{\partial x} = 0$$

$$u\frac{\partial v}{\partial x} + v\frac{\partial v}{\partial y} + g\frac{\partial h}{\partial y} = 0$$

$$\frac{\partial}{\partial x}(uh) + \frac{\partial}{\partial y}(vh) = 0$$

where u and v represent the velocity components, and h represents the surface elevation.

(a) Define the unknown vector Q as

$$Q = \begin{bmatrix} u \\ v \\ h \end{bmatrix}$$

and rewrite the system of equations in a vector form similar to (1-23).

(b) Classify the system.

1.12 Classify the following system of PDEs:

$$a_1\frac{\partial u}{\partial x} + a_2\frac{\partial v}{\partial y} = g_1$$

$$b_1\frac{\partial v}{\partial x} + b_2\frac{\partial u}{\partial y} = g_2$$

Consider three cases where

(a) $a_1 = b_1 = a_2 = b_2 = 1$

(b) $a_1 = b_2 = 1 \ , \quad b_1 = 0 \ , \quad a_2 = -1$

(c) $a_1 = b_1 = b_2 = 1 \ , \quad a_2 = -1$

1.13 Consider the system

$$\frac{\partial u}{\partial x} + \frac{\partial v}{\partial y} = 0$$

$$u\frac{\partial u}{\partial x} + v\frac{\partial u}{\partial y} = \frac{1}{Re}\frac{\partial^2 u}{\partial y^2}$$

(a) Reduce the system to a first-order system.

(b) Write the system in a vector form.

(c) Show that the system is parabolic.

1.14 A system of PDEs is given by the following:

$$\frac{\partial u}{\partial x} + \frac{\partial v}{\partial y} = 0$$

$$u\frac{\partial u}{\partial x} + v\frac{\partial u}{\partial y} = -\frac{1}{\rho}\frac{\partial p}{\partial x} + \nu\left(\frac{\partial^2 u}{\partial x^2} + \frac{\partial^2 u}{\partial y^2}\right)$$

$$u\frac{\partial v}{\partial x} + v\frac{\partial v}{\partial y} = -\frac{1}{\rho}\frac{\partial p}{\partial y} + \nu\left(\frac{\partial^2 v}{\partial x^2} + \frac{\partial^2 v}{\partial y^2}\right)$$

where the kinematic viscosity ν and the density ρ are constants.

(a) Reduce the system to a first-order system.

(b) Write the system in a vector form.

(c) Classify the system.

1.15 Classify the following system of equations:

(a) $(x+y)\dfrac{\partial u}{\partial x}+\dfrac{\partial v}{\partial y}=0$

$$(x-y)\frac{\partial v}{\partial x}+\frac{\partial u}{\partial y}=0$$

(b) $\dfrac{\partial u}{\partial y}-\dfrac{\partial v}{\partial x}-\dfrac{\partial w}{\partial x}=0$

$$\frac{\partial u}{\partial x}-\frac{\partial v}{\partial y}-\frac{\partial w}{\partial y}=0$$

$$\frac{\partial v}{\partial y}-\frac{\partial w}{\partial x}=0$$

Chapter 2

Finite Difference Formulations

2.1 Introductory Remarks

The differentials of the dependent variables appearing in partial differential equations must be expressed as approximate expressions, so that a digital computer (which can perform only standard arithmetic and logical operations) can be employed to obtain a solution. Two methods for approximating the differentials of a function f are considered in this chapter. One method of approximation often used is the Taylor series expansion of the function f. A second method is the use of a polynomial of degree n. The Taylor series expansion will be considered first, and subsequently some examples using the polynomial representation of the function f are given.

2.2 Taylor Series Expansion

Given a function $f(x)$, which is analytical, $f(x+\Delta x)$ can be expanded in a Taylor series about x as

$$\begin{aligned} f(x+\Delta x) &= f(x) + (\Delta x)\frac{\partial f}{\partial x} + \frac{(\Delta x)^2}{2!}\frac{\partial^2 f}{\partial x^2} + \frac{(\Delta x)^3}{3!}\frac{\partial^3 f}{\partial x^3} + \cdots \\ &= f(x) + \sum_{n=1}^{\infty}\frac{(\Delta x)^n}{n!}\frac{\partial^n f}{\partial x^n} \end{aligned} \tag{2-1}$$

Solving for $\partial f/\partial x$, one obtains

$$\frac{\partial f}{\partial x} = \frac{f(x+\Delta x) - f(x)}{\Delta x} - \frac{\Delta x}{2!}\frac{\partial^2 f}{\partial x^2} - \frac{(\Delta x)^2}{3!}\frac{\partial^3 f}{\partial x^3} + \cdots \tag{2-2}$$

Summing all the terms with factors of Δx and higher and representing them as $O(\Delta x)$ (that is read as terms of order Δx) yields

$$\frac{\partial f}{\partial x} = \frac{f(x+\Delta x) - f(x)}{\Delta x} + O(\Delta x), \tag{2-3}$$

which is an approximation for the first partial derivative of f with respect to x. Graphically, as shown in Figure 2-1, this approximation may be interpreted as the slope of the function at point B, using the values of the function at points B and C. If the subscript index i is used to represent the discrete points in the x-direction, Equation (2-3) is written as

$$\left.\frac{\partial f}{\partial x}\right|_i = \frac{f_{i+1} - f_i}{\Delta x} + O(\Delta x) \tag{2-4}$$

This equation is known as the first forward difference approximation of $\partial f/\partial x$ of order (Δx). It is obvious that as the step size decreases, the error term is reduced and therefore the accuracy of the approximation is increased. Now consider the Taylor series expansion of $f(x - \Delta x)$ about x.

$$f(x-\Delta x) = f(x) - \Delta x \frac{\partial f}{\partial x} + \frac{(\Delta x)^2}{2!}\frac{\partial^2 f}{\partial x^2} - \frac{(\Delta x)^3}{3!}\frac{\partial^3 f}{\partial x^3} + \cdots \tag{2-5}$$

$$= f(x) + \sum_{n=1}^{\infty}\left[(-1)^n \frac{(\Delta x)^n}{n!}\right]\frac{\partial^n f}{\partial x^n}$$

Solving for $\partial f/\partial x$,

$$\frac{\partial f}{\partial x} = \frac{f(x) - f(x-\Delta x)}{\Delta x} + O(\Delta x)$$

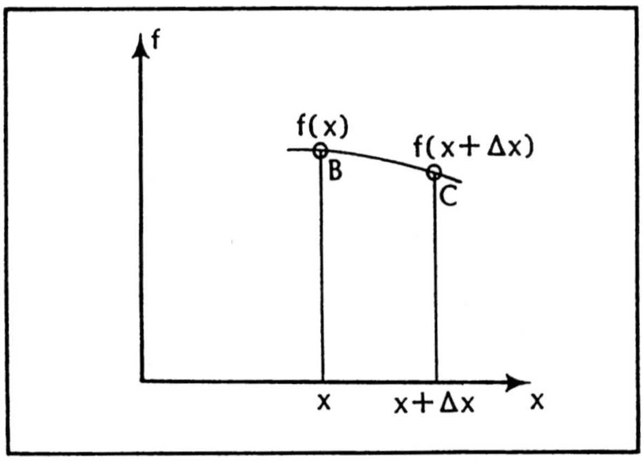

Figure 2-1. Illustration of grid points used in Equation (2-3).

or

$$\left.\frac{\partial f}{\partial x}\right|_i = \frac{f_i - f_{i-1}}{\Delta x} + O(\Delta x) \tag{2-6}$$

which represents the slope of the function at B using the values of the function at points A and B, as shown in Figure 2-2. Equation (2-6) is the first backward difference approximation of $\partial f/\partial x$ of order (Δx). Now, consider the Taylor series expansions (2-1) and (2-5), which are repeated here:

$$f(x + \Delta x) = f(x) + \Delta x \frac{\partial f}{\partial x} + \frac{(\Delta x)^2}{2!}\frac{\partial^2 f}{\partial x^2} + \frac{(\Delta x)^3}{3!}\frac{\partial^3 f}{\partial x^3} + \cdots \tag{2-7}$$

and

$$f(x - \Delta x) = f(x) - \Delta x \frac{\partial f}{\partial x} + \frac{(\Delta x)^2}{2!}\frac{\partial^2 f}{\partial x^2} - \frac{(\Delta x)^3}{3!}\frac{\partial^3 f}{\partial x^3} + \cdots \tag{2-8}$$

Subtracting Equation (2-8) from Equation (2-7), one obtains

$$f(x + \Delta x) - f(x - \Delta x) = 2\Delta x \frac{\partial f}{\partial x} + 2\frac{(\Delta x)^3}{3!}\frac{\partial^3 f}{\partial x^3} + \cdots \tag{2-9}$$

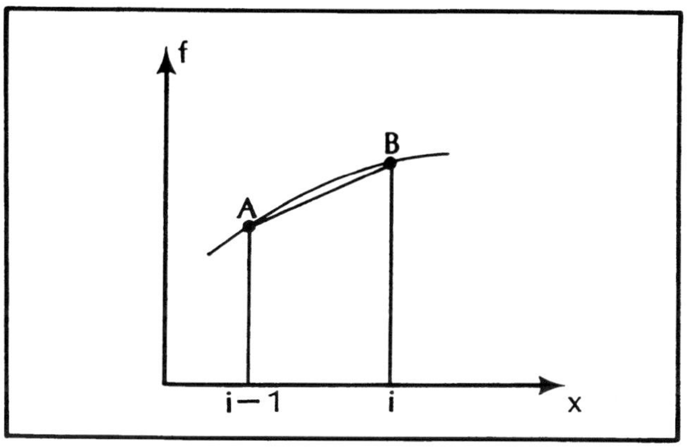

Figure 2-2. Illustration of grid points used in Equation (2-6).

Solving for $\partial f/\partial x$,

$$\frac{\partial f}{\partial x} = \frac{f(x + \Delta x) - f(x - \Delta x)}{2\Delta x} + O(\Delta x)^2$$

or

$$\left.\frac{\partial f}{\partial x}\right|_i = \frac{f_{i+1} - f_{i-1}}{2\Delta x} + O(\Delta x)^2 \tag{2-10}$$

which represents the slope of the function f at point B using the values of the function at points A and C, as shown in Figure 2-3. This representation of $\partial f/\partial x$

is known as the central difference approximation of order $(\Delta x)^2$. Thus far, three approximations for the first derivative $\partial f/\partial x$ have been introduced. Now, the derivations of approximate expressions for the higher order derivatives are considered. Again, consider the Taylor series expansion

$$f(x+\Delta x) = f(x) + (\Delta x)\frac{\partial f}{\partial x} + \frac{(\Delta x)^2}{2!}\frac{\partial^2 f}{\partial x^2} + \frac{(\Delta x)^3}{3!}\frac{\partial^3 f}{\partial x^3} + \cdots \tag{2-11}$$

Expanding by a Taylor series $f(x+2\Delta x)$ about x produces the expansion

$$f(x+2\Delta x) = f(x) + (2\Delta x)\frac{\partial f}{\partial x} + \frac{(2\Delta x)^2}{2!}\frac{\partial^2 f}{\partial x^2} + \frac{(2\Delta x)^3}{3!}\frac{\partial^3 f}{\partial x^3} + \cdots \tag{2-12}$$

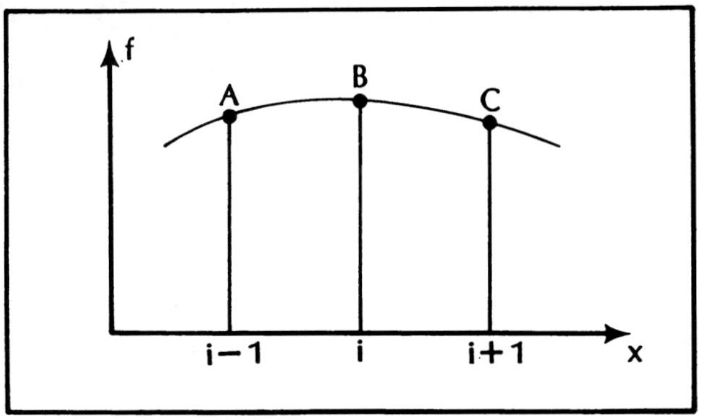

Figure 2-3. Illustration of grid points used in Equation (2-10).

Multiply Equation (2-11) by 2 and subtract it from Equation (2-12), and the result is

$$-2f(x+\Delta x) + f(x+2\Delta x) = -f(x) + (\Delta x)^2\frac{\partial^2 f}{\partial x^2} + (\Delta x)^3\frac{\partial^3 f}{\partial x^3} + \cdots \tag{2-13}$$

Solving for $\partial^2 f/\partial x^2$,

$$\frac{\partial^2 f}{\partial x^2} = \frac{f(x+2\Delta x) - 2f(x+\Delta x) + f(x)}{(\Delta x)^2} + O(\Delta x)$$

or

$$\left.\frac{\partial^2 f}{\partial x^2}\right|_i = \frac{f_{i+2} - 2f_{i+1} + f_i}{(\Delta x)^2} + O(\Delta x) \tag{2-14}$$

This equation represents the forward difference approximation for the second derivative of f with respect to x and is of the order (Δx). A similar approximation for the second derivative can be produced using the Taylor series expansions of $f(x-\Delta x)$ and $f(x-2\Delta x)$. The result is

$$\left.\frac{\partial^2 f}{\partial x^2}\right|_i = \frac{f_i - 2f_{i-1} + f_{i-2}}{(\Delta x)^2} + O(\Delta x) \tag{2-15}$$

This equation is the backward difference approximation of $\partial^2 f/\partial x^2$. The derivation of Equation (2-15) is left as an exercise. To obtain a central difference approximation of the second derivative, simply add Equations (2-7) and (2-8). Thus,

$$\frac{\partial^2 f}{\partial x^2} = \frac{f(x+\Delta x) - 2f(x) + f(x-\Delta x)}{(\Delta x)^2} + O(\Delta x)^2$$

or

$$\left.\frac{\partial^2 f}{\partial x^2}\right|_i = \frac{f_{i+1} - 2f_i + f_{i-1}}{(\Delta x)^2} + O(\Delta x)^2 \tag{2-16}$$

Approximations for higher order derivatives of f with respect to x can be found using the same procedure (see example (2.2)). For convenience, define the first forward difference $f_{i+1} - f_i$ as $\Delta_x f_i$ and the first backward difference $f_i - f_{i-1}$ as $\nabla_x f_i$. In general, first order forward and backward differences can be expressed as

$$\Delta_x^n f_i = \Delta_x^{n-1}(\Delta_x f_i) \quad \text{and} \quad \nabla_x^n f_i = \nabla_x^{n-1}(\nabla_x f_i)$$

Various central difference operators can be similarly defined. Some typical operators are:

$$\begin{aligned}
\delta_x^* f_i &= f_{i+1} - f_{i-1} = \Delta_x f_i + \nabla_x f_i \\
\delta_x f_i &= f_{i+\frac{1}{2}} - f_{i-\frac{1}{2}} \\
\delta_x^2 f_i &= \delta_x(\delta_x f_i) = \delta_x \left(f_{i+\frac{1}{2}} - f_{i-\frac{1}{2}}\right) \\
&= (f_{i+1} - f_i) - (f_i - f_{i-1}) \\
&= f_{i+1} - 2f_i + f_{i-1} \\
(\Delta_x \nabla_x)\, f_i &= \Delta_x f_i - \nabla_x f_i = f_{i+1} - 2f_i + f_{i-1} \\
(\Delta_x \nabla_x)^2 f_i &= f_{i+2} - 4f_{i+1} + 6f_i - 4f_{i-1} + f_{i-2}
\end{aligned}$$

Using the operators just defined, the approximations of the higher derivatives by forward, backward, and central differencing may be expressed as

$$\left.\frac{\partial^n f}{\partial x^n}\right|_i = \frac{\Delta_x^n f_i}{(\Delta x)^n} + O(\Delta x) \tag{2-17}$$

$$\left.\frac{\partial^n f}{\partial x^n}\right|_i = \frac{\nabla_x^n f_i}{(\Delta x)^n} + O(\Delta x) \tag{2-18}$$

$$\left.\frac{\partial^n f}{\partial x^n}\right|_i = \frac{\Delta_x^n f_{i-\frac{n}{2}} + \nabla_x^n f_{i+\frac{n}{2}}}{2(\Delta x)^n} + O(\Delta x)^2 \qquad \text{for } n \text{ even} \tag{2-19}$$

$$\left.\frac{\partial^n f}{\partial x^n}\right|_i = \frac{\Delta_x^n f_{i-\frac{n-1}{2}} + \nabla_x^n f_{i+\frac{n-1}{2}}}{2(\Delta x)^n} + O(\Delta x)^2 \qquad \text{for } n \text{ odd} \tag{2-20}$$

These expressions are tabulated in Tables 2.1, 2.2, and 2.3 for derivatives of up to the fourth order.

So far the first and higher order derivatives have been expressed using forward and backward differencing of order (Δx) and central differencing of order $(\Delta x)^2$. By considering additional terms in the Taylor series expansions, a more accurate approximation of the derivatives is produced. Consider the Taylor series expansion,

$$f(x+\Delta x) = f(x) + (\Delta x)\frac{\partial f}{\partial x} + \frac{(\Delta x)^2}{2!}\frac{\partial^2 f}{\partial x^2} + \frac{(\Delta x)^3}{3!}\frac{\partial^3 f}{\partial x^3} + \cdots \tag{2-21}$$

Solving for $\partial f/\partial x$,

$$\frac{\partial f}{\partial x} = \frac{f(x+\Delta x) - f(x)}{(\Delta x)} - \frac{\Delta x}{2}\frac{\partial^2 f}{\partial x^2} - \frac{(\Delta x)^2}{6}\frac{\partial^3 f}{\partial x^3} + \cdots \tag{2-22}$$

Substitute a forward difference expression for $\partial^2 f/\partial x^2$, i.e.,

$$\frac{\partial^2 f}{\partial x^2} = \frac{f(x) - 2f(x+\Delta x) + f(x+2\Delta x)}{(\Delta x)^2} + O(\Delta x)$$

and one obtains

$$\frac{\partial f}{\partial x} = \frac{f(x+\Delta x) - f(x)}{(\Delta x)} - \frac{\Delta x}{2}\left[\frac{f(x+2\Delta x) - 2f(x+\Delta x) + f(x)}{(\Delta x)^2} + O(\Delta x)\right]$$

$$- \frac{(\Delta x)^2}{6}\frac{\partial^3 f}{\partial x^3} + \cdots \quad \text{or}$$

$$\frac{\partial f}{\partial x} = \frac{-f(x+2\Delta x) + 4f(x+\Delta x) - 3f(x)}{2\Delta x} + O(\Delta x)^2 \tag{2-23}$$

Thus, a second-order accurate finite difference approximation for $\partial f/\partial x$ has been obtained. A similar expression for the backward difference approximation can be generated by replacing the second-order derivative with a first order accurate backward approximation. In general, higher order forward, backward, and central approximations are obtained by replacing more terms in the Taylor series by forward, backward, and central difference representations of the derivatives. In practice, approximations of order three or more are rarely used because they require greater computation time, since computation time increases as *(Nodes)*^{3} in most machines; however, with sufficient convergence criteria, a good approximation with a more reasonable computation time can be obtained with second-order differencing. Tables 2.4, 2.5, and 2.6 present the forward, backward, and central difference approximations of first and higher order derivatives (up to fourth order) with error orders of $(\Delta x)^2$ (for forward and backward differencing) and $(\Delta x)^4$ (for central differencing).

2.3 Finite Difference by Polynomials

The second procedure for approximating a derivative is to represent the function as a polynomial. The coefficients of the polynomial are computed by substitution of data (dependent variable) from a series of usually equally spaced points of the independent variable. The approximate values of the derivatives are computed from the polynomial. For example, consider a second-order polynomial,

$$f(x) = Ax^2 + Bx + C \tag{2-24}$$

which is shown in Figure 2-4. Select the origin at x_i. Thus, $x_i = 0$, $x_{i+1} = \Delta x$, and $x_{i+2} = 2\Delta x$ and the values of the function f at these locations are, $f(x_i) = f_i$, $f(x_{i+1}) = f_{i+1}$, and $f(x_{i+2}) = f_{i+2}$. Thus,

$$f_i = Ax_i^2 + Bx_i + C = C$$

$$f_{i+1} = Ax_{i+1}^2 + Bx_{i+1} + C = A(\Delta x)^2 + B(\Delta x) + C$$

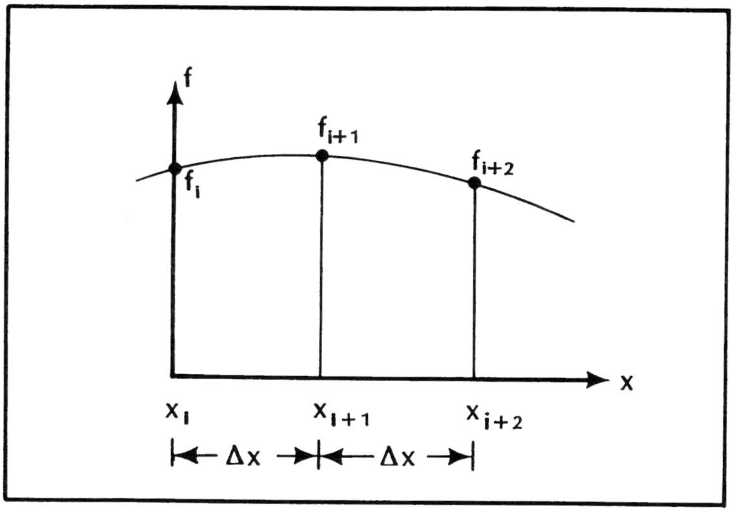

Figure 2-4. A second-order polynomial approximating the actual function.

and

$$f_{i+2} = Ax_{i+2}^2 + Bx_{i+2} + C = A(2\Delta x)^2 + B(2\Delta x) + C$$

From which it follows that

$$C = f_i$$

$$B = \frac{-f_{i+2} + 4f_{i+1} - 3f_i}{2(\Delta x)}$$

and

$$A = \frac{f_{i+2} - 2f_{i+1} + f_i}{2(\Delta x)^2}$$

Now computing the first derivative of f, one has

$$\frac{\partial f}{\partial x} = 2Ax + B$$

or at $x_i = 0$,

$$\left.\frac{\partial f}{\partial x}\right|_i = B$$

Therefore,

$$\frac{\partial f}{\partial x} = \frac{-f_{i+2} + 4f_{i+1} - 3f_i}{2\Delta x}$$

which is identical to the second-order accurate forward difference expression obtained from Taylor series expansion. Note that this approximation is classified as second-order accurate for $\partial f/\partial x$, since $\partial^3 f/\partial x^3$ vanishes just as in the accuracy analysis of the Taylor series expansion. The second derivative of f may be determined as

$$\frac{\partial^2 f}{\partial x^2} = 2A$$

from which

$$\frac{\partial^2 f}{\partial x^2} = \frac{f_{i+2} - 2f_{i+1} + f_i}{(\Delta x)^2}$$

and is consistent with the first-order finite difference approximation given by Equation (2-14). If the spacing of points i, $i+1$, and $i+2$ is not identical, as shown in Figure 2-5, a finite difference approximation of the derivative is found by the same procedure. Assume $x_i = 0$, $x_{i+1} = \Delta x$, and $x_{i+2} = (1+\alpha)\Delta x$. Then,

$$f_i = C$$

$$f_{i+1} = A(\Delta x)^2 + B(\Delta x) + C$$

and

$$f_{i+2} = A(1+\alpha)^2(\Delta x)^2 + B(1+\alpha)\Delta x + C$$

Consequently,

$$C = f_i$$

$$B = \frac{-f_{i+2} + (1+\alpha)^2 f_{i+1} - (\alpha^2 + 2\alpha)f_i}{\alpha(1+\alpha)\Delta x}$$

and

$$A = \frac{f_{i+2} - (1+\alpha)f_{i+1} + \alpha f_i}{\alpha(1+\alpha)(\Delta x)^2}$$

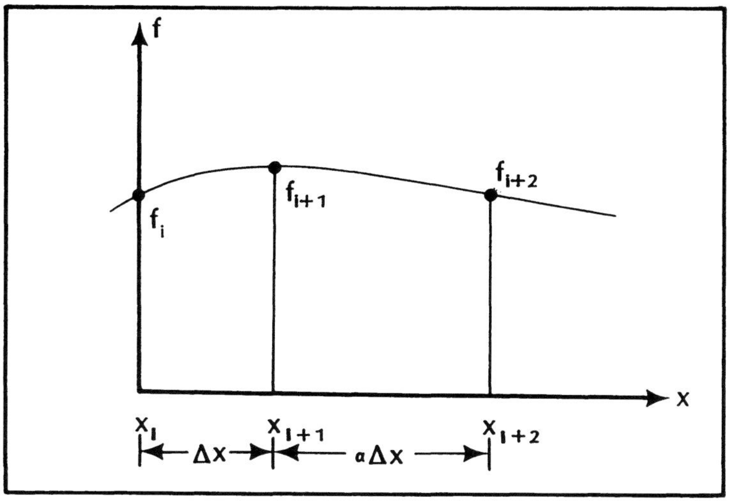

Figure 2-5. A second-order polynomial with unequal stepsizes.

Therefore,

$$\left.\frac{\partial f}{\partial x}\right|_i = \frac{-f_{i+2} + (1+\alpha)^2 f_{i+1} - \alpha(\alpha+2)f_i}{\alpha(1+\alpha)\Delta x} \tag{2-25}$$

which is a second-order accurate approximation. The second derivative of f is obtained as

$$\frac{\partial^2 f}{\partial x^2} = 2A$$

Hence,

$$\frac{\partial^2 f}{\partial x^2} = 2\left[\frac{f_{i+2} - (1+\alpha)f_{i+1} + \alpha f_i}{\alpha(1+\alpha)(\Delta x)^2}\right]$$

which is a first-order accurate expression. Similar relations for backward and central difference approximations may be obtained by this procedure.

2.4 Finite Difference Equations

The finite difference approximations just discussed are used to replace the derivatives that appear in the PDEs. Consider an example involving time (t) and two spatial coordinates (x, y); i.e., the dependent variable f is $f = f(t, x, y)$. A governing PDE of the form

$$\frac{\partial f}{\partial t} = \alpha\left(\frac{\partial^2 f}{\partial x^2} + \frac{\partial^2 f}{\partial y^2}\right) \tag{2-26}$$

where α is assumed constant, is used for illustration purposes.

It is required to approximate the PDE by a finite difference equation in a domain with equal grid spacing. The subscript indices i and j are used to represent the Cartesian coordinates x and y, and the superscript index n is used to represent time. It is specified that a first-order finite difference approximation in time and central differencing of second-order accuracy in space be used. The spatial grid spacings are Δx and Δy, whereas Δt designates the time step. The grid system is shown in Figure 2-6.

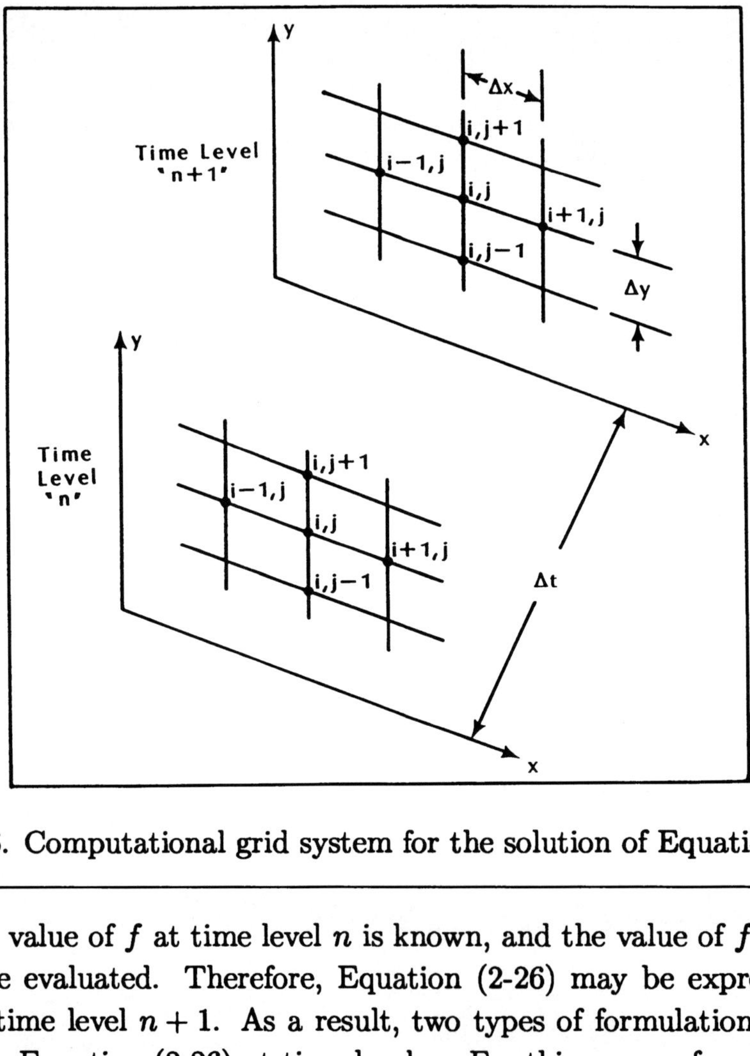

Figure 2-6. Computational grid system for the solution of Equation (2-26).

Note that the value of f at time level n is known, and the value of f at time level $n+1$ is to be evaluated. Therefore, Equation (2-26) may be expressed at time level n or at time level $n+1$. As a result, two types of formulation are possible. First, consider Equation (2-26) at time level n. For this case, a forward difference approximation which is first-order accurate is used. Hence,

$$\frac{\partial f}{\partial t} = \frac{f_{i,j}^{n+1} - f_{i,j}^{n}}{\Delta t} + O(\Delta t)$$

From Equation (2-16)

$$\frac{\partial^2 f}{\partial x^2} = \frac{f^n_{i+1,j} - 2f^n_{i,j} + f^n_{i-1,j}}{(\Delta x)^2} + O(\Delta x)^2$$

and

$$\frac{\partial^2 f}{\partial y^2} = \frac{f^n_{i,j+1} - 2f^n_{i,j} + f^n_{i,j-1}}{(\Delta y)^2} + O(\Delta y)^2$$

Therefore, the finite difference formulation of the partial differential equation (2-26) is:

$$\frac{f^{n+1}_{i,j} - f^n_{i,j}}{\Delta t} = \alpha \left[\frac{f^n_{i+1,j} - 2f^n_{i,j} + f^n_{i-1,j}}{(\Delta x)^2} + \frac{f^n_{i,j+1} - 2f^n_{i,j} + f^n_{i,j-1}}{(\Delta y)^2} \right] + O\left[\Delta t,\ (\Delta x)^2,\ (\Delta y)^2\right] \tag{2-27a}$$

Note that in this formulation, the spatial approximations are applied at time level n. For the second case, Equation (2-26) is evaluated at $n+1$ time level. Therefore, a first-order backward difference approximation in time is employed, and the spatial approximations are at time level $n+1$. Hence, the finite difference formulation takes the form:

$$\frac{f^{n+1}_{i,j} - f^n_{i,j}}{\Delta t} = \alpha \left[\frac{f^{n+1}_{i+1,j} - 2f^{n+1}_{i,j} + f^{n+1}_{i-1,j}}{(\Delta x)^2} + \frac{f^{n+1}_{i,j+1} - 2f^{n+1}_{i,j} + f^{n+1}_{i,j-1}}{(\Delta y)^2} \right] + O\left[\Delta t,\ (\Delta x)^2,\ (\Delta y)^2\right] \tag{2-27b}$$

The resulting finite difference equations, (2-27a) and (2-27b), are classified as explicit and implicit formulations, respectively.

An obvious distinction between two finite difference equations is the number of unknowns appearing in each equation. Close examination of Equation (2-27a) reveals that it involves only one unknown, $f^{n+1}_{i,j}$, whereas Equation (2-27b) involves five unknowns. Thus, the solution procedures based on explicit and implicit formulations are different. In the explicit formulation, only one unknown appears and may therefore be solved for directly at each grid point. In the implicit formulation, more than one unknown exists and therefore the finite difference equation must be written for all the spatial grid points at $n+1$ time level to provide the same number of equations as there are unknowns and solved simultaneously. Obviously, the solution of explicit formulation is simpler than the implicit equation. However, as will be seen shortly, implicit formulations are more stable than explicit formulations. Other differences between explicit and implicit formulations are discussed in future chapters.

Before some applications are considered, the finite difference equation (2-27a) is rearranged as

$$\frac{f_{i,j}^{n+1} - f_{i,j}^{n}}{\Delta t} - \alpha\left[\frac{f_{i+1,j}^{n} - 2f_{i,j}^{n} + f_{i-1,j}^{n}}{(\Delta x)^2} + \frac{f_{i,j+1}^{n} - 2f_{i,j}^{n} + f_{i,j-1}^{n}}{(\Delta y)^2}\right]$$
$$= O[\Delta t,\ (\Delta x)^2,\ (\Delta y)^2]$$

where the right-hand side represents the truncation error. Since the lowest term on the right-hand side is of order one, the formulation is classified as a first-order accurate method. If the approximation of the original PDE was such that the truncation error was of the order $[(\Delta t)^2, (\Delta x)^2, (\Delta y)^2]$, in which case the lowest term is of order two, then it would be classified as a second-order accurate method.

2.5 Applications

Example 2.1. Find a forward difference approximation of $O(\Delta x)$ for $\frac{\partial^4 f}{\partial x^4}$.

Solution. From Equation (2-17),

$$\left.\frac{\partial^n f}{\partial x^n}\right|_i = \frac{\Delta_x^n f_i}{(\Delta x)^n} + O(\Delta x)$$

For $n = 4$, one has

$$\left.\frac{\partial^4 f}{\partial x^4}\right|_i = \frac{1}{(\Delta x)^4}\Delta_x^4 f_i + O(\Delta x)$$

But,

$$\begin{aligned}
\Delta_x^4 f_i &= \Delta_x^3(\Delta_x f_i) = \Delta_x^3(f_{i+1} - f_i) = \Delta_x^2(\Delta_x f_{i+1} - \Delta_x f_i) = \\
&= \Delta_x^2\Big[(f_{i+2} - f_{i+1}) - (f_{i+1} - f_i)\Big] = \Delta_x^2(f_{i+2} - 2f_{i+1} + f_i) \\
&= \Delta_x(\Delta_x f_{i+2} - 2\Delta_x f_{i+1} + \Delta_x f_i) = \Delta_x\Big[(f_{i+3} - f_{i+2}) - 2(f_{i+2} - f_{i+1}) \\
&\quad + (f_{i+1} - f_i)\Big] = \Delta_x[f_{i+3} - 3f_{i+2} + 3f_{i+1} - f_i] \\
&= \Delta_x f_{i+3} - 3\Delta_x f_{i+2} + 3\Delta_x f_{i+1} - \Delta_x f_i \\
&= (f_{i+4} - f_{i+3}) - 3(f_{i+3} - f_{i+2}) + 3(f_{i+2} - f_{i+1}) - (f_{i+1} - f_i) \\
&= f_{i+4} - 4f_{i+3} + 6f_{i+2} - 4f_{i+1} + f_i
\end{aligned}$$

Therefore,

$$\left.\frac{\partial^4 f}{\partial x^4}\right|_i = \frac{1}{(\Delta x)^4}\left[f_{i+4} - 4f_{i+3} + 6f_{i+2} - 4f_{i+1} + f_i\right]$$

Example 2.2. Determine the approximate forward difference representation for $\partial^3 f/\partial x^3$ which is of the order (Δx), given evenly spaced grid points as shown in Figure 2-7 by means of:

(a) Taylor series expansion

(b) Forward difference recurrence formula

(c) A third-degree polynomial passing through the four points.

Solution.

(a) The Taylor series expansions of $f(x+\Delta x)$, $f(x+2\Delta x)$, and $f(x+3\Delta x)$ about x are

$$f(x+\Delta x) = f(x) + \Delta x \frac{\partial f}{\partial x} + \frac{(\Delta x)^2}{2!}\frac{\partial^2 f}{\partial x^2} + \frac{(\Delta x)^3}{3!}\frac{\partial^3 f}{\partial x^3} + O(\Delta x)^4 \quad (2\text{-}28)$$

$$f(x+2\Delta x) = f(x) + 2\Delta x \frac{\partial f}{\partial x} + \frac{(2\Delta x)^2}{2!}\frac{\partial^2 f}{\partial x^2} + \frac{(2\Delta x)^3}{3!}\frac{\partial^3 f}{\partial x^3} + O(2\Delta x)^4 \quad (2\text{-}29)$$

$$f(x+3\Delta x) = f(x) + 3\Delta x \frac{\partial f}{\partial x} + \frac{(3\Delta x)^2}{2!}\frac{\partial^2 f}{\partial x^2} + \frac{(3\Delta x)^3}{3!}\frac{\partial^3 f}{\partial x^3} + O(3\Delta x)^4 \quad (2\text{-}30)$$

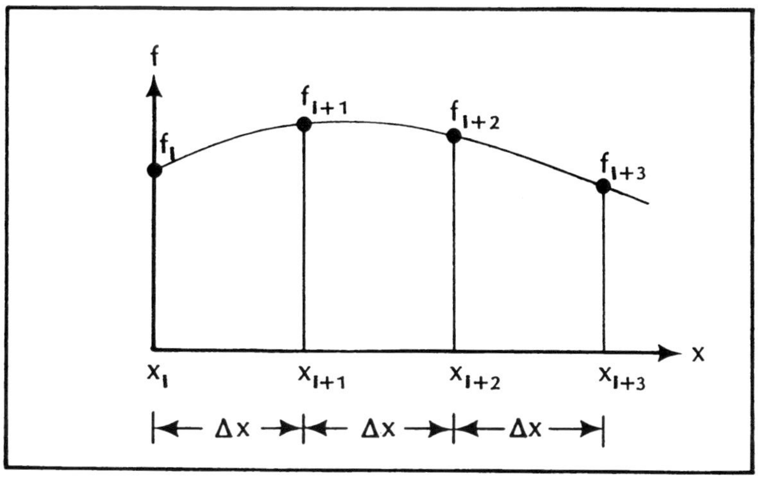

Figure 2-7. Grid points used for the solution of example 2.2.

The three simultaneous equations can be solved for $\partial^3 f/\partial x^3$; for example, from Equations (2-28) and (2-30),

$$3f_{i+1} - f_{i+3} = 2f_i - 3(\Delta x)^2 \frac{\partial^2 f}{\partial x^2} - 4(\Delta x)^3 \frac{\partial^3 f}{\partial x^3} + O(\Delta x)^4 \tag{2-31}$$

and, from Equations (2-28) and (2-29),

$$f_{i+2} - 2f_{i+1} = -f_i + (\Delta x)^2 \frac{\partial^2 f}{\partial x^2} + (\Delta x)^3 \frac{\partial^3 f}{\partial x^3} + O(\Delta x)^4 \tag{2-32}$$

Now Equations (2-31) and (2-32) can be solved for $\partial^3 f/\partial x^3$, resulting in

$$\frac{\partial^3 f}{\partial x^3} = \frac{f_{i+3} - 3f_{i+2} + 3f_{i+1} - f_i}{(\Delta x)^3} + O(\Delta x)$$

(b) From Equation (2-17),

$$\frac{\partial^3 f}{\partial x^3} = \frac{\Delta_x^3 f_i}{(\Delta x)^3} + O(\Delta x)$$

where

$$\begin{aligned}
\Delta_x^3 f_i &= \Delta_x^2(\Delta_x f_i) = \Delta_x^2(f_{i+1} - f_i) = \Delta_x(\Delta_x f_{i+1} - \Delta_x f_i) \\
&= \Delta_x\Big[(f_{i+2} - f_{i+1}) - (f_{i+1} - f_i)\Big] = \Delta_x(f_{i+2} - 2f_{i+1} + f_i) \\
&= \Delta_x f_{i+2} - 2\Delta_x f_{i+1} + \Delta_x f_i = (f_{i+3} - f_{i+2}) - 2(f_{i+2} - f_{i+1}) + (f_{i+1} - f_i) \\
&= f_{i+3} - 3f_{i+2} + 3f_{i+1} - f_i
\end{aligned}$$

Thus,

$$\frac{\partial^3 f}{\partial x^3} = \frac{f_{i+3} - 3f_{i+2} + 3f_{i+1} - f_i}{(\Delta x)^3} + O(\Delta x)$$

(c) Fitting the third-degree polynomial $f(x) = Ax^3 + Bx^2 + Cx + D$ through the four equally spaced points at x_{i+3}, x_{i+2}, x_{i+1}, and x_i, one obtains

$$f_{i+3} = A(3\Delta x)^3 + B(3\Delta x)^2 + C(3\Delta x) + D$$

$$f_{i+2} = A(2\Delta x)^3 + B(2\Delta x)^2 + C(2\Delta x) + D$$

$$f_{i+1} = A(\Delta x)^3 + B(\Delta x)^2 + C(\Delta x) + D$$

and

$$f_i = D$$

(Note that the selection of $x_i = 0$ does not affect the generality of the solution.) From which it follows that

$$D = f_i$$

$$C = \frac{2f_{i+3} - 9f_{i+2} + 18f_{i+1} - 11f_i}{6\Delta x}$$

$$B = \frac{-3f_{i+3} + 12f_{i+2} - 15f_{i+1} + 6f_i}{6(\Delta x)^2}$$

and

$$A = \frac{f_{i+3} - 3f_{i+2} + 3f_{i+1} - f_i}{6(\Delta x)^3}$$

Now, the derivatives of $f(x)$ are determined as

$$\frac{\partial f}{\partial x} = 3Ax^2 + 2Bx + C\ , \quad \frac{\partial^2 f}{\partial x^2} = 6Ax + 2B\ , \quad \text{and} \quad \frac{\partial^3 f}{\partial x^3} = 6A$$

Therefore,

$$\frac{\partial^3 f}{\partial x^3} = \frac{f_{i+3} - 3f_{i+2} + 3f_{i+1} - f_i}{(\Delta x)^3} + O(\Delta x)$$

In addition, the first and second derivatives at i, where x is zero, are obtained easily as

$$\frac{\partial^2 f}{\partial x^2} = 2B \quad \text{and} \quad \frac{\partial f}{\partial x} = C$$

Hence,

$$\frac{\partial^2 f}{\partial x^2} = \frac{-3f_{i+3} + 12f_{i+2} - 15f_{i+1} + 6f_i}{3(\Delta x)^2} + O(\Delta x)^2$$

$$= \frac{-f_{i+3} + 4f_{i+2} - 5f_{i+1} + 2f_i}{(\Delta x)^2} + O(\Delta x)^2$$

and

$$\frac{\partial f}{\partial x} = \frac{2f_{i+3} - 9f_{i+2} + 18f_{i+1} - 11f_i}{6(\Delta x)} + O(\Delta x)^3$$

Example 2.3. Determine a backward difference approximation for $\partial f/\partial x$ which is of the order $(\Delta x)^3$.

Solution. Consider the Taylor series expansion,

$$f(x - \Delta x) = f(x) - \Delta x \frac{\partial f}{\partial x} + \frac{(\Delta x)^2}{2!}\frac{\partial^2 f}{\partial x^2} - \frac{(\Delta x)^3}{3!}\frac{\partial^3 f}{\partial x^3} + O(\Delta x)^4$$

from which

$$\Delta x \frac{\partial f}{\partial x} = f(x) - f(x - \Delta x) + \frac{(\Delta x)^2}{2} \frac{\partial^2 f}{\partial x^2} - \frac{(\Delta x)^3}{6} \frac{\partial^3 f}{\partial x^3} + O(\Delta x)^4 \tag{2-33}$$

Now substitute the backward difference approximations for $\partial^2 f/\partial x^2$ and $\partial^3 f/\partial x^3$. It is important to pay attention to the order of accuracy of the approximations for $\partial^2 f/\partial x^2$ and $\partial^3 f/\partial x^3$. Since we are interested in $\partial f/\partial x$ which is of the order $(\Delta x)^3$, we must select a second-order accurate formulation for $\partial^2 f/\partial x^2$ and a first-order accurate formulation for $\partial^3 f/\partial x^3$, i.e.,

$$\frac{\partial^2 f}{\partial x^2} = \frac{-f_{i-3} + 4f_{i-2} - 5f_{i-1} + 2f_i}{(\Delta x)^2} + O(\Delta x)^2$$

and

$$\frac{\partial^3 f}{\partial x^3} = \frac{-f_{i-3} + 3f_{i-2} - 3f_{i-1} + f_i}{(\Delta x)^3} + O(\Delta x)$$

Substituting the expressions above into (2-33) produces

$$\begin{aligned} \Delta x \frac{\partial f}{\partial x} &= f_i - f_{i-1} + \frac{(\Delta x)^2}{2} \left[\frac{-f_{i-3} + 4f_{i-2} - 5f_{i-1} + 2f_i}{(\Delta x)^2} + O(\Delta x)^2 \right] \\ &\quad - \frac{(\Delta x)^3}{6} \left[\frac{-f_{i-3} + 3f_{i-2} - 3f_{i-1} + f_i}{(\Delta x)^3} + O(\Delta x) \right] + O(\Delta x)^4 \end{aligned}$$

or

$$6\Delta x \frac{\partial f}{\partial x} = -2f_{i-3} + 9f_{i-2} - 18f_{i-1} + 11f_i + O(\Delta x)^4$$

from which

$$\frac{\partial f}{\partial x} = \frac{-2f_{i-3} + 9f_{i-2} - 18f_{i-1} + 11f_i}{6\Delta x} + O(\Delta x)^3$$

Example 2.4. Using the Taylor series expansion, find a second-order forward difference approximation for $\partial f/\partial x$ with unequally spaced grid points, as shown in Figure 2-8.

Solution: Expand $f(x + \Delta x)$ and $f[x + (1 + \alpha)\Delta x]$ about x

$$f(x + \Delta x) = f(x) + \Delta x \frac{\partial f}{\partial x} + \frac{(\Delta x)^2}{2!} \frac{\partial^2 f}{\partial x^2} + \frac{(\Delta x)^3}{3!} \frac{\partial^3 f}{\partial x^3} + O(\Delta x)^4 \tag{2-34}$$

and

$$\begin{aligned} f[x + (1 + \alpha)\Delta x] &= f(x) + (1 + \alpha)\Delta x \frac{\partial f}{\partial x} + \frac{(1 + \alpha)^2 (\Delta x)^2}{2!} \frac{\partial^2 f}{\partial x^2} \\ &\quad + \frac{(1 + \alpha)^3 (\Delta x)^3}{3!} \frac{\partial^3 f}{\partial x^3} + O(\Delta x)^4 \end{aligned} \tag{2-35}$$

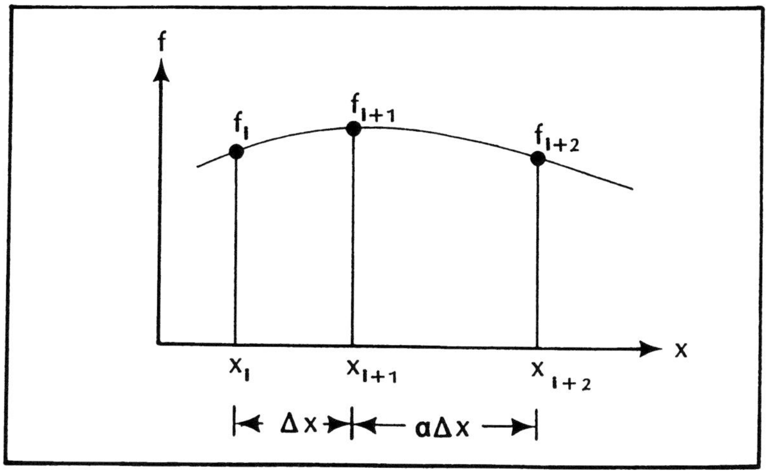

Figure 2-8. Unequally spaced grid points for the solution of example 2.4.

Multiply Equation (2-34) by $-(1+\alpha)$ and add it to Equation (2-35), so that

$$f_{i+2} - (1+\alpha)f_{i+1} + \alpha f_i = (1+\alpha)(\alpha)\frac{(\Delta x)^2}{2}\frac{\partial^2 f}{\partial x^2} + O(\Delta x)^3$$

or, solving for $\partial^2 f/\partial x^2$,

$$\frac{\partial^2 f}{\partial x^2} = \frac{f_{i+2} - (1+\alpha)f_{i+1} + \alpha f_i}{\frac{1}{2}\alpha(1+\alpha)(\Delta x)^2} + O(\Delta x) \tag{2-36}$$

Substitute Equation (2-36) into Equation (2-34):

$$f_{i+1} = f_i + \Delta x\frac{\partial f}{\partial x} + \frac{(\Delta x)^2}{2!}\frac{f_{i+2} - (1+\alpha)f_{i+1} + \alpha f_i}{\frac{1}{2}\alpha(1+\alpha)(\Delta x)^2} + O(\Delta x)^3$$

Solving for $\partial f/\partial x$, one obtains

$$\frac{\partial f}{\partial x} = \frac{-f_{i+2} + (1+\alpha)^2 f_{i+1} - \alpha(\alpha+2)f_i}{\alpha(1+\alpha)\Delta x} + O(\Delta x)^2$$

Note that this expression was obtained (as Equation (2-25)) by using the polynomial technique for finite differencing.

Example 2.5. Determine a central difference approximation of $\partial f/\partial x$ for the unequally spaced grid points by the polynomial technique. Refer to Figure 2-9.

Solution. The second-order polynomial $f(x) = Ax^2 + Bx + C$ is passed through the points f_{i-1}, f_i, and f_{i+1}, where $x_{i-1} = -\Delta x$, $x_i = 0$, and $x_{i+1} = \alpha\Delta x$. Thus,

$$f_{i-1} = A(-\Delta x)^2 + B(-\Delta x) + C$$

$$f_i = C$$

and

$$f_{i+1} = A(\alpha\Delta x)^2 + B(\alpha\Delta x) + C$$

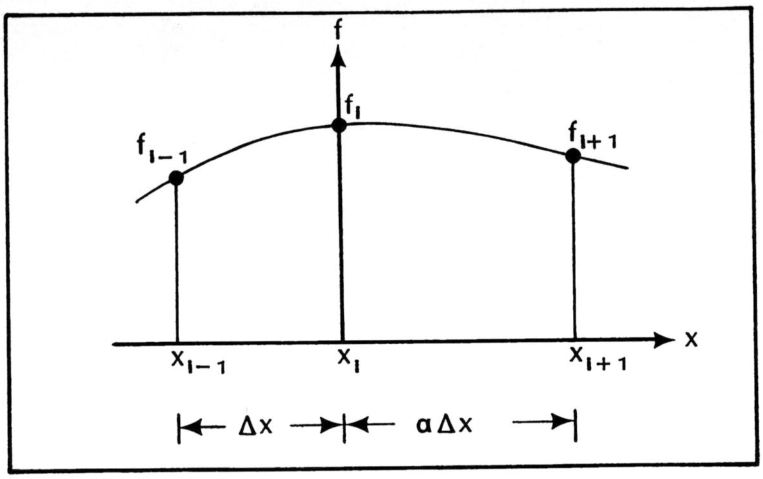

Figure 2-9. Unequally spaced grid points for example 2.5.

Solving for the coefficients A, B, and C, one finds

$$A = \frac{f_{i+1} - (\alpha+1)f_i + \alpha f_{i-1}}{\alpha(\alpha+1)(\Delta x)^2}$$

$$B = \frac{f_{i+1} + (\alpha^2-1)f_i - \alpha^2 f_{i-1}}{\alpha(\alpha+1)(\Delta x)}$$

and

$$C = f_i$$

The first derivative of the function is $\partial f/\partial x = 2Ax + B$, which at point x_i reduces to $\partial f/\partial x = B$. Therefore,

$$\frac{\partial f}{\partial x} = \frac{f_{i+1} + (\alpha^2-1)f_i - \alpha^2 f_{i-1}}{\alpha(\alpha+1)(\Delta x)} + O(\Delta x)^2$$

Note that for $\alpha = 1$, this expression reduces to

$$\frac{\partial f}{\partial x} = \frac{f_{i+1} - f_{i-1}}{2\Delta x} + O(\Delta x)^2$$

i.e., Equation (2-10).

Example 2.6. Given the function $f(x) = \frac{1}{4}x^2$, compute the first derivative of f at $x = 2$ using forward and backward differencing of order (Δx). Compare the results with a central differencing of $O(\Delta x)^2$ and the exact analytical value. Use a step size of $\Delta x = 0.1$. Repeat the computations for a step size of 0.4.

Solution. From Equation (2-4), the forward difference approximation of order (Δx) is

$$\frac{\partial f}{\partial x} = \frac{f_{i+1} - f_i}{\Delta x} + O(\Delta x)$$

With step size of $\Delta x = 0.1$,

$$\frac{\partial f}{\partial x} = \frac{f(2.1) - f(2)}{0.1} + O(0.1)$$

$$\frac{\partial f}{\partial x} = \frac{\frac{4.41}{4} - \frac{4}{4}}{0.1} + O(0.1) = 1.025 + O(0.1)$$

From Equation (2-6), the backward difference approximation which is of order Δx is:

$$\frac{\partial f}{\partial x} = \frac{f_i - f_{i-1}}{\Delta x} + O(\Delta x)$$

For $\Delta x = 0.1$,

$$\frac{\partial f}{\partial x} = \frac{f(2) - f(1.9)}{0.1} + O(0.1)$$

and

$$\frac{\partial f}{\partial x} = \frac{\frac{4}{4} - \frac{3.61}{4}}{0.1} + O(0.1) = 0.975 + O(0.1)$$

From Equation (2-10), the central differencing of $O(\Delta x)^2$ is

$$\frac{\partial f}{\partial x} = \frac{f_{i+1} - f_{i-1}}{2\Delta x} + O(\Delta x)^2$$

For step size of $\Delta x = 0.1$,

$$\frac{\partial f}{\partial x} = \frac{\frac{(2.1)^2}{4} - \frac{(1.9)^2}{4}}{2(0.1)} + O(0.1)^2 = \frac{\frac{4.41}{4} - \frac{3.61}{4}}{0.2} + O(0.1)^2$$

and

$$\frac{\partial f}{\partial x} = 1 + O(0.01)$$

The exact value is $\partial f/\partial x = \frac{1}{4}(2x)$, which at $x = 2$ is $\partial f/\partial x = 1$. Repeating the computations for $\Delta x = 0.4$, the results are

$$\frac{\partial f}{\partial x} = \frac{f_{i+1} - f_i}{\Delta x} + O(\Delta x) = \frac{f(2.4) - f(2)}{0.4} + O(0.4)$$

$$= \frac{\frac{5.76}{4} - \frac{4}{4}}{0.4} + O(0.4) = 1.1 + O(0.4) \ .$$

$$\frac{\partial f}{\partial x} = \frac{f_i - f_{i-1}}{\Delta x} + O(\Delta x) = \frac{f(2) - f(1.6)}{0.4} + O(0.4)$$

$$= \frac{\frac{4}{4} - \frac{2.56}{4}}{0.4} + O(0.4) = 0.9 + O(0.4) \ .$$

$$\frac{\partial f}{\partial x} = \frac{f_{i+1} - f_{i-1}}{2\Delta x} + O(\Delta x)^2 = \frac{f(2.4) - f(1.6)}{2(0.4)} + O(0.4)^2$$

$$\frac{\partial f}{\partial x} = \frac{\frac{5.76}{4} - \frac{2.56}{4}}{2(0.4)} + O(0.4)^2 = 1 + O(0.4)^2$$

Note that the results obtained from backward and forward differencing deviate from the exact value when a larger step size is used. Selection of the step size is extremely important in numerical analysis. As it is shown in the next example, selecting a step size without careful consideration can result in meaningless values.

Example 2.7. Given the function $f(x) = \sin(2\pi x)$, shown in Figure 2-10, determine $\partial f/\partial x$ at $x = 0.375$ using central difference representation of order $(\Delta x)^2$ and order $(\Delta x)^4$. Use step sizes of 0.01, 0.1, and 0.25. Compare and discuss the results.

Solution. The central difference approximations of order $(\Delta x)^2$ and $(\Delta x)^4$ are

$$\frac{\partial f}{\partial x} = \frac{f_{i+1} - f_{i-1}}{2\Delta x} + O(\Delta x)^2$$

and

$$\frac{\partial f}{\partial x} = \frac{-f_{i+2} + 8f_{i+1} - 8f_{i-1} + f_{i-2}}{12(\Delta x)} + O(\Delta x)^4$$

For the step size of 0.01,

$$\frac{\partial f}{\partial x} = \frac{\sin(2.4190) - \sin(2.2934)}{2(0.01)} = \frac{0.6613 - 0.7501}{0.02} = -4.4399602 + O(0.01)^2$$

and

$$\frac{\partial f}{\partial x} = \frac{-\sin(2.4819) + 8\sin(2.4190) - 8\sin(2.2934) + \sin(2.2305)}{12(0.01)}$$

$$= \frac{-0.6129 + 8(0.6613) - 8(0.7501) + 0.7902}{0.12}$$

$$= -4.4428806 + O(0.01)^4$$

For the step size of 0.1,

$$\frac{\partial f}{\partial x} = \frac{\sin(2.9845) - \sin(1.7279)}{2(0.1)} = \frac{0.1564 - 0.9877}{0.2} = -4.1562694 + O(0.1)^2$$

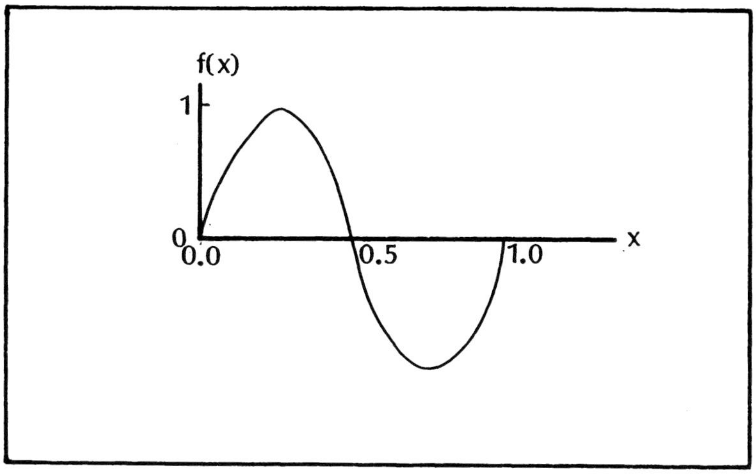

Figure 2-10. The function $f(x) = \sin(2\pi x)$.

and

$$\begin{aligned}\frac{\partial f}{\partial x} &= \frac{-\sin(3.6128) + 8\sin(2.9845) - 8\sin(1.7279) + \sin(1.0996)}{12(0.1)} \\ &= \frac{-(-0.4540) + 8(0.1564) - 8(0.9877) + 0.8910}{1.2} \\ &= -4.4211667 + O(0.1)^4\end{aligned}$$

For the step size 0.25,

$$\frac{\partial f}{\partial x} = \frac{\sin(2.9270) - \sin(0.7854)}{2(0.25)} = \frac{-0.7071 - 0.7071}{0.5} = -2.8284271 + O(0.25)^2$$

and

$$\begin{aligned}\frac{\partial f}{\partial x} &= \frac{-\sin(5.4978) + 8\sin(3.9270) - 8\sin(0.7854) + \sin(-0.7854)}{12(0.25)} \\ &= \frac{-(-0.7071) + 8(-0.7071) - 8(0.7071) + (-0.7071)}{3} \\ &= -3.7712362 + O(0.25)^4\end{aligned}$$

The analytical solution yields:

$$\frac{\partial f}{\partial x} = 2\pi \cos 2\pi x = 2\pi \cos[2\pi(0.375)] = -4.4428829$$

The calculations are summarized in Table 2.7.

Step size	$O(\Delta x)^2$	error (%)	$O(\Delta x)^4$	error (%)	Exact
0.01	−4.4399602	0.0658	−4.4428806	0.000052	−4.4428829
0.1	−4.1562694	6.4511	−4.4211667	0.4888	−4.4428829
0.25	−2.8284271	36.3380	−3.7712362	15.1174	−4.4428829

Table 2.7

The error is determined as follows:

$$\% \text{ error} = \left(\frac{\text{Approximate value - Analytical value}}{\text{Analytical value}}\right) 100$$

This problem clearly indicates the importance of step size in the computation of derivatives. It illustrates that a large step size yields inaccurate results for the derivatives. Before any computations of the derivatives are attempted, the behavior of the given function must be examined, and the selection of the step size reviewed with care.

Example 2.8. Given the following data, compute $f'(5)$, $f'(7)$, and $f'(9)$. Use finite differencing of order (Δx). Compare the results to the values obtained by finite differencing of order $(\Delta x)^2$.

x	5	6	7	8	9
$f(x)$	25	36	49	64	81

Solution. Only a forward difference approximation can be used to compute $f'(5)$. Similarly, only a backward differencing may be applied for the computation of $f'(9)$. Thus, from Equation (2-4),

$$f'(5) = \frac{\partial f}{\partial x} = \frac{f_{i+1} - f_i}{\Delta x} + O(\Delta x) = \frac{36 - 25}{1} = 11 + O(1)$$

and from Equation (2-23),

$$\frac{\partial f}{\partial x} = \frac{-f_{i+2} + 4f_{i+1} - 3f_i}{2\Delta x} + O(\Delta x)^2$$

$$\frac{\partial f}{\partial x} = \frac{-49 + 4(36) - 3(25)}{2(1)} = 10 + O(1)^2$$

The exact value of $f'(5)$ is 10. [Note that the data may be represented by $f(x) = x^2$; thus, $f'(x) = 2x$.] By forward or backward differencing of order Δx, $f'(7)$ may be calculated. Using forward differencing,

$$f'(7) = \frac{\partial f}{\partial x} = \frac{f_{i+1} - f_i}{\Delta x} = \frac{64 - 49}{1} = 15 + O(1)$$

Using backward differencing,

$$f'(7) = \frac{\partial f}{\partial x} = \frac{f_i - f_{i-1}}{\Delta x} = \frac{49 - 36}{1} = 13 + O(1)$$

Using a central difference approximation of order $(\Delta x)^2$, one obtains

$$f'(7) = \frac{\partial f}{\partial x} = \frac{f_{i+1} - f_{i-1}}{2\Delta x} = \frac{64 - 36}{2} = 14 + O(1)^2$$

The exact value of $f'(7)$ is 14; $f'(9)$ can only be computed using a backward representation:

$$f'(9) = \frac{\partial f}{\partial x} = \frac{f_i - f_{i-1}}{\Delta x} = \frac{81 - 64}{1} = 17 + O(1)$$

Using a second-order approximation,

$$f'(9) = \frac{\partial f}{\partial x} = \frac{f_{i-2} - 4f_{i-1} + 3f_i}{2\Delta x} = \frac{49 - 4(64) + 3(81)}{2} = 18 + O(1)^2$$

The exact value is 18.

2.6 Finite Difference Approximation of Mixed Partial Derivatives

Approximating mixed partial derivatives can be performed by two procedures. One method is to use the Taylor series expansion for the two variables. A second and easier method is to use the approximation of partial derivatives discussed earlier, in which only one independent variable is involved. Procedures for both methods are presented.

2.6.1 Taylor Series Expansion

To illustrate the use of the Taylor series expansion to compute approximate expressions for mixed partial derivatives, consider $\partial^2 f/(\partial x \partial y)$. The Taylor series expansion for two variables x and y, for $f(x + \Delta x,\ y + \Delta y)$, is

$$\begin{aligned} f(x + \Delta x,\ y + \Delta y) &= f(x, y) + \Delta x \frac{\partial f}{\partial x} + \Delta y \frac{\partial f}{\partial y} + \frac{(\Delta x)^2}{2!} \frac{\partial^2 f}{\partial x^2} \\ &+ \frac{(\Delta y)^2}{2!} \frac{\partial^2 f}{\partial y^2} + 2 \frac{\Delta x \Delta y}{2!} \frac{\partial^2 f}{\partial x \partial y} + O\left[(\Delta x)^3, (\Delta y)^3\right] \end{aligned}$$

Using indices i and j to represent a grid point at x, y,

$$\begin{aligned} f_{i+1,j+1} &= f_{i,j} + \Delta x \frac{\partial f}{\partial x} + \Delta y \frac{\partial f}{\partial y} + \Delta x \Delta y \frac{\partial^2 f}{\partial x \partial y} \\ &+ \frac{(\Delta x)^2}{2} \frac{\partial^2 f}{\partial x^2} + \frac{(\Delta y)^2}{2} \frac{\partial^2 f}{\partial y^2} + O\left[(\Delta x)^3, (\Delta y)^3\right] \end{aligned} \tag{2-37}$$

Similarly, the expansions of $f(x - \Delta x,\ y - \Delta y)$, $f(x + \Delta x,\ y - \Delta y)$, and $f(x - \Delta x,\ y + \Delta y)$ yield:

$$\begin{aligned} f_{i-1,j-1} &= f_{i,j} - \Delta x \frac{\partial f}{\partial x} - \Delta y \frac{\partial f}{\partial y} + \Delta x \Delta y \frac{\partial^2 f}{\partial x \partial y} \\ &+ \frac{(\Delta x)^2}{2} \frac{\partial^2 f}{\partial x^2} + \frac{(\Delta y)^2}{2} \frac{\partial^2 f}{\partial y^2} + O\left[(\Delta x)^3, (\Delta y)^3\right] \end{aligned} \tag{2-38}$$

$$\begin{aligned} f_{i+1,j-1} &= f_{i,j} + \Delta x \frac{\partial f}{\partial x} - \Delta y \frac{\partial f}{\partial y} - \Delta x \Delta y \frac{\partial^2 f}{\partial x \partial y} \\ &+ \frac{(\Delta x)^2}{2} \frac{\partial^2 f}{\partial x^2} + \frac{(\Delta y)^2}{2} \frac{\partial^2 f}{\partial y^2} + O\left[(\Delta x)^3, (\Delta y)^3\right] \end{aligned} \tag{2-39}$$

and

$$\begin{aligned} f_{i-1,j+1} &= f_{i,j} - \Delta x \frac{\partial f}{\partial x} + \Delta y \frac{\partial f}{\partial y} - \Delta x \Delta y \frac{\partial^2 f}{\partial x \partial y} \\ &+ \frac{(\Delta x)^2}{2} \frac{\partial^2 f}{\partial x^2} + \frac{(\Delta y)^2}{2} \frac{\partial^2 f}{\partial y^2} + O\left[(\Delta x)^3, (\Delta y)^3\right] \end{aligned} \tag{2-40}$$

From Equations (2-37) through (2-40),

$$\frac{\partial^2 f}{\partial x \partial y} = \frac{f_{i+1,j+1} - f_{i+1,j-1} - f_{i-1,j+1} + f_{i-1,j-1}}{4(\Delta x)(\Delta y)} + O\left[(\Delta x)^2, (\Delta y)^2\right]$$

Finite difference approximation of higher order derivatives may be obtained by following the same procedure.

2.6.2 The Use of Partial Derivatives with Respect to One Independent Variable

Approximate expressions for partial derivatives with respect to one independent variable have already been developed. These expressions can now be used to compute mixed partial derivatives. Again, consider the partial derivative

$$\frac{\partial^2 f}{\partial x \partial y} = \frac{\partial}{\partial x}\left(\frac{\partial f}{\partial y}\right)$$

Using central differencing of order $(\Delta y)^2$ for $\partial f/\partial y$, one may write

$$\frac{\partial f}{\partial y} = \frac{f_{i,j+1} - f_{i,j-1}}{2\Delta y} + O(\Delta y)^2$$

Therefore,

$$\frac{\partial^2 f}{\partial x \partial y} = \frac{\partial}{\partial x}\left[\frac{f_{i,j+1} - f_{i,j-1}}{2\Delta y}\right] + O(\Delta y)^2 = \frac{1}{2\Delta y}\left[\left.\frac{\partial f}{\partial x}\right|_{i,j+1} - \left.\frac{\partial f}{\partial x}\right|_{i,j-1}\right] + O(\Delta y)^2$$

Now apply a central differencing of order $(\Delta x)^2$ for $\partial f/\partial x$:

$$\frac{\partial^2 f}{\partial x \partial y} = \frac{1}{2\Delta y}\left[\frac{f_{i+1,j+1} - f_{i-1,j+1}}{2\Delta x} - \frac{f_{i+1,j-1} - f_{i-1,j-1}}{2\Delta x}\right] + O\left[(\Delta x)^2, (\Delta y)^2\right]$$

Hence,

$$\frac{\partial^2 f}{\partial x \partial y} = \frac{f_{i+1,j+1} - f_{i-1,j+1} - f_{i+1,j-1} + f_{i-1,j-1}}{4(\Delta x)(\Delta y)} + O\left[(\Delta x)^2, (\Delta y)^2\right]$$

As a second example, consider calculation of the approximate finite difference expression for the mixed partial derivative of $\partial^2 f/(\partial x \partial y)$ which is of order $(\Delta x, \Delta y)$. In this particular example, forward differencing is used for all the derivatives.

$$\frac{\partial^2 f}{\partial x \partial y} = \frac{\partial}{\partial x}\left(\frac{\partial f}{\partial y}\right) = \frac{\partial}{\partial x}\left[\frac{f_{i,j+1} - f_{i,j}}{\Delta y}\right] + O(\Delta y)$$

$$= \frac{1}{\Delta y}\left[\left.\frac{\partial f}{\partial x}\right|_{i,j+1} - \left.\frac{\partial f}{\partial x}\right|_{i,j}\right] + O(\Delta y)$$

$$= \frac{1}{\Delta y}\left[\frac{f_{i+1,j+1} - f_{i,j+1}}{\Delta x} - \frac{f_{i+1,j} - f_{i,j}}{\Delta x}\right] + O(\Delta x, \Delta y)$$

$$= \frac{f_{i+1,j+1} - f_{i,j+1} - f_{i+1,j} + f_{i,j}}{\Delta x \Delta y} + O(\Delta x, \Delta y)$$

Similar approximations can be obtained by using backward differencing for the derivatives, or by using forward differencing for the x derivatives and backward differencing for the y derivatives, or vice versa.

The finite difference approximations derived in this chapter will be used in the following chapters to formulate various FDEs of model PDEs. Subsequently, various numerical solutions of the FDEs will be investigated.

2.7 Summary Objectives

At this point, you should be able to do the following:

1. Use Taylor series expansion to approximate partial derivatives with various orders of accuracy.
2. Use polynomials to approximate partial derivatives.
3. Approximate partial derivatives with finite difference expressions with variable step sizes.
4. Approximate mixed partial derivatives.
5. Solve the problems for Chapter Two.

2.8 Problems

2.1 Derive a central difference approximation for $\partial^3 f/\partial x^3$ which is of order $(\Delta x)^2$.

2.2 Determine an approximate backward difference representation for $\partial^3 f/\partial x^3$ which is of order (Δx), given evenly spaced grid points f_i, f_{i-1}, f_{i-2}, and f_{i-3} by means of:

(a) Taylor series expansions.

(b) A backward difference recurrence formula.

(c) A third-degree polynomial passing through the four points.

2.3 Find a forward difference approximation of the order (Δx) for $\partial^6 f/\partial x^6$.

2.4 Derive a backward difference approximation of $\partial f/\partial x$ with the use of a second order polynomial. Use unequally spaced grid points as shown in Figure (P2.4).

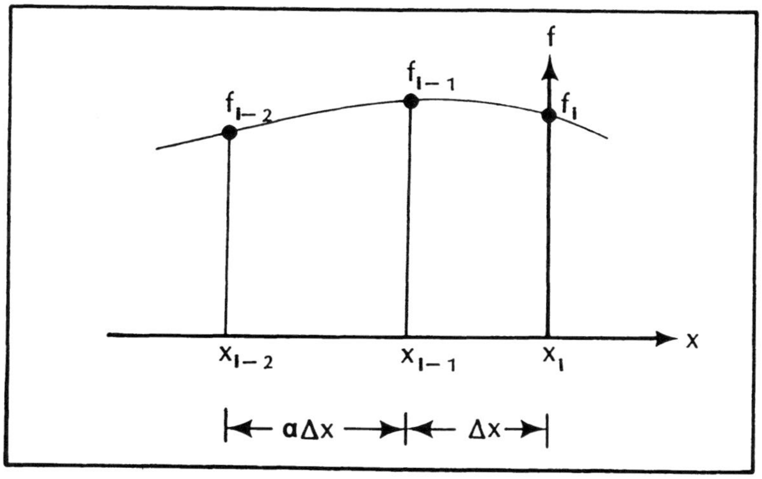

Figure P2.4. Grid points for problem 2.4.

2.5 Derive a first-order backward finite difference approximation for the mixed partial derivative $\dfrac{\partial^2 f}{\partial x \partial y}$.

2.6 Derive a third-order accurate, forward difference approximation for $\dfrac{\partial f}{\partial x}$.

2.7 Given the function $f(x) = \cos \pi x$, find $f'(0.25)$ using forward and backward difference representations of order $(\Delta x)^2$. Use step sizes of 0.01, 0.1, and 0.25. Compare and discuss your findings.

2.8 Solve problem 2.7 using a second-order accurate central difference approximation.

2.9 Compute the first derivative of the function $f(x) = \tan(\pi x/4)$ at $x = 1.5$, using first-order forward and backward approximations. Use step sizes of 0.01, 0.1, 0.5, and 0.8. Discuss the results.

2.10 Use the second-order accurate central difference approximation and the first-order forward difference approximation to evaluate $\frac{\partial}{\partial x}(e^x)$ at $x = 1$. A step size of $\Delta x = 0.1$ is to be employed. Recall that $e = 2.71828$.

2.11 Write a computer program to compute $\partial f/\partial x$, $\partial^2 f/\partial x^2$, $\partial^3 f/\partial x^3$, and $\partial^4 f/\partial x^4$ at $x = 1.5$ for $f = \sin(\pi x/2)$. Use central differencing of second order with the following step sizes: $\Delta x = 0.0005$, 0.001, 0.01, 0.1, 0.2, 0.3, and 0.4. Also determine the % of error for each computation by

$$\% \text{ Error} = 100\left(\frac{\text{Numerical value} - \text{Analytical value}}{\text{Analytical value}}\right)$$

2.12 Given the function $f(x) = x^3 - 5x$, write a program to compute $\partial f/\partial x$ and $\partial^2 f/\partial x^2$ at $x = 0.5$ and 1.5 by second-order central, backward and forward differencing. Use step sizes of 0.00001, 0.0001, 0.001, 0.01, 0.1, 0.2, and 0.3. Determine the numerical error for each computation.

2.13 Compute $f'(1)$, $f'(3)$, and $f'(4)$ for a function represented by the following data. Use finite differencing of order $(\Delta x)^2$.

x	0	1	2	3	3.6	4	5
$f(x)$	0	1	4	3	2.4	2	1

	f_i	f_{i+1}	f_{i+2}	f_{i+3}	f_{i+4}
$(\Delta x)\dfrac{\partial f}{\partial x}$	-1	1			
$(\Delta x)^2\dfrac{\partial^2 f}{\partial x^2}$	1	-2	1		
$(\Delta x)^3\dfrac{\partial^3 f}{\partial x^3}$	-1	3	-3	1	
$(\Delta x)^4\dfrac{\partial^4 f}{\partial x^4}$	1	-4	6	-4	1

Table 2.1 Forward difference representations of $O(\Delta x)$.

	f_{i-4}	f_{i-3}	f_{i-2}	f_{i-1}	f_i
$(\Delta x)\dfrac{\partial f}{\partial x}$				-1	1
$(\Delta x)^2\dfrac{\partial^2 f}{\partial x^2}$			1	-2	1
$(\Delta x)^3\dfrac{\partial^3 f}{\partial x^3}$		-1	3	-3	1
$(\Delta x)^4\dfrac{\partial^4 f}{\partial x^4}$	1	-4	6	-4	1

Table 2.2 Backward difference representations of $O(\Delta x)$.

	f_{i-2}	f_{i-1}	f_i	f_{i+1}	f_{i+2}
$2(\Delta x)\frac{\partial f}{\partial x}$		-1	0	1	
$(\Delta x)^2\frac{\partial^2 f}{\partial x^2}$		1	-2	1	
$2(\Delta x)^3\frac{\partial^3 f}{\partial x^3}$	-1	2	0	-2	1
$(\Delta x)^4\frac{\partial^4 f}{\partial x^4}$	1	-4	6	-4	1

Table 2.3 Central difference representations of $O(\Delta x)^2$.

	f_i	f_{i+1}	f_{i+2}	f_{i+3}	f_{i+4}	f_{i+5}
$2(\Delta x)\frac{\partial f}{\partial x}$	-3	4	-1			
$(\Delta x)^2\frac{\partial^2 f}{\partial x^2}$	2	-5	4	-1		
$2(\Delta x)^3\frac{\partial^3 f}{\partial x^3}$	-5	18	-24	14	-3	
$(\Delta x)^4\frac{\partial^4 f}{\partial x^4}$	3	-14	26	-24	11	-2

Table 2.4 Forward difference representations of $O(\Delta x)^2$.

	f_{i-5}	f_{i-4}	f_{i-3}	f_{i-2}	f_{i-1}	f_i
$2(\Delta x)\frac{\partial f}{\partial x}$				1	−4	3
$(\Delta x)^2\frac{\partial^2 f}{\partial x^2}$			−1	4	−5	2
$2(\Delta x)^3\frac{\partial^3 f}{\partial x^3}$		3	−14	24	−18	5
$(\Delta x)^4\frac{\partial^4 f}{\partial x^4}$	−2	11	−24	26	−14	3

Table 2.5 Backward difference representations of $O(\Delta x)^2$.

	f_{i-3}	f_{i-2}	f_{i-1}	f_i	f_{i+1}	f_{i+2}	f_{i+3}
$12(\Delta x)\frac{\partial f}{\partial x}$		1	−8	0	8	−1	
$12(\Delta x)^2\frac{\partial^2 f}{\partial x^2}$		−1	16	−30	16	−1	
$8(\Delta x)^3\frac{\partial^3 f}{\partial x^3}$	1	−8	13	0	−13	8	−1
$6(\Delta x)^4\frac{\partial^4 f}{\partial x^4}$	−1	12	−39	56	−39	12	−1

Table 2.6 Central difference representations of $O(\Delta x)^4$.

Chapter 3

Parabolic Partial Differential Equations

3.1 Introductory Remarks

Equations of motion in fluid mechanics are frequently reduced to parabolic formulations. Boundary layer equations and Parabolized Navier-Stokes (PNS) equations are examples of such formulations. In addition, the unsteady heat conduction equation is also parabolic.

In this chapter, various finite difference formulations of the model parabolic differential equation will be investigated. Each of the resulting FDEs has its own merit as far as accuracy, consistency, and stability are concerned. Although for simple model equations some of the methods behave in much the same way, they are investigated here to familiarize the reader with each of the methods. Obviously, all the possible methods of solutions for parabolic equations cannot be presented here; however, the most commonly used techniques are discussed. These methods are applied to the model equations, and the resulting solutions for given initial and boundary conditions are presented.

3.2 Finite Difference Formulations

A typical parabolic second-order PDE is the unsteady heat conduction equation, which is considered first in one-space dimension, and later in higher space dimensions. The model equation under consideration has the following form

$$\frac{\partial u}{\partial t} = \alpha \frac{\partial^2 u}{\partial x^2} \tag{3-1}$$

where α is assumed constant.

Various finite difference approximations can be used to represent the derivatives in Equation (3-1). The resulting finite difference equations will be discussed in detail shortly. For now, $\partial u/\partial t$ will be represented by a forward difference approximation which is of order Δt:

$$\frac{\partial u}{\partial t} = \frac{u_i^{n+1} - u_i^n}{\Delta t} + O(\Delta t) \tag{3-2}$$

Using the second-order central differencing of order $(\Delta x)^2$ for the diffusion term, Equation (3-1) can be approximated by the following difference equation:

$$\frac{u_i^{n+1} - u_i^n}{\Delta t} = \alpha \frac{u_{i+1}^n - 2u_i^n + u_{i-1}^n}{(\Delta x)^2} \tag{3-3}$$

In this equation, u_i^{n+1} is the only unknown and, therefore, it can be computed from the following:

$$u_i^{n+1} = u_i^n + \frac{\alpha(\Delta t)}{(\Delta x)^2}(u_{i+1}^n - 2u_i^n + u_{i-1}^n) \tag{3-4}$$

Thus, the second-order PDE has been replaced by an algebraic equation. Graphical representation of the grid points in Equation (3-4) is shown in Figure 3-1.

Note that the value of the dependent variable at time level n is known from a previous solution or given as initial data; i.e., the computed values at $n+1$ depend only on the past history. To start the solution, an initial condition and two boundary conditions must be specified. The formulation of a continuum equation in a finite difference equation (such as Equation (3-4)), which expresses one unknown in terms of the known values, is known as the explicit method. Since each finite difference equation involves only one unknown, the resulting equations at time level $n + 1$ are solved independently to provide the values of the unknowns. Now assume that the solution of a PDE using explicit finite differencing has progressed and that the unknowns at time level $n + 4$ are being computed (Figure 3-2).

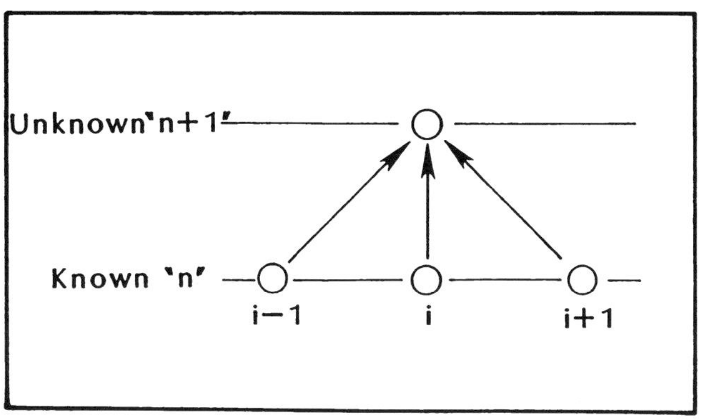

Figure 3-1. Grid points for the explicit formulation.

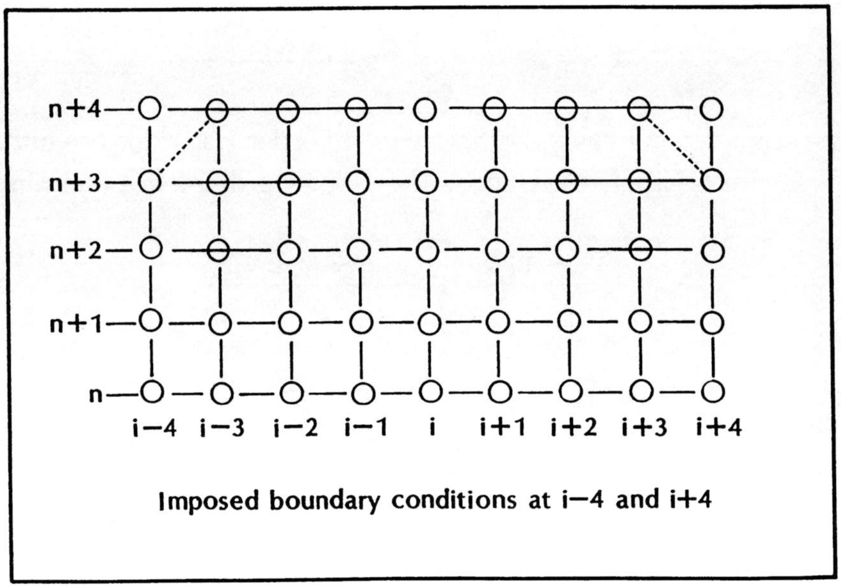

Figure 3-2. Effect of the boundary conditions for the explicit formulation.

Note that the information at the boundaries at the same time level $(n+4)$ does not feed into the computation of the unknowns at $n+4$. That is contrary to the physics of the problem, since the characteristic lines for this parabolic equation are lines of constant t. Thus, in an explicit formulation, the boundary conditions lag behind computation by one step. Next, a technique for which the formulation may include the boundary conditions at every time level for the computations will be reviewed.

When a first-order backward difference approximation for the time derivative and a second-order central difference approximation for the spatial derivative is used, the discretized equation takes the form

$$\frac{u_i^{n+1} - u_i^n}{\Delta t} = \alpha \frac{u_{i+1}^{n+1} - 2u_i^{n+1} + u_{i-1}^{n+1}}{(\Delta x)^2} \tag{3-5}$$

In this equation, there are three unknowns: u_{i-1}^{n+1}, u_i^{n+1}, and u_{i+1}^{n+1}. Figure 3-3 shows the grid points involved in Equation (3-5).

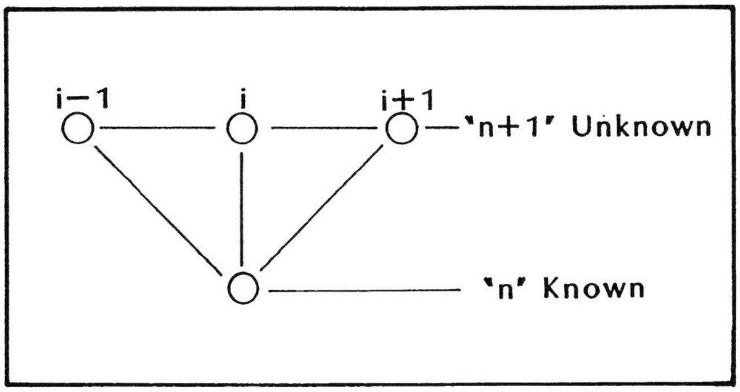

Figure 3-3. Grid points for the implicit formulation.

The computation of the unknowns would require a set of coupled finite difference equations, which are found by writing finite difference equations for all the grid points. In order to apply a standard solution procedure for Equation (3-5), it is rearranged as

$$\frac{\alpha\Delta t}{(\Delta x)^2}u_{i-1}^{n+1} - \left[1 + 2\frac{\alpha\Delta t}{(\Delta x)^2}\right]u_i^{n+1} + \frac{\alpha\Delta t}{(\Delta x)^2}u_{i+1}^{n+1} = -u_i^n \tag{3-6}$$

A formulation of this type, which includes more than one unknown in each FDE, is known as an implicit method. Equation (3-6) may be expressed in a general form by defining the coefficients of u_{i-1}^{n+1}, u_i^{n+1}, and u_{i+1}^{n+1} as a_i^n, b_i^n, and c_i^n, and by defining the right-hand side of the equation as D_i^n; therefore,

$$a_i^n u_{i-1}^{n+1} + b_i^n u_i^{n+1} + c_i^n u_{i+1}^{n+1} = D_i^n \tag{3-7}$$

This finite difference equation is written for all grid points at the advanced time level, resulting in a set of algebraic equations. When these equations are put in a matrix form, the coefficient matrix is tridiagonal. The solution procedure for the tridiagonal system is presented in Appendix B.

Having identified two techniques for the discretization of the continuum equation, we must yet learn which method is superior for a particular application. What about the step sizes Δt and Δx? How does the selection of the step size affect the solution? Will the solution be stable, and how accurate will it be? These important issues are addressed in the upcoming sections and are illustrated by various examples. The very important issue of stability is addressed in Chapter 4, which covers stability analysis of finite difference formulations. Various formulations using explicit and implicit methods are presented in the following sections.

3.3 Explicit Methods

This section introduces some of the commonly used explicit methods for solving parabolic equations.

3.3.1 *The forward time/central space (FTCS) method.* As discussed earlier, using forward difference approximation for the time derivative and central differencing for the space derivative in Equation (3-1), one obtains

$$u_i^{n+1} = u_i^n + \frac{\alpha(\Delta t)}{(\Delta x)^2}(u_{i+1}^n - 2u_i^n + u_{i-1}^n) \tag{3-8}$$

which is of order $[(\Delta t), (\Delta x)^2]$. It will be shown that the solution is stable for $\alpha\Delta t/(\Delta x)^2 \leq 1/2$. The grid points involved in Equation (3-8) were shown in Figure 3-1.

3.3.2 *The Richardson method.* In this approximation, central differencing is used for both time and space derivatives. For the model Equation (3-1), the resulting FDE is

$$\frac{u_i^{n+1} - u_i^{n-1}}{2\Delta t} = \alpha \frac{u_{i+1}^n - 2u_i^n + u_{i-1}^n}{(\Delta x)^2}$$

which is of order $[(\Delta t)^2, (\Delta x)^2]$. It turns out that this method is unconditionally unstable and, therefore, has no practical value.

3.3.3 *The DuFort-Frankel method.* In this formulation the time derivative $\partial u/\partial t$ is approximated by a central differencing which is of order $(\Delta t)^2$. The second-order space derivative is also approximated by a central differencing of order $(\Delta x)^2$; however, due to stability considerations, u_i^n in the diffusion term is replaced by the average value of u_i^{n+1} and u_i^{n-1}. This formulation is a modification of the Richardson method. The resulting FDE is

$$\frac{u_i^{n+1} - u_i^{n-1}}{2\Delta t} = \alpha \frac{u_{i+1}^n - 2\dfrac{u_i^{n+1} + u_i^{n-1}}{2} + u_{i-1}^n}{(\Delta x)^2} \tag{3-9}$$

From which,

$$u_i^{n+1} = u_i^{n-1} + \frac{2\alpha(\Delta t)}{(\Delta x)^2}\left[u_{i+1}^n - u_i^{n+1} - u_i^{n-1} + u_{i-1}^n\right] \tag{3-10}$$

Even though the $n+1$ value appears on the right-hand side, it is at i location only, so that the equation can be solved explicitly for the unknown u_i at the time level $n+1$. Thus,

$$\left[1 + \frac{2\alpha(\Delta t)}{(\Delta x)^2}\right]u_i^{n+1} = \left[1 - 2\frac{\alpha(\Delta t)}{(\Delta x)^2}\right]u_i^{n-1} + \frac{2\alpha(\Delta t)}{(\Delta x)^2}[u_{i+1}^n + u_{i-1}^n] \tag{3-11}$$

This method is of order $[(\Delta t)^2, (\Delta x)^2, (\Delta t/\Delta x)^2]$. Interestingly enough, this formulation is unconditionally stable! The additional term $(\Delta t/\Delta x)^2$ is included in the error term as a result of consistency analysis, which will be considered shortly.

The values of u_i at time levels n and $n-1$ are required to start the computation. Therefore, either two sets of data must be specified, or, from a practical point of view, a one-step method can be used as a starter. Of course, for the one-step (in Δt) starter solution, only one set of initial data, say at $n-1$, is required to generate the solution at n. With the values of u_i at $n-1$ and n specified, the DuFort-Frankel method can be used. Two points about this scheme must be kept in mind. First, the accuracy of the solution provided by the DuFort-Frankel method is affected by the accuracy of the starter solution. Second, since the solution at the unknown station requires data from two previous stations, computer storage requirements will increase. The grid points involved in Equation (3-11) are shown in Figure 3-4.

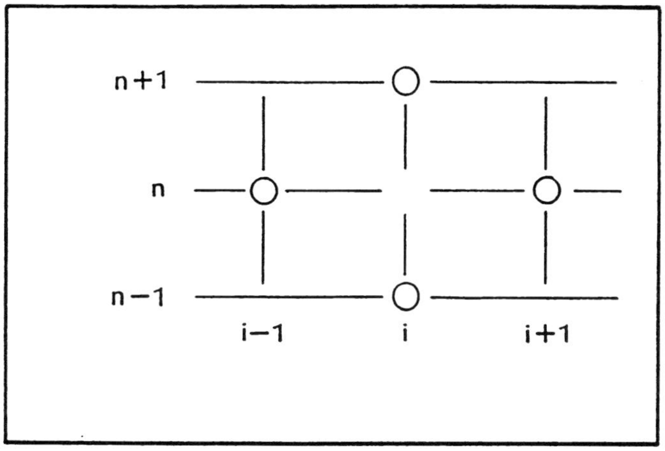

Figure 3-4. Grid points for the DuFort-Frankel method.

3.4 Implicit Methods

When model Equation (3-1) is discretized as

$$\frac{u_i^{n+1} - u_i^n}{(\Delta t)} = \alpha \frac{u_{i+1}^{n+1} - 2u_i^{n+1} + u_{i-1}^{n+1}}{(\Delta x)^2} \tag{3-12}$$

it is defined as being implicit, since more than one unknown appears in the finite difference equation. As a result, a set of simultaneous equations needs to be solved, which requires more computation time per time step. Implicit methods offer great advantage on the stability of the finite difference equations, since most are unconditionally stable. Therefore, a larger step size in time is permitted; however, the selection of a larger time step is limited due to accuracy consideration because an increase in time step will increase the truncation error of the finite difference equation. In this section some commonly used implicit formulations are described.

3.4.1 ***The Laasonen method.*** The simple formulation of Equation (3-12) is known as the Laasonen implicit method. Applying the formulation to all grid points would lead to a set of linear algebraic equations, for which a solution procedure is described in Appendix B. The grid points were shown in Figure 3-3.

3.4.2 ***The Crank-Nicolson method.*** If the diffusion term in Equation (3-1) is replaced by the average of the central differences at time levels n and $n+1$, the discretized equation would be of the form

$$\frac{u_i^{n+1}-u_i^n}{\Delta t}=\alpha\left(\frac{1}{2}\right)\left[\frac{u_{i+1}^{n+1}-2u_i^{n+1}+u_{i-1}^{n+1}}{(\Delta x)^2}+\frac{u_{i+1}^n-2u_i^n+u_{i-1}^n}{(\Delta x)^2}\right] \tag{3-13}$$

Note that the left side of the equation is a central difference of step $\Delta t/2$, i.e.,

$$\frac{\partial u}{\partial t}=\frac{u_i^{n+1}-u_i^n}{2\left(\frac{\Delta t}{2}\right)}$$

which is of order $(\Delta t)^2$.

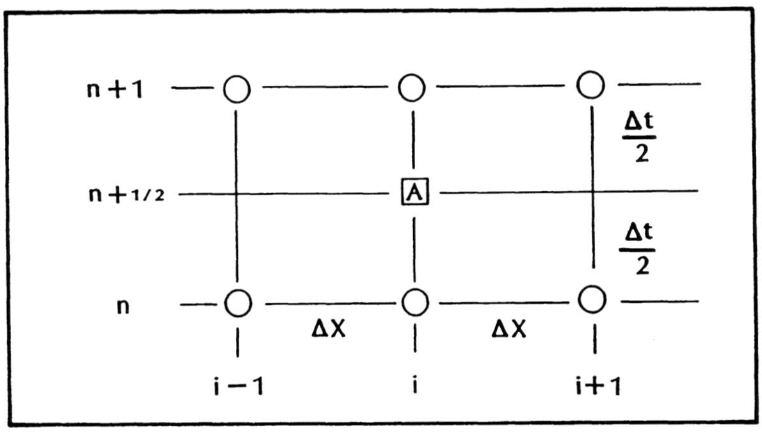

Figure 3-5. Grid points for the Crank-Nicolson implicit method.

In terms of the grid points (see Figure 3-5) the left side can be interpreted as the central difference representation of $\partial u/\partial t$ at point A, while the right side is the average of the diffusion term at the same point. The method may be thought of as the addition of two step computations as follows. Using the explicit method,

$$\frac{u_i^{n+\frac{1}{2}}-u_i^n}{\frac{\Delta t}{2}}=\alpha\frac{u_{i-1}^n-2u_i^n+u_{i+1}^n}{(\Delta x)^2} \tag{3-14}$$

while using the implicit method,

$$\frac{u_i^{n+1} - u_i^{n+\frac{1}{2}}}{\frac{\Delta t}{2}} = \alpha \frac{u_{i-1}^{n+1} - 2u_i^{n+1} + u_{i+1}^{n+1}}{(\Delta x)^2} \tag{3-15}$$

Adding Equations (3-14) and (3-15), one obtains

$$\frac{u_i^{n+1} - u_i^n}{\Delta t} = \frac{1}{2}\alpha \left[\frac{u_{i+1}^{n+1} - 2u_i^{n+1} + u_{i-1}^{n+1}}{(\Delta x)^2} + \frac{u_{i+1}^n - 2u_i^n + u_{i-1}^n}{(\Delta x)^2} \right] \tag{3-16}$$

Note that by this analogy it is difficult to recognize the order $(\Delta t)^2$ of the time derivative. This implicit method is unconditionally stable and is of order $[(\Delta t)^2, (\Delta x)^2]$, i.e., a second-order scheme.

3.4.3 *The Beta Formulation.* A general form of the finite difference equation for model Equation (3-1) can be written as

$$\frac{u_i^{n+1} - u_i^n}{\Delta t} = \alpha \left[\beta \frac{u_{i+1}^{n+1} - 2u_i^{n+1} + u_{i-1}^{n+1}}{(\Delta x)^2} + (1-\beta) \frac{u_{i+1}^n - 2u_i^n + u_{i-1}^n}{(\Delta x)^2} \right] \tag{3-17}$$

For $1/2 \leq \beta \leq 1$, the method is unconditionally stable. Note that for $\beta = 1/2$, the formulation is Crank-Nicolson implicit. For $0 \leq \beta < 1/2$, the formulation is conditionally stable. For $\beta = 0$, the formulation is FTCS explicit.

3.5 Applications

Various finite difference equations were used to represent the parabolic model Equation (3-1) in the previous section. It is extremely important to experiment with the application of these numerical techniques. It is hoped that by writing computer codes and analyzing the results, additional insights into the solution procedures are gained. Therefore, this section proposes an example and presents solutions by various methods. In addition, readers are encouraged to work out the problems proposed at the end of the chapter, since one gains valuable experience by writing computer codes and overcoming difficulties in the process.

As a first example, consider a fluid bounded by two parallel plates extended to infinity such that no end effects are encountered. The planar walls and the fluid are initially at rest. Now, the lower wall is suddenly accelerated in the x-direction, as illustrated in Figure 3-6. A spatial coordinate system is selected such that the lower wall includes the xz plane to which the y-axis is perpendicular. The spacing between two plates is denoted by h.

The Navier-Stokes equations for this problem may be expressed as

$$\frac{\partial u}{\partial t} = \nu \frac{\partial^2 u}{\partial y^2}$$

where ν is the kinematic viscosity of the fluid. It is required to compute the velocity profile $u = u(t, y)$. The initial and boundary conditions for this problem are stated as follows:

(a) Initial condition $\quad t = 0\,, \quad u = U_0 \quad \text{for} \quad y = 0$

$$u = 0 \quad \text{for} \quad 0 < y \leq h$$

(b) Boundary conditions $\quad t \geq 0\,, \quad u = U_0 \quad \text{for} \quad y = 0$

$$u = 0 \quad \text{for} \quad y = h$$

The fluid is oil with a kinematic viscosity of 0.000217 m^2/s, and the spacing between plates is 40 mm. The velocity of the lower wall is specified as $U_0 = 40$ m/s. A solution for the velocity is to be obtained up to 1.08 seconds.

A grid system with $\Delta y = 0.001$ m and various values of time steps is to be used to investigate the numerical schemes and the effect of time step on stability and accuracy. By selecting $j = 1$ at the lower surface and spatial step size of 0.001, j at the upper surface would be 41. JM and NM will be used to denote the number of steps in the y-direction and time, respectively. Note that $n = 1$ is used for $t = 0$, i.e., initial condition. The grid system is illustrated in Figure 3-6.

An attempt is made to solve the stated problem subject to the imposed initial and boundary conditions by the following:

(a) The FTCS explicit method with

(I) $\Delta t = 0.002\,, \quad NM = 541$

(II) $\Delta t = 0.00232\,, \quad NM = 541$

(b) The DuFort-Frankel explicit method with

(I) $\Delta t = 0.002\,, \quad NM = 541$

(II) $\Delta t = 0.003\,, \quad NM = 361$

(c) The Laasonen implicit method with

(I) $\Delta t = 0.002\,, \quad NM = 541$

(II) $\Delta t = 0.01\,, \quad NM = 109$

(d) The Crank-Nicolson method with

(I) $\Delta t = 0.002\,, \quad NM = 541$

(II) $\Delta t = 0.01\,, \quad NM = 109$

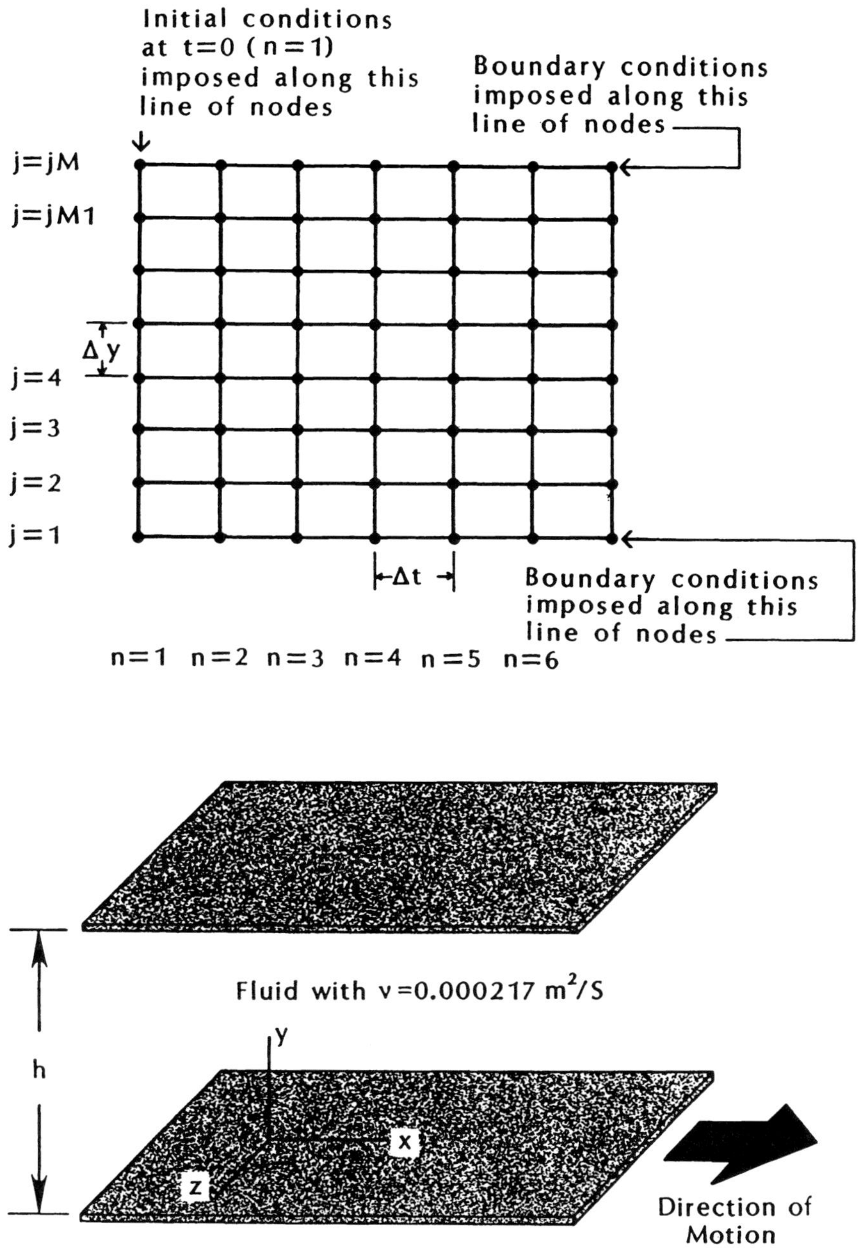

Figure 3-6. Physical space and the computational grid system for a suddenly accelerated plane wall.

Solutions:

Case a.I. In this case, the FTCS explicit method is to be used. As stated previously, the stability requirement of this explicit method is $\nu\Delta t/(\Delta y)^2 \leq 0.5$. (The term $\nu\Delta t/(\Delta y)^2 = d$ is known as the diffusion number.) For this particular application, the diffusion number is

$$d = \nu\frac{\Delta t}{(\Delta y)^2} = 0.000217\frac{(0.002)}{(0.001)^2} = 0.434$$

Therefore the stability condition is satisfied, and a stable solution is expected. The tabulated solution and the velocity profiles at various time levels are shown in Table 3-1 and Figure 3-7.

Case a.II. When the time step is increased to $\Delta t = 0.00232$, which is only a fraction of an increase over Case a.I., the diffusion number exceeds the stability requirement. In this case,

$$d = \nu\frac{\Delta t}{(\Delta y)^2} = 0.000217\frac{(0.00232)}{(0.001)^2} = 0.50344$$

With the step sizes indicated, an unstable solution is developed. The velocity profiles are given in Table 3-2 and shown in Figure 3-8. The oscillatory behavior of the unstable solution, in which the value of the dependent variable (in this problem, the velocity) at a grid point changes sign for each step, is known as dynamic instability. This concept is discussed further in Chapter 4.

Case b. The DuFort-Frankel explicit method behaves exceptionally well with regard to stability, since it turns out to be unconditionally stable. Therefore, it allows larger steps in time. However, large time steps should be selected carefully, since an increase in the time step also increases the truncation error. The computed values of velocity profiles are given in Table 3-3. Note that for the given values of step sizes in Case b.II., $d = \nu\Delta t/(\Delta y)^2 = (0.000217)(0.003)/(0.001)^2 = 0.651$, which exceeds the stability limit of the FTCS explicit method. Therefore, the DuFort-Frankel has the advantage of providing a stable solution with a large time step and thus shorter computational time. Its disadvantage is the starting procedure it requires. Since two sets of values are required, usually a one-step procedure is used to provide the necessary information as the second set of initial data.

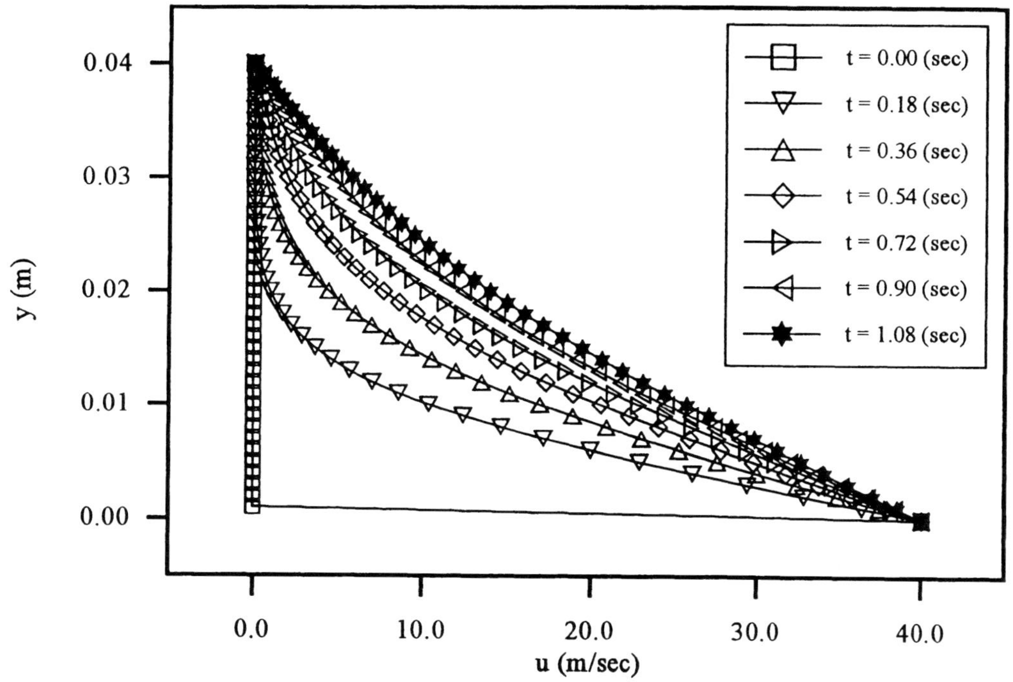

Figure 3-7. Velocity profiles obtained by the FTCS explicit method, $\Delta y = 0.001$, $\Delta t = 0.002$.

To generate the second set of data required for the DuFort-Frankel method, various schemes may be considered. If the FTCS explicit method is employed for this purpose, stability requirements must be considered. For this example, where $d = 0.651$, the FTCS explicit will result in an unstable solution. Therefore, if it is used to start the DuFort-Frankel method, some instabilities are introduced into the solution. To overcome this difficulty, one may use the Laasonen or Crank-Nicolson scheme, which provides stable solutions for the specified step size. Or, if the FTCS explicit is used, a step size which would satisfy the stability condition must be employed. For example, in this problem, a time step of 0.001 may be used and solution proceeds to $t = 0.003$. Now this stable solution is used as the second plane of data.

Case c. The Laasonen implicit method is unconditionally stable. Therefore, a larger time step is allowed as long as the truncation error is within the accuracy criteria for the specific problem. Note that the time step of $\Delta t = 0.01$ is larger by a factor of five than the time step used for the stable solution of the FTCS explicit method. This large step reduces the total computation time by decreasing the total number of grid points. However, the computation time per step is larger than that of the FTCS explicit method since a system of tridiagonal simultaneous equations

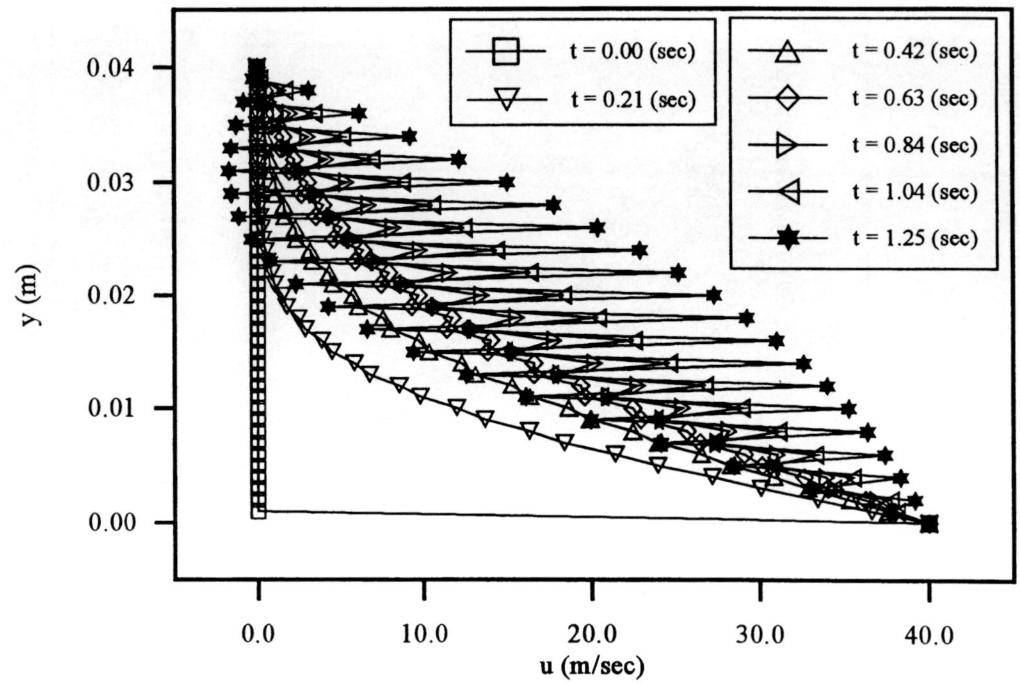

Figure 3-8. Velocity profiles obtained by the FTCS explicit method, $\Delta y = 0.001$, $\Delta t = 0.00232$.

must be solved. A routine for solving the system of simultaneous equations should be selected with care. Since a tridiagonal system is to be solved, a scheme specifically designed for the tridiagonal system of equations, which takes advantage of the zeros of the coefficient matrix, is used. Various routines are available; one such method is presented in Appendix B. The solution for the velocity profiles at various times are shown in Table 3-4.

Case d. Since the Crank-Nicolson implicit method is unconditionally stable, a time step of $\Delta t = 0.01$ does not encounter stability restrictions. The formulation results in a tridiagonal system of equations, which is solved using the method described in Appendix B. The solution is presented in Table 3-5.

3.6 Analysis

In the preceding section, various finite difference formulations were applied to the reduced form of the Navier-Stokes equation and the solutions were presented. The effect of the stability imposed by the diffusion number on the FTCS explicit method was clearly indicated. Therefore, for this method the selection of step sizes is limited due to the stability requirement. On the other hand, the DuFort-Frankel explicit method, the Laasonen implicit, and the Crank-Nicolson implicit methods are unconditionally stable and allow larger time steps. However, the accuracy requirement limits the use of large time steps, since an increase in time steps will increase the truncation errors introduced in the approximation process of the PDE. This point will be elaborated on shortly.

For the simple problem under consideration, an analytical solution may be obtained. The partial differential equation $\partial u/\partial t = \nu(\partial^2 u/\partial y^2)$ is transformed to an ordinary differential equation by defining $\eta = y/(2\sqrt{\nu t})$. For the imposed initial and boundary conditions, the solution is given in the form of a series of complementary error functions as

$$u = U_0 \left\{ \sum_{n=0}^{\infty} erfc[2n\eta_1 + \eta] - \sum_{n=0}^{\infty} erfc[2(n+1)\eta_1 - \eta] \right\}$$

$$= U_0 \left\{ erfc(\eta) - erfc(2\eta_1 - \eta) + erfc(2\eta_1 + \eta) - erfc(4\eta_1 - \eta) + erfc(4\eta_1 + \eta) - \cdots + \cdots \right\}$$

where $\eta_1 = \dfrac{h}{2\sqrt{\nu t}}$.

The analytical solution is given in Table 3-6.

The analytical result is used for code validation and for comparison of various methods. In addition, it is used to study the effect of step size on the accuracy of solutions.

An error term is defined as

$$\text{ER} = \left(\frac{\text{Analytical value} - \text{Computed value}}{\text{Analytical value}} \right) 100$$

A comparison of various methods used is illustrated in Figures 3-9 and 3-10.

The results shown in Figure 3-9 are the error term as defined above at time level of 0.18 seconds for the solutions obtained by a time step of 0.002, whereas Figure 3-10 represents time level of 1.08 seconds.

Two points to emphasize with regard to Figures 3-9 and 3-10 are: (1) For this application, the Crank-Nicolson scheme has minimum error in comparison with other schemes, and (2) the amount of error is decreased for all schemes as the solution is marched in time. This error reduction is due to a decrease in the influence of the initial data. This effect is generally true for both time marching as well as space marching schemes, i.e., the effect of initial data is "washed out" after few time (space) steps.

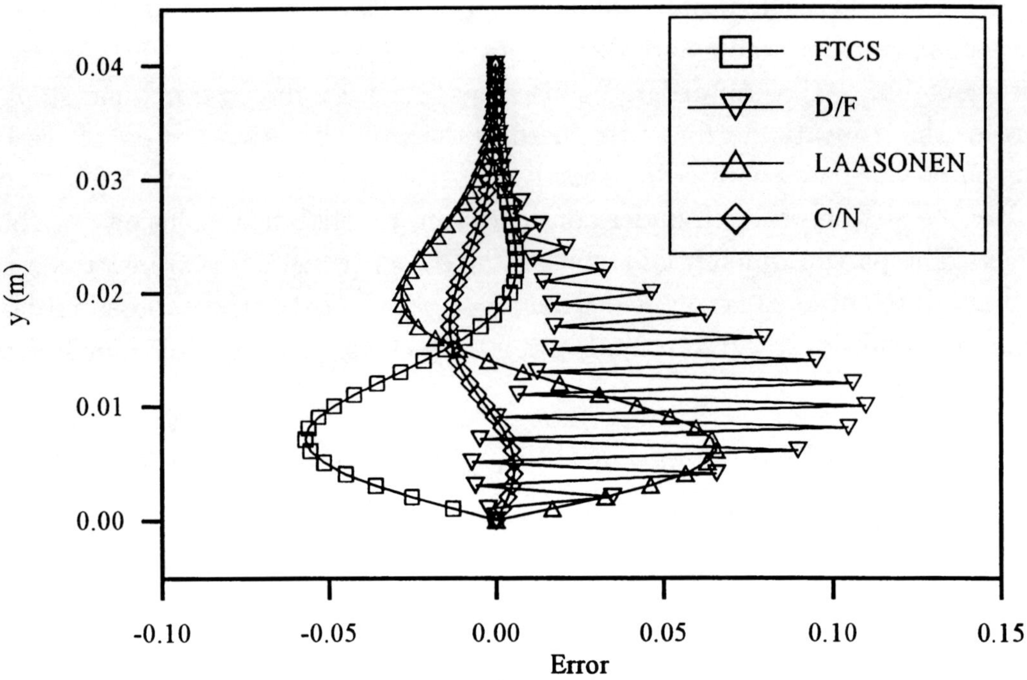

Figure 3-9. Comparison of error distributions for various schemes at $t = 0.18$ sec.

To study the effect of step size on the accuracy of the solution, the Laasonen implicit method is used to generate solutions using different time steps. Naturally, as the value of the time step increases, the total grid points in the computational domain decreases and, as a result, computation time is decreased. However, these advantages are accompanied by an increase in error. Accuracy comparison is illustrated in Figure 3-11.

Clearly, increasing the step size increases the error as shown. It should be noted that selecting a very small step size should also be avoided, since in addition to the enormous amount of computer time required for a solution, the accuracy of the solution will be dominated by round-off errors.

This example clearly illustrates the factors involved in the selection of step size. First of all, stability analysis imposes limitations on some of the numerical methods. Second, accuracy of the solution and the computation time required to generate the solution play an important role in selecting step sizes. Once a solution is obtained, it should be compared to other solutions, analytical or numerical, and to experimental data if available. As one gains experience with numerical methods and their behaviors, one develops a feeling for choosing the right step sizes and numerical techniques.

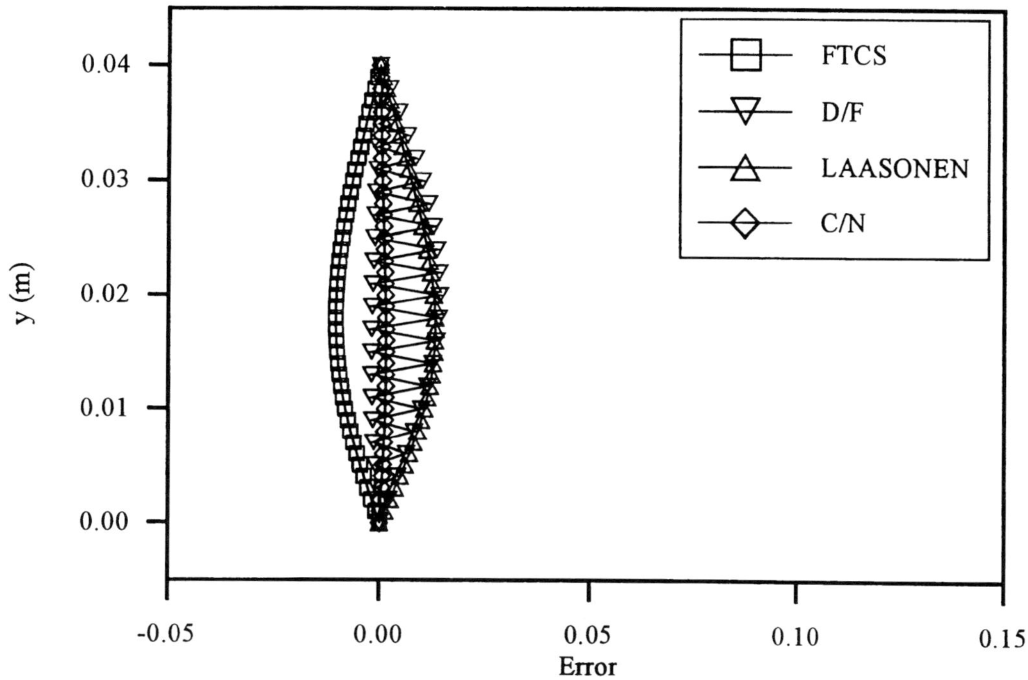

Figure 3-10. Comparison of error distributions for various schemes at $t = 1.08$ sec.

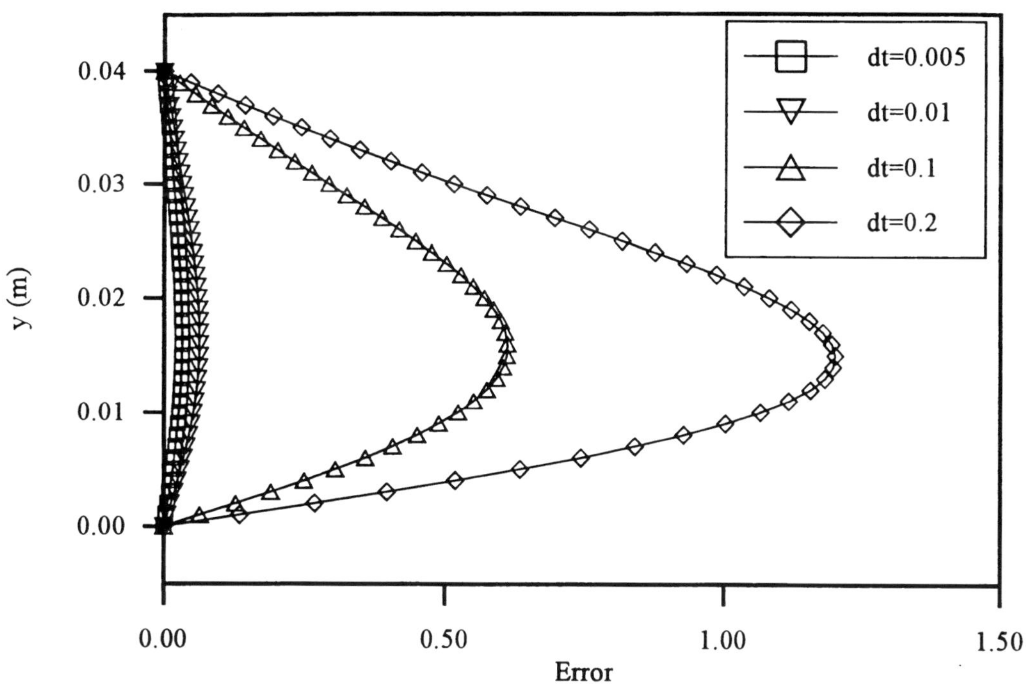

Figure 3-11. Effect of time steps on the solution obtained by the Laasonen implicit method at $t = 1.0$ sec.

Selecting a numerical technique depends on the problem represented by the governing PDE and the imposed initial and boundary conditions. Each of the methods described have their own advantages and disadvantages. Thus, when a problem is posed, the advantages and disadvantages of the available numerical methods should be carefully weighed before selecting a particular algorithm.

3.7 Parabolic Equations in Two-Space Dimensions

So far, various finite difference formulations of parabolic PDEs have been discussed by considering a model equation which was limited to unsteady, one-space dimension. In this section, the space dimension is extended to two, and an efficient method of solution is presented. Consider the model equation

$$\frac{\partial u}{\partial t} = \alpha \left[\frac{\partial^2 u}{\partial x^2} + \frac{\partial^2 u}{\partial y^2} \right] \tag{3-18}$$

where α is considered to be a constant.

An explicit finite difference equation using forward differencing for the time derivative and central differencing for the space derivatives is

$$\frac{u_{i,j}^{n+1} - u_{i,j}^n}{\Delta t} = \alpha \left[\frac{u_{i+1,j}^n - 2u_{i,j}^n + u_{i-1,j}^n}{(\Delta x)^2} + \frac{u_{i,j+1}^n - 2u_{i,j}^n + u_{i,j-1}^n}{(\Delta y)^2} \right] \tag{3-19}$$

which is of order $[(\Delta t), (\Delta x)^2, (\Delta y)^2]$. Stability analysis indicates that the method is stable for

$$\left[\frac{\alpha \Delta t}{(\Delta x)^2} + \frac{\alpha \Delta t}{(\Delta y)^2} \right] \leq \frac{1}{2}$$

Define the diffusion numbers

$$d_x = \frac{\alpha \Delta t}{(\Delta x)^2}$$

and

$$d_y = \frac{\alpha \Delta t}{(\Delta y)^2}$$

Then, the stability requirement is expressed as

$$(d_x + d_y) \leq \frac{1}{2}$$

To make an easy comparison with the stability restriction of the one-dimensional FDE (i.e., Equation (3-8)), select equal step sizes in space such that $\Delta x = \Delta y$. Then $d_x = d_y = d$, and the stability requirement of the FDE (3-19) is $d \leq 0.25$, which is twice as restrictive as the one-dimensional case. Such a severe restriction on the step sizes makes the explicit formulation given by (3-19) an inefficient procedure for some applications.

Instead, consider an implicit formulation for which the FDE is

$$\frac{u_{i,j}^{n+1} - u_{i,j}^{n}}{(\Delta t)} = \alpha\left[\frac{u_{i+1,j}^{n+1} - 2u_{i,j}^{n+1} + u_{i-1,j}^{n+1}}{(\Delta x)^2} + \frac{u_{i,j+1}^{n+1} - 2u_{i,j}^{n+1} + u_{i,j-1}^{n+1}}{(\Delta y)^2}\right]$$

From which it follows that

$$d_x u_{i+1,j}^{n+1} + d_x u_{i-1,j}^{n+1} - (2d_x + 2d_y + 1)u_{i,j}^{n+1} + d_y u_{i,j-1}^{n+1} + d_y u_{i,j+1}^{n+1} = -u_{ij}^{n} \tag{3-20}$$

By defining the coefficients of the unknowns as a, b, c, d, and e, and the right-hand side by f, Equation (3-20) may be written as

$$a_{i,j}u_{i+1,j}^{n+1} + b_{i,j}u_{i-1,j}^{n+1} + c_{i,j}u_{i,j}^{n+1} + d_{i,j}u_{i,j-1}^{n+1} + e_{i,j}u_{i,j+1}^{n+1} = f_{i,j}^{n}$$

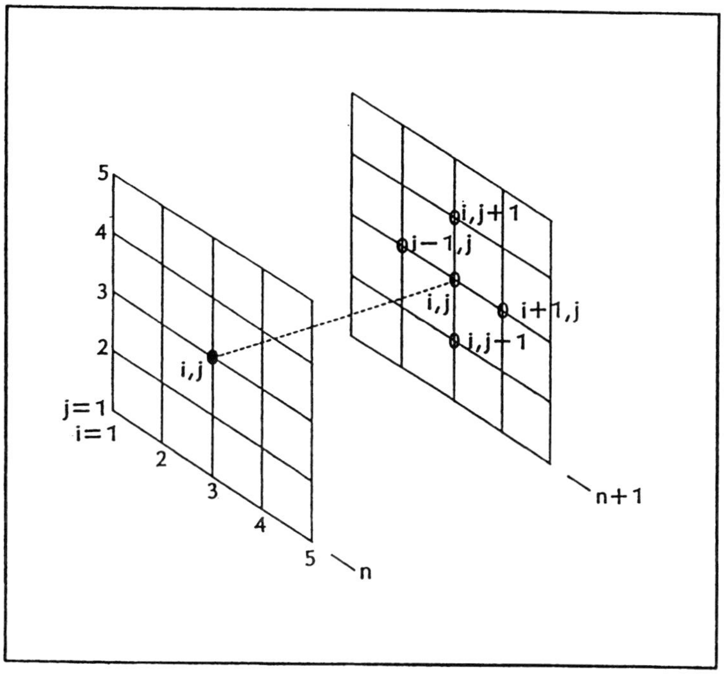

Figure 3-12. The 5 by 5 grid system for Equation (3-20).

For discussion purposes, consider the 5 by 5 grid system as shown in Figure 3-12. There are a total of nine unknowns at time level $n+1$. Therefore, a total of nine simultaneous equations must be solved. The implicit finite difference equations for the grid system of Figure 3-12 are

$$a_{2,2}u_{3,2} + c_{2,2}u_{2,2} + e_{2,2}u_{2,3} = f_{2,2} - b_{2,2}u_{1,2} - d_{2,2}u_{2,1}$$

$$a_{2,3}u_{3,3} + c_{2,3}u_{2,3} + d_{2,3}u_{2,2} + e_{2,3}u_{2,4} = f_{2,3} - b_{2,3}u_{1,3}$$

$$a_{2,4}u_{3,4} + c_{2,4}u_{2,4} + d_{2,4}u_{2,3} = f_{2,4} - b_{2,4}u_{1,4} - e_{2,4}u_{2,5}$$

$$a_{3,2}u_{4,2} + b_{3,2}u_{2,2} + c_{3,2}u_{3,2} + e_{3,2}u_{3,3} = f_{3,2} + d_{3,2}u_{3,1}$$

$$a_{3,3}u_{4,3} + b_{3,3}u_{2,3} + c_{3,3}u_{3,3} + d_{3,3}u_{3,2} + e_{3,3}u_{3,4} = f_{3,3}$$

$$a_{3,4}u_{4,4} + b_{3,4}u_{2,4} + c_{3,4}u_{3,4} + d_{3,4}u_{3,3} = f_{3,4} - e_{3,4}u_{3,5}$$

$$b_{4,2}u_{3,2} + c_{4,2}u_{4,2} + e_{4,2}u_{4,3} = f_{4,2} - a_{4,2}u_{5,2} - d_{4,2}u_{4,1}$$

$$b_{4,3}u_{3,3} + c_{4,3}u_{4,3} + d_{4,3}u_{4,2} + e_{4,3}u_{4,4} = f_{4,3} - a_{4,3}u_{5,3}$$

$$b_{4,4}u_{3,4} + c_{4,4}u_{4,4} + d_{4,4}u_{4,3} = f_{4,4} - a_{4,4}u_{5,4} - e_{4,4}u_{4,5}$$

where all the known quantities from the imposed boundary conditions have been moved to the right-hand side and added to the known quantities from the previous n time level. The data at time level n are provided from the imposed initial condition for the first level of computation and, subsequently, from the solution at the previous station. The set of equations can be written in a matrix form as

$$\begin{bmatrix} c_{2,2} & e_{2,2} & 0 & a_{2,2} & 0 & 0 & 0 & 0 & 0 \\ d_{2,3} & c_{2,3} & e_{2,3} & 0 & a_{2,3} & 0 & 0 & 0 & 0 \\ 0 & d_{2,4} & c_{2,4} & 0 & 0 & a_{2,4} & 0 & 0 & 0 \\ b_{3,2} & 0 & 0 & c_{3,2} & e_{3,2} & 0 & a_{3,2} & 0 & 0 \\ 0 & b_{3,3} & 0 & d_{3,3} & c_{3,3} & e_{3,3} & 0 & a_{3,3} & 0 \\ 0 & 0 & b_{3,4} & 0 & d_{3,4} & c_{3,4} & 0 & 0 & a_{3,4} \\ 0 & 0 & 0 & b_{4,2} & 0 & 0 & c_{4,2} & e_{4,2} & 0 \\ 0 & 0 & 0 & 0 & b_{4,3} & 0 & d_{4,3} & c_{4,3} & e_{4,3} \\ 0 & 0 & 0 & 0 & 0 & b_{4,4} & 0 & d_{4,4} & c_{4,4} \end{bmatrix} \begin{bmatrix} u_{2,2} \\ u_{2,3} \\ u_{2,4} \\ u_{3,2} \\ u_{3,3} \\ u_{3,4} \\ u_{4,2} \\ u_{4,3} \\ u_{4,4} \end{bmatrix} = \begin{bmatrix} f_{2,2} - b_{2,2}u_{1,2} - d_{2,2}u_{2,1} \\ f_{2,3} - b_{2,3}u_{1,3} \\ f_{2,4} - b_{2,4}u_{1,4} - e_{2,4}u_{2,5} \\ f_{3,2} - d_{3,2}u_{3,1} \\ f_{3,3} \\ f_{3,4} - e_{3,4}u_{3,5} \\ f_{4,2} - a_{4,2}u_{5,2} - d_{4,2}u_{4,1} \\ f_{4,3} - a_{4,3}u_{5,3} \\ f_{4,4} - a_{4,4}u_{5,4} - e_{4,4}u_{4,5} \end{bmatrix}$$

The coefficient matrix is pentadiagonal. The solution procedure for a pentadiagonal system of equations is also very time-consuming. One way to overcome the shortcomings and inefficiency of the method described above is to use a splitting method. This method is known as the alternating direction implicit method or ADI. The algorithm produces two sets of tridiagonal simultaneous equations to be solved in sequence. Earlier, an efficient method of solution for tridiagonal systems of equations was introduced. The finite difference equations of model Equation (3-18) in the ADI formulation are

$$\frac{u_{i,j}^{n+\frac{1}{2}} - u_{i,j}^{n}}{\left(\frac{\Delta t}{2}\right)} = \alpha \left[\frac{u_{i+1,j}^{n+\frac{1}{2}} - 2u_{i,j}^{n+\frac{1}{2}} + u_{i-1,j}^{n+\frac{1}{2}}}{(\Delta x)^2} + \frac{u_{i,j+1}^{n} - 2u_{i,j}^{n} + u_{i,j-1}^{n}}{(\Delta y)^2} \right] \tag{3-21a}$$

and

$$\frac{u_{i,j}^{n+1} - u_{i,j}^{n+\frac{1}{2}}}{\left(\frac{\Delta t}{2}\right)} = \alpha \left[\frac{u_{i+1,j}^{n+\frac{1}{2}} - 2u_{i,j}^{n+\frac{1}{2}} + u_{i-1,j}^{n+\frac{1}{2}}}{(\Delta x)^2} + \frac{u_{i,j+1}^{n+1} - 2u_{i,j}^{n+1} + u_{i,j-1}^{n+1}}{(\Delta y)^2} \right] \tag{3-21b}$$

The method is of order $[(\Delta t)^2, (\Delta x)^2, (\Delta y)^2]$ and is unconditionally stable. Equations (3-21a) and (3-21b) are written in the tridiagonal form as

$$-d_1 u_{i-1,j}^{n+\frac{1}{2}} + (1 + 2d_1) u_{i,j}^{n+\frac{1}{2}} - d_1 u_{i+1,j}^{n+\frac{1}{2}} = $$
$$= d_2 u_{i,j+1}^{n} + (1 - 2d_2) u_{i,j}^{n} + d_2 u_{i,j-1}^{n} \tag{3-22a}$$

and

$$-d_2 u_{i,j-1}^{n+1} + (1 + 2d_2) u_{i,j}^{n+1} - d_2 u_{i,j+1}^{n+1}$$
$$= d_1 u_{i+1,j}^{n+\frac{1}{2}} + (1 - 2d_1) u_{i,j}^{n+\frac{1}{2}} + d_1 u_{i-1,j}^{n+\frac{1}{2}} \tag{3-22b}$$

where

$$d_1 = \frac{1}{2} d_x = \frac{1}{2} \frac{\alpha \Delta t}{(\Delta x)^2}$$

and

$$d_2 = \frac{1}{2} d_y = \frac{1}{2} \frac{\alpha \Delta t}{(\Delta y)^2}$$

The solution procedure starts with the solution of the tridiagonal system (3-22a). The formulation of Equation (3-22a) is implicit in the x-direction and explicit in the y-direction; thus the solution at this stage is referred to as the x sweep. Solving the tridiagonal system of (3-22a) provides the necessary data for the right-hand side of Equation (3-22b) to solve the tridiagonal system of (3-22b). In this equation, the FDE is implicit in the y-direction and explicit in the x-direction, and it is referred to as the y sweep. Graphical presentation of the method is shown in Figure 3-13.

For application purposes, consider the unsteady two-dimensional heat conduction equation

$$\frac{\partial T}{\partial t} = \alpha \left[\frac{\partial^2 T}{\partial x^2} + \frac{\partial^2 T}{\partial y^2} \right]$$

where α, the thermal diffusivity, is assumed constant. It is required to determine the temperature distribution in a long bar with a rectangular cross-section. The rectangular bar is shown in Figure 3-14. Assume the bar is composed of chrome steel, which has cross-sectional dimensions of 3.5 ft. by 3.5 ft., i.e., $b = h = 3.5'$. The thermal diffusivity is provided as $\alpha = 0.645$ ft^2/hr. The following initial and boundary conditions are imposed (see Figure 3-14).

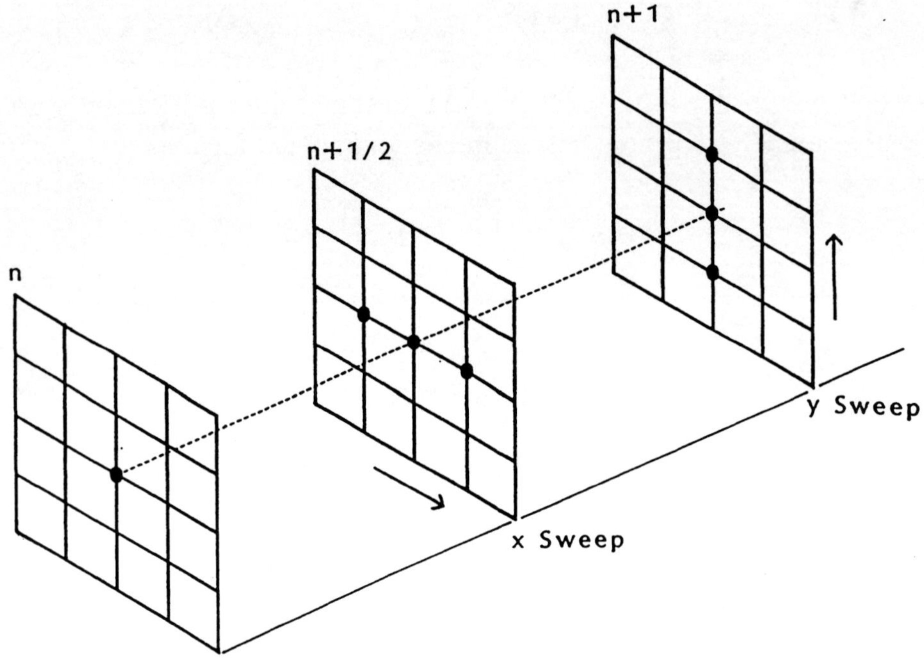

Figure 3-13. Illustration of the grid system for the ADI method.

Initial condition: $t = 0$ $\quad T = T_0 = 0.0$

Boundary conditions: $t \geq 0$ $\quad T(x,0) = T_1 = 200.00$

$$T(0,y) = T_2 = 200.00$$

$$T(x,h) = T_3 = 0.0$$

$$T(b,y) = T_4 = 0.0$$

Application of the ADI method yields the following set of equations for the x sweep

$$-d_1 T_{i-1,j}^{n+\frac{1}{2}} + (2d_1 + 1) T_{i,j}^{n+\frac{1}{2}} - d_1 T_{i+1,j}^{n+\frac{1}{2}}$$

$$= d_2 T_{i,j+1}^{n} + (1 - 2d_2) T_{i,j}^{n} + d_2 T_{i,j-1}^{n} \tag{3-23}$$

where $d_1 = \frac{1}{2} d_x = \frac{1}{2} \frac{\alpha \Delta t}{(\Delta x)^2}$ and $d_2 = \frac{1}{2} d_y = \frac{1}{2} \frac{\alpha \Delta t}{(\Delta y)^2}$

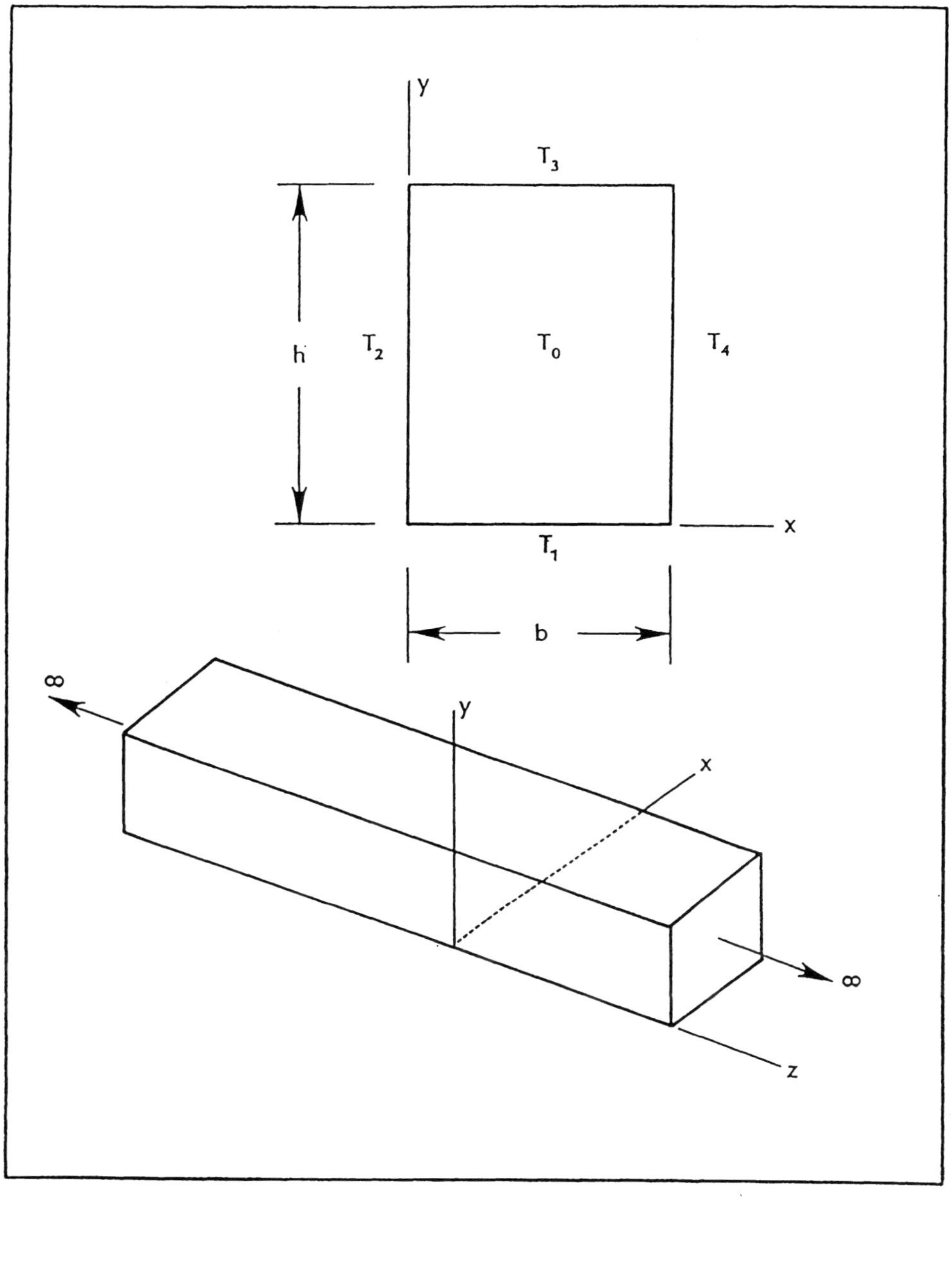

Figure 3-14. Nomenclature for the rectangular plate.

By defining the coefficients of the unknowns in Equation (3-23) so that they are consistent with the general form of the tridiagonal system of equations, one has

$$\begin{aligned} a_1 &= -d_1 , \\ b_1 &= (1+2d_1) , \\ c_1 &= -d_1 , \end{aligned}$$

and

$$D_1 = d_2 T^n_{i,j+1} + (1-2d_2)T^n_{i,j} + d_2 T^n_{i,j-1}$$

Note that in order to rearrange the equations as a tridiagonal system, D_1 must be modified at $i = 2$ and $i = IMM1$ ($IMM1 = IM-1$), where boundary conditions at $i = 1$ and $i = IM$ enter the equation. Thus, at $i = 2$,

$$D_1 = d_2 T^n_{2,j+1} + (1-2d_2)T^n_{2,j} + d_2 T^n_{2,j-1} + d_1 T_{1,j}$$

where the last term includes the specified boundary condition $T_{1,j} = T_2$. Similarly, at $i = IMM1$,

$$D_1 = d_2 T^n_{IMM1,j+1} + (1-2d_2)T^n_{IMM1,j} + d_2 T^n_{IMM1,j-1} + d_1 T_{IM,j}$$

where $T_{IM,j}$ is specified as T_4 for this application. Following the procedure described in Appendix B, the imposed boundary conditions for this problem yield

$$\begin{aligned} H_{1x} &= 0 \\ \text{and} \qquad & \\ G_{1x} &= T_2 \end{aligned} \tag{3-24}$$

The FDE (3-23), along with the boundary conditions specified by (3-24), completes the x sweep computations.

For the y sweep, the FDE becomes

$$\begin{aligned} -d_2 T^{n+1}_{i,j-1} &+ (1+2d_2)T^{n+1}_{i,j} - d_2 T^{n+1}_{i,j+1} \\ &= d_1 T^{n+\frac{1}{2}}_{i+1,j} + (1-2d_1)T^{n+\frac{1}{2}}_{i,j} + d_1 T^{n+\frac{1}{2}}_{i-1,j} \end{aligned} \tag{3-25}$$

The coefficients are

$$\begin{aligned} a_2 &= -d_2 , \\ b_2 &= (1+2d_2) , \\ c_2 &= -d_2 , \end{aligned}$$

and

$$D_2 = d_1 T^{n+\frac{1}{2}}_{i+1,j} + (1-2d_1)T^{n+\frac{1}{2}}_{i,j} + d_1 T^{n+\frac{1}{2}}_{i-1,j}$$

Modifications of D_2 at the boundaries are as follows. At $j = 2$,

$$D_2 = d_1 T_{i+1,2}^{n+\frac{1}{2}} + (1 - 2d_1) T_{i,2}^{n+\frac{1}{2}} + d_1 T_{i-1,2}^{n+\frac{1}{2}} + d_2 T_{i,1}$$

where $T_{i,1}$ is specified as T_1. At $j = JMM1$ $(JMM1 = JM - 1)$

$$D_2 = d_1 T_{i+1,JMM1}^{n+\frac{1}{2}} + (1 - 2d_1) T_{i,JMM1}^{n+\frac{1}{2}} + d_1 T_{i-1,JMM1}^{n+\frac{1}{2}} + d_2 T_{i,JM}$$

where $T_{i,JM}$ is specified as T_3. The imposed boundary conditions yield

$$H_{1y} = 0$$

and

$$G_{1y} = T_1 \tag{3-26}$$

Solving Equation (3-25) subject to the boundary conditions specified by (3-26) would advance the computation to the $n + 1$ time level. The temperature distributions at time $t = 0.1$ hr. and $t = 0.4$ hr. are presented in Tables 3-7 and 3-8. The computed values are printed for all y locations and selected x locations on intervals of 0.5 ft. The symmetry of the initial and boundary conditions produces the symmetry in the solution. The temperature distributions are presented in Figures 3-15 through 3-17. Figure 3-15 shows the initial and boundary conditions, while Figures 3-16 and 3-17 show the temperature distributions at $t = 0.1$ hr. and $t = 0.4$ hr.

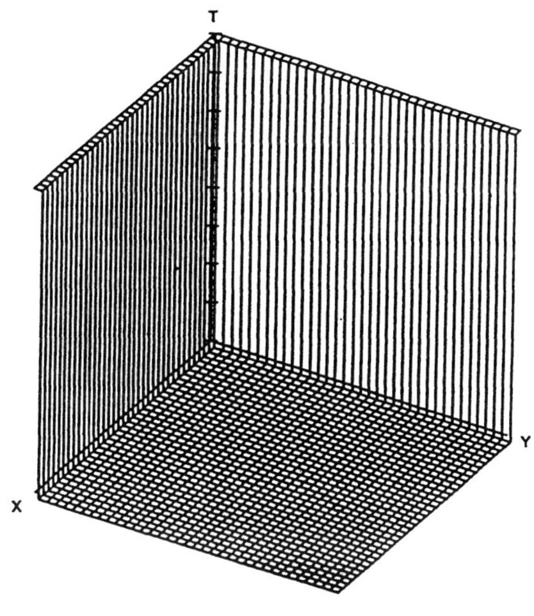

Figure 3-15. Initial temperature distribution for the rectangular bar.

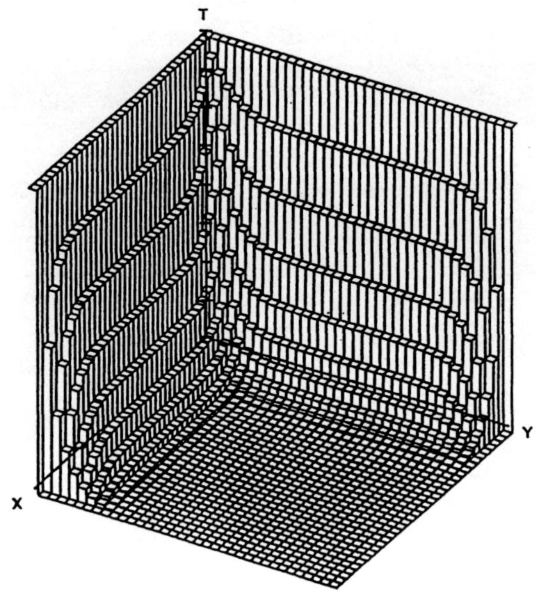

Figure 3-16. Temperature distribution at $t = 0.1$ hr.

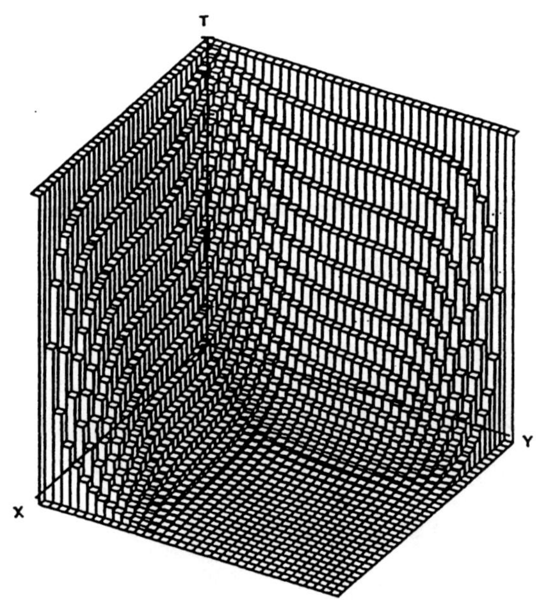

Figure 3-17. Temperature distribution at $t = 0.4$ hr.

3.8 Approximate Factorization

The ADI scheme just investigated belongs to a class of methods known as "Approximate Factorization." In these methods, the original multidimensional FDEs are replaced by a series of finite difference equations which can be represented as tridiagonal formulations. The purpose, as the ADI technique demonstrated, is to simplify the solution procedure and permit it to be more rapidly executed. To see the mathematical development and details of this procedure, consider the model equation

$$\frac{\partial u}{\partial t} = \alpha \left[\frac{\partial^2 u}{\partial x^2} + \frac{\partial^2 u}{\partial y^2} \right]$$

Applying the Crank-Nicolson scheme, one obtains

$$\begin{aligned} \frac{u_{i,j}^{n+1} - u_{i,j}^{n}}{\Delta t} &= \frac{1}{2}\alpha \left[\frac{u_{i+1,j}^{n+1} - 2u_{i,j}^{n+1} + u_{i-1,j}^{n+1}}{(\Delta x)^2} + \frac{u_{i+1,j}^{n} - 2u_{i,j}^{n} + u_{i-1,j}^{n}}{(\Delta x)^2} \right. \\ &\left. + \frac{u_{i,j+1}^{n+1} - 2u_{i,j}^{n+1} + u_{i,j-1}^{n+1}}{(\Delta y)^2} + \frac{u_{i,j+1}^{n} - 2u_{i,j}^{n} + u_{i,j-1}^{n}}{(\Delta y)^2} \right] \end{aligned} \tag{3-27}$$

which is of order $[(\Delta t)^2, (\Delta x)^2, (\Delta y)^2]$. Define the following operators such that Equation (3-27) may be expressed in a compact form:

$$\delta_x^2 u_{i,j} = u_{i+1,j} - 2u_{i,j} + u_{i-1,j} \tag{3-28}$$

and

$$\delta_y^2 u_{i,j} = u_{i,j+1} - 2u_{i,j} + u_{i,j-1} \tag{3-29}$$

Therefore (3-27) is represented as

$$\frac{u_{i,j}^{n+1} - u_{i,j}^{n}}{\Delta t} = \frac{1}{2}\alpha \left[\frac{\delta_x^2 u_{i,j}^{n+1}}{(\Delta x)^2} + \frac{\delta_x^2 u_{i,j}^{n}}{(\Delta x)^2} + \frac{\delta_y^2 u_{i,j}^{n+1}}{(\Delta y)^2} + \frac{\delta_y^2 u_{i,j}^{n}}{(\Delta y)^2} \right]$$

or

$$u_{i,j}^{n+1} - \frac{\alpha \Delta t}{2} \left[\frac{\delta_x^2}{(\Delta x)^2} + \frac{\delta_y^2}{(\Delta y)^2} \right] u_{i,j}^{n+1} = u_{i,j}^{n} + \frac{\alpha \Delta t}{2} \left[\frac{\delta_x^2}{(\Delta x)^2} + \frac{\delta_y^2}{(\Delta y)^2} \right] u_{i,j}^{n}$$

Hence

$$\left[1 - \frac{1}{2} \left[d_x \delta_x^2 + d_y \delta_y^2 \right] \right] u_{i,j}^{n+1} = \left[1 + \frac{1}{2} \left[d_x \delta_x^2 + d_y \delta_y^2 \right] \right] u_{i,j}^{n} \tag{3-30}$$

where $d_x = \dfrac{\alpha \Delta t}{(\Delta x)^2}$ and $d_y = \dfrac{\alpha \Delta t}{(\Delta y)^2}$.

Now, turn back to the ADI formulation given by (3-21a) and (3-21b), and rewrite them with the operators defined by (3-28) and (3-29), to provide

$$\frac{u_{i,j}^{n+\frac{1}{2}} - u_{i,j}^{n}}{\frac{\Delta t}{2}} = \alpha \left[\frac{\delta_x^2 u_{i,j}^{n+\frac{1}{2}}}{(\Delta x)^2} + \frac{\delta_y^2 u_{i,j}^{n}}{(\Delta y)^2} \right]$$

and

$$\frac{u_{i,j}^{n+1} - u_{i,j}^{n+\frac{1}{2}}}{\frac{\Delta t}{2}} = \alpha \left[\frac{\delta_x^2 u_{i,j}^{n+\frac{1}{2}}}{(\Delta x)^2} + \frac{\delta_y^2 u_{i,j}^{n+1}}{(\Delta y)^2} \right]$$

or

$$u_{i,j}^{n+\frac{1}{2}} - \frac{\alpha \Delta t}{2(\Delta x)^2} \delta_x^2 u_{i,j}^{n+\frac{1}{2}} = u_{i,j}^{n} + \frac{\alpha \Delta t}{2(\Delta y)^2} \delta_y^2 u_{i,j}^{n}$$

and

$$u_{i,j}^{n+1} - \frac{\alpha \Delta t}{2(\Delta y)^2} \delta_y^2 u_{i,j}^{n+1} = u_{i,j}^{n+\frac{1}{2}} + \frac{\alpha \Delta t}{2(\Delta x)^2} \delta_x^2 u_{i,j}^{n+\frac{1}{2}}$$

Then, in terms of the diffusion numbers, one may write

$$\left[1 - \frac{1}{2} d_x \delta_x^2\right] u_{i,j}^{n+\frac{1}{2}} = \left[1 + \frac{1}{2} d_y \delta_y^2\right] u_{i,j}^{n} \tag{3-31}$$

and

$$\left[1 - \frac{1}{2} d_y \delta_y^2\right] u_{i,j}^{n+1} = \left[1 + \frac{1}{2} d_x \delta_x^2\right] u_{i,j}^{n+\frac{1}{2}} \tag{3-32}$$

These two equations can be combined by eliminating $u_{i,j}^{n+\frac{1}{2}}$, resulting in

$$(1 - \frac{1}{2} d_x \delta_x^2)(1 - \frac{1}{2} d_y \delta_y^2) u_{i,j}^{n+1} = (1 + \frac{1}{2} d_x \delta_x^2)(1 + \frac{1}{2} d_y \delta_y^2) u_{i,j}^{n} \tag{3-33}$$

In order to compare (3-33) with the Crank-Nicolson formulation of (3-30), multiply the terms in Equation (3-33). Thus,

$$\left[1 - \frac{1}{2}(d_x \delta_x^2 + d_y \delta_y^2) + \frac{1}{4} d_x d_y \delta_x^2 \delta_y^2\right] u_{i,j}^{n+1}$$
$$= \left[1 + \frac{1}{2}(d_x \delta_x^2 + d_y \delta_y^2) + \frac{1}{4} d_x d_y \delta_x^2 \delta_y^2\right] u_{i,j}^{n} \tag{3-34}$$

Compared to (3-30), Equation (3-34) has the additional term $\frac{1}{4} d_x d_y \delta_x^2 \delta_y^2 (u_{i,j}^{n+1} - u_{i,j}^{n})$, which is smaller than the truncation error of (3-30). Therefore, it is seen that the Crank-Nicolson formulation of (3-30) can be approximated by (3-34), which in turn can be factored as (3-33) and split as (3-31) and (3-32). Equation (3-33) is known as the approximate factorization of (3-30).

The procedure described above may be applied to any multidimensional problem to provide the approximate factorization of the PDE.

3.9 Fractional Step Methods

An approximation of multidimensional problems similar to ADI (or, in general, approximate factorization schemes) is the method of fractional step. This method splits the multidimensional equation into a series of one-space dimensional equations and solves them sequentially. For the two-dimensional model equation

$$\frac{\partial u}{\partial t} = \alpha \left[\frac{\partial^2 u}{\partial x^2} + \frac{\partial^2 u}{\partial y^2} \right]$$

the method provides the following finite difference equations: (Note that the Crank-Nicolson scheme is used.)

$$\frac{u_{i,j}^{n+\frac{1}{2}} - u_{i,j}^n}{\frac{\Delta t}{2}} = \alpha \frac{1}{2} \left[\frac{u_{i+1,j}^{n+\frac{1}{2}} - 2u_{i,j}^{n+\frac{1}{2}} + u_{i-1,j}^{n+\frac{1}{2}}}{(\Delta x)^2} + \frac{u_{i+1,j}^n - 2u_{i,j}^n + u_{i-1,j}^n}{(\Delta x)^2} \right]$$

and

$$\frac{u_{i,j}^{n+1} - u_{i,j}^{n+\frac{1}{2}}}{\frac{\Delta t}{2}} = \alpha \frac{1}{2} \left[\frac{u_{i,j+1}^{n+1} - 2u_{i,j}^{n+1} + u_{i,j-1}^{n+1}}{(\Delta y)^2} + \frac{u_{i,j+1}^{n+\frac{1}{2}} - 2u_{i,j}^{n+\frac{1}{2}} + u_{i,j-1}^{n+\frac{1}{2}}}{(\Delta y)^2} \right]$$

The scheme is unconditionally stable and is of order $[(\Delta t)^2, (\Delta x)^2, (\Delta y)^2]$.

3.10 Extension to Three-Space Dimensions

The ADI method just investigated for the unsteady two-space dimensional parabolic equation can be extended to three-space dimensions, which is accomplished by considering time intervals of n, $n + \frac{1}{3}$, $n + \frac{2}{3}$, and $n + 1$. The resulting equations for the model equation

$$\frac{\partial u}{\partial t} = \alpha \left[\frac{\partial^2 u}{\partial x^2} + \frac{\partial^2 u}{\partial y^2} + \frac{\partial^2 u}{\partial z^2} \right] \tag{3-35}$$

are:

$$\frac{u_{i,j,k}^{n+\frac{1}{3}} - u_{i,j,k}^n}{\frac{\Delta t}{3}} = \alpha \left[\frac{\delta_x^2 u_{i,j,k}^{n+\frac{1}{3}}}{(\Delta x)^2} + \frac{\delta_y^2 u_{i,j,k}^n}{(\Delta y)^2} + \frac{\delta_z^2 u_{i,j,k}^n}{(\Delta z)^2} \right]$$

$$\frac{u_{i,j,k}^{n+\frac{2}{3}} - u_{i,j,k}^{n+\frac{1}{3}}}{\frac{\Delta t}{3}} = \alpha \left[\frac{\delta_x^2 u_{i,j,k}^{n+\frac{1}{3}}}{(\Delta x)^2} + \frac{\delta_y^2 u_{i,j,k}^{n+\frac{2}{3}}}{(\Delta y)^2} + \frac{\delta_z^2 u_{i,j,k}^{n+\frac{1}{3}}}{(\Delta z)^2} \right]$$

and

$$\frac{u_{i,j,k}^{n+1} - u_{i,j,k}^{n+\frac{2}{3}}}{\frac{\Delta t}{3}} = \alpha \left[\frac{\delta_x^2 u_{i,j,k}^{n+\frac{2}{3}}}{(\Delta x)^2} + \frac{\delta_y^2 u_{i,j,k}^{n+\frac{2}{3}}}{(\Delta y)^2} + \frac{\delta_z^2 u_{i,j,k}^{n+1}}{(\Delta z)^2} \right]$$

The method is of order $[(\Delta t), (\Delta x)^2, (\Delta y)^2, (\Delta z)^2]$ and is only conditionally stable with the requirement of $(d_x + d_y + d_z) \leq (3/2)$. As a result of this requirement, the method is not very attractive. A method that is unconditionally stable and is second-order accurate uses the Crank-Nicolson scheme. The finite difference equations of the model Equation (3-35) are

$$\frac{u^*_{i,j,k} - u^n_{i,j,k}}{\Delta t} = \alpha \left[\frac{1}{2}\frac{\delta^2_x u^*_{i,j,k} + \delta^2_x u^n_{i,j,k}}{(\Delta x)^2} + \frac{\delta^2_y u^n_{i,j,k}}{(\Delta y)^2} + \frac{\delta^2_z u^n_{i,j,k}}{(\Delta z)^2}\right],$$

$$\frac{u^{**}_{i,j,k} - u^n_{i,j,k}}{\Delta t} = \alpha \left[\frac{1}{2}\frac{\delta^2_x u^*_{i,j,k} + \delta^2_x u^n_{i,j,k}}{(\Delta x)^2} + \frac{1}{2}\frac{\delta^2_y u^{**}_{i,j,k} + \delta^2_y u^n_{i,j,k}}{(\Delta y)^2} + \frac{\delta^2_z u^n_{i,j,k}}{(\Delta z)^2}\right],$$

and

$$\frac{u^{n+1}_{i,j,k} - u^n_{i,j,k}}{\Delta t} = \alpha \left[\frac{1}{2}\frac{\delta^2_x u^*_{i,j,k} + \delta^2_x u^n_{i,j,k}}{(\Delta x)^2} + \frac{1}{2}\frac{\delta^2_y u^{**}_{i,j,k} + \delta^2_y u^n_{i,j,k}}{(\Delta y)^2} + \frac{1}{2}\frac{\delta^2_z u^{n+1}_{i,j,k} + \delta^2_z u^n_{i,j,k}}{(\Delta z)^2}\right]$$

3.11 Consistency Analysis of the Finite Difference Equations

By previous definition, an FDE approximation of a PDE is consistent if the FDE reduces to the original PDE as the step sizes approach zero. In this section, the consistency of some of the methods discussed earlier will be investigated. Since the procedure is simple and straightforward, only a couple of examples are illustrated. As a first example, consider model Equation (3-1), i.e.,

$$\frac{\partial u}{\partial t} = \alpha \frac{\partial^2 u}{\partial x^2}$$

where the FDE approximation by the FTCS explicit method is

$$\frac{u^{n+1}_i - u^n_i}{\Delta t} = \alpha \frac{u^n_{i+1} - 2u^n_i + u^n_{i-1}}{(\Delta x)^2} \tag{3-36}$$

Expand each u in a Taylor series expansion about u^n_i; therefore,

$$u^{n+1}_i = u^n_i + \frac{\partial u}{\partial t}(\Delta t) + \frac{\partial^2 u}{\partial t^2}\frac{(\Delta t)^2}{2!} + O(\Delta t)^3 \tag{3-37}$$

$$u^n_{i+1} = u^n_i + \frac{\partial u}{\partial x}(\Delta x) + \frac{\partial^2 u}{\partial x^2}\frac{(\Delta x)^2}{2!} + \frac{\partial^3 u}{\partial x^3}\frac{(\Delta x)^3}{3!} + O(\Delta x)^4 \tag{3-38}$$

and

$$u^n_{i-1} = u^n_i - \frac{\partial u}{\partial x}(\Delta x) + \frac{\partial^2 u}{\partial x^2}\frac{(\Delta x)^2}{2!} - \frac{\partial^3 u}{\partial x^3}\frac{(\Delta x)^3}{3!} + O(\Delta x)^4 \tag{3-39}$$

Substituting Equations (3-37), (3-38), and (3-39) into Equation (3-36) yields

$$\frac{1}{\Delta t}\left[u_i^n + \frac{\partial u}{\partial t}(\Delta t) + \frac{\partial^2 u}{\partial t^2}\frac{(\Delta t)^2}{2!} + O(\Delta t)^3 - u_i^n\right] =$$

$$\frac{\alpha}{(\Delta x)^2}\left[u_i^n + \frac{\partial u}{\partial x}\Delta x + \frac{\partial^2 u}{\partial x^2}\frac{(\Delta x)^2}{2!} + \frac{\partial^3 u}{\partial x^3}\frac{(\Delta x)^3}{3!} + O(\Delta x)^4 - 2u_i^n\right.$$

$$\left. + u_i^n - \frac{\partial u}{\partial x}\Delta x + \frac{\partial^2 u}{\partial x^2}\frac{(\Delta x)^2}{2!} - \frac{\partial^3 u}{\partial x^3}\frac{(\Delta x)^3}{3!} + O(\Delta x)^4\right]$$

from which

$$\left[\frac{\partial u}{\partial t} + \frac{\partial^2 u}{\partial t^2}\frac{\Delta t}{2} + O(\Delta t)^2\right] = \alpha\left[\frac{\partial^2 u}{\partial x^2} + O(\Delta x)^2\right]$$

or

$$\frac{\partial u}{\partial t} = \alpha\frac{\partial^2 u}{\partial x^2} - \left[\frac{\Delta t}{2}\right]\frac{\partial^2 u}{\partial t^2} + O\left[(\Delta t)^2 , (\Delta x)^2\right]$$

Consistency requires that as the step sizes Δx and Δt approach zero, the FDE must reduce to the original PDE. In this example, it is obvious that as Δx, $\Delta t \to 0$, the original PDE is recovered, i.e., $\partial u/\partial t = \alpha(\partial^2 u/\partial x^2)$. Therefore, the method is consistent.

As a second example, consider the DuFort-Frankel method. The finite difference formulation of the model equation is

$$(1 + 2d)u_i^{n+1} = (1 - 2d)u_i^{n-1} + 2d(u_{i+1}^n + u_{i-1}^n) \tag{3-40}$$

where

$$d = \alpha\frac{\Delta t}{(\Delta x)^2} \tag{3-41}$$

Following the procedure illustrated for the previous example, expand u_i^{n+1}, u_i^{n-1}, u_{i-1}^n, and u_{i+1}^n in a Taylor series about u_i^n, and substitute the results into (3-40) to obtain

$$(1 + 2d)\left[u_i^n + \frac{\partial u}{\partial t}\Delta t + \frac{\partial^2 u}{\partial t^2}\frac{(\Delta t)^2}{2} + \frac{\partial^3 u}{\partial t^3}\frac{(\Delta t)^3}{3!} + O(\Delta t)^4\right] =$$

$$(1 - 2d)\left[u_i^n - \frac{\partial u}{\partial t}\Delta t + \frac{\partial^2 u}{\partial t^2}\frac{(\Delta t)^2}{2} - \frac{\partial^3 u}{\partial t^3}\frac{(\Delta t)^3}{3!} + O(\Delta t)^4\right]$$

$$+ 2d\left\{\left[u_i^n + \frac{\partial u}{\partial x}\Delta x + \frac{\partial^2 u}{\partial x^2}\frac{(\Delta x)^2}{2} + \frac{\partial^3 u}{\partial x^3}\frac{(\Delta x)^3}{3!} + O(\Delta x)^4\right]\right.$$

$$\left. + \left[u_i^n - \frac{\partial u}{\partial x}\Delta x + \frac{\partial^2 u}{\partial x^2}\frac{(\Delta x)^2}{2} - \frac{\partial^3 u}{\partial x^3}\frac{(\Delta x)^3}{3!} + O(\Delta x)^4\right]\right\}$$

This equation is reduced to:

$$\frac{\partial u}{\partial t}\Delta t + (d)\frac{\partial^2 u}{\partial t^2}(\Delta t)^2 + O(\Delta t)^3 = (d)\frac{\partial^2 u}{\partial x^2}(\Delta x)^2 + (d)[O(\Delta x)^4]$$

Substitution of (3-41) yields

$$\frac{\partial u}{\partial t}\Delta t + \frac{\alpha \Delta t}{(\Delta x)^2}\frac{\partial^2 u}{\partial t^2}(\Delta t)^2 + O(\Delta t)^3 =$$

$$\frac{\alpha \Delta t}{(\Delta x)^2}\left(\frac{\partial^2 u}{\partial x^2}\right)(\Delta x)^2 + \alpha\frac{\Delta t}{(\Delta x)^2}\left[O(\Delta x)^4\right] ,$$

or

$$\frac{\partial u}{\partial t} + \alpha\frac{\partial^2 u}{\partial t^2}\left(\frac{\Delta t}{\Delta x}\right)^2 = \alpha\frac{\partial^2 u}{\partial x^2} + O\left[(\Delta t)^2 \ , \ (\Delta x)^2\right]$$

or

$$\frac{\partial u}{\partial t} = \alpha\frac{\partial^2 u}{\partial x^2} + O\left[(\Delta t)^2 \ , \ (\Delta x)^2 \ , \ \left(\frac{\Delta t}{\Delta x}\right)^2\right]$$

The method is consistent if only Δt and Δx approach zero and if $(\Delta t/\Delta x) \rightarrow 0$, so that $\partial u/\partial t = \alpha(\partial^2 u/\partial x^2)$, i.e., the original PDE, is recovered. Note that if Δx approaches zero faster than Δt, then $(\Delta t/\Delta x) \rightarrow K$, and the equation becomes

$$\frac{\partial u}{\partial t} + \alpha K^2\frac{\partial^2 u}{\partial t^2} = \alpha\frac{\partial^2 u}{\partial x^2}$$

which represents a hyperbolic equation!

3.12 Linearization

The model equations investigated so far have been linear. In general, the equations of interest in fluid mechanics and heat transfer are nonlinear. In this section various methods to linearize the finite difference equations will be reviewed. To show the linearization procedures, consider a nonlinear term such as $u(\partial u/\partial x)$ (which is a convection term in the Navier-Stokes equation). In the illustration, a steady two-dimensional flow is assumed and the partial derivative is approximated by a forward differencing scheme. The properties at all j nodes, for a given i location, are known, whereas the properties at $i+1$ are to be computed.

Method I. Lagging. In this linearization, the coefficient is used at the known value, i.e., at station i. Therefore, the finite difference representation becomes

$$u_{i,j}\frac{u_{i+1,j} - u_{i,j}}{\Delta x}$$

There is one unknown, $u_{i+1,j}$, in this expression and the finite difference representation is linear.

Method II. Iterative. In this procedure, the lagged value is updated until a specified convergence criterion is reached. Denoting by k the iteration level, the formulation is

$$u_{i+1,j}^{k}\frac{u_{i+1,j}^{k+1}-u_{i,j}}{\Delta x}$$

For the first iteration $k = 1$, $u_{i+1,j}$ is the value at the previous station, i.e., $u_{i,j}$. Once $u_{i+1,j}^{k+1}$ has been computed, the coefficient $u_{i+1,j}^{k}$ is updated and a new solution is sought. The procedure continues until a convergence criterion is met. The convergence criterion can be specified by either of the following forms (usually depending on the particular problem):

$$\left|u_{i+1,j}^{k+1}-u_{i+1,j}^{k}\right| \leq \epsilon$$

or

$$\left|\frac{u_{i+1,j}^{k+1}-u_{i+1,j}^{k}}{u_{i+1,j}^{k+1}}\right| \leq \epsilon$$

Method III. Newton's iterative linearization. To illustrate how this procedure works, assume a nonlinear term represented by $(A)(B)$. Define by δA the change in the variable A between two consecutive iterations, then

$$\delta A = A^{k+1} - A^{k}$$

and, similarly,

$$\delta B = B^{k+1} - B^{k} \; ;$$

or

$$A^{k+1} = A^{k} + \delta A \; ; \tag{3-42}$$

$$B^{k+1} = B^{k} + \delta B \tag{3-43}$$

Return to the nonlinear term $(A)(B)$, and substitute relations (3-42) and (3-43) to get

$$\begin{aligned} A^{k+1}B^{k+1} &= (A^{k}+\delta A)(B^{k}+\delta B) \\ &= A^{k}B^{k} + B^{k}\delta A + A^{k}\delta B + (\delta A)(\delta B) \end{aligned}$$

After dropping the second-order term $(\delta A)(\delta B)$, the equation is rearranged as

$$\begin{aligned} A^{k+1}B^{k+1} &= A^{k}B^{k} + B^{k}(A^{k+1}-A^{k}) + A^{k}(B^{k+1}-B^{k}) \\ &= A^{k}B^{k} + A^{k+1}B^{k} - A^{k}B^{k} + A^{k}B^{k+1} - A^{k}B^{k} \\ &= A^{k}B^{k+1} + A^{k+1}B^{k} - A^{k}B^{k} \end{aligned}$$

Now, for the proposed problem where it is required to linearize $u\partial u/\partial x$, one has

$$u\frac{\partial u}{\partial x} = u^k\left(\frac{\partial u}{\partial x}\right)^{k+1} + u^{k+1}\left(\frac{\partial u}{\partial x}\right)^{k} - u^k\left(\frac{\partial u}{\partial x}\right)^{k}$$

With forward differencing,

$$u\frac{\partial u}{\partial x} = u^k_{i+1,j}\frac{u^{k+1}_{i+1,j} - u_{i,j}}{\Delta x} + u^{k+1}_{i+1,j}\frac{u^{k}_{i+1,j} - u_{i,j}}{\Delta x} - u^k_{i+1,j}\frac{u^{k}_{i+1,j} - u_{i,j}}{\Delta x} =$$

$$\frac{1}{\Delta x}\Big[u^k_{i+1,j}u^{k+1}_{i+1,j} - u^k_{i+1,j}u_{i,j} + u^{k+1}_{i+1,j}u^k_{i+1,j} - u^{k+1}_{i+1,j}u_{i,j}$$

$$- u^k_{i+1,j}u^k_{i+1,j} + u^k_{i+1,j}u_{i,j}\Big]$$

Thus,

$$u\frac{\partial u}{\partial x} = \frac{1}{\Delta x}\left[2u^k_{i+1,j}u^{k+1}_{i+1,j} - (u^k_{i+1,j})^2 - u_{i,j}u^{k+1}_{i+1,j}\right]$$

The linearization procedures just described are some of those most commonly used in numerical methods. These procedures will be used in various applications in the upcoming sections.

3.13 Irregular Boundaries

The problems investigated so far have all had nice rectangular boundaries, on which a computational grid system was superimposed. Therefore, the physical and computational spaces were identical. Simple rectangular boundaries rarely occur in nature; indeed, most boundaries are irregular. Irregular boundaries create tremendous difficulties in implementation of the boundary conditions. Various schemes are available to treat irregular boundaries. One may consider using variable step sizes in the neighborhood of the irregular boundaries, as illustrated in Figure 3-18. This procedure is cumbersome and, in most cases, very inefficient.

A popular method is grid generation. Broadly speaking, grid generation schemes can be categorized into two camps. One is the so-called *structured grids* and the other is *unstructured grids*. For structured grids, a transformation from the physical space to the computational space is performed. The computational plane has rectangular boundaries and constant step sizes. A typical grid is illustrated graphically in Figure 3-19. There are three schemes for structured grid generation: 1) complex variable methods, 2) algebraic methods, and 3) differential equations methods. Some of these methods will be described and applied to problems of interest in Chapter 9.

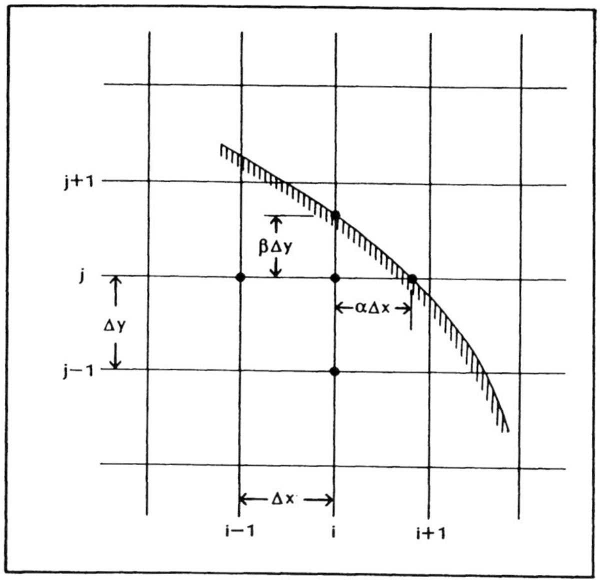

Figure 3-18. Illustration of the variable step sizes for irregular boundaries.

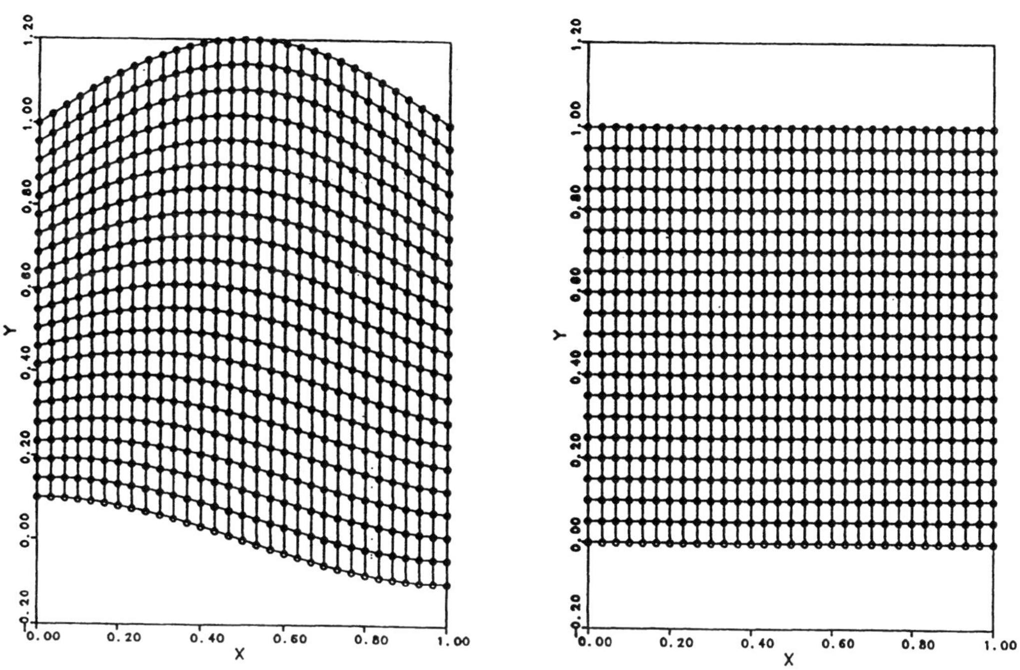

Figure 3-19. Physical space and the transformed computational space.

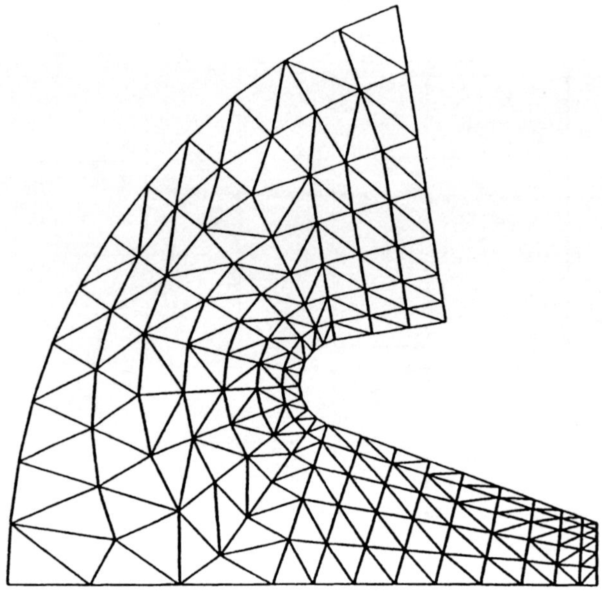

Figure 3-20. Illustration of an unstructured grid with triangular elements.

The second category includes the unstructured grids. This type grid is directly imposed on the physical space. Various types of elements may be used to construct the grid system. Among many choices available, triangular elements are the most popular. A typical unstructured grid generated by triangular elements is shown in Figure 3-20. The unstructured grids will be addressed in Chapter 15.

In the remaining applications of various numerical methods in the first seven chapters, only rectangular boundaries are considered. Therefore, specification of initial and boundary conditions will not pose additional difficulties. After all, at this point it is desirable to learn the procedures of the numerical methods in the simplest forms and avoid any additional complexity.

3.14 Summary Objectives

After studying the material in this chapter, you should be able to do the following:

1. Define and give examples of

    a. explicit formulation

    b. implicit formulation

    c. the FTCS explicit method

   d. Richardson's method
   e. the DuFort-Frankel method
   f. the Laasonen method
   g. the Crank-Nicolson method
   h. Beta formulation
   i. the ADI (alternating direction implicit) method
   j. linearization by iteration
   k. irregular boundaries

2. Define

   a. diffusion number
   b. approximate factorization

3. Solve the problems for Chapter Three.

3.15 Problems

3.1 A wall 1 ft. thick and infinite in other directions (see Figure P3-1) has an initial uniform temperature (T_i) of 100.0°F. The surface temperatures (T_s) at the two sides are suddenly increased and maintained at 300.0°F. The wall is composed of nickel steel (40% Ni) with a diffusivity of $\alpha = 0.1$ ft ^{2}/hr. We are interested in computing the temperature distribution within the wall as a function of time.

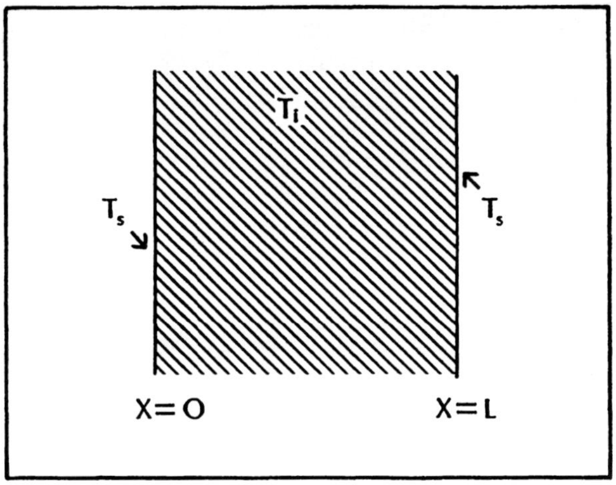

Figure P3-1. Nomenclature for problem 3.1.

The governing equation to be solved is the unsteady one-space dimensional heat conduction equation, which in Cartesian coordinates is

$$\frac{\partial T}{\partial t} = \alpha \frac{\partial^2 T}{\partial x^2}$$

Use the following techniques with the specified step sizes to solve the problem.

a. FTCS explicit

b. DuFort-Frankel

c. Laasonen

d. Crank-Nicolson

For each method, two sets of step sizes are to be used.

I. $\Delta x = 0.05, \quad \Delta t = 0.01$

II. $\Delta x = 0.05, \quad \Delta t = 0.05$

The analytical solution of this problem, subject to the imposed initial and boundary conditions, is

$$T = T_s + 2(T_i - T_s) \sum_{m=1}^{\infty} e^{-(m\pi/L)^2 \alpha t} \, \frac{1-(-1)^m}{m\pi} \sin\left(\frac{m\pi x}{L}\right)$$

where T_s denotes the equal surface temperatures at the two sides and T_i is the initial temperature distribution within the wall. This solution is to be used to validate the numerical solutions.

In all cases the solution is to be printed and plotted for all x locations at each 0.1 hr. time intervals from 0.0 to 0.5 hr.

e. To investigate the effect of step size on the accuracy of the solution and required computation time of an implicit technique, use the Laasonen method with the following step sizes.

I. $\Delta t = 0.01$

II. $\Delta t = 0.025$

III. $\Delta t = 0.05$

IV. $\Delta t = 0.1$

For all cases, $\Delta x = 0.05$.

3.2 Consider the wall described in problem 3.1. Assume the left surface is insulated and the right surface is subject to a constant temperature of 500.0°F, as shown in Figure P3-2.

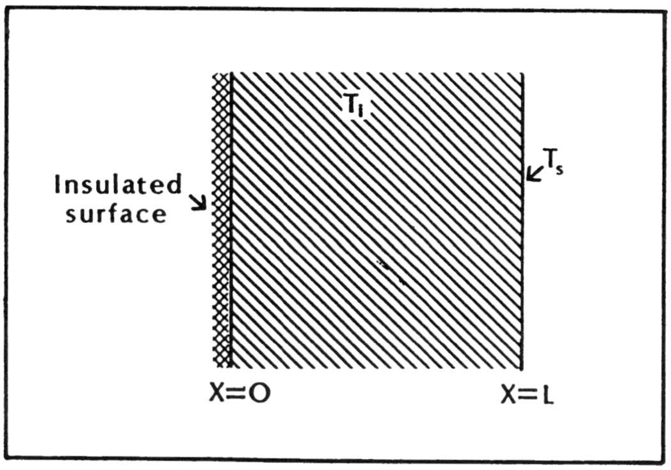

Figure P3-2. Nomenclature for problem 3.2.

Obtain the solution by the following methods:

I. FTCS explicit

II. Laasonen implicit

The step sizes to be employed are $\Delta x = 0.05$, $\Delta t = 0.01$. The analytical solution provided in the previous problem may be used for code validation and accuracy investigation. For this purpose, a plate of thickness 2L subject to surface temperatures of T_s is considered. Note that, due to symmetry, the heat transfer rate at the center line is zero, providing zero temperature gradient, i.e., insulated conditions.

3.3 Consider two infinite parallel plates, a distance of 0.3 cm apart. The upper plate is stationary and the lower plate oscillates according to

$$u(0,t) = u_0 \cos(1000t) = u_0 \cos(1000 n \Delta t)$$

where n represents the computational time level, selected to be 1 at $t = 0.0$.

The governing equation is obtained from the Navier-Stokes equation as

$$\frac{\partial u}{\partial t} = \nu \frac{\partial^2 u}{\partial y^2}$$

Assume the kinematic viscosity ν is constant and has a value of 0.000217 m^2/s, and $u_0 = 40$ m/s. Select 31 equally spaced grid points with $j = 1$ located at the lower plate. A time step of $\Delta t = 0.00002$ sec. is specified. Use the FTCS explicit scheme to obtain the solution within the domain up to $t = 0.00632$ sec. Print the solution for all spatial locations at time levels of 0.0, 0.00158, 0.00316, 0.00474, and 0.00632 seconds. Plot the velocity profiles at the time levels indicated above.

3.4 Two parallel plates extended to infinity are a distance of h apart. The fluid within the plates has a kinematic viscosity of 0.000217 m^2/s and density of 800 kg/m^3. The upper plate is stationary and the lower plate is suddenly set in motion with a constant velocity of 40 m/s. The spacing h is 4 cm. A constant streamwise pressure gradient of dp/dx is imposed within the domain at the instant motion starts. A spatial size of 0.001 m is specified. Recall that the governing equation is reduced from the Navier-Stokes equation and is given by

$$\frac{\partial u}{\partial t} = \nu \frac{\partial^2 u}{\partial y^2} - \frac{1}{\rho} \frac{\partial p}{\partial x}$$

(a) Use the FTCS explicit scheme with a time step of 0.002 sec. to compute the velocity within the domain for
(I) $dp/dx = 0.0$, (II) $dp/dx = 20000.0$ N/m²/m,
(III) $dp/dx = -30000.0$ N/m²/m
Print the solutions at time levels of 0.0, 0.18, 0.36, 0.54, 0.72, 0.9, and 1.08 seconds. Plot the velocity profiles at time levels of 0.0, 0.18, and 1.08 seconds.

(b) Use the Laasonen implicit scheme to compute the velocity profiles and print them at the specified time levels as that of (a) for

(I) $\Delta t = 0.01$, $dp/dx = 0.0$

(II) $\Delta t = 0.01$, $dp/dx = 20000.0$ N/m²/m

(III) $\Delta t = 0.002$, $dp/dx = 20000.0$ N/m²/m

3.5 A long, rectangular bar has dimensions of L by W, as shown in Figure P3-5. The bar is initially heated to a temperature of T_o. Subsequently, its surfaces are subjected to the constant temperatures of T_1, T_2, T_3, and T_4, as depicted in Figure P3-5. It is required to compute the transient solution where the governing equation is

$$\frac{\partial T}{\partial t} = \alpha \left(\frac{\partial^2 T}{\partial x^2} + \frac{\partial^2 T}{\partial y^2} \right)$$

The bar is composed of copper with a thermal conductivity of 380 W/(m°C) and a thermal diffusivity of 11.234×10^{-5} m²/sec, both assumed constant for this problem. The rectangular bar has dimensions of $L = 0.3$ m. and $W = 0.4$ m. The computational grid is specified by IMAX= 31 and JMAX= 41.

Use the FTCS explicit scheme with time steps of 0.2 sec. and 1.0 sec. to compute the transient solution. The initial and boundary conditions are specified as: $T_o =$ 0.0°C, $T_1 = 40.0$°C, $T_2 = 0.0$°C, $T_3 = 10.0$°C, and $T_4 = 0.0$°C.

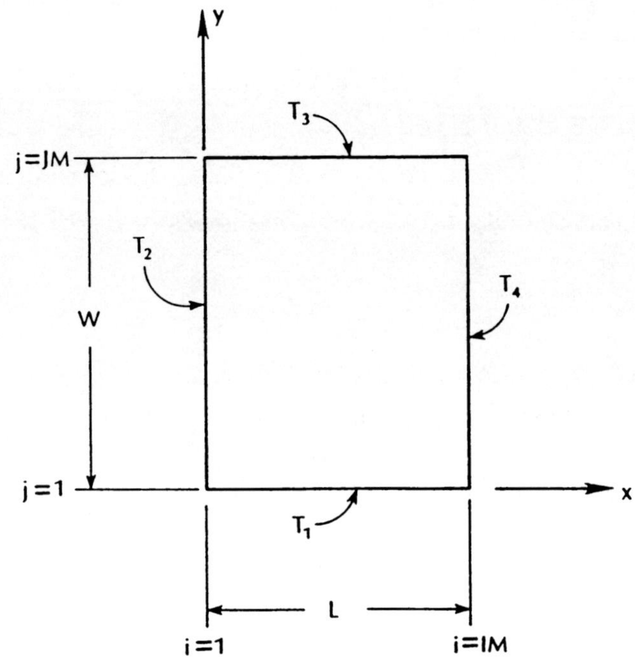

Figure P3-5. Nomenclature for Problem 3.5.

(I) Print the solutions at intervals of 0.05 m in the x-direction and all y-locations at $t = 10.0$ sec., $t = 40.0$ sec., and steady state. Assume the solution has reached steady state if the total variation in temperature from one time level to the next is less than CONSS, where CONSS is specified as 0.01°C. The total variation is determined as

$$TV = \sum_{\substack{i=2\\j=2}}^{\substack{j=JMM1\\i=IMM1}} ABS\left(T_{i,j}^{n+1} - T_{i,j}^{n}\right)$$

(II) Compare the steady-state solution obtained in (I) to the analytical solution. Recall that the analytical solution is

$$T = T_A + T_B$$

where

$$T_A = T_1 * 2.0 \sum_{m=1}^{\infty} \frac{1-\cos(m\pi)}{m\pi} \, \frac{\sinh\left(\dfrac{m\pi(W-y)}{L}\right)}{\sinh\dfrac{m\pi W}{L}} \sin\frac{m\pi x}{L}$$

and

$$T_B = T_3 * 2.0 \sum_{m=1}^{\infty} \frac{1-\cos(m\pi)}{m\pi} \frac{\sinh\dfrac{m\pi y}{L}}{\sinh\dfrac{m\pi W}{L}} \sin\frac{m\pi x}{L}$$

(III) Plot the transient solution for the following locations: (0.1, 0.05), (0.15, 0.10), and (0.1, 0.3).

(IV) Print and plot the heat transfer distributions along the sides $y = 0.0$, $y = W$, and $x = 0.0$ for the steady state solution.

3.6 Repeat problem 3.5 using the ADI scheme.

3.7 Repeat problem 3.5 using the fractional step method.

3.8 Compare the accuracy, stability, and efficiency of the FTCS explicit, ADI, and FSM from the results of problems 3.5 through 3.7.

3.9 Consider a fluid with a temperature of T_c and a constant velocity of u_o travelling from left to right in a channel. The temperature at the end of the channel is suddenly changed to T_h and is maintained at that constant value. It is required to compute the steady state temperature distribution within the channel. The governing equation is given by

$$u\frac{\partial T}{\partial x} = \alpha\frac{\partial^2 T}{\partial x^2}$$

where α is the thermal diffusivity. The boundary conditions are specified as follows:

$$x = 0.0 \quad , \quad T = T_c$$

$$x = L \quad , \quad T = T_h$$

(a) Nondimensionalize the equation by

$$x^* = \frac{x}{L} \quad , \quad u^* = \frac{u}{u_o} \quad , \quad T^* = \frac{T - T_c}{T_h - T_c}$$

and define

$$\alpha^* = \frac{\alpha}{Lu_o}$$

For the following, use the nondimensional equation.

(b) Obtain the analytical solution.

(c) Use second-order central difference approximations and write an implicit formulation.

(d) Use second-order central difference approximations with nonequal grid spacing and write an implicit formulation. Denote the variable spatial steps as shown in Figure P3-9a.

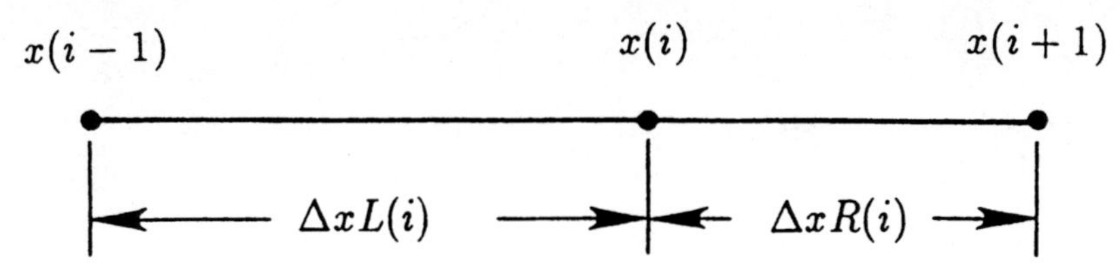

Figure P3-9a. Illustration of nonequal grid spacing used in Problem 3.9.

where

$$\Delta xL(i) = x(i) - x(i-1)$$

and

$$\Delta xR(i) = x(i+1) - x(i)$$

Define the ratio of stepsizes as

$$\gamma_i = \frac{\Delta xR(i)}{\Delta xL(i)}$$

To simplify the formulation, use

$$\Delta xL(i) = \Delta x \quad , \qquad \Delta xR(i) = \gamma \Delta x$$

However, bear in mind that Δx and γ will change from grid point to grid point. The grid point clustering near the boundary at $x = L$ is desired to better resolve the temperature gradient within that region. A procedure for grid point clustering is suggested at the end of the problem.

(e) Obtain the numerical solution by the scheme developed in (c) for the following set of data:

$$T_c = 20°\text{C} \quad , \qquad T_h = 100°\text{C} \quad , \qquad u_o = 0.2 \text{ m/sec}$$

$$\alpha = 0.04 \text{ m}^2/\text{sec} \quad , \qquad L = 2.0 \text{ m}$$

(I) IM = 21

(II) IM = 41

Print the temperature and error distributions for each case. Determine the error by the following:

$$\text{Error} \quad = \quad \text{ABS (TA} - \text{TN)}$$

where

$$\text{TA} \quad \Longrightarrow \quad \text{Analytical solution}$$

$$\text{TN} \quad \Longrightarrow \quad \text{Numerical solution}$$

(f) Use the formulation developed in (d) with the data set specified above and obtain the solution for the clustering parameter of $\beta = 1.1$. Print the temperature and error distributions for each case.

HINT on grid point clustering: Various schemes are available for grid point clustering. Detailed discussion is given in Chapter 9. For now, the following function is suggested

$$C_i = \frac{\beta\left[\left(\dfrac{\beta+1}{\beta-1}\right)^{\theta_i} - 1\right]}{\left[\left(\dfrac{\beta+1}{\beta-1}\right)^{\theta_i} + 1\right]}$$

where $\theta_i = \xi_i/L$, and β is the clustering parameter between 1 and ∞. More clustering of grid points is enforced as β approaches 1. The length of the domain is L, which would be 1.0 when it is nondimensionalized. The variable ξ represents the location of grid points in an arbitrary domain, where they are equally spaced. The physical locations of the grid points are related to ξ locations by

$$x_i = C_i * L$$

The graphical illustration of ξ and its relation to x is shown in Figure P3-9b.

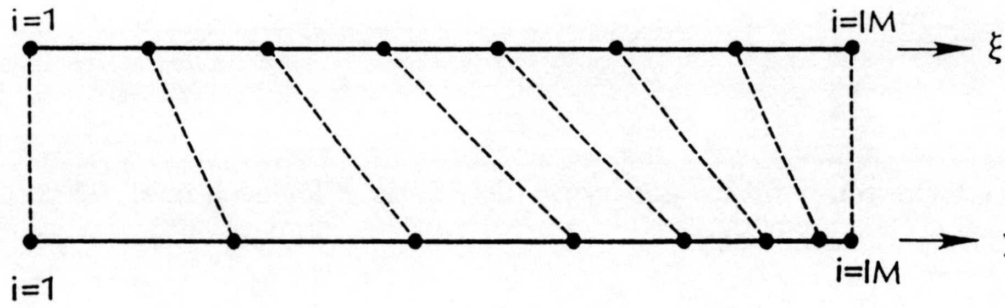

Figure P3-9b. Physical and the corresponding computational grid points distributions.

y	t=0.00	t=0.18	t=0.36	t=0.54	t=0.72	t=0.90	t=1.08
0.000	40.000	40.000	40.000	40.000	40.000	40.000	40.000
0.001	0.000	36.410	37.454	37.919	38.197	38.386	38.524
0.002	0.000	32.864	34.924	35.847	36.400	36.776	37.051
0.003	0.000	29.408	32.426	33.793	34.614	35.175	35.584
0.004	0.000	26.079	29.976	31.764	32.845	33.585	34.127
0.005	0.000	22.915	27.586	29.770	31.099	32.012	32.681
0.006	0.000	19.945	25.272	27.818	29.380	30.458	31.250
0.007	0.000	17.192	23.044	25.915	27.693	28.927	29.836
0.008	0.000	14.672	20.914	24.067	26.044	27.424	28.443
0.009	0.000	12.395	18.889	22.281	24.436	25.950	27.072
0.010	0.000	10.364	16.976	20.563	22.874	24.509	25.727
0.011	0.000	8.574	15.182	18.916	21.360	23.104	24.408
0.012	0.000	7.018	13.509	17.344	19.899	21.737	23.118
0.013	0.000	5.682	11.959	15.850	18.492	20.411	21.859
0.014	0.000	4.550	10.532	14.436	17.142	19.127	20.633
0.015	0.000	3.602	9.227	13.104	15.850	17.887	19.440
0.016	0.000	2.820	8.041	11.854	14.619	16.691	18.281
0.017	0.000	2.182	6.970	10.687	13.448	15.543	17.159
0.018	0.000	1.669	6.009	9.601	12.338	14.440	16.073
0.019	0.000	1.261	5.153	8.594	11.289	13.386	15.023
0.020	0.000	0.942	4.394	7.666	10.300	12.378	14.011
0.021	0.000	0.695	3.726	6.813	9.372	11.417	13.035
0.022	0.000	0.506	3.142	6.032	8.502	10.503	12.097
0.023	0.000	0.364	2.634	5.321	7.689	9.634	11.194
0.024	0.000	0.259	2.196	4.676	6.931	8.810	10.327
0.025	0.000	0.182	1.820	4.092	6.226	8.029	9.494
0.026	0.000	0.126	1.500	3.566	5.572	7.290	8.695
0.027	0.000	0.086	1.229	3.094	4.965	6.591	7.928
0.028	0.000	0.058	1.000	2.672	4.404	5.929	7.191
0.029	0.000	0.039	0.809	2.295	3.885	5.304	6.484
0.030	0.000	0.025	0.651	1.960	3.405	4.711	5.804
0.031	0.000	0.016	0.519	1.661	2.961	4.149	5.148
0.032	0.000	0.011	0.411	1.396	2.549	3.616	4.516
0.033	0.000	0.007	0.322	1.160	2.167	3.107	3.905
0.034	0.000	0.004	0.249	0.948	1.809	2.621	3.312
0.035	0.000	0.003	0.190	0.758	1.474	2.154	2.735
0.036	0.000	0.002	0.140	0.586	1.157	1.704	2.172
0.037	0.000	0.001	0.098	0.427	0.855	1.266	1.619
0.038	0.000	0.000	0.063	0.279	0.564	0.839	1.075
0.039	0.000	0.000	0.030	0.138	0.280	0.418	0.536
0.040	0.000	0.000	0.000	0.000	0.000	0.000	0.000

Table 3-1. Solution of the accelerated lower plate by the FCTS explicit scheme, $\Delta y = 0.001$, $\Delta t = 0.002$.

y	t=0.00	t=0.21	t=0.42	t=0.63	t=0.84	t=1.04	t=1.25
0.000	40.000	40.000	40.000	40.000	40.000	40.000	40.000
0.001	0.000	36.634	37.597	37.996	38.165	38.114	37.656
0.002	0.000	33.430	35.361	36.287	36.975	37.772	39.184
0.003	0.000	30.049	32.845	34.018	34.519	34.365	33.002
0.004	0.000	27.134	30.820	32.623	33.975	35.548	38.336
0.005	0.000	23.888	28.252	30.133	30.944	30.687	28.453
0.006	0.000	21.352	26.467	29.057	31.026	33.331	37.424
0.007	0.000	18.372	23.911	26.398	27.482	27.126	24.075
0.008	0.000	16.265	22.381	25.632	28.151	31.124	36.419
0.009	0.000	13.649	19.901	22.862	24.175	23.721	19.928
0.010	0.000	11.976	18.627	22.386	25.371	28.932	35.294
0.011	0.000	9.781	16.279	19.568	21.055	20.510	16.066
0.012	0.000	8.512	15.250	19.352	22.706	26.759	34.029
0.013	0.000	6.753	13.080	16.547	18.150	17.522	12.536
0.014	0.000	5.834	12.277	16.554	20.173	24.609	32.604
0.015	0.000	4.486	10.320	13.820	15.481	14.780	9.374
0.016	0.000	3.851	9.713	14.007	17.784	22.488	31.008
0.017	0.000	2.864	7.991	11.395	13.061	12.299	6.606
0.018	0.000	2.445	7.550	11.721	15.549	20.400	29.233
0.019	0.000	1.756	6.071	9.274	10.894	10.088	4.248
0.020	0.000	1.492	5.764	9.694	13.473	18.349	27.276
0.021	0.000	1.032	4.522	7.447	8.979	8.148	2.304
0.022	0.000	0.874	4.319	7.921	11.557	16.339	25.141
0.023	0.000	0.581	3.303	5.896	7.307	6.473	0.769
0.024	0.000	0.491	3.176	6.388	9.800	14.372	22.833
0.025	0.000	0.314	2.363	4.600	5.865	5.051	-0.372
0.026	0.000	0.265	2.290	5.078	8.196	12.449	20.363
0.027	0.000	0.162	1.656	3.530	4.633	3.864	-1.144
0.028	0.000	0.136	1.618	3.969	6.733	10.571	17.745
0.029	0.000	0.080	1.135	2.658	3.590	2.889	-1.579
0.030	0.000	0.067	1.117	3.035	5.399	8.734	14.996
0.031	0.000	0.037	0.758	1.952	2.710	2.100	-1.718
0.032	0.000	0.032	0.750	2.248	4.177	6.936	12.135
0.033	0.000	0.017	0.490	1.381	1.965	1.467	-1.606
0.034	0.000	0.014	0.482	1.581	3.048	5.171	9.183
0.035	0.000	0.007	0.299	0.913	1.327	0.958	-1.293
0.036	0.000	0.006	0.284	1.004	1.992	3.432	6.161
0.037	0.000	0.003	0.160	0.519	0.765	0.539	-0.835
0.038	0.000	0.002	0.131	0.487	0.983	1.711	3.092
0.039	0.000	0.001	0.050	0.168	0.250	0.174	-0.288
0.040	0.000	0.000	0.000	0.000	0.000	0.000	0.000

Table 3-2. Solution of the accelerated lower plate by the FCTS explicit scheme, $\Delta y = 0.001$, $\Delta t = 0.00232$.

y	t=0.00	t=0.18	t=0.36	t=0.54	t=0.72	t=0.90	t=1.08
0.000	40.000	40.000	40.000	40.000	40.000	40.000	40.000
0.001	0.000	36.405	37.452	37.918	38.196	38.386	38.524
0.002	0.000	32.795	34.899	35.834	36.391	36.770	37.047
0.003	0.000	29.392	32.421	33.790	34.612	35.173	35.584
0.004	0.000	25.948	29.927	31.738	32.828	33.573	34.118
0.005	0.000	22.887	27.577	29.765	31.095	32.009	32.680
0.006	0.000	19.766	25.203	27.779	29.354	30.440	31.237
0.007	0.000	17.151	23.030	25.907	27.689	28.924	29.835
0.008	0.000	14.463	20.828	24.018	26.011	27.400	28.426
0.009	0.000	12.341	18.870	22.272	24.430	25.946	27.070
0.010	0.000	10.143	16.878	20.505	22.835	24.481	25.706
0.011	0.000	8.510	15.159	18.903	21.353	23.099	24.405
0.012	0.000	6.805	13.402	17.279	19.855	21.706	23.095
0.013	0.000	5.613	11.931	15.835	18.483	20.405	21.856
0.014	0.000	4.358	10.423	14.366	17.093	19.092	20.607
0.015	0.000	3.534	9.196	13.087	15.840	17.880	19.436
0.016	0.000	2.659	7.933	11.782	14.567	16.654	18.254
0.017	0.000	2.120	6.937	10.667	13.436	15.535	17.155
0.018	0.000	1.542	5.907	9.528	12.285	14.402	16.044
0.019	0.000	1.209	5.118	8.573	11.276	13.378	15.019
0.020	0.000	0.849	4.300	7.595	10.248	12.339	13.982
0.021	0.000	0.654	3.692	6.791	9.358	11.409	13.031
0.022	0.000	0.442	3.058	5.965	8.450	10.464	12.068
0.023	0.000	0.335	2.602	5.299	7.674	9.626	11.189
0.024	0.000	0.217	2.125	4.613	6.882	8.773	10.299
0.025	0.000	0.162	1.791	4.070	6.212	8.021	9.489
0.026	0.000	0.100	1.441	3.510	5.526	7.256	8.669
0.027	0.000	0.074	1.203	3.074	4.951	6.583	7.923
0.028	0.000	0.044	0.953	2.623	4.363	5.898	7.168
0.029	0.000	0.031	0.788	2.277	3.872	5.296	6.480
0.030	0.000	0.018	0.614	1.918	3.370	4.684	5.783
0.031	0.000	0.013	0.502	1.645	2.950	4.143	5.145
0.032	0.000	0.007	0.384	1.362	2.520	3.593	4.499
0.033	0.000	0.005	0.309	1.146	2.157	3.102	3.902
0.034	0.000	0.002	0.230	0.923	1.787	2.604	3.299
0.035	0.000	0.002	0.181	0.748	1.467	2.150	2.733
0.036	0.000	0.001	0.128	0.569	1.142	1.692	2.163
0.037	0.000	0.001	0.093	0.421	0.851	1.264	1.618
0.038	0.000	0.000	0.057	0.271	0.556	0.833	1.070
0.039	0.000	0.000	0.029	0.136	0.279	0.417	0.536
0.040	0.000	0.000	0.000	0.000	0.000	0.000	0.000

Table 3-3. Solution of the accelerated lower plate by the Dufort-Frankel explicit scheme, $\Delta y = 0.001$, $\Delta t = 0.003$.

y	t=0.00	t=0.18	t=0.36	t=0.54	t=0.72	t=0.90	t=1.08
0.000	40.000	40.000	40.000	40.000	40.000	40.000	40.000
0.001	0.000	36.318	37.422	37.902	38.186	38.378	38.517
0.002	0.000	32.687	34.861	35.813	36.377	36.760	37.038
0.003	0.000	29.157	32.334	33.742	34.580	35.150	35.565
0.004	0.000	25.771	29.857	31.698	32.801	33.553	34.101
0.005	0.000	22.568	27.446	29.690	31.045	31.972	32.649
0.006	0.000	19.580	25.113	27.726	29.317	30.411	31.212
0.007	0.000	16.829	22.872	25.812	27.623	28.874	29.793
0.008	0.000	14.329	20.733	23.956	25.967	27.365	28.395
0.009	0.000	12.086	18.705	22.165	24.353	25.886	27.020
0.010	0.000	10.101	16.795	20.442	22.786	24.441	25.670
0.011	0.000	8.364	15.006	18.793	21.269	23.032	24.348
0.012	0.000	6.863	13.343	17.221	19.806	21.662	23.055
0.013	0.000	5.581	11.807	15.730	18.398	20.334	21.793
0.014	0.000	4.498	10.396	14.321	17.048	19.048	20.565
0.015	0.000	3.595	9.109	12.994	15.758	17.807	19.370
0.016	0.000	2.848	7.942	11.752	14.528	16.612	18.211
0.017	0.000	2.238	6.891	10.593	13.360	15.464	17.088
0.018	0.000	1.744	5.949	9.516	12.254	14.363	16.002
0.019	0.000	1.349	5.111	8.519	11.210	13.310	14.953
0.020	0.000	1.035	4.370	7.601	10.227	12.305	13.942
0.021	0.000	0.788	3.717	6.758	9.304	11.347	12.967
0.022	0.000	0.595	3.147	5.988	8.439	10.435	12.030
0.023	0.000	0.447	2.652	5.287	7.632	9.570	11.130
0.024	0.000	0.333	2.223	4.650	6.880	8.750	10.265
0.025	0.000	0.246	1.855	4.075	6.181	7.973	9.435
0.026	0.000	0.181	1.540	3.557	5.532	7.238	8.639
0.027	0.000	0.132	1.273	3.092	4.932	6.543	7.875
0.028	0.000	0.096	1.046	2.675	4.375	5.885	7.142
0.029	0.000	0.069	0.856	2.303	3.861	5.264	6.439
0.030	0.000	0.050	0.696	1.971	3.385	4.675	5.762
0.031	0.000	0.035	0.562	1.674	2.945	4.118	5.111
0.032	0.000	0.025	0.451	1.410	2.537	3.588	4.483
0.033	0.000	0.018	0.358	1.174	2.157	3.083	3.875
0.034	0.000	0.012	0.281	0.963	1.802	2.601	3.287
0.035	0.000	0.009	0.217	0.771	1.469	2.138	2.714
0.036	0.000	0.006	0.162	0.597	1.153	1.691	2.155
0.037	0.000	0.004	0.115	0.436	0.852	1.257	1.607
0.038	0.000	0.002	0.074	0.285	0.562	0.832	1.066
0.039	0.000	0.001	0.036	0.141	0.279	0.414	0.532
0.040	0.000	0.000	0.000	0.000	0.000	0.000	0.000

Table 3-4. Solution of the accelerated lower plate by the Laasonen scheme, $\Delta y = 0.001$, $\Delta t = 0.01$.

y	t=0.00	t=0.18	t=0.36	t=0.54	t=0.72	t=0.90	t=1.08
0.000	40.000	40.000	40.000	40.000	40.000	40.000	40.000
0.001	0.000	36.396	37.449	37.916	38.195	38.385	38.523
0.002	0.000	32.839	34.914	35.842	36.396	36.774	37.049
0.003	0.000	29.371	32.412	33.785	34.609	35.171	35.581
0.004	0.000	26.035	29.957	31.754	32.838	33.580	34.122
0.005	0.000	22.864	27.565	29.757	31.090	32.005	32.676
0.006	0.000	19.890	25.247	27.803	29.370	30.450	31.244
0.007	0.000	17.137	23.017	25.898	27.682	28.919	29.829
0.008	0.000	14.619	20.885	24.049	26.032	27.414	28.435
0.009	0.000	12.346	18.860	22.263	24.423	25.939	27.064
0.010	0.000	10.321	16.947	20.543	22.860	24.498	25.717
0.011	0.000	8.538	15.154	18.896	21.346	23.092	24.398
0.012	0.000	6.990	13.482	17.324	19.884	21.725	23.108
0.013	0.000	5.662	11.934	15.830	18.477	20.398	21.848
0.014	0.000	4.538	10.510	14.417	17.126	19.114	20.621
0.015	0.000	3.598	9.208	13.086	15.835	17.874	19.428
0.016	0.000	2.821	8.025	11.838	14.604	16.678	18.270
0.017	0.000	2.189	6.957	10.671	13.433	15.530	17.147
0.018	0.000	1.679	5.999	9.586	12.324	14.428	16.061
0.019	0.000	1.274	5.145	8.582	11.276	13.373	15.012
0.020	0.000	0.956	4.389	7.655	10.288	12.366	13.999
0.021	0.000	0.710	3.724	6.803	9.361	11.405	13.024
0.022	0.000	0.521	3.142	6.025	8.491	10.492	12.086
0.023	0.000	0.378	2.636	5.315	7.679	9.624	11.183
0.024	0.000	0.272	2.200	4.671	6.922	8.800	10.316
0.025	0.000	0.193	1.826	4.089	6.218	8.020	9.484
0.026	0.000	0.136	1.506	3.564	5.565	7.281	8.686
0.027	0.000	0.094	1.236	3.094	4.959	6.583	7.919
0.028	0.000	0.065	1.008	2.672	4.399	5.922	7.183
0.029	0.000	0.044	0.817	2.296	3.881	5.297	6.476
0.030	0.000	0.030	0.658	1.961	3.402	4.705	5.797
0.031	0.000	0.020	0.526	1.663	2.958	4.144	5.142
0.032	0.000	0.013	0.418	1.398	2.547	3.611	4.511
0.033	0.000	0.009	0.328	1.162	2.165	3.103	3.900
0.034	0.000	0.006	0.255	0.950	1.808	2.618	3.308
0.035	0.000	0.004	0.194	0.760	1.473	2.152	2.732
0.036	0.000	0.002	0.144	0.587	1.157	1.702	2.169
0.037	0.000	0.001	0.101	0.429	0.855	1.265	1.617
0.038	0.000	0.001	0.064	0.280	0.564	0.838	1.074
0.039	0.000	0.000	0.031	0.138	0.280	0.417	0.535
0.040	0.000	0.000	0.000	0.000	0.000	0.000	0.000

Table 3-5. Solution of the accelerated lower plate by the Crank-Nicolson scheme, $\Delta y = 0.001$, $\Delta t = 0.01$.

y	t=0.00	t=0.18	t=0.36	t=0.54	t=0.72	t=0.90	t=1.08
0.000	40.000	40.000	40.000	40.000	40.000	40.000	40.000
0.001	0.000	36.397	37.449	37.917	38.195	38.385	38.523
0.002	0.000	32.839	34.915	35.842	36.396	36.774	37.049
0.003	0.000	29.372	32.413	33.785	34.609	35.171	35.582
0.004	0.000	26.035	29.958	31.755	32.839	33.580	34.123
0.005	0.000	22.864	27.566	29.759	31.091	32.006	32.676
0.006	0.000	19.889	25.249	27.804	29.371	30.451	31.245
0.007	0.000	17.135	23.019	25.900	27.683	28.920	29.830
0.008	0.000	14.616	20.886	24.051	26.033	27.415	28.436
0.009	0.000	12.342	18.860	22.264	24.424	25.941	27.065
0.010	0.000	10.315	16.948	20.544	22.861	24.499	25.719
0.011	0.000	8.532	15.154	18.897	21.347	23.094	24.399
0.012	0.000	6.983	13.482	17.324	19.885	21.726	23.109
0.013	0.000	5.654	11.933	15.831	18.477	20.399	21.850
0.014	0.000	4.528	10.508	14.417	17.127	19.115	20.623
0.015	0.000	3.587	9.205	13.086	15.836	17.875	19.430
0.016	0.000	2.810	8.021	11.837	14.604	16.679	18.271
0.017	0.000	2.177	6.953	10.670	13.433	15.531	17.149
0.018	0.000	1.668	5.994	9.585	12.324	14.429	16.062
0.019	0.000	1.263	5.140	8.580	11.276	13.374	15.013
0.020	0.000	0.946	4.383	7.653	10.288	12.366	14.001
0.021	0.000	0.700	3.718	6.801	9.360	11.406	13.025
0.022	0.000	0.512	3.136	6.022	8.491	10.492	12.087
0.023	0.000	0.370	2.630	5.312	7.678	9.624	11.184
0.024	0.000	0.265	2.194	4.668	6.921	8.800	10.317
0.025	0.000	0.187	1.819	4.085	6.217	8.020	9.485
0.026	0.000	0.131	1.500	3.561	5.563	7.281	8.686
0.027	0.000	0.090	1.230	3.090	4.958	6.583	7.920
0.028	0.000	0.061	1.002	2.669	4.397	5.922	7.184
0.029	0.000	0.041	0.812	2.293	3.879	5.297	6.477
0.030	0.000	0.028	0.653	1.958	3.400	4.705	5.797
0.031	0.000	0.018	0.522	1.660	2.957	4.144	5.143
0.032	0.000	0.012	0.414	1.395	2.546	3.611	4.511
0.033	0.000	0.008	0.325	1.159	2.163	3.103	3.900
0.034	0.000	0.005	0.252	0.948	1.807	2.617	3.308
0.035	0.000	0.003	0.192	0.758	1.472	2.151	2.732
0.036	0.000	0.002	0.142	0.586	1.156	1.701	2.169
0.037	0.000	0.001	0.100	0.427	0.854	1.265	1.617
0.038	0.000	0.001	0.063	0.279	0.563	0.838	1.074
0.039	0.000	0.000	0.031	0.138	0.280	0.417	0.535
0.040	0.000	0.000	0.000	0.000	0.000	0.000	0.000

Table 3-6. Analytical solution of the accelerated lower plate.

	0.00	0.50	1.00	1.50	2.00	2.50	3.00	3.50
0	200.00	200.00	200.00	200.00	200.00	200.00	200.00	200.00
.10	200.00	163.57	156.64	156.37	156.36	156.32	153.82	0
.20	200.00	129.52	116.11	115.59	115.58	115.51	111.57	0
.30	200.00	100.69	81.81	81.07	81.06	80.97	76.75	0
.40	200.00	77.51	54.21	53.30	53.28	53.21	49.61	0
.50	200.00	60.58	34.07	33.03	33.02	32.96	30.29	0
.60	200.00	49.06	20.36	19.24	19.22	19.18	17.43	0
.70	200.00	41.80	11.71	10.54	10.52	10.49	9.45	0
.80	200.00	37.56	6.67	5.46	5.44	5.43	4.85	0
.90	200.00	35.25	3.92	2.70	2.68	2.67	2.38	0
1.00	200.00	34.07	2.52	1.28	1.26	1.26	1.12	0
1.10	200.00	33.50	1.83	.59	.57	.57	.51	0
1.20	200.00	33.23	1.51	.27	.25	.25	.22	0
1.30	200.00	33.11	1.37	.13	.11	.11	.09	0
1.40	200.00	33.06	1.31	.06	.05	.05	.04	0
1.50	200.00	33.03	1.28	.04	.02	.02	.02	0
1.60	200.00	33.02	1.27	.03	.01	.01	.01	0
1.70	200.00	33.02	1.27	.02	.00	.00	.00	0
1.80	200.00	33.02	1.26	.02	.00	.00	.00	0
1.90	200.00	33.02	1.26	.02	.00	.00	.00	0
2.00	200.00	33.02	1.26	.02	.00	.00	.00	0
2.10	200.00	33.02	1.26	.02	.00	.00	.00	0
2.20	200.00	33.01	1.26	.02	.00	.00	.00	0
2.30	200.00	33.01	1.26	.02	.00	.00	.00	0
2.40	200.00	32.99	1.26	.02	.00	.00	.00	0
2.50	200.00	32.96	1.26	.02	.00	.00	.00	0
2.60	200.00	32.88	1.25	.02	.00	.00	.00	0
2.70	200.00	32.71	1.24	.02	.00	.00	.00	0
2.80	200.00	32.35	1.22	.02	.00	.00	.00	0
2.90	200.00	31.64	1.18	.02	.00	.00	.00	0
3.00	200.00	30.29	1.12	.02	.00	.00	.00	0
3.10	200.00	27.90	1.00	.01	.00	.00	.00	0
3.20	200.00	23.95	.83	.01	.00	.00	.00	0
3.30	200.00	17.94	.60	.01	.00	.00	.00	0
3.40	200.00	9.75	.32	.00	.00	.00	.00	0
3.50	200.00	0	0	0	0	0	0	0

Table 3-7. Temperature distribution at $t = 0.1$ hr.

	0.00	0.50	1.00	1.50	2.00	2.50	3.00	3.50
0	200.00	200.00	200.00	200.00	200.00	200.00	200.00	200.00
.10	200.00	188.62	181.48	178.66	177.82	176.65	168.67	0
.20	200.00	177.46	163.32	157.73	156.08	153.85	139.56	0
.30	200.00	166.73	145.86	137.60	135.19	132.10	113.90	0
.40	200.00	156.60	129.38	118.61	115.50	111.81	91.98	0
.50	200.00	147.23	114.13	101.04	97.31	93.28	73.62	0
.60	200.00	138.72	100.29	85.10	80.82	76.69	58.41	0
.70	200.00	131.15	87.98	70.91	66.18	62.14	45.91	0
.80	200.00	124.55	77.23	58.53	53.42	49.61	35.73	0
.90	200.00	118.89	68.02	47.93	42.51	39.02	27.50	0
1.00	200.00	114.13	60.28	39.02	33.37	30.24	20.92	0
1.10	200.00	110.21	53.91	31.68	25.85	23.08	15.72	0
1.20	200.00	107.04	48.75	25.75	19.78	17.35	11.66	0
1.30	200.00	104.52	44.66	21.05	14.97	12.85	8.53	0
1.40	200.00	102.55	41.47	17.39	11.24	9.38	6.15	0
1.50	200.00	101.04	39.02	14.59	8.39	6.75	4.37	0
1.60	200.00	99.89	37.18	12.49	6.26	4.79	3.06	0
1.70	200.00	99.01	35.80	10.93	4.69	3.35	2.11	0
1.80	200.00	98.34	34.77	9.80	3.56	2.32	1.44	0
1.90	200.00	97.79	33.98	8.98	2.75	1.60	.96	0
2.00	200.00	97.31	33.37	8.39	2.19	1.09	.64	0
2.10	200.00	96.82	32.83	7.95	1.80	.75	.42	0
2.20	200.00	96.26	32.32	7.60	1.52	.52	.27	0
2.30	200.00	95.55	31.76	7.31	1.34	.36	.17	0
2.40	200.00	94.60	31.09	7.04	1.20	.26	.11	0
2.50	200.00	93.28	30.24	6.75	1.09	.20	.07	0
2.60	200.00	91.43	29.15	6.42	1.00	.15	.04	0
2.70	200.00	88.85	27.74	6.03	.92	.12	.03	0
2.80	200.00	85.26	25.94	5.56	.83	.10	.02	0
2.90	200.00	80.33	23.69	5.01	.74	.08	.01	0
3.00	200.00	73.62	20.92	4.37	.64	.07	.01	0
3.10	200.00	64.62	17.62	3.64	.53	.05	.01	0
3.20	200.00	52.83	13.80	2.81	.40	.04	.00	0
3.30	200.00	37.90	9.50	1.92	.27	.03	.00	0
3.40	200.00	19.95	4.85	.97	.14	.01	.00	0
3.50	200.00	0	0	0	0	0	0	0

Table 3-8. Temperature distribution at $t = 0.4$ hr.

Chapter 4

Stability Analysis

4.1 Introductory Remarks

In general, two types of errors are introduced in the solution of finite difference equations. These errors may be caused by round-off error, which is a property of the computer, or by the application of a particular numerical method, i.e., a discretization error. If the errors introduced into the FDE are not controlled, the growth of errors with the solution of the FDE will result in an unstable solution. Understanding and controlling these errors by stability analysis is essential for a successful solution of an FDE. In this chapter, stability analyses of scalar model equations are introduced. The results will establish valuable guidelines for the stability requirement of complicated FDEs. In some cases, stability analysis becomes so cumbersome that numerical experimentation is a more realistic approach. Stability analysis of model equations, along with numerical experimentation, should provide insight into the limitations on stepsizes that are needed to obtain a stable solution.

In this chapter, two methods of stability analysis and their applications to various model equations are investigated. The two methods are discrete perturbation stability analysis and von Neumann (Fourier) stability analysis.

The von Neumann stability analysis is more commonly used and is less cumbersome mathematically. However, discrete perturbation stability analysis is presented here to familiarize the reader with the method and to illustrate graphically the effect of a disturbance as it grows or decays with the solution.

4.2 Discrete Perturbation Stability Analysis

In this method, a disturbance is introduced at a point, and its effect on neighboring points is investigated. If the disturbance dies out as the solution proceeds, then the numerical technique used is indeed stable. However, if the disturbance grows with the solution, the method is unstable. To illustrate this analysis, consider the following parabolic model equation

$$\frac{\partial u}{\partial t} = \alpha \frac{\partial^2 u}{\partial x^2} \tag{4-1}$$

where α is assumed constant. An explicit finite difference equation using second-order central differencing for the space derivative and first-order forward differencing for the time derivative is

$$\frac{u_i^{n+1} - u_i^n}{\Delta t} = \alpha \frac{u_{i+1}^n - 2u_i^n + u_{i-1}^n}{(\Delta x)^2} \tag{4-2}$$

Two procedures may be used to approach the mathematical work. The first procedure assumes that a solution $u_i^n = 0$ at all i has been obtained. Then a disturbance ϵ at node i at the time level n is introduced and the solution at the time level $n+1$ for all i nodes is sought. Therefore,

$$\frac{u_i^{n+1} - (u_i^n + \epsilon)}{\Delta t} = \alpha \frac{u_{i+1}^n - 2\,(u_i^n + \epsilon) + u_{i-1}^n}{(\Delta x)^2}$$

from which it follows that

$$\frac{u_i^{n+1} - \epsilon}{\Delta t} = \alpha \frac{-2\epsilon}{(\Delta x)^2} \qquad \text{or}$$

$$u_i^{n+1} = -\alpha \frac{2\epsilon}{(\Delta x)^2} \Delta t + \epsilon = \epsilon \left[1 - 2\alpha \left(\frac{\Delta t}{(\Delta x)^2}\right)\right]$$

The expression $\alpha \Delta t/(\Delta x)^2$ is known as the diffusion number and will be denoted by d. Thus,

$$\frac{u_i^{n+1}}{\epsilon} = (1 - 2d) \tag{4-3}$$

Before proceeding with the effect of disturbance on other nodes, the second procedure is considered. In this procedure, a disturbance ϵ is introduced at node i at time level n, and the disturbance at $(i, n+1)$ is computed as follows. The FDE is

$$\frac{u_i^{n+1} - u_i^n}{\Delta t} = \alpha \frac{u_{i+1}^n - 2u_i^n + u_{i-1}^n}{(\Delta x)^2} \tag{4-4}$$

A disturbance ϵ_i^n is introduced at (i, n), and the disturbance to be calculated at $(i, n+1)$ is ϵ_i^{n+1}. Thus, from Equation (4-4),

$$\frac{\left(u_i^{n+1} + \epsilon_i^{n+1}\right) - \left(u_i^n + \epsilon_i^n\right)}{\Delta t} = \alpha \frac{u_{i+1}^n - 2\left(u_i^n + \epsilon_i^n\right) + u_{i-1}^n}{(\Delta x)^2} \tag{4-5}$$

Subtracting Equation (4-4) from (4-5) produces

$$\frac{\epsilon_i^{n+1} - \epsilon_i^n}{\Delta t} = \alpha \frac{-2\epsilon_i^n}{(\Delta x)^2} \qquad \text{or}$$

$$\epsilon_i^{n+1} = \epsilon_i^n - 2\frac{\alpha \Delta t}{(\Delta x)^2}\epsilon_i^n = (1 - 2d)\epsilon_i^n$$

hence,

$$\frac{\epsilon_i^{n+1}}{\epsilon_i^n} = (1 - 2d) \tag{4-6}$$

This result is identical to that obtained by the first procedure, i.e., Equation (4-3). Since less mathematical labor is involved in the first procedure, it will be used to continue the analysis. One important point to note is that the assumed solution at time level n does not effect the outcome of the analysis.

Returning to Equation (4-3), we have

$$\frac{u_i^{n+1}}{\epsilon} = (1 - 2d)$$

which represents propagation of the error to grid point i at time level $n + 1$. In order to prevent its growing with the solution, the error must be bounded. For this purpose, the absolute value of error propagation is set to be less than or equal to one. Mathematically, this requirement is expressed as

$$\left|\frac{u_i^{n+1}}{\epsilon}\right| \leq 1$$

This limitation requires that $1 - 2d \leq 1$, which is always satisfied since d has a positive value, and

$$1 - 2d \geq -1\,, \qquad \text{so that} \quad d \leq 1 \tag{4-7}$$

To calculate the effect of the disturbance at time $n + 1$ at point $i + 1$, consider

$$\frac{u_{i+1}^{n+1} - u_{i+1}^n}{\Delta t} = \alpha \frac{u_{i+2}^n - 2u_{i+1}^n + u_i^n}{(\Delta x)^2} \tag{4-8}$$

or $u_{i+1}^{n+1} - u_{i+1}^n = d(u_{i+2}^n - 2u_{i+1}^n + u_i^n)$. With a disturbance at i, n

$$u_{i+1}^{n+1} - u_{i+1}^n = d\left[u_{i+2}^n - 2u_{i+1}^n + (u_i^n + \epsilon)\right] \qquad \text{or}$$

$$u_{i+1}^{n+1} = d\epsilon \tag{4-9a}$$

Similarly, at $i-1$

$$u_{i-1}^{n+1} = d\epsilon \tag{4-9b}$$

Now for the time level $(n+2)$,

$$\frac{u_i^{n+2} - u_i^{n+1}}{\Delta t} = \alpha \frac{u_{i+1}^{n+1} - 2u_i^{n+1} + u_{i-1}^{n+1}}{(\Delta x)^2} \quad \text{or}$$

$$u_i^{n+2} = u_i^{n+1} + d\left(u_{i+1}^{n+1} - 2u_i^{n+1} + u_{i-1}^{n+1}\right) \tag{4-10}$$

Substituting Equations (4-3) and (4-9) will yield

$$u_i^{n+2} = \epsilon(1-2d) + d\Big[\epsilon d - 2\epsilon(1-2d) + \epsilon d\Big] \quad \text{or}$$

$$u_i^{n+2} = \epsilon(6d^2 - 4d + 1) \tag{4-11}$$

For a stable solution, $\left|\dfrac{u_i^{n+2}}{\epsilon}\right| = \left|6d^2 - 4d + 1\right| \leq 1$

This inequality includes two possible cases, which are $6d^2 - 4d + 1 \leq 1$ and $6d^2 - 4d + 1 \geq -1$. The first condition requires that $2d(3d-2) \leq 0$. Thus,

$$d \leq \frac{2}{3} \tag{4-12}$$

From the second condition, it follows that $2d - 3d^2 \leq 1$, a requirement that is always satisfied. Note that the analysis indicates a more severe restriction on d at this time level. Now, consider the disturbance at point $i+2$ at the time level $(n+2)$, for which

$$\frac{u_{i+2}^{n+2} - u_{i+2}^{n+1}}{\Delta t} = \alpha \frac{u_{i+3}^{n+1} - 2u_{i+2}^{n+1} + u_{i+1}^{n+1}}{(\Delta x)^2} \quad \text{or} \tag{4-13}$$

$$u_{i+2}^{n+2} = u_{i+2}^{n+1} + d\left(u_{i+3}^{n+1} - 2u_{i+2}^{n+1} + u_{i+1}^{n+1}\right)$$

Hence,

$$u_{i+2}^{n+2} = d(\epsilon d) = \epsilon d^2 \tag{4-14}$$

Similarly, at point $i+1$

$$u_{i+1}^{n+2} - u_{i+1}^{n+1} = d\left(u_{i+2}^{n+1} - 2u_{i+1}^{n+1} + u_i^{n+1}\right)$$

$$u_{i+1}^{n+2} - \epsilon d = d\left[-2\epsilon d + \epsilon(1-2d)\right]$$

$$u_{i+1}^{n+2} = 2\epsilon d(1-2d) \tag{4-15}$$

The analysis may be continued to higher time levels, in which case a more restrictive condition on d is imposed at each time level.

Eventually the error will reach all grid points after many time steps with approximately the same magnitude. Two possibilities may be considered. In the first case, the errors at the time m have the same sign; for example,

$$u_{i+1}^m = \epsilon^m$$

$$u_i^m = \epsilon^m$$

$$u_{i-1}^m = \epsilon^m$$

Then,

$$\frac{u_i^{m+1} - \epsilon^m}{\Delta t} = \alpha \frac{\epsilon^m - 2\epsilon^m + \epsilon^m}{(\Delta x)^2}$$

or

$$u_i^{m+1} = \epsilon^m$$

Thus, this case does not impose a stability restriction. For the second case, the errors alternate signs, i.e.,

$$u_{i+1}^m = -\epsilon^m$$

$$u_i^m = \epsilon^m$$

$$u_{i-1}^m = -\epsilon^m$$

This oscillatory error distribution is illustrated in Figure 4-1. Now the FDE is

$$\frac{u_i^{m+1} - \epsilon^m}{\Delta t} = \alpha \frac{-\epsilon^m - 2\epsilon^m - \epsilon^m}{(\Delta x)^2}$$

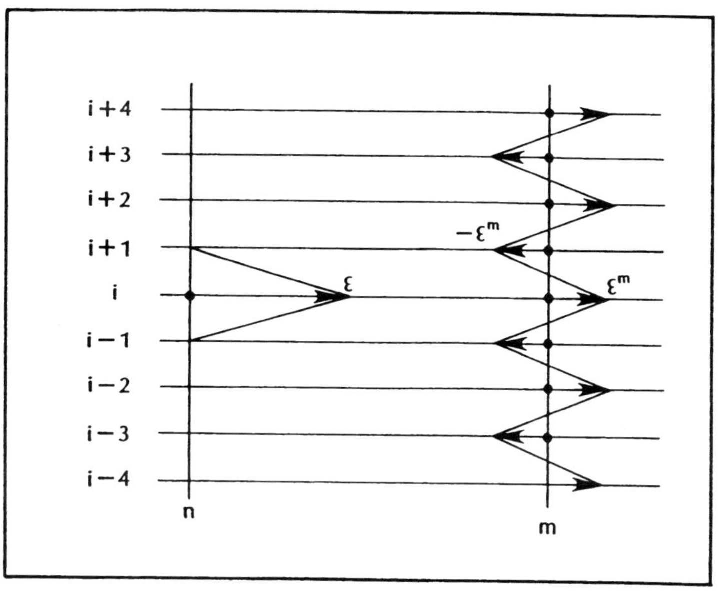

Figure 4-1. Distribution of error at time level m.

Hence,

$$u_i^{m+1} = \epsilon^m + d(-\epsilon^m - 2\epsilon^m - \epsilon^m) = (1-4d)\epsilon^m \tag{4-16}$$

The solution will be stable if

$$\left|\frac{u_i^{m+1}}{\epsilon^m}\right| \le 1 \quad \text{or} \quad \left|1-4d\right| \le 1$$

Hence, the two requirements are

$$1-4d \le 1 \quad \text{and} \quad 1-4d \ge -1$$

The first condition will yield $-4d \le 0$, a requirement that is always satisfied. The second condition will result in

$$d \le \frac{1}{2} \tag{4-17}$$

or, in terms of the step sizes Δx and Δt,

$$\Delta t \le \frac{1}{2\alpha}(\Delta x)^2 \tag{4-18}$$

This requirement imposes limitations on the step sizes and provides a valuable guide for the selection of step sizes. It is concluded that, for a stable solution of the FDE given by Equation (4-2) and for a specified α and Δx, a Δt must be selected that is less than or equal to $(\Delta x)^2/2\alpha$.

Decay or growth of a disturbance for two values of d are now considered graphically. For the first case, assume that $d = 0.25$, which should cause the disturbance ϵ, introduced at (i, n), to decay as illustrated in Figure 4-2.

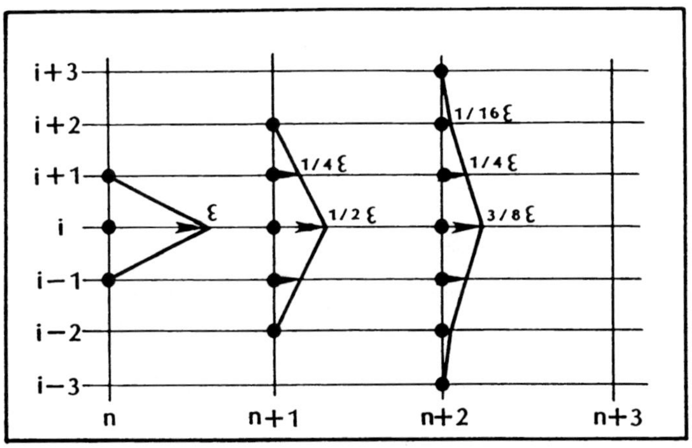

Figure 4-2. Decay of error for FTCS explicit formulation of model Equation (4-1) for $d = 0.25$.

For $d = 1.5$, which exceeds the stability requirement, the disturbance grows as shown in Figure 4-3.

A second example illustrating the application of discrete perturbation stability analysis involves a first-order hyperbolic equation. The relevant model equation is the one-dimensional wave equation:

$$\frac{\partial u}{\partial t} = -a\frac{\partial u}{\partial x} \quad \text{where} \quad a > 0 \tag{4-19}$$

To approximate the PDE, the derivatives are replaced with a forward differencing of the first order in time and a backward differencing of the first order in space. Then the explicit formulation is

$$\frac{u_i^{n+1} - u_i^n}{\Delta t} = -a\frac{u_i^n - u_{i-1}^n}{\Delta x} \tag{4-20}$$

When a disturbance ϵ is introduced into the ith node at the time n, the FDE is

$$\frac{u_i^{n+1} - (u_i^n + \epsilon)}{\Delta t} = -a\frac{(u_i^n + \epsilon) - u_{i-1}^n}{\Delta x}$$

from which $u_i^{n+1} = u_i^n + \epsilon - c(u_i^n + \epsilon - u_{i-1}^n)$, where $c = a\Delta t/\Delta x$ is known as the Courant number. Therefore the solution at $(i, n+1)$ is

$$u_i^{n+1} = (1 - c)\epsilon \tag{4-21}$$

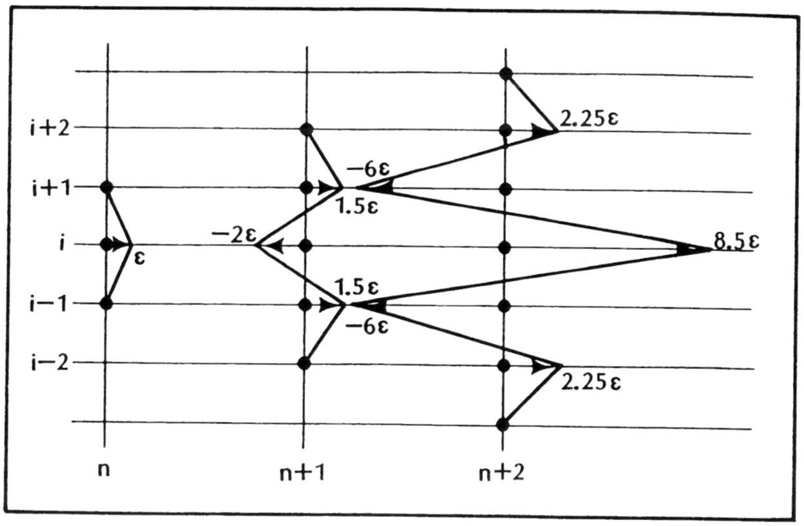

Figure 4-3. Growth of error for FTCS explicit formulation of model Equation (4-1) for $d = 1.5$.

For a stable solution,

$$\left|\frac{u_i^{n+1}}{\epsilon}\right| \le 1 \quad \text{or} \quad \left|(1-c)\right| \le 1$$

from which

$$1 - c \le 1 \tag{4-22a}$$

and

$$1 - c \ge -1 \tag{4-22b}$$

Condition (4-22a) requires that $-c \le 0$; since c is positive, this requirement is always satisfied. Condition (4-22b) requires that $c \le 2$.

Now consider the propagation of the disturbance to $(i+1, n+1)$. The FDE is

$$\frac{u_{i+1}^{n+1} - u_{i+1}^n}{\Delta t} = -a\frac{u_{i+1}^n - u_i^n}{\Delta x} \quad \text{or} \quad u_{i+1}^{n+1} = u_{i+1}^n - c\left[u_{i+1}^n - (u_i^n + \epsilon)\right]$$

from which $u_{i+1}^{n+1} = c\epsilon$. For a stable solution,

$$\left|\frac{u_{i+1}^{n+1}}{\epsilon}\right| \le 1$$

Thus, $|c| \le 1$; so that $c \le 1$ and $c \ge -1$. The second condition is always satisfied; therefore, for a stable solution, c must be ≤ 1. At $(i-1, n+1)$

$$\frac{u_{i-1}^{n+1} - u_{i-1}^n}{\Delta t} = -a\frac{u_{i-1}^n - u_{i-2}^n}{\Delta x}$$

Therefore, $u_{i-1}^{n+1} = 0$. The procedure may be continued to the next time level at $(n+2)$ with the following solution:

$$u_i^{n+2} = (1-c)^2\epsilon \tag{4-23a}$$

$$u_{i+1}^{n+2} = 2c(1-c)\epsilon \tag{4-23b}$$

$$u_{i+2}^{n+2} = c^2\epsilon \tag{4-23c}$$

$$u_{i+3}^{n+2} = 0 \tag{4-23d}$$

$$u_{i-1}^{n+2} = 0 \tag{4-23e}$$

Imposing the stability requirement indicates that the solution will be stable if $c \le 1$.

Now the decay or growth of the disturbance ϵ for $c = 0.75$ and $c = 1.5$ is investigated. When $c = 0.75$, stability analysis indicates a decay of disturbance, which is illustrated in Figure 4-4.

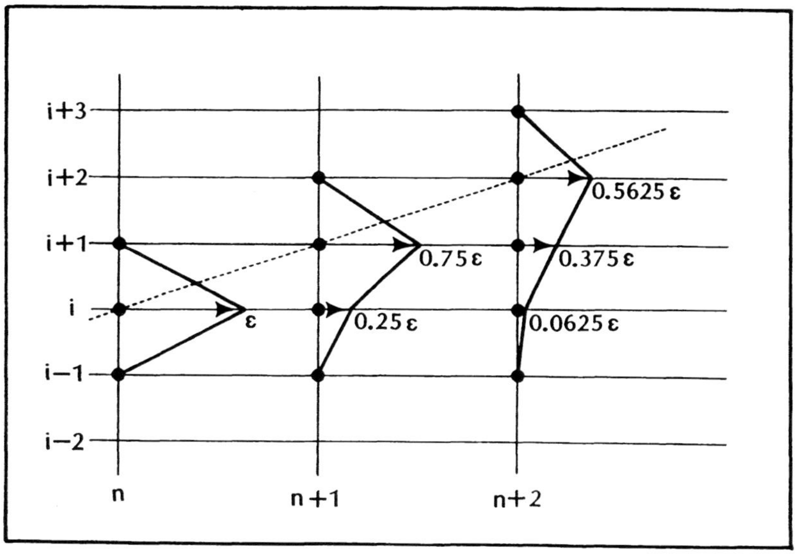

Figure 4-4. Decay of error for the finite difference Equation (4-20) for $c = 0.75$.

Note that for the upper limit of the stability requirement, i.e., at $c = 1$, the solution is exact. The disturbance propagates along the characteristic line, as shown in Figure 4-5.

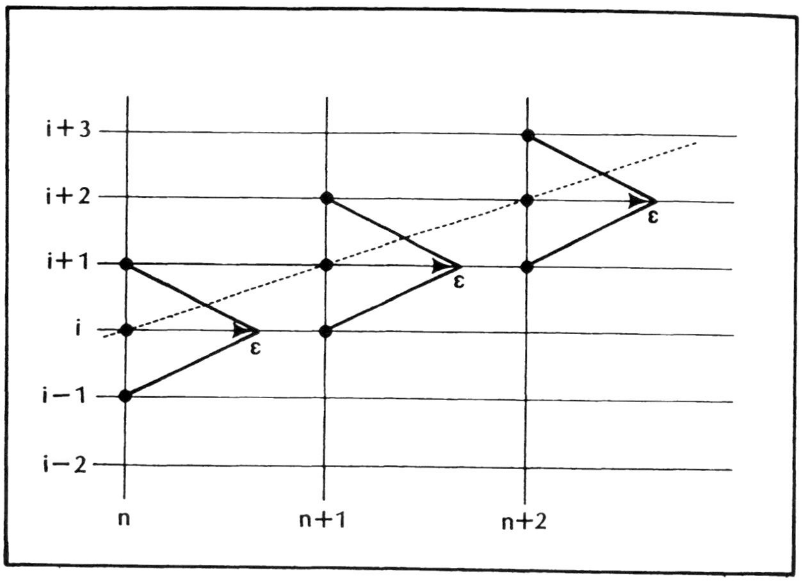

Figure 4-5. Propagation of disturbance along the characteristic line.

An unstable solution of Equation (4-20) results when $c = 1.5$. The resulting propagation of disturbance (for the first two time levels) is shown in Figure 4-6.

A comment on instability: Recall the unstable solution presented in Figure 4-3. The instability of solution for node i at various time levels is repeated in Figure 4-7. This figure clearly indicates the oscillatory behavior of an unstable solution, which is known as "dynamic instability". The error propagations for node i at various time levels are

$$u_i^n = \epsilon$$

$$u_i^{n+1} = (1 - 2d)\epsilon$$

$$u_i^{n+2} = (6d^2 - 4d + 1)\epsilon$$

Note that for $d > \frac{1}{2}$, the unstable solution is indeed oscillatory; for example, when $d = 1.5$,

$$u_i^n = \epsilon > 0$$

$$u_i^{n+1} = -2\epsilon < 0$$

$$u_i^{n+2} = +8.5\epsilon > 0$$

That is, the solution changes sign at each time step as it proceeds forward in time.

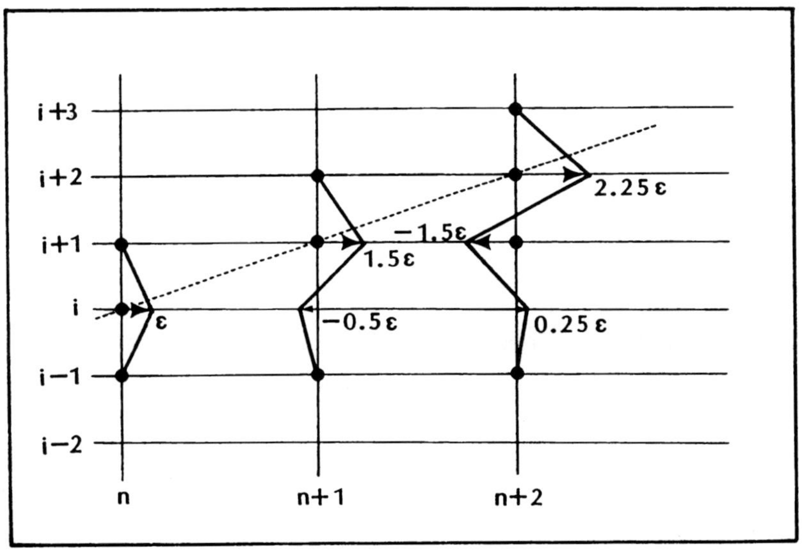

Figure 4-6. Growth of error for the finite difference Equation (4-20) for $c = 1.5$.

If the instability is such that the amplitude grows without oscillation, it is classified as a static instability. A typical result is shown in Figure 4-8.

Applying discrete perturbation analysis to a model equation becomes cumbersome when it contains both convection and diffusion terms, i.e., when

$$\frac{\partial u}{\partial t} = -a\frac{\partial u}{\partial x} + \alpha\frac{\partial^2 u}{\partial x^2}$$

Indeed, in such instances the von Neumann stability analysis, which is discussed in the next section, would be more practical.

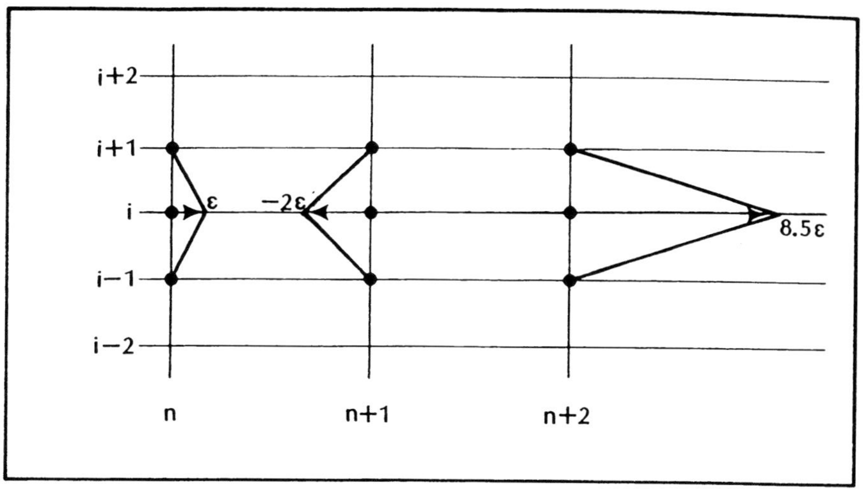

Figure 4-7. Sketch illustrating dynamic instability.

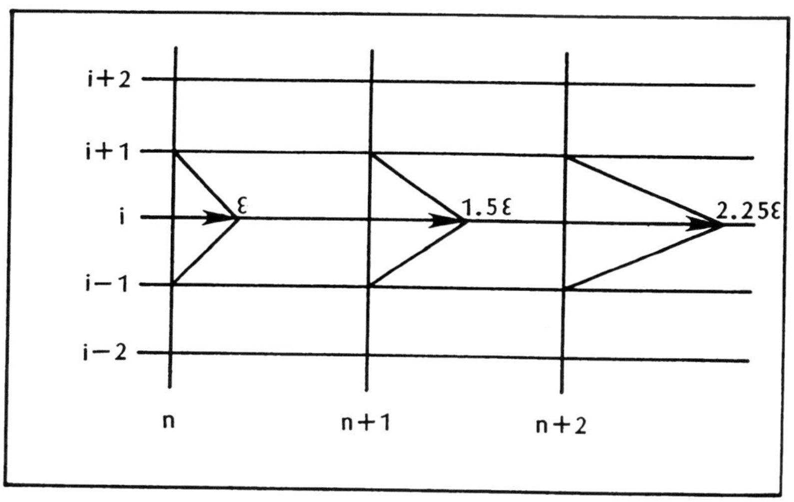

Figure 4-8. Sketch illustrating static instability.

4.3 Von Neumann Stability Analysis

Von Neumann stability analysis is a commonly used procedure for determining the stability requirements of finite difference equations. In this method, a solution of the finite difference equation is expanded in a Fourier series. The decay or growth of the amplification factor indicates whether or not the numerical algorithm is stable.

Recall that, for a linear equation, various solutions may be added. Therefore, when the FDE under investigation is linear, it is sufficient to investigate only one component of the Fourier series. In fact, the linearity of the equation is a general requirement for the application of the von Neumann stability analysis. Furthermore, the effect of the boundary condition on the stability of the solution is not included with this procedure. To overcome these limitations, one may locally linearize the nonlinear equation and subsequently apply the von Neumann stability analysis. However, note that the resulting stability requirement is satisfied locally. Therefore, the actual stability requirement may be more restrictive than the one obtained from the von Neumann stability analysis. Nevertheless, the results will provide very useful information on stability requirements.

To illustrate the procedure, assume a Fourier component for u_i^n as

$$u_i^n = U^n e^{IP(\Delta x)i} \tag{4-24}$$

where $I = \sqrt{-1}$, U^n is the amplitude at time level n, and P is the wave number in the x-direction, i.e., $\lambda_x = 2\pi/P$, where λ_x is the wavelength. Similarly,

$$u_i^{n+1} = U^{n+1} e^{IP(\Delta x)i} \quad \text{and} \quad u_{i\pm1}^n = U^n e^{IP(\Delta x)(i\pm1)}$$

If a phase angle $\theta = P\Delta x$ is defined, then

$$u_i^n = U^n e^{I\theta i} \tag{4-25}$$

$$u_i^{n+1} = U^{n+1} e^{I\theta i} \tag{4-26}$$

and

$$u_{i\pm1}^n = U^n e^{I\theta(i\pm1)} \tag{4-27}$$

To proceed with the application of this method, consider the explicit representation of Stokes' first problem. Recall that the partial differential equation is Equation (3-1), and its FTCS explicit formulation is

$$\frac{u_i^{n+1} - u_i^n}{\Delta t} = \alpha \frac{u_{i+1}^n - 2u_i^n + u_{i-1}^n}{(\Delta x)^2}$$

or, in terms of the diffusion number

$$u_i^{n+1} = u_i^n + d\left(u_{i+1}^n - 2u_i^n + u_{i-1}^n\right) \tag{4-28}$$

Substituting Equations (4-25), (4-26), and (4-27) into the finite difference equation (4-28), one obtains

$$U^{n+1}e^{I\theta i} = U^n e^{I\theta i} + d\left[U^n e^{I\theta(i+1)} - 2U^n e^{I\theta i} + U^n e^{I\theta(i-1)}\right]$$

After canceling the common factor $e^{I\theta i}$,

$$U^{n+1} = U^n + d\left(U^n e^{I\theta} - 2U^n + U^n e^{-I\theta}\right) \quad \text{or}$$

$$U^{n+1} = U^n\left[1 + d\left(e^{I\theta} + e^{-I\theta} - 2\right)\right]$$

Utilizing the relation

$$\cos\theta = \frac{e^{I\theta} + e^{-I\theta}}{2}$$

$$U^{n+1} = U^n\left[1 + 2d(\cos\theta - 1)\right]$$

Introducing an amplification factor such that $U^{n+1} = GU^n$, then

$$G = 1 - 2d(1 - \cos\theta) \tag{4-29}$$

For a stable solution, the absolute value of G must be bounded for all values of θ. Mathematically, it is expressed as

$$\left|G\right| \le 1 \quad \text{or} \quad \left|1 - 2d(1 - \cos\theta)\right| \le 1$$

so that

$$1 - 2d(1 - \cos\theta) \le 1 \tag{4-30}$$

and

$$1 - 2d(1 - \cos\theta) \ge -1 \tag{4-31}$$

Inequality (4-30) is satisfied for all values of θ. With the maximum value of $(1 - \cos\theta) = 2$, the left-hand side of (4-31) is $(1 - 4d)$, which must be larger than or equal to -1; thus, $1 - 4d \ge -1$ or $4d \le 2$. So the stability condition is that

$$d \le \frac{1}{2} \tag{4-32}$$

This result was expected because this stability condition was determined by the discrete perturbation analysis in the previous section. On some occasions, the amplification factor may become a complicated mathematical expression. Once the stability requirement ($|G| \le 1$) is imposed, one may not be able to clearly conclude a stability condition(s). In such instances, a graphical representation of the amplification factor along with some numerical experimentation may facilitate the analysis. Furthermore, experience gained by the stability analysis of simple model equations should also provide some very useful guidelines. To illustrate the graphical representation, the amplification factor given by (4-29) is considered.

The graphical solution may be either in a polar coordinate or a Cartesian coordinate. They are shown in Figures 4-9a and 4-9b, respectively. Note that in Figure 4-9a, when $d = 0.625$, some values of G have fallen outside the circle of radius 1; and in Figure 4-9b, it has exceeded the stability limit, indicating an unstable solution for this value of d.

(a) Polar

(b) Cartesian

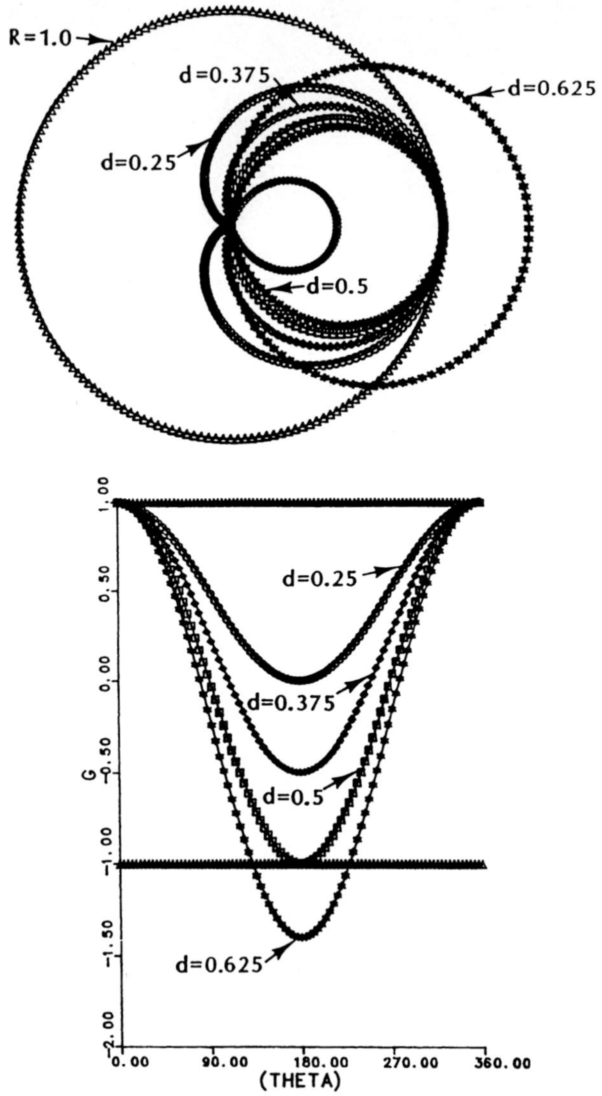

Figure 4-9. Amplification factor given by (4-29) illustrated for various values of d.

For a second application of the von Neumann stability analysis, consider a one-dimensional equation with both convection and diffusion terms, i.e.,

$$\frac{\partial u}{\partial t} = -a\frac{\partial u}{\partial x} + \alpha\frac{\partial^2 u}{\partial x^2} \tag{4-33}$$

The FTCS explicit formulation is expressed as

$$\frac{u_i^{n+1} - u_i^n}{\Delta t} = -a\frac{u_{i+1}^n - u_{i-1}^n}{2\Delta x} + \alpha\frac{u_{i+1}^n - 2u_i^n + u_{i-1}^n}{(\Delta x)^2} \quad \text{or}$$

$$u_i^{n+1} = u_i^n - \frac{c}{2}\left(u_{i+1}^n - u_{i-1}^n\right) + d\left(u_{i+1}^n - 2u_i^n + u_{i-1}^n\right) \tag{4-34}$$

Following the von Neumann stability analysis,

$$U^{n+1}e^{I\theta i} = U^n e^{I\theta i} - \frac{c}{2}\left[U^n e^{I\theta(i+1)} - U^n e^{I\theta(i-1)}\right]$$
$$+ d\left[U^n e^{I\theta(i+1)} - 2U^n e^{I\theta i} + U^n e^{I\theta(i-1)}\right]$$

Eliminating $e^{I\theta i}$,

$$U^{n+1} = U^n - \frac{c}{2}\left[U^n e^{I\theta} - U^n e^{-I\theta}\right] + d\left[U^n e^{I\theta} - 2U^n + U^n e^{-I\theta}\right] \quad \text{or}$$

$$U^{n+1} = U^n\left\{\left[1 - \frac{c}{2}\left(e^{I\theta} - e^{-I\theta}\right)\right] + d\left[\left(e^{I\theta} + e^{-I\theta}\right) - 2\right]\right\} \tag{4-35}$$

With the identities

$$\cos\theta = \frac{e^{I\theta} + e^{-I\theta}}{2} \tag{4-36a}$$

and

$$\sin\theta = \frac{e^{I\theta} - e^{-I\theta}}{2I} \tag{4-36b}$$

Equation (4-35) is written as

$$U^{n+1} = U^n\left[1 - c(I\sin\theta) + 2d(\cos\theta - 1)\right]$$

from which it follows that

$$G = \left[1 - 2d(1 - \cos\theta)\right] - I\left[c\sin\theta\right] \tag{4-37}$$

Note that, for this particular problem, the amplification factor has real and imaginary parts.

A stable solution requires that the modulus of the amplification factor must be bounded. Thus, a formal requirement may be expressed as

$$\left|G\right|^2 \le 1$$

Before mathematical arguments are considered, a graphical presentation is studied. When Equation (4-37) is written in the form

$$G = X + IY = [(1 - 2d) + 2d\cos\theta] - I\,[c\sin\theta] \tag{4-38}$$

it becomes the equation of an ellipse which is illustrated in Fig. 4-10. For a stable solution, the requirement of $|G| \leq 1$ indicates that the ellipse should fall inside the circle of unit radius. In the two most extreme cases, $1 - 2d = 0$ or $d = 0.5$ (Figure 4-11a), or $c = 1$ (Figure 4-11b). Thus, it appears that the stability requirements are $d \leq 0.5$ and $c \leq 1$. However, a more restrictive condition is found by considering the formal requirement that the modulus of (4-38) must be bounded, i.e., $|G|^2 \leq 1$. Note that

$$\left|G\right|^2 = G\bar{G} = \{[(1 - 2d) + 2d\cos\theta] - I(c\sin\theta)\}\ \{[(1 - 2d) + 2d\cos\theta] + I(c\sin\theta)\}$$

or

$$\left|G\right|^2 = [(1 - 2d) + 2d\cos\theta]^2 + (c\sin\theta)^2 \tag{4-39}$$

When this equation is written as a quadratic, it follows that

$$\left|G\right|^2 = (1 - 2d)^2 + (4d^2)\cos^2\theta + 4d(1 - 2d)\cos\theta + c^2(1 - \cos^2\theta)$$

or

$$\left|G\right|^2 = \left(4d^2 - c^2\right)\cos^2\theta + 4d(1 - 2d)\cos\theta + 4d(d - 1) + c^2 + 1 \tag{4-40}$$

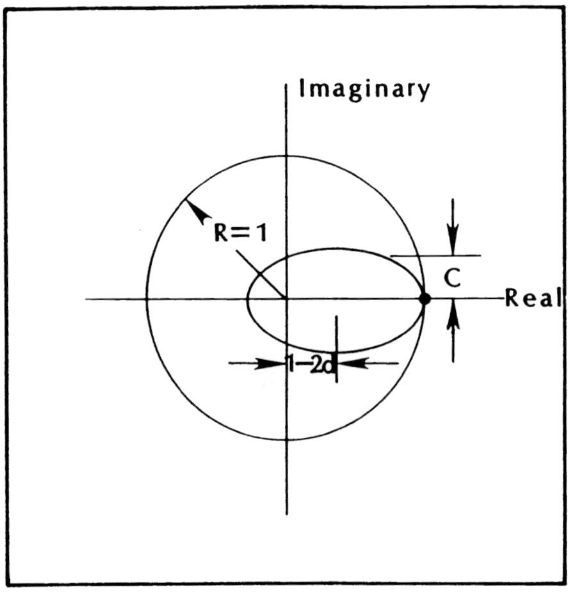

Figure 4-10. Sketch of amplification factor G given by Equation (4-38).

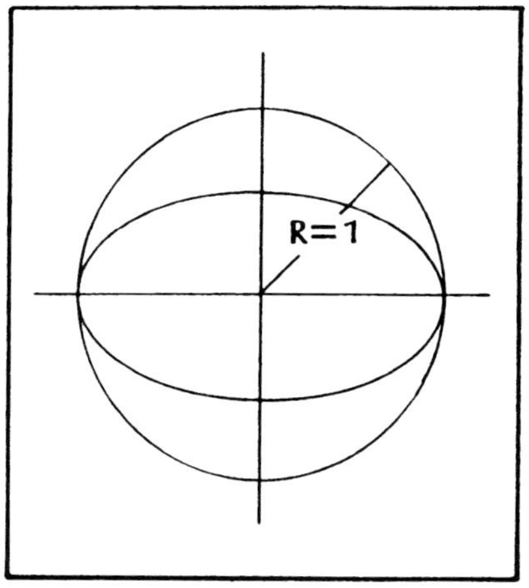

Figure 4-11a. Amplification factor for $d = 0.5$.

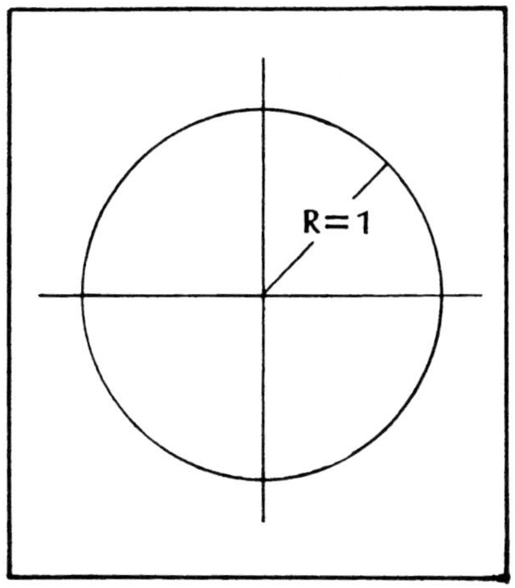

Figure 4-11b. Amplification factor for $c = 1$.

This quadratic equation represents either a convex curve (with a minimum) or a concave curve (with a maximum) (see Figure 4-12).

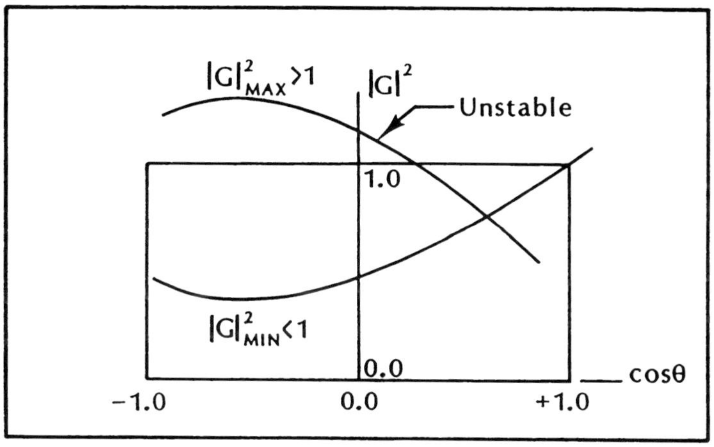

Figure 4-12. Quadratic function representing Equation (4-40).

It can be shown that for a stable solution, this quadratic function may not have a global maximum, i.e., the curve cannot be concave. The mathematical procedure is as follows. The first and second derivatives of the function $|G|^2$ with respect to $\cos\theta$ are

$$\frac{d|G|^2}{d(\cos\theta)} = (4d^2 - c^2)(2\cos\theta) + 4d(1 - 2d) \tag{4-41}$$

and

$$\frac{d^2|G|^2}{d(\cos\theta)^2} = 2(4d^2 - c^2) \tag{4-42}$$

The function has a maximum, and it would be a concave curve if the second derivative is negative, that is, if $(4d^2 - c^2) < 0$ or $c > 2d$. Now from Equation (4-41),

$$(\cos\theta)_{\max} = -\frac{4d(1-2d)}{2(4d^2-c^2)} = \frac{2d(2d-1)}{4d^2-c^2} \tag{4-43}$$

and, from Equation (4-40), with the substitution of (4-43)

$$\left|G\right|^2_{\max} = (4d^2 - c^2)\left[\frac{2d(2d-1)}{4d^2-c^2}\right]^2$$

$$+ 4d(1-2d)\frac{2d(2d-1)}{4d^2-c^2} + 4d(d-1) + c^2 + 1$$

From which it follows that

$$\left|G\right|^2_{\max} = \frac{c^2(c^2 - 4d + 1)}{c^2 - 4d^2} \tag{4-44}$$

Requiring that $|G|^2 \leq 1$ will result in

$$c^2(c^2 - 4d + 1) \leq c^2 - 4d^2$$

which is reduced to

$$(c^2 - 2d)^2 \leq 0$$

Since $(c^2 - 2d)^2$ is always a positive quantity, the condition for stability cannot be achieved; and the quadratic function represented by (4-40), which corresponds to a concave curve is not allowed. Thus, it is concluded that one must first enforce

$$c \leq 2d \tag{4-45}$$

Enforcing of (4-45) would ensure that a maximum would not occur. However, the extreme values of $\cos\theta$, namely, ± 1, still need to be explored. For $\cos\theta = 1$, Equation (4-40) yields $|G|^2 = 1$, and the stability requirement is satisfied. For $\cos\theta = -1$, $|G|^2 = 16d^2 - 8d + 1$ and, imposing the requirement of $|G|^2 \leq 1$, yields

$$d \leq \frac{1}{2} \tag{4-46}$$

which was also obtained previously as Equation (4-32). Note that the combination of requirements (4-45) and (4-46) includes the requirement of

$$c \leq 1 \tag{4-47}$$

A more restrictive requirement than (4-45) is obtained by the combination of (4-45) and (4-47), which provides

$$c^2 \leq 2d \tag{4-48}$$

Thus, the stability requirements for the FTCS explicit scheme given by Equation (4-34) are

$$d \leq \frac{1}{2} \quad \text{and} \quad c^2 \leq 2d$$

Using the definitions of the Courant number and the diffusion number, the requirement (4-45) can be written as

$$a\frac{\Delta t}{\Delta x} \leq 2\alpha\frac{\Delta t}{(\Delta x)^2}$$

or

$$a\frac{\Delta x}{\alpha} \leq 2 \tag{4-49}$$

The expression $a\Delta x/\alpha$ is called the "Cell Reynolds number Re_c." Therefore, (4-49) can be written as

$$Re_c \leq 2 \tag{4-50}$$

Similarly, the more restrictive requirement of (4-48) can be written as

$$Re_c \leq (2/c) \tag{4-51}$$

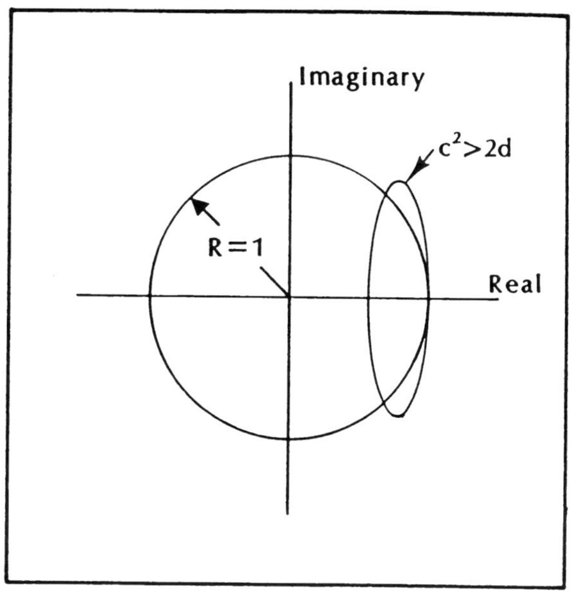

Figure 4-13. Sketch of amplification factor indicating instability for $c^2 > 2d$.

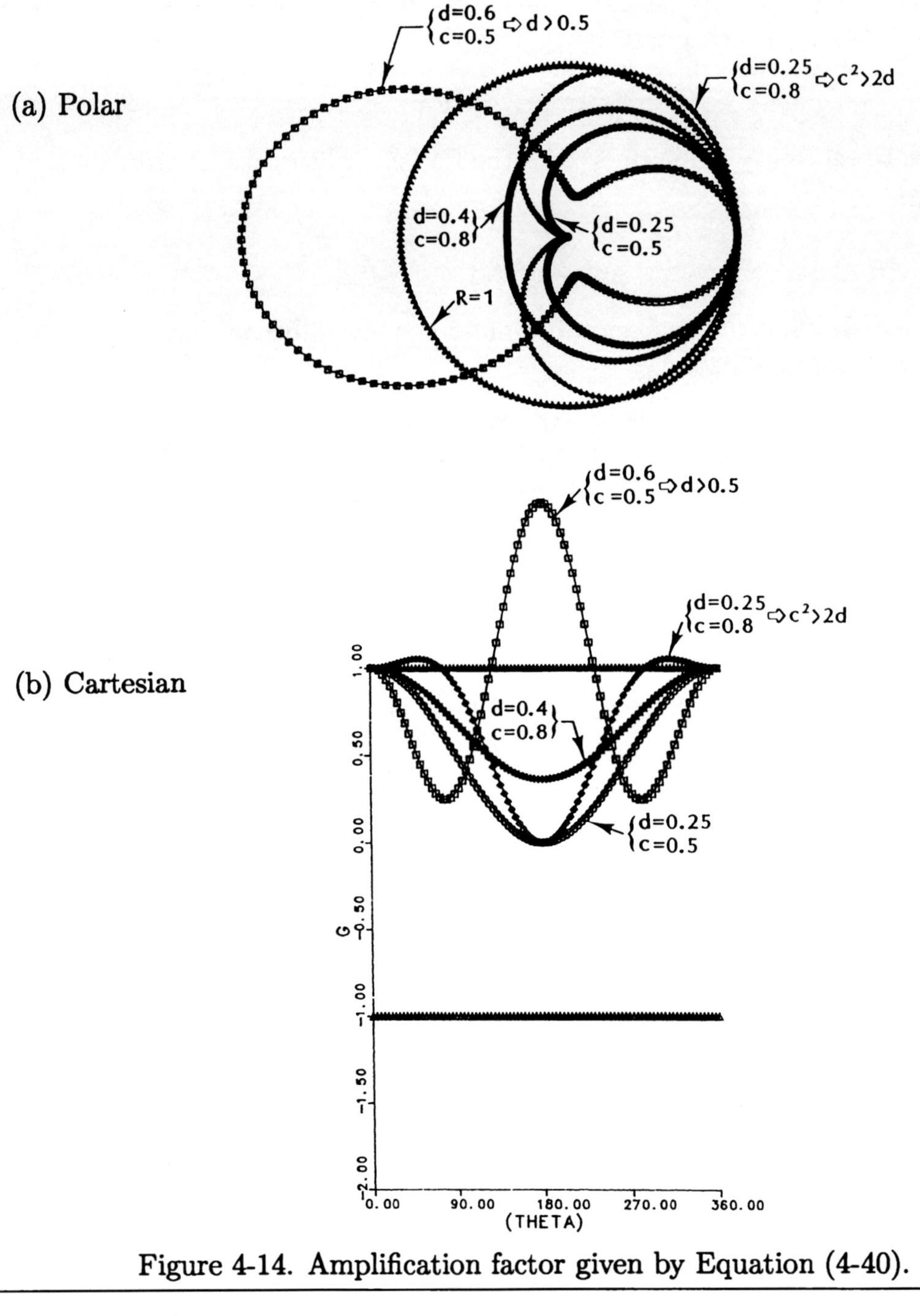

Figure 4-14. Amplification factor given by Equation (4-40).

Graphical representation of the amplification factor for various values of c and d is illustrated in Figures 4-14a and 4-14b.

The finite difference equations discussed so far in this chapter have involved two time levels, i.e., time levels n and $n+1$. Some of the methods described in Chapter 3 involve three time levels, so that time levels $n-1$, n, and $n+1$ appear in the finite difference equations. To illustrate the application of stability analysis to the three-level FDEs, the following example is proposed.

Consider the wave equation,

$$\frac{\partial u}{\partial t} = -a\frac{\partial u}{\partial x} \tag{4-52}$$

and approximate the PDE using central differencing. The resulting equation (known as *the midpoint leapfrog method*)

$$\frac{u_i^{n+1} - u_i^{n-1}}{2\Delta t} = -a\frac{u_{i+1}^n - u_{i-1}^n}{2\Delta x} \tag{4-53}$$

is an explicit equation of order $[(\Delta t)^2, (\Delta x)^2]$. Using the definition of Courant number $c = a\Delta t/\Delta x$, rewrite Equation (4-53) as

$$u_i^{n+1} = u_i^{n-1} - c\,(u_{i+1}^n - u_{i-1}^n)$$

Application of the von Neumann stability analysis yields

$$U^{n+1}e^{I\theta i} = U^{n-1}e^{I\theta i} - c\left(U^n e^{I\theta(i+1)} - U^n e^{I\theta(i-1)}\right) \quad \text{or}$$

$$U^{n+1} = U^{n-1} - cU^n\left(e^{I\theta} - e^{-I\theta}\right)$$

from which

$$U^{n+1} = U^{n-1} - c(2I\sin\theta)U^n$$

Let $-2Ic\sin\theta = A$; then

$$U^{n+1} = AU^n + U^{n-1} \tag{4-54}$$

Furthermore, the following identity may be written:

$$U^n = U^n + (0)U^{n-1} \tag{4-55}$$

Equations (4-54) and (4-55) in a matrix form are expressed as

$$\begin{vmatrix} U^{n+1} \\ U^n \end{vmatrix} = \begin{vmatrix} A & 1 \\ 1 & 0 \end{vmatrix} \begin{vmatrix} U^n \\ U^{n-1} \end{vmatrix}$$

Now, the amplification factor is a matrix:

$$G = \begin{vmatrix} A & 1 \\ 1 & 0 \end{vmatrix}$$

For stability purposes, the eigenvalues of G (call them λ) must obey the condition $|\lambda| \le 1$. Proceeding with the determination of the eigenvalues,

$$\begin{vmatrix} A-\lambda & 1 \\ 1 & 0-\lambda \end{vmatrix} = \lambda^2 - \lambda A - 1 = 0$$

Hence,

$$\lambda_{1,2} = \frac{A \pm \sqrt{A^2 - 4(-1)}}{2} = \frac{A \pm \sqrt{A^2 + 4}}{2}$$

In terms of the Courant number

$$\lambda_{1,2} = \frac{1}{2}\left[-2Ic\sin\theta \pm \sqrt{4 - 4c^2\sin^2\theta}\ \right] \quad \text{or}$$

$$\lambda_{1,2} = -Ic\sin\theta \pm \sqrt{1 - c^2\sin^2\theta} \tag{4-56}$$

Here, two cases must be investigated. If $c^2\sin^2\theta \leq 1$, then

$$\left|\lambda_{1,2}\right|^2 = c^2\sin^2\theta + (1 - c^2\sin^2\theta) = 1$$

So, when $c^2\sin^2\theta \leq 1$, the stability requirement is satisfied. The most restrictive condition corresponds to the maximum value of $\sin^2\theta$, in which case $c \leq 1$.

For the second case, when $c^2\sin^2\theta > 1$, one may write

$$\lambda_{1,2} = I\left[-c\sin\theta \pm \sqrt{c^2\sin^2\theta - 1}\ \right] \quad \text{or}$$

$$\left|\lambda_{1,2}\right|^2 = c^2\sin^2\theta + (c^2\sin^2\theta - 1) \pm 2c\sin\theta\sqrt{c^2\sin^2\theta - 1}$$

$$= 2c^2\sin^2\theta \pm 2c\sin\theta\sqrt{c^2\sin^2\theta - 1} - 1$$

The stability condition requires that

$$\left|\lambda_{1,2}\right|^2 \leq 1$$

or that

$$2c^2\sin^2\theta \pm 2c\sin\theta\sqrt{c^2\sin^2\theta - 1} - 1 \leq 1$$

When the "+" is used, one obtains

$$c^2\sin^2\theta + c\sin\theta\sqrt{c^2\sin^2\theta - 1} < 1$$

which is never satisfied for $c^2\sin^2\theta > 1$. Thus, the stability requirement for the finite difference equation (4-53) is $c \leq 1$.

Another example of stability analysis and its interpretation is proposed as follows. Consider the second-order wave equation,

$$\frac{\partial^2 u}{\partial t^2} = a^2\frac{\partial^2 u}{\partial x^2} \tag{4-57}$$

In this equation, u is the fluid speed and a is the speed of sound, which is assumed constant, resulting in a linearized equation.

It can be shown that the analytical solution of Equation (4-57) may be expressed as

$$u(x,t) = f(x-at) + g(x+at)$$

for which the characteristic equations are:

$$x - at = c_1$$

and

$$x + at = c_2$$

These characteristics have slopes of $dx/dt = \pm a$ in the $x-t$ plane (lines AB and AC in Figure 4-15). Recall from Chapter 1 that information at point A is influenced by the data within the region BAC, known as the domain of dependence of point A.

Now, consider the finite difference representation of Equation (4-57), which is obtained using second-order central difference approximations for the derivatives; thus,

$$\frac{u_i^{n+1} - 2u_i^n + u_i^{n-1}}{(\Delta t)^2} = a^2 \frac{u_{i+1}^n - 2u_i^n + u_{i-1}^n}{(\Delta x)^2} \tag{4-58}$$

or

$$u_i^{n+1} = 2u_i^n - u_i^{n-1} + \left(\frac{a\Delta t}{\Delta x}\right)^2 (u_{i+1}^n - 2u_i^n + u_{i-1}^n) \ .$$

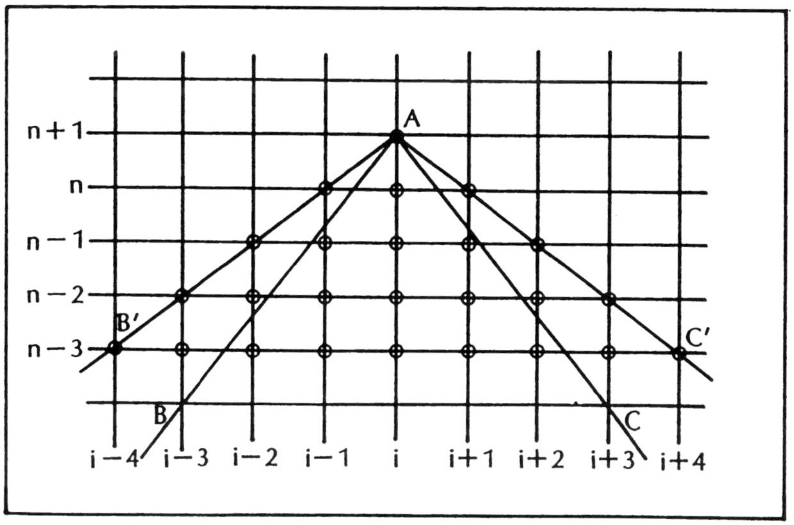

Figure 4-15. Physical and computational characteristic lines.

In terms of the Courant number,

$$u_i^{n+1} = 2u_i^n - u_i^{n-1} + c^2\left(u_{i+1}^n - 2u_i^n + u_{i-1}^n\right) \tag{4-59}$$

Equation (4-59) indicates that a newly computed value at $(i, n+1)$ depends on the previous values at i and $i \pm 1$ at time level n and node i at time level $n-1$. Similarly, properties at $i, n+1$ will propagate to the points i, $i \pm 1$ at time level $n+2$.

The von Neumann stability analysis indicates that $c^2 \leq 1$ will result in a stable solution. (Verifying this statement is left as an exercise.) Therefore, the stability condition of the numerical solution is

$$\left(\frac{a\Delta t}{\Delta x}\right)^2 \leq 1 \qquad \text{or} \qquad \frac{\Delta t}{\Delta x} \leq \pm\frac{1}{a}$$

This relation indicates that, for a stable solution, the zone of dependence in the finite difference domain shown as $B'AC'$ in Figure 4-15 must include the zone of dependence of the continuum domain shown as region BAC. This result is inaccurate because it includes some unnecessary information propagating from regions $B'AB$ and $C'AC$. It is clear that when $c = 1$, i.e., when the continuum and computational domains are identical, the numerical solution is exact. On the other hand, if the numerical domain does not include all of the continuum domain, some of the vital information necessary for the computation is deleted and an unstable numerical solution results.

It is appropriate at this time to discuss the stability analysis of a system of equations. Consider the model equations

$$\frac{\partial u}{\partial t} + a_2\frac{\partial v}{\partial x} = 0 \tag{4-60}$$

and

$$\frac{\partial v}{\partial t} + b_1\frac{\partial u}{\partial x} = 0 \tag{4-61}$$

Expressing the set of equations as a vector equation, as in Chapter 1, one obtains

$$\frac{\partial \phi}{\partial t} + [A]\frac{\partial \phi}{\partial x} = 0 \quad \text{where} \quad \phi = \begin{bmatrix} u \\ v \end{bmatrix} \quad \text{and}$$

$$A = \begin{bmatrix} 0 & a_2 \\ b_1 & 0 \end{bmatrix}$$

Explicit finite differencing produces

$$\frac{\phi_i^{n+1} - \phi_i^n}{\Delta t} + [A]^n\frac{\phi_i^n - \phi_{i-1}^n}{\Delta x} = 0$$

so that

$$\phi_i^{n+1} = \phi_i^n - [A]\frac{\Delta t}{\Delta x}(\phi_i^n - \phi_{i-1}^n)$$

Application of the von Neumann stability analysis yields:

$$\Phi^{n+1}e^{I\theta i} = \Phi^n e^{I\theta i} - [A]\frac{\Delta t}{\Delta x}\left[\Phi^n e^{I\theta i} - \Phi^n e^{I\theta(i-1)}\right]$$

After canceling the common term $e^{I\theta i}$, one obtains

$$\Phi^{n+1} = \Phi^n - [A]\frac{\Delta t}{\Delta x}\left[1 - e^{-I\theta}\right]\Phi^n$$

It follows that the amplification matrix is

$$[G] = [ID] - \frac{\Delta t}{\Delta x}(1 - \cos\theta + I\sin\theta)\,[A] \qquad \text{or}$$

$$[G] = \left\{[ID] - (1 - \cos\theta)\left(\frac{\Delta t}{\Delta x}\right)[A]\right\} - I\left\{\frac{\Delta t}{\Delta x}\sin\theta\right\}[A]$$

where $[ID]$ is the identity matrix. Stability requires that the largest eigenvalue of $[G]$, say $|\lambda G_{\max}| \leq 1$. An equivalent of the stability condition is that $|\lambda_{\max}\Delta t/\Delta x| \leq 1$, where $\lambda_{\max}$ is the largest eigenvalue of matrix $[A]$. Thus, for our example, where

$$[A] = \begin{bmatrix} 0 & a_2 \\ b_1 & 0 \end{bmatrix}$$

$\lambda_{\max} = \sqrt{b_1 a_2}$, and the stability requirement is

$$\left|\frac{\Delta t}{\Delta x}\sqrt{b_1 a_2}\right| \leq 1 \tag{4-62}$$

4.4 Multidimensional Problems

The von Neumann stability analysis can be easily extended to multidimensional problems. Even though discrete perturbation and other stability analyses can be used to analyze multidimensional problems, an extension of the von Neumann method is the most straightforward and the most commonly used. Thus, this technique for the stability analysis of multidimensional problems will be considered in this section. As a first example, consider the unsteady, two-dimensional heat conduction equation

$$\frac{\partial u}{\partial t} = \alpha\left(\frac{\partial^2 u}{\partial x^2} + \frac{\partial^2 u}{\partial y^2}\right) \tag{4-63}$$

Approximating the derivatives by forward differencing in time and central differencing in space, one obtains the explicit formulation

$$\frac{u_{i,j}^{n+1} - u_{i,j}^{n}}{\Delta t} = \alpha \left[\frac{u_{i+1,j}^{n} - 2u_{i,j}^{n} + u_{i-1,j}^{n}}{(\Delta x)^2} + \frac{u_{i,j+1}^{n} - 2u_{i,j}^{n} - u_{i,j-1}^{n}}{(\Delta y)^2} \right] \tag{4-64}$$

Define the diffusion numbers $d_x = \alpha \Delta t / (\Delta x)^2$ and $d_y = \alpha \Delta t / (\Delta y)^2$, then

$$u_{i,j}^{n+1} = u_{i,j}^{n} + d_x \left(u_{i+1,j}^{n} - 2u_{i,j}^{n} + u_{i-1,j}^{n} \right) + d_y \left(u_{i,j+1}^{n} - 2u_{i,j}^{n} + u_{i,j-1}^{n} \right) \tag{4-65}$$

Assume a Fourier component of the form

$$u_{i,j}^{n} = U^n e^{IP(\Delta x)i} e^{Iq(\Delta y)j}$$

where U^n is the amplitude at time level n, and P and q are the wave numbers in the x and y directions. The phase angles $\theta = P\Delta x$ and $\phi = q\Delta y$ are defined as in the previous case. Then,

$$u_{i,j}^{n} = U^n e^{I\theta i} e^{I\phi j} = U^n e^{I(\theta i + \phi j)} \tag{4-66}$$

Similarly,

$$u_{i\pm 1,j}^{n} = U^n e^{I(\theta(i\pm 1) + \phi j)} \tag{4-67}$$

$$u_{i,j\pm 1}^{n} = U^n e^{I(\theta i + \phi(j\pm 1))} \tag{4-68}$$

Substituting Equations (4-66) through (4-68) into Equation (4-65), one obtains

$$U^{n+1} e^{I(\theta i + \phi j)} = U^n e^{I(\theta i + \phi j)} + d_x \left[U^n e^{I(\theta(i+1) + \phi j)} - 2U^n e^{I(\theta i + \phi j)} + U^n e^{I(\theta(i-1) + \phi j)} \right]$$
$$+ d_y \left[U^n e^{I(\theta i + \phi(j+1))} - 2U^n e^{I(\theta i + \phi j)} + U^n e^{I(\theta i + \phi(j-1))} \right]$$

Canceling terms and factoring U^n,

$$U^{n+1} = U^n \left[1 + d_x \left(e^{I\theta} - 2 + e^{-I\theta} \right) + d_y \left(e^{I\phi} - 2 + e^{-I\phi} \right) \right]$$

Using the identities (4-36a) and (4-36b), it follows that

$$U^{n+1} = U^n \left[1 + 2d_x(\cos\theta - 1) + 2d_y(\cos\phi - 1) \right] \quad \text{or}$$

$$G = 1 + 2d_x(\cos\theta - 1) + 2d_y(\cos\phi - 1)$$

For a stable solution, $|G| \le 1$, or

$$\left| 1 + 2d_x(\cos\theta - 1) + 2d_y(\cos\phi - 1) \right| \le 1$$

that is,

$$2d_x(\cos\theta - 1) + 2d_y(\cos\phi - 1) \le 0 \tag{4-69}$$

and

$$2d_x(\cos\theta - 1) + 2d_y(\cos\phi - 1) \ge -2 \tag{4-70}$$

The first condition, given by (4-69), is always satisfied. The second condition, (4-70), specifies that $d_x(1-\cos\theta) + d_y(1-\cos\phi) \le 1$. At extreme values of $\cos\theta$ and $\cos\phi$,

$$d_x(2) + d_y(2) \le 1 \qquad \text{or}$$

$$d_x + d_y \le \frac{1}{2} \tag{4-71}$$

In terms of the step sizes Δt, Δx, and Δy,

$$\alpha\Delta t\left[\frac{1}{(\Delta x)^2} + \frac{1}{(\Delta y)^2}\right] \le \frac{1}{2}$$

Note that, in the one-dimensional case, $d \le 0.5$. In the two-dimensional case just considered, if $\Delta x = \Delta y$, then $d \le 0.25$, i.e., the stability condition is twice as restrictive as in the one-dimensional case.

The procedure described above can be used to analyze three-dimensional problems as well. For the FTCS explicit formulation of the three-dimensional heat conduction equation,

$$\frac{\partial u}{\partial t} = \alpha\left(\frac{\partial^2 u}{\partial x^2} + \frac{\partial^2 u}{\partial y^2} + \frac{\partial^2 u}{\partial z^2}\right) \tag{4-72}$$

stability analysis indicates that a numerical solution will be stable if

$$d_x + d_y + d_z \le \frac{1}{2}$$

The analysis is left as an exercise in problem (4-5).

As a second example of a multidimensional problem, consider an unsteady, two-dimensional convection equation with positive constant coefficients a and b, i.e.,

$$\frac{\partial u}{\partial t} = -a\frac{\partial u}{\partial x} - b\frac{\partial u}{\partial y} \tag{4-73}$$

An explicit finite difference representation of the partial differential equation may be obtained by using a first-order accurate forward difference approximation of the time derivative and a first-order accurate backward difference approximation of the spatial derivatives. Thus,

$$\frac{u_{i,j}^{n+1} - u_{i,j}^{n}}{\Delta t} = -a\frac{u_{i,j}^{n} - u_{i-1,j}^{n}}{\Delta x} - b\frac{u_{i,j}^{n} - u_{i,j-1}^{n}}{\Delta y} \tag{4-74}$$

which can be written as

$$u_{i,j}^{n+1} = u_{i,j}^{n} - c_x \left(u_{i,j}^{n} - u_{i-1,j}^{n}\right) - c_y \left(u_{i,j}^{n} - u_{i,j-1}^{n}\right)$$

where $c_x = a\Delta t/\Delta x$ and $c_y = b\Delta t/\Delta y$. Substituting the Fourier components defined in (4-66) through (4-68) yields

$$U^{n+1} e^{I(\theta i+\phi j)} = U^n e^{I(\theta i+\phi j)} - c_x U^n \left[e^{I(\theta i+\phi j)} - e^{I(\theta(i-1)+\phi j)}\right]$$
$$- c_y U^n \left[e^{I(\theta i+\phi j)} - e^{I(\theta i+\phi(j-1))}\right]$$

After eliminating common terms,

$$U^{n+1} = U^n - c_x U^n \left[1 - e^{-I\theta}\right] - c_y U^n \left[1 - e^{-I\phi}\right]$$

hence,

$$G = 1 - c_x \left[1 - (\cos\theta - I\sin\theta)\right] - c_y \left[1 - (\cos\phi - I\sin\phi)\right] \quad \text{or}$$

$$G = \left[1 - c_x(1-\cos\theta) - c_y(1-\cos\phi)\right] - I\left[c_x \sin\theta + c_y \sin\phi\right]$$

Therefore,

$$\left|G\right|^2 = \left[1 - c_x(1-\cos\theta) - c_y(1-\cos\phi)\right]^2 + \left[c_x \sin\theta + c_y \sin\phi\right]^2$$

For a stable solution, $|G|^2 \leq 1$, or

$$\left[c_x(1-\cos\theta) + c_y(1-\cos\phi)\right]^2 - 2\left[c_x(1-\cos\theta) + c_y(1-\cos\phi)\right]$$
$$+ (c_x \sin\theta + c_y \sin\phi)^2 \leq 0 \tag{4-75}$$

For the maximum values of $(1-\cos\theta)$ and $(1-\cos\phi)$, one has

$$4c_x^2 + 4c_y^2 + 8c_x c_y - 4c_x - 4c_y \leq 0$$

or $(c_x+c_y)^2 - (c_x+c_y) \leq 0$. From which $(c_x+c_y)(c_x+c_y-1) \leq 0$. Since (c_x+c_y) is positive, $(c_x+c_y-1) \leq 0$ or $c_x+c_y \leq 1$. When Equation (4-75) is investigated for the maximum values of $\sin\theta$ and $\sin\phi$, the same requirement will result. Therefore, the solution of the FDE given by Equation (4-74) is stable for $c_x + c_y \leq 1$.

Before proceeding to the next section, a summary on application and limitations of the von Neumann stability analysis is considered.

1. The von Neumann stability analysis can be applied to linear equations only.
2. The influence of the boundary conditions on the stability of solution is not included.

3. For a scalar PDE which is approximated by a two-level FDE, the mathematical requirement is imposed on the amplification factor as follows:

   (a) if G is real, then $|G| \leq 1$

   (b) if G is complex, then $|G|^2 \leq 1$, where $|G|^2 = G\bar{G}$

4. For a scalar PDE which is approximated by a three-level FDE, the amplification factor is a matrix. In this case, the requirement is imposed on the eigenvalues of G as follows:

   (a) if λ is real, then $|\lambda| \leq 1$

   (b) if λ is complex, then $|\lambda|^2 \leq 1$

5. The method can be easily extended to multidimensional problems.

6. The procedure can be used for stability analysis of a system of linear PDEs. The requirement is imposed on the largest eigenvalue of the amplification matrix.

7. Benchmark values for the stability of unsteady one-dimensional problems may be established as follows:

   (a) For most explicit formulations:

      I. Courant number, $c \leq 1$

      II. Diffusion number, $d \leq 0.5$

      III. Cell Reynolds number, $Re_c \leq (2/c)$

   (b) For implicit formulation, most are unconditionally stable.

8. For multi-dimensional problems with equal grid sizes in all spatial directions, the stated benchmark values are usually adjusted by dividing them by the number of spatial dimensions.

9. On occasions where the amplification factor is a difficult expression to analyze, graphical solution along with some numerical experimentation will facilitate the analysis.

4.5 Error Analysis

Methods for which the lowest order term in the truncation error is of order one are classified as first-order accurate methods. An example of a first-order accurate method for the model equation

$$\frac{\partial u}{\partial t} = -a\frac{\partial u}{\partial x} \tag{4-76}$$

is

$$\frac{u_i^{n+1} - u_i^n}{\Delta t} = -a\frac{u_i^n - u_{i-1}^n}{\Delta x} \tag{4-77}$$

Note that, in the approximation process, the Taylor series expansion was truncated at an even (second) derivative, i.e.,

$$\frac{\partial u}{\partial t} = \frac{u_i^{n+1} - u_i^n}{\Delta t} \quad \Big| \quad + \left(\frac{\Delta t}{2}\right)\frac{\partial^2 u}{\partial t^2} + \frac{(\Delta t)^2}{3!}\frac{\partial^3 u}{\partial t^3} + \cdots$$

$$\Big| \rightarrow \text{truncated terms represented as } O(\Delta t)\ .$$

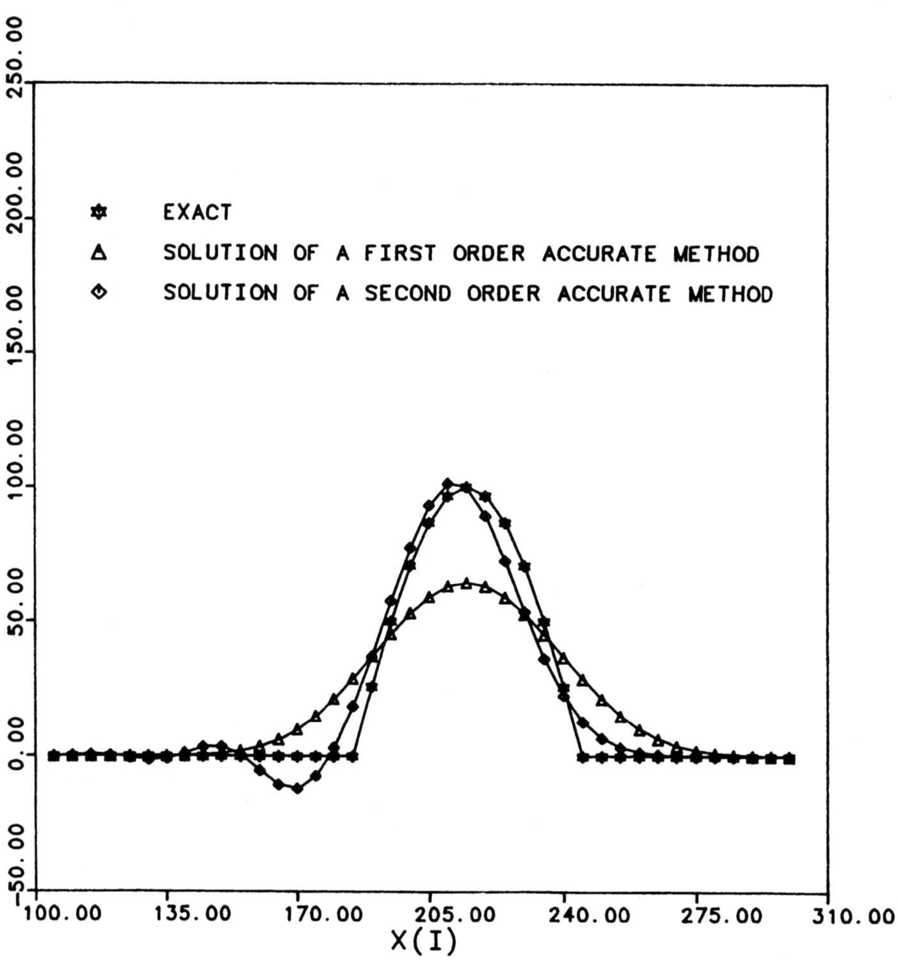

Figure 4-16. Comparison of solutions obtained by a first-order accurate method and a second-order accurate method.

For a second-order accurate formulation, the truncation of the error term is at an odd (third) derivative; for example,

$$\frac{\partial u}{\partial t} = \frac{u_i^{n+1} - u_i^{n-1}}{2\Delta t} \quad \Bigg| \quad -\frac{2(\Delta t)^2}{3!}\frac{\partial^3 u}{\partial t^3} + \cdots$$

$$\Bigg| \rightarrow \text{truncated terms represented as } O(\Delta t)^2 \ .$$

The errors associated with first-order accurate methods are known as dissipation errors. These errors tend to decrease the gradients within the solution domain. Errors associated with second-order accurate methods are known as dispersion errors. Typical dissipation and dispersion errors are shown in Figure 4-16. The dissipation error produced by a first-order accurate method is indicated by a decrease in the amplitude of the wave, while the dispersion error created by a second-order accurate method is indicated by an oscillation in the solution.

4.6 Modified Equation

In order to determine the dominant error term of a finite difference equation, Taylor series expansions are substituted back into the finite difference equation and, after some algebraic manipulation, the so-called modified equation is obtained. To illustrate the procedure, two examples will be considered. The selected examples represent a first-order accurate and a second-order accurate finite difference formulation of a model equation. These examples will show the dominant error terms of these methods and their relation to dissipation and dispersion errors.

A first-order finite difference equation for model equation (4-76) is given in (4-77), which can be rearranged as

$$u_i^{n+1} = u_i^n - c(u_i^n - u_{i-1}^n) \tag{4-78}$$

The terms u_i^{n+1} and u_{i-1}^n are expanded in Taylor series as follows:

$$u_i^{n+1} = u_i^n + \frac{\partial u}{\partial t}\Delta t + \frac{\partial^2 u}{\partial t^2}\frac{(\Delta t)^2}{2!} + \frac{\partial^3 u}{\partial t^3}\frac{(\Delta t)^3}{3!} + O(\Delta t)^4 \tag{4-79}$$

$$u_{i-1}^n = u_i^n - \frac{\partial u}{\partial x}\Delta x + \frac{\partial^2 u}{\partial x^2}\frac{(\Delta x)^2}{2!} - \frac{\partial^3 u}{\partial x^3}\frac{(\Delta x)^3}{3!} + O(\Delta x)^4 \tag{4-80}$$

Substitution of (4-79) and (4-80) into (4-78) yields:

$$u_i^n + \frac{\partial u}{\partial t}\Delta t + \frac{\partial^2 u}{\partial t^2}\frac{(\Delta t)^2}{2!} + \frac{\partial^3 u}{\partial t^3}\frac{(\Delta t)^3}{3!} + O(\Delta t)^4 =$$

$$u_i^n - c\left\{u_i^n - \left[u_i^n - \frac{\partial u}{\partial x}\Delta x + \frac{\partial^2 u}{\partial x^2}\frac{(\Delta x)^2}{2!} - \frac{\partial^3 u}{\partial x^3}\frac{(\Delta x)^3}{3!} + O(\Delta x)^4\right]\right\}$$

Canceling terms and factoring produces:

$$\left[\frac{\partial u}{\partial t} + \frac{\partial^2 u}{\partial t^2}\frac{\Delta t}{2} + \frac{\partial^3 u}{\partial t^3}\frac{(\Delta t)^2}{6} + O(\Delta t)^3\right](\Delta t) =$$

$$- a\frac{\Delta t}{\Delta x}\left[\frac{\partial u}{\partial x} - \frac{\partial^2 u}{\partial x^2}\frac{\Delta x}{2} + \frac{\partial^3 u}{\partial x^3}\frac{(\Delta x)^2}{6} + O(\Delta x)^3\right](\Delta x)$$

This equation is divided by Δt and rearranged as

$$\frac{\partial u}{\partial t} = -a\frac{\partial u}{\partial x} - \frac{\partial^2 u}{\partial t^2}\frac{\Delta t}{2} + a\frac{\partial^2 u}{\partial x^2}\frac{\Delta x}{2} - \frac{\partial^3 u}{\partial t^3}\frac{(\Delta t)^2}{6}$$

$$- a\frac{\partial^3 u}{\partial x^3}\frac{(\Delta x)^2}{6} + O\left[(\Delta t)^3, (\Delta x)^3\right] \tag{4-81}$$

In order to analyze Equation (4-81) and to compare it to the original partial differential equation, the higher order derivatives in time must be replaced by spatial derivatives. This substitution requires determination of $\partial^2 u/\partial t^2$ with $O[(\Delta t)^2, (\Delta x)^2]$ and $\partial^3 u/\partial t^3$ with $O[(\Delta t), (\Delta x)]$ such that the order of accuracy of (4-81) is not altered, i.e., it remains third order. These calculations, presented in Appendix C, yield the equations

$$\frac{\partial^2 u}{\partial t^2} = a^2\frac{\partial^2 u}{\partial x^2} + \left[a^3\Delta t - a^2\Delta x\right]\frac{\partial^3 u}{\partial x^3} + O\left[(\Delta t)^2,\ (\Delta x\Delta t),\ (\Delta x)^2\right] \tag{4-82}$$

and

$$\frac{\partial^3 u}{\partial t^3} = -a^3\frac{\partial^3 u}{\partial x^3} + O\left[(\Delta t),\ (\Delta x)\right] \tag{4-83}$$

Substituting (4-82) and (4-83) into (4-81) produces

$$\frac{\partial u}{\partial t} = -a\frac{\partial u}{\partial x} - \frac{a^2\Delta t}{2}\frac{\partial^2 u}{\partial x^2} - \left(a^3\Delta t - a^2\Delta x\right)\frac{\Delta t}{2}\frac{\partial^3 u}{\partial x^3}$$

$$+ \frac{a\Delta x}{2}\frac{\partial^2 u}{\partial x^2} + \frac{a^3(\Delta t)^2}{6}\frac{\partial^3 u}{\partial x^3} - \frac{a(\Delta x)^2}{6}\frac{\partial^3 u}{\partial x^3}$$

$$+ O\left[(\Delta t)^3,\ (\Delta x)(\Delta t)^2,\ (\Delta x)^2(\Delta t),\ (\Delta x)^3\right]$$

Rewrite the equation as

$$\frac{\partial u}{\partial t} = -a\frac{\partial u}{\partial x} + \frac{a\Delta x}{2}\left(1 - \frac{a\Delta t}{\Delta x}\right)\frac{\partial^2 u}{\partial x^2}$$

$$- \frac{a(\Delta x)^2}{6}\left[2\frac{a^2(\Delta t)^2}{(\Delta x)^2} - 3\frac{a\Delta t}{\Delta x} + 1\right]\frac{\partial^3 u}{\partial x^3}$$

$$+ O\left[(\Delta t)^3,\ (\Delta x)(\Delta t)^2,\ (\Delta x)^2(\Delta t),\ (\Delta x)^3\right]$$

or, in terms of the Courant number,

$$\frac{\partial u}{\partial t} = -a\frac{\partial u}{\partial x} + \frac{a\Delta x}{2}(1-c)\frac{\partial^2 u}{\partial x^2} - \frac{a(\Delta x)^2}{6}(2c^2 - 3c + 1)\frac{\partial^3 u}{\partial x^3} + O\left[(\Delta t)^3,\ (\Delta x)(\Delta t)^2,\ (\Delta x)^2(\Delta t),\ (\Delta x)^3\right] \tag{4-84}$$

This equation is known as the modified equation. When compared to the original PDE given by (4-76), the error introduced in the approximation process is clearly indicated. Note that the dominating term in the error is the second term on the right-hand side of Equation (4-84), which includes the second derivative. Equation (4-84) and its associated error will be further investigated in the next section when artificial viscosity is introduced.

At this time, consider a second-order method in which the dominant error term includes a third-order (odd) derivative. Such a formulation is the midpoint leapfrog method given by (4-53). The equation may be rearranged as

$$u_i^{n+1} = u_i^{n-1} - c(u_{i+1}^n - u_{i-1}^n) \tag{4-85}$$

Substituting the Taylor series expansions,

$$u_i^{n+1} = u_i^n + \frac{\partial u}{\partial t}\Delta t + \frac{\partial^2 u}{\partial t^2}\frac{(\Delta t)^2}{2!} + \frac{\partial^3 u}{\partial t^3}\frac{(\Delta t)^3}{3!} + O(\Delta t)^4 \tag{4-86}$$

$$u_i^{n-1} = u_i^n - \frac{\partial u}{\partial t}\Delta t + \frac{\partial^2 u}{\partial t^2}\frac{(\Delta t)^2}{2!} - \frac{\partial^3 u}{\partial t^3}\frac{(\Delta t)^3}{3!} + O(\Delta t)^4 \tag{4-87}$$

$$u_{i+1}^n = u_i^n + \frac{\partial u}{\partial x}\Delta x + \frac{\partial^2 u}{\partial x^2}\frac{(\Delta x)^2}{2!} + \frac{\partial^3 u}{\partial x^3}\frac{(\Delta x)^3}{3!} + O(\Delta x)^4 \tag{4-88}$$

and

$$u_{i-1}^n = u_i^n - \frac{\partial u}{\partial x}\Delta x + \frac{\partial^2 u}{\partial x^2}\frac{(\Delta x)^2}{2!} - \frac{\partial^3 u}{\partial x^3}\frac{(\Delta x)^3}{3!} + O(\Delta x)^4 \tag{4-89}$$

into Equation (4-85), one obtains

$$\frac{\partial u}{\partial t}\Delta t + \frac{\partial^3 u}{\partial t^3}\frac{(\Delta t)^3}{3!} = -\frac{\partial u}{\partial t}\Delta t - \frac{\partial^3 u}{\partial t^3}\frac{(\Delta t)^3}{3!} + - c\left[2\frac{\partial u}{\partial x}\Delta x + 2\frac{\partial^3 u}{\partial x^3}\frac{(\Delta x)^3}{3!}\right] + O\left[(\Delta t)^5,\ (\Delta x)^5\right]$$

from which it follows that

$$\frac{\partial u}{\partial t} = -a\frac{\partial u}{\partial x} - \frac{\partial^3 u}{\partial t^3}\frac{(\Delta t)^2}{3!} - a\frac{\partial^3 u}{\partial x^3}\frac{(\Delta x)^2}{3!} + O\left[(\Delta t)^4,\ (\Delta x)^4\right] \tag{4-90}$$

From Appendix C, Equation (C-6) is

$$\frac{\partial^3 u}{\partial t^3} = -a^3 \frac{\partial^3 u}{\partial x^3} + O(\Delta t,\ \Delta x)$$

After substituting into (4-90) and factoring terms one obtains

$$\frac{\partial u}{\partial t} = -a\frac{\partial u}{\partial x} + \frac{a(\Delta x)^2}{6}\left[\frac{a^2(\Delta t)^2}{(\Delta x)^2} - 1\right]\frac{\partial^3 u}{\partial x^3} + O\left[(\Delta t)^4,\ (\Delta x)^3(\Delta t),\ (\Delta x)^4\right] \tag{4-91}$$

This is the modified equation, clearly indicating the dominant error term. Note that this term includes the third-order (odd) derivative. Additional error terms, such as some factor times $\partial^5 u/\partial x^5$, may have been included in Equation (4-91); however, to avoid cumbersome mathematical work, they were lumped and represented by higher order terms in $O\,[(\Delta t)^4,\ (\Delta x)^3(\Delta t),\ (\Delta x)^4]$.

4.7 Artificial Viscosity

Recall the modified Equation (4-84), repeated here for convenience:

$$\frac{\partial u}{\partial t} = -a\frac{\partial u}{\partial x} + \frac{a\Delta x}{2}(1-c)\frac{\partial^2 u}{\partial x^2} - \frac{a(\Delta x)^2}{6}\left(2c^2 - 3c + 1\right)\frac{\partial^3 u}{\partial x^3} + \cdots$$

The error term of this first-order accurate method is clearly indicated, with the second term on the right-hand side being the dominant term in error. The coefficient of $\partial^2 u/\partial x^2$ in the error is known as the numerical or artificial viscosity. This is a nonphysical coefficient introduced by a particular approximation of the PDE. We may use α_e to denote the artificial viscosity, in which case

$$\alpha_e = \frac{1}{2}a\Delta x(1-c)$$

Note that for $c = 1$, $\alpha_e = 0$, the solution is exact. Additional investigation and analysis of dissipation and dispersion errors will be studied in Chapter 6. The effect of artificial viscosity is to dissipate the solution; as a result, the gradients in the solution domain are reduced.

Note that the artificial viscosity just defined must be distinguished from a damping or smoothing term that is sometimes added intentionally to an FDE. An example of that will be illustrated in Chapter 6 when a fourth-order damping term is added to an FDE in order to reduce the oscillations in the solution.

In closing, another procedure for determining the dominant error term is investigated. Consider Equation (4-81) where

$$\frac{\partial u}{\partial t} = -a\frac{\partial u}{\partial x} - \frac{\partial^2 u}{\partial t^2}\frac{\Delta t}{2} + a\frac{\partial^2 u}{\partial x^2}\frac{\Delta x}{2} + O\left[(\Delta t)^2,\ (\Delta x)^2\right] \tag{4-92}$$

To eliminate the time derivative on the right-hand side of Equation (4-92), the original PDE will be used, i.e.,

$$\frac{\partial u}{\partial t} = -a\frac{\partial u}{\partial x} \tag{4-93}$$

Taking the time derivative, one obtains

$$\frac{\partial^2 u}{\partial t^2} = -a\frac{\partial}{\partial t}\left(\frac{\partial u}{\partial x}\right) = -a\frac{\partial}{\partial x}\left(\frac{\partial u}{\partial t}\right) \tag{4-94}$$

Substitute (4-93) into (4-94) to obtain

$$\frac{\partial^2 u}{\partial t^2} = -a\frac{\partial}{\partial x}\left(-a\frac{\partial u}{\partial x}\right) = a^2\frac{\partial^2 u}{\partial x^2} \tag{4-95}$$

Substituting (4-95) into (4-92) yields:

$$\frac{\partial u}{\partial t} = -a\frac{\partial u}{\partial x} - a^2\frac{\partial^2 u}{\partial x^2}\frac{\Delta t}{2} + a\frac{\Delta x}{2}\frac{\partial^2 u}{\partial x^2}$$

Hence,

$$\frac{\partial u}{\partial t} = -a\frac{\partial u}{\partial x} + \left(a\frac{\Delta x}{2} - a^2\frac{\Delta t}{2}\right)\frac{\partial^2 u}{\partial x^2}$$

or, in terms of the Courant number,

$$\frac{\partial u}{\partial t} = -a\frac{\partial u}{\partial x} + \frac{a\Delta x}{2}(1-c)\frac{\partial^2 u}{\partial x^2}$$

where the second term on the right-hand side represents the error term, and the coefficient $(a\Delta x/2)(1-c)$ represents artificial viscosity.

This procedure, though relatively simple mathematically, is not precise, since the original PDE is being used in the approximate FDE. Furthermore, this procedure does not provide the additional terms shown in Equation (4-84). However, this method has been used and presented in various literature.

4.8 Summary Objectives

After studying this chapter, you should be able to do the following:

1. Define:
    a. Stability
    b. Discrete perturbation stability analysis
    c. Von Neumann stability analysis
    d. The Courant number
    e. The cell Reynolds number
    f. Static instability
    g. Dynamic instability
    h. Dissipation error
    i. Dispersion error
    j. A modified equation
    k. Artificial viscosity
2. Solve the problems for Chapter Four.

4.9 Problems

4.1 Given the model equation

$$\frac{\partial u}{\partial t} = \alpha \frac{\partial^2 u}{\partial x^2}$$

use the DuFort-Frankel method for this PDE and apply the von Neumann stability analysis.

4.2 Consider the model equation

$$\frac{\partial u}{\partial t} = a \frac{\partial u}{\partial x}$$

A numerical technique with the following formulation has been suggested for the FDE:

$$u_i^{n+\frac{1}{2}} = u_i^n + \frac{a\Delta t}{\Delta x}\left(u_{i+1}^n - u_i^n\right)$$

$$u_i^{n+1} = \frac{1}{2}\left(u_i^n + u_i^{n+\frac{1}{2}}\right) + \frac{a\Delta t}{2\Delta x}\left(u_i^{n+\frac{1}{2}} - u_{i-1}^{n+\frac{1}{2}}\right)$$

Use the von Neumann stability analysis to obtain the stability condition for this technique. *Hint: Eliminate "$n+\frac{1}{2}$" time level in the second equation by the substitution of the first equation. The resulting equation will have time levels of "n" and "$n+1$" only. Now apply the von Neumann stability analysis.*

4.3 Using the von Neumann stability analysis, determine the stability requirement of the following FDE:

$$\frac{u_i^{n+1} - u_i^n}{\Delta t} = -a\frac{u_i^n - u_{i-1}^n}{\Delta x}$$

$$a > 0$$

Plot the amplification factor for Courant numbers of 0.5, 0.75, 1.0, and 1.1.

4.4 Consider the model equation

$$a\frac{\partial u}{\partial x} = \nu\frac{\partial^2 u}{\partial y^2}$$

Define $\nu/a = K$.

(a) Write an explicit formulation using a first-order forward differencing in x and a second-order central differencing in y.

(b) Use von Neumann stability analysis to determine the stability requirement of the scheme.

4.5 Consider the unsteady three-dimensional heat conduction equation

$$\frac{\partial T}{\partial t} = \alpha\left(\frac{\partial^2 T}{\partial x^2} + \frac{\partial^2 T}{\partial y^2} + \frac{\partial^2 T}{\partial z^2}\right)$$

(a) Obtain an explicit FDE using forward time and central differencing of $O[\Delta t, (\Delta x)^2, (\Delta y)^2, (\Delta z)^2]$.

(b) Apply the von Neumann stability analysis to the FDE.

4.6 Determine the modified equation for the Laplace's equation,

$$\frac{\partial^2 u}{\partial x^2} + \frac{\partial^2 u}{\partial y^2} = 0$$

Use second-order central difference approximation for the derivatives. The modified equation should include terms up to fourth-order derivatives.

4.7 Write an implicit finite difference formulation for the wave equation which is first-order in time and second-order (central) in space. Determine the modified equation.

4.8 Consider the model equation,

$$\frac{\partial u}{\partial t} + a\frac{\partial u}{\partial x} = \alpha\frac{\partial^2 u}{\partial x^2}$$

(a) Write an explicit formulation using a first-order forward differencing in time, a first-order backward differencing in space for the convective term, and a second-order central differencing for the diffusion term.

(b) Determine the amplification factor G.

4.9 Consider the wave equation,

$$\frac{\partial u}{\partial t} + a\frac{\partial u}{\partial x} = 0 \qquad a > 0$$

(a) Write an explicit finite difference equation which employs a first-order forward differencing in time and a second-order central differencing in space.

(b) Determine the stability requirement of the scheme.

4.10 Repeat problem 4.9 using the implicit Crank-Nicolson scheme.

4.11 Consider the diffusion equation given by

$$\frac{\partial u}{\partial t} = \alpha\,\frac{\partial^2 u}{\partial x^2}$$

(a) Write an implicit Crank-Nicolson formulation.

(b) Use von Neumann stability analysis to determine the stability requirement of the scheme.

4.12 Show that the application of explicit FTCS differencing to the model equation

$$\frac{\partial u}{\partial t} = -a\frac{\partial u}{\partial x}$$

introduces an artificial viscosity $\alpha_e = -a^2\Delta t/2$.

4.13 For the model equation

$$\frac{\partial u}{\partial t} = -a\frac{\partial u}{\partial x}$$

the following finite difference equation (known as the Lax method) has been proposed:

$$u_i^{n+1} = \frac{1}{2}\left(u_{i+1}^n + u_{i-1}^n\right) - \frac{a\Delta t}{2\Delta x}\left(u_{i+1}^n - u_{i-1}^n\right)$$

Determine the artificial viscosity of the method.

Chapter 5

Elliptic Equations

5.1 Introductory Remarks

The governing equations in fluid mechanics and heat transfer can be reduced to elliptic form for particular applications. Such examples are the steady-state heat conduction equation, velocity potential equation for incompressible, inviscid flow, and the stream function equation. Typical elliptic equations in a two-dimensional Cartesian system are Laplace's equations,

$$\frac{\partial^2 u}{\partial x^2} + \frac{\partial^2 u}{\partial y^2} = 0 \tag{5-1}$$

and Poisson's equation,

$$\frac{\partial^2 u}{\partial x^2} + \frac{\partial^2 u}{\partial y^2} = f(x, y) \tag{5-2}$$

These model equations are used to investigate a variety of solution procedures.

5.2 Finite Difference Formulations

Of the various existing finite difference formulations, the so-called "five-point formula" is the most commonly used. In this representation of the PDE, central differencing which is second-order accurate is utilized. Therefore, model Equation (5-1) is approximated as

$$\frac{u_{i+1,j} - 2u_{i,j} + u_{i-1,j}}{(\Delta x)^2} + \frac{u_{i,j+1} - 2u_{i,j} + u_{i,j-1}}{(\Delta y)^2} = 0 \tag{5-3}$$

The corresponding grid points are shown in Figure (5-1).

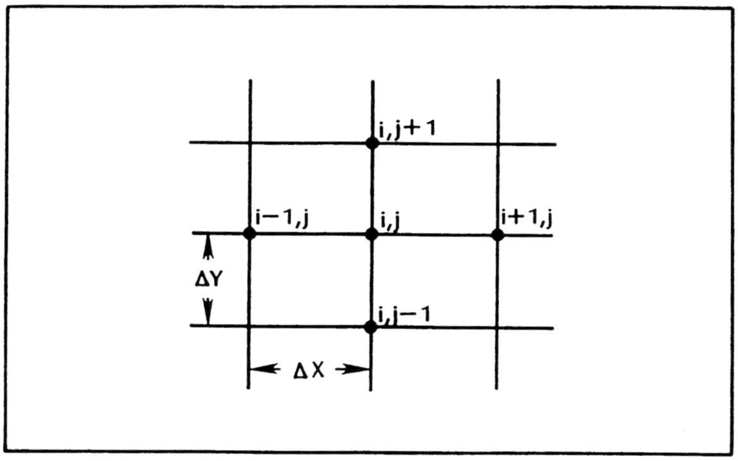

Figure 5-1. Grid points for a five-point formula.

A higher order formulation is the nine-point formula, which uses a fourth-order approximation for the derivatives. With this formulation, the FDE of model Equation (5-1) is:

$$\frac{-u_{i-2,j} + 16u_{i-1,j} - 30u_{i,j} + 16u_{i+1,j} - u_{i+2,j}}{12(\Delta x)^2}$$

$$+ \frac{-u_{i,j-2} + 16u_{i,j-1} - 30u_{i,j} + 16u_{i,j+1} - u_{i,j+2}}{12(\Delta y)^2} = 0 \qquad (5\text{-}4)$$

The grid points involved in Equation (5-4) are shown in Figure (5-2).

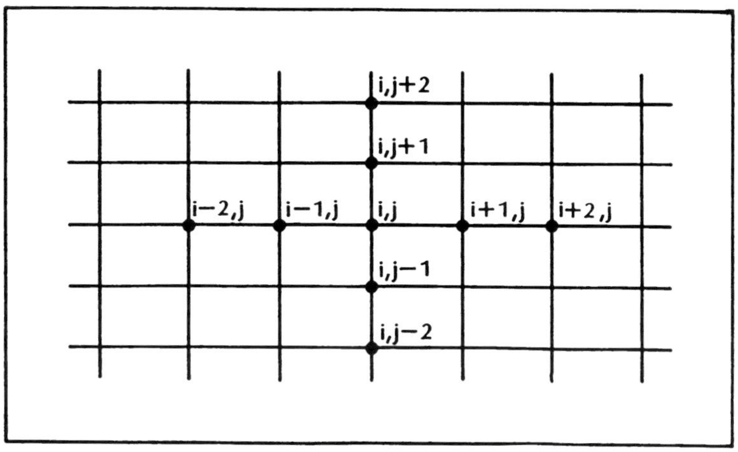

Figure 5-2. Grid points for a nine-point formula.

One obvious difficulty with the application of this formula is the implementation of the boundary conditions. Thus, for problems where higher accuracy is required, it is easier to use the five-point formula with small grid sizes than the fourth-order accurate nine-point formula. Due to its simplicity, the five-point formula represented by Equation (5-3) will be considered in this chapter. Rewrite Equation (5-3) as

$$u_{i+1,j} - 2u_{i,j} + u_{i-1,j} + \left(\frac{\Delta x}{\Delta y}\right)^2 (u_{i,j+1} - 2u_{i,j} + u_{i,j-1}) = 0 \tag{5-5}$$

Define the ratio of step sizes as β, so that $\beta = \Delta x/\Delta y$. By rearranging the terms in Equation (5-5), one obtains

$$u_{i+1,j} + u_{i-1,j} + \beta^2 u_{i,j+1} + \beta^2 u_{i,j-1} - 2(1+\beta^2)u_{i,j} = 0 \tag{5-6}$$

In order to explore various solution procedures, first consider a square domain with Dirichlet boundary conditions. For instance, a simple 6×6 grid system (see Figure 5-3) subject to the following boundary conditions is considered:

$$x = 0 \quad u = u_2 \quad , \quad y = 0 \quad u = u_1$$

$$x = L \quad u = u_4 \quad , \quad y = H \quad u = u_3$$

Applying Equation (5-6) to the interior grid points produces sixteen equations with sixteen unknowns. The equations are:

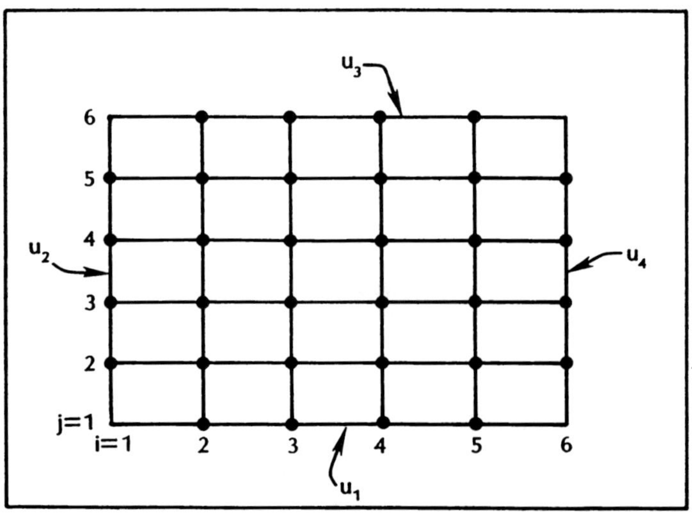

Figure 5-3. Grid system used for solution of Equation (5-6).

$$u_{3,2} + u_{1,2} + \beta^2 u_{2,3} + \beta^2 u_{2,1} - 2(1+\beta^2)u_{2,2} = 0$$

$$u_{4,2} + u_{2,2} + \beta^2 u_{3,3} + \beta^2 u_{3,1} - 2(1+\beta^2)u_{3,2} = 0$$

$$u_{5,2} + u_{3,2} + \beta^2 u_{4,3} + \beta^2 u_{4,1} - 2(1+\beta^2)u_{4,2} = 0$$

$$u_{6,2} + u_{4,2} + \beta^2 u_{5,3} + \beta^2 u_{5,1} - 2(1+\beta^2)u_{5,2} = 0$$

$$u_{3,3} + u_{1,3} + \beta^2 u_{2,4} + \beta^2 u_{2,2} - 2(1+\beta^2)u_{2,3} = 0$$

$$u_{4,3} + u_{2,3} + \beta^2 u_{3,4} + \beta^2 u_{3,2} - 2(1+\beta^2)u_{3,3} = 0$$

$$u_{5,3} + u_{3,3} + \beta^2 u_{4,4} + \beta^2 u_{4,2} - 2(1+\beta^2)u_{4,3} = 0$$

$$u_{6,3} + u_{4,3} + \beta^2 u_{5,4} + \beta^2 u_{5,2} - 2(1+\beta^2)u_{5,3} = 0$$

$$u_{3,4} + u_{1,4} + \beta^2 u_{2,5} + \beta^2 u_{2,3} - 2(1+\beta^2)u_{2,4} = 0$$

$$u_{4,4} + u_{2,4} + \beta^2 u_{3,5} + \beta^2 u_{3,3} - 2(1+\beta^2)u_{3,4} = 0$$

$$u_{5,4} + u_{3,4} + \beta^2 u_{4,5} + \beta^2 u_{4,3} - 2(1+\beta^2)u_{4,4} = 0$$

$$u_{6,4} + u_{4,4} + \beta^2 u_{5,5} + \beta^2 u_{5,3} - 2(1+\beta^2)u_{5,4} = 0$$

$$u_{3,5} + u_{1,5} + \beta^2 u_{2,6} + \beta^2 u_{2,4} - 2(1+\beta^2)u_{2,5} = 0$$

$$u_{4,5} + u_{2,5} + \beta^2 u_{3,6} + \beta^2 u_{3,4} - 2(1+\beta^2)u_{3,5} = 0$$

$$u_{5,5} + u_{3,5} + \beta^2 u_{4,6} + \beta^2 u_{4,4} - 2(1+\beta^2)u_{4,5} = 0$$

$$u_{6,5} + u_{4,5} + \beta^2 u_{5,6} + \beta^2 u_{5,4} - 2(1+\beta^2)u_{5,5} = 0$$

These equations are expressed in a matrix form as

$$
\begin{bmatrix}
\alpha & 1 & 0 & 0 & \beta^2 & 0 & 0 & 0 & 0 & 0 & 0 & 0 & 0 & 0 & 0 & 0 \\
1 & \alpha & 1 & 0 & 0 & \beta^2 & 0 & 0 & 0 & 0 & 0 & 0 & 0 & 0 & 0 & 0 \\
0 & 1 & \alpha & 1 & 0 & 0 & \beta^2 & 0 & 0 & 0 & 0 & 0 & 0 & 0 & 0 & 0 \\
0 & 0 & 1 & \alpha & 0 & 0 & 0 & \beta^2 & 0 & 0 & 0 & 0 & 0 & 0 & 0 & 0 \\
\beta^2 & 0 & 0 & 0 & \alpha & 1 & 0 & 0 & \beta^2 & 0 & 0 & 0 & 0 & 0 & 0 & 0 \\
0 & \beta^2 & 0 & 0 & 1 & \alpha & 1 & 0 & 0 & \beta^2 & 0 & 0 & 0 & 0 & 0 & 0 \\
0 & 0 & \beta^2 & 0 & 0 & 1 & \alpha & 1 & 0 & 0 & \beta^2 & 0 & 0 & 0 & 0 & 0 \\
0 & 0 & 0 & \beta^2 & 0 & 0 & 1 & \alpha & 0 & 0 & 0 & \beta^2 & 0 & 0 & 0 & 0 \\
0 & 0 & 0 & 0 & \beta^2 & 0 & 0 & 0 & \alpha & 1 & 0 & 0 & \beta^2 & 0 & 0 & 0 \\
0 & 0 & 0 & 0 & 0 & \beta^2 & 0 & 0 & 1 & \alpha & 1 & 0 & 0 & \beta^2 & 0 & 0 \\
0 & 0 & 0 & 0 & 0 & 0 & \beta^2 & 0 & 0 & 1 & \alpha & 1 & 0 & 0 & \beta^2 & 0 \\
0 & 0 & 0 & 0 & 0 & 0 & 0 & \beta^2 & 0 & 0 & 1 & \alpha & 0 & 0 & 0 & \beta^2 \\
0 & 0 & 0 & 0 & 0 & 0 & 0 & 0 & \beta^2 & 0 & 0 & 0 & \alpha & 1 & 0 & 0 \\
0 & 0 & 0 & 0 & 0 & 0 & 0 & 0 & 0 & \beta^2 & 0 & 0 & 1 & \alpha & 1 & 0 \\
0 & 0 & 0 & 0 & 0 & 0 & 0 & 0 & 0 & 0 & \beta^2 & 0 & 0 & 1 & \alpha & 1 \\
0 & 0 & 0 & 0 & 0 & 0 & 0 & 0 & 0 & 0 & 0 & \beta^2 & 0 & 0 & 1 & \alpha
\end{bmatrix}
\begin{bmatrix}
u_{2,2} \\ u_{3,2} \\ u_{4,2} \\ u_{5,2} \\ u_{2,3} \\ u_{3,3} \\ u_{4,3} \\ u_{5,3} \\ u_{2,4} \\ u_{3,4} \\ u_{4,4} \\ u_{5,4} \\ u_{2,5} \\ u_{3,5} \\ u_{4,5} \\ u_{5,5}
\end{bmatrix}
=
\begin{bmatrix}
-u_{1,2} - \beta^2 u_{2,1} \\ -\beta^2 u_{3,1} \\ -\beta^2 u_{4,1} \\ -u_{6,2} - \beta^2 u_{5,1} \\ -u_{1,3} \\ 0 \\ 0 \\ -u_{6,3} \\ -u_{1,4} \\ 0 \\ 0 \\ -u_{6,4} \\ -u_{1,5} - \beta^2 u_{3,6} \\ -\beta^2 u_{3,6} \\ -\beta^2 u_{4,6} \\ -u_{6,5} - \beta^2 u_{5,6}
\end{bmatrix}
\tag{5-7}
$$

where $\alpha = -2(1 + \beta^2)$.

The matrix formulation has two noteworthy features. First, it is a pentadiagonal matrix with nonadjacent diagonals; and second, the elements in the main diagonal in each row are the largest. These features are important when developing solution procedures.

5.3 Solution Algorithms

In general, there are two methods of solution for the system of simultaneous linear algebraic equations given by (5-7). These schemes are classified as direct or iterative methods.

Some familiar direct methods are Cramer's rule and Gaussian elimination. The major disadvantage of these methods is the enormous amount of arithmetic operations required to produce a solution. Some advanced direct methods have been proposed which require moderate computation time, but almost all of them have disadvantages. Usually these methods are limited by one or more restrictions such as the Cartesian coordinate system, a rectangular domain, the size of the coefficient matrix, a large storage requirement, boundary conditions, or difficulty of programming. Nevertheless, some of the advanced direct methods show promise for applications and indeed are useful tools. The discussion of these methods, however, is beyond the scope of this text. Since the general application of solution procedures is of more interest, the investigation will be confined to the simple and easily understood iterative methods. Iterative procedures for solving a system of linear algebraic equations are simple and easy to program. The idea behind these methods is to obtain the solution by iteration. Usually an initial solution is guessed and new values are computed; based on the newly computed values, a newer solution is sought, and the procedure is repeated until a specified convergence criterion has been reached.

The various formulations of the iterative method can be divided into two categories. If a formulation results in only one unknown, then it is called a point iterative method. This formulation is similar to the explicit methods of parabolic equations. On the other hand, if the formulation involves more than one unknown (usually three unknowns, resulting in a tridiagonal matrix coefficient), it is classified as a line iterative method, which is similar to an implicit formulation of a parabolic equation. Some of these iterative methods will now be discussed.

5.3.1 The Jacobi Iteration Method

In this method, the dependent variable at each grid point is solved, using initial guessed values of the neighboring points or previously computed values. Therefore, Equation (5-6) is used to compute a new value of $u_{i,j}$ at the new iteration $k+1$ level as

$$u_{i,j}^{k+1} = \frac{1}{2(1+\beta^2)} \left[u_{i+1,j}^k + u_{i-1,j}^k + \beta^2 (u_{i,j+1}^k + u_{i,j-1}^k) \right] \tag{5-8}$$

where k corresponds to the previously computed values (or, to initial guesses for the first round of computations). The computation is carried out until a specified convergence criterion is met. One immediately notices an easy way to improve the solution procedure. Why not use the newly computed values of the dependent variable to compute the neighboring points when available? Indeed, implementation of this idea gives us the Gauss-Seidel iteration method, which increases the convergence rate and, as a result, involves much less computation time. The Jacobi method is rarely (if ever) used for the solution of elliptic equations. It is reviewed here to show the step-by-step improvements of various iteration methods.

Before proceeding with other iterative methods, the analogy between the iterative method just discussed and a time-dependent parabolic equation is investigated. Consider the following parabolic equation:

$$\frac{\partial u}{\partial t} = \frac{\partial^2 u}{\partial x^2} + \frac{\partial^2 u}{\partial y^2} \tag{5-9}$$

The explicit formulation of the finite difference equation (FTCS) is:

$$\frac{u_{i,j}^{n+1}-u_{i,j}^{n}}{\Delta t}=\frac{u_{i+1,j}^{n}-2u_{i,j}^{n}+u_{i-1,j}^{n}}{(\Delta x)^2}+\frac{u_{i,j+1}^{n}-2u_{i,j}^{n}+u_{i,j-1}^{n}}{(\Delta y)^2}$$

or

$$u_{i,j}^{n+1}=u_{i,j}^{n}+\frac{\Delta t}{(\Delta x)^2}\left(u_{i+1,j}^{n}-2u_{i,j}^{n}+u_{i-1,j}^{n}\right)+\frac{\Delta t}{(\Delta y)^2}\left(u_{i,j+1}^{n}-2u_{i,j}^{n}+u_{i,j-1}^{n}\right)$$

For simplicity, assume that $\Delta x = \Delta y$; then

$$u_{i,j}^{n+1}=u_{i,j}^{n}+\frac{\Delta t}{(\Delta x)^2}\left[u_{i+1,j}^{n}+u_{i-1,j}^{n}-4u_{i,j}^{n}+u_{i,j+1}^{n}+u_{i,j-1}^{n}\right]$$

If $\Delta t/(\Delta x)^2 \leq 0.25$, a stable solution is obtained. So for the upper limit of $\Delta t/(\Delta x)^2 = 0.25$, one has

$$u_{i,j}^{n+1}=u_{i,j}^{n}+\frac{1}{4}\left[u_{i+1,j}^{n}+u_{i-1,j}^{n}-4u_{i,j}^{n}+u_{i,j+1}^{n}+u_{i,j-1}^{n}\right]$$

or

$$u_{i,j}^{n+1}=\frac{1}{4}\left(u_{i+1,j}^{n}+u_{i-1,j}^{n}+u_{i,j+1}^{n}+u_{i,j-1}^{n}\right) \tag{5-10}$$

Now apply the Jacobi iterative method to the two-dimensional elliptic equation (5-1). With $\beta = \Delta x/\Delta y = 1$, the result is

$$u_{i,j}^{k+1}=\frac{1}{4}\left(u_{i+1,j}^{k}+u_{i-1,j}^{k}+u_{i,j+1}^{k}+u_{i,j-1}^{k}\right) \tag{5-11}$$

Comparison of Equation (5-10), which is the FTCS approximation of a parabolic equation, and Equation (5-11), which is the Jacobi iteration method for an elliptic equation, indicates that the two equations are identical. Even though the two equations represent different formulations and completely different physical phenomenon, mathematically (and to the computer) they are equal. This analogy suggests that some of the techniques used for the solution of parabolic equations can be extended or modified to provide efficient methods for the solution of elliptic equations. Indeed, the ADI method described earlier for the solution of parabolic equations is such an example. The ADI method for solving elliptic equations will be discussed shortly.

The discussion above may be presented graphically as follows. Assume that Equation (5-10) is solved for the time dependent values of $u_{i,j}$. A steady-state solution is approached after a sufficiently large time, which is illustrated in Figure 5-4.

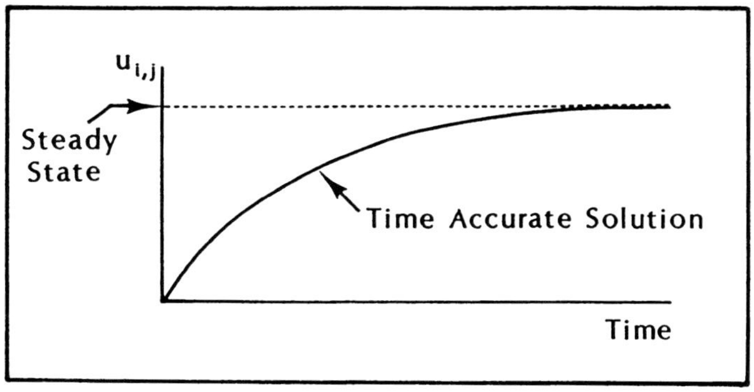

Figure 5-4. A time accurate solution.

Now consider the solution of the steady-state equation (5-1). Assume that an iterative method is used to obtain a solution. For instance, Equation (5-11) is employed. A typical solution is illustrated in Figure 5-5.

An important difference between the two solutions is emphasized at this point. The time dependent solution obtained from Equation (5-10) is a valid solution at any intermediate time level. On the other hand, the intermediate solution of Equation (5-11) has no physical significance; it is only a path to the steady-state solution.

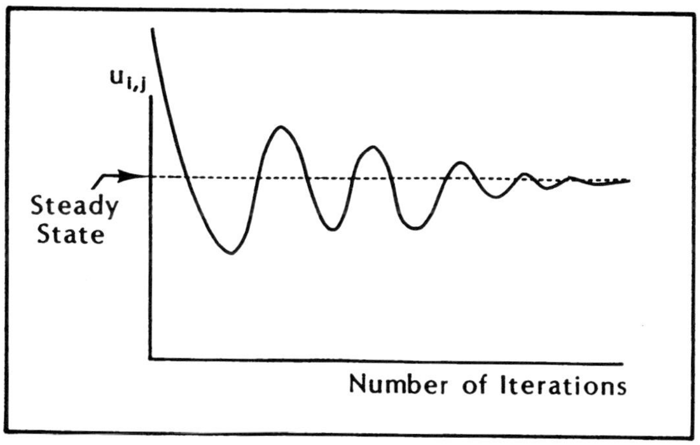

Figure 5-5. An iterative solution for steady-state problem.

Another point to address is as follows. In order to start a solution procedure for an unsteady problem, an initial set of data is required, as was illustrated in Chapter Three. When the imposed initial data and time step correctly represent the physics of the problem, then the solution is time accurate. But, what about problems where an accurate initial data set is not available? A hypersonic flow field around a complex configuration is such an example. In such instances, an arbitrary initial condition is imposed. Hence, the required equations may be solved for a converged solution which represents the steady-state solution. Thus, a time dependent problem with an arbitrary initial data set is solved to reach steady-state. The intermediate steps are not physically correct solutions. In many instances where such a procedure is used, the expressions *converged solution* and *steady-state solution* are used interchangeably. Similarly, time step and iteration step may be used interchangeably. These issues will be further explored in Chapter Twelve, when Euler equations are investigated. For now, return to the iterative procedures for the solution of elliptic equations.

5.3.2 The Point Gauss-Seidel Iteration Method

In this method, the current values of the dependent variable are used to compute the neighboring points as soon as they are available. This will certainly increase the convergence rate dramatically over the Jacobi method (about 100%). The method is convergent if the largest elements are located in the main diagonal of the coefficient matrix, as in the case of the formulation that produced (5-7). The formal requirement (sufficient condition) for the convergence of the method is

$$|a_{ii}| \geq \sum_{\substack{j=1 \\ j\neq i}}^{n} |a_{ij}|$$

and, at least for one row,

$$|a_{ii}| > \sum_{\substack{j=1 \\ j\neq i}}^{n} |a_{ij}|$$

Since this is a sufficient condition, the method may converge even though the condition is not met for all rows. Now the formulation of the method is considered. The finite difference equation is given by Equation (5-6), which is repeated here:

$$u_{i,j} = \frac{1}{2(1+\beta^2)}\left[u_{i+1,j} + u_{i-1,j} + \beta^2(u_{i,j+1} + u_{i,j-1})\right] \tag{5-12}$$

In order to solve for the value of u at grid point i, j, the values of u on the right-hand side must be provided. This procedure is easy to understand if one considers the application of Equation (5-12) to a few grid points. For the computation of the first point, say (2,2), as shown in Figure 5-6, it follows that

$$u_{2,2}^{k+1} = \frac{1}{2(1+\beta^2)}\left[u_{3,2} + \underline{u_{1,2}} + \beta^2(u_{2,3} + \underline{u_{2,1}})\right]$$

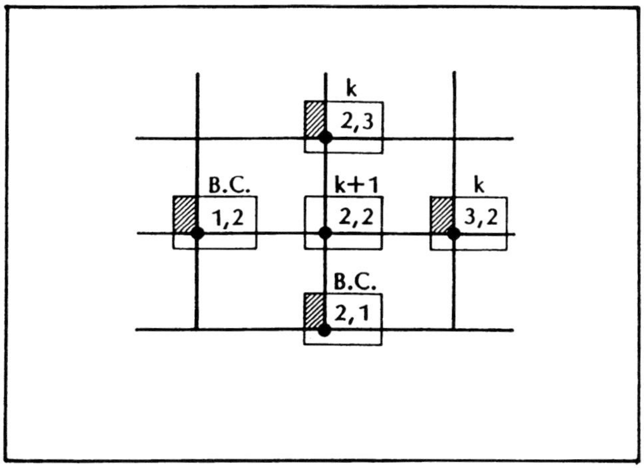

Figure 5-6. Grid points for Equation (5-13).

In this equation, $u_{2,1}$ and $u_{1,2}$ are provided by the boundary conditions (underlined in the equation above). Only two values, namely $u_{3,2}$ and $u_{2,3}$, use the values from the previous iteration at k. Thus, in terms of the iteration levels,

$$u_{2,2}^{k+1} = \frac{1}{2(1+\beta^2)}\left[u_{3,2}^k + \underline{u_{1,2}} + \beta^2(u_{2,3}^k + \underline{u_{2,1}})\right] \tag{5-13}$$

Now, for point (3,2), one has

$$u_{3,2} = \frac{1}{2(1+\beta^2)}\left[u_{4,2} + u_{2,2} + \beta^2(u_{3,3} + \underline{u_{3,1}})\right]$$

In this equation, $u_{3,1}$ is provided by the boundary condition, and $u_{4,2}$ and $u_{3,3}$ are taken from the previous computation; but $u_{2,2}$ is given by Equation (5-13). Thus, (see Figure 5-7)

$$u_{3,2}^{k+1} = \frac{1}{2(1+\beta^2)}\left[u_{4,2}^k + u_{2,2}^{k+1} + \beta^2(u_{3,3}^k + \underline{u_{3,1}})\right] \tag{5-14}$$

Finally, the general formulation provides the equation

$$u_{i,j}^{k+1} = \frac{1}{2(1+\beta^2)}\left[u_{i+1,j}^k + u_{i-1,j}^{k+1} + \beta^2(u_{i,j+1}^k + u_{i,j-1}^{k+1})\right] \tag{5-15}$$

This is a point iteration method, since only one unknown is being sought. The grid points involved in Equation (5-15) are shown in Figure 5-8.

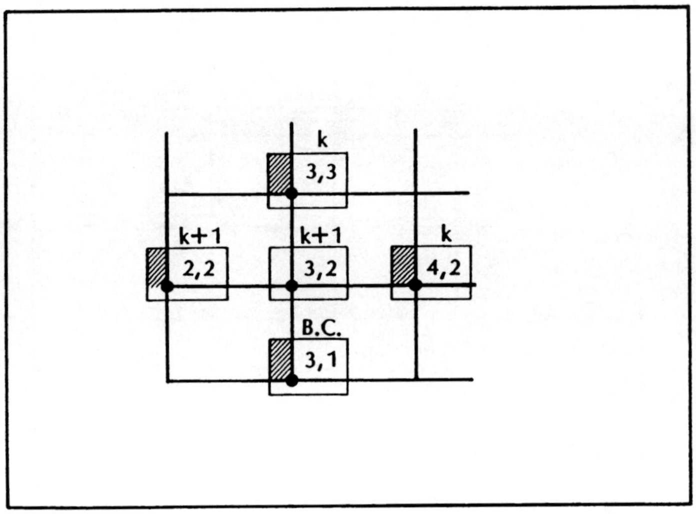

Figure 5-7. Grid points for Equation (5-14).

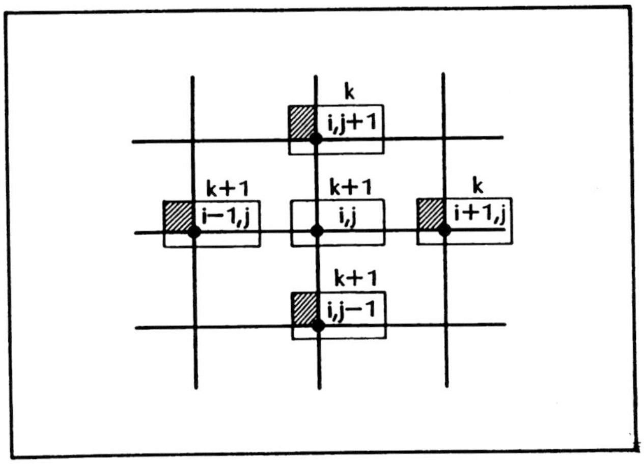

Figure 5-8. Grid points employed in Equation (5-15).

5.3.3 The Line Gauss-Seidel Iteration Method

In this formulation, Equation (5-6) results in three unknowns at points $(i-1,j)$, (i,j), and $(i+1,j)$. The formulation becomes

$$u_{i-1,j}^{k+1} - 2(1+\beta^2)u_{i,j}^{k+1} + u_{i+1,j}^{k+1} = -\beta^2 u_{i,j+1}^{k} - \beta^2 u_{i,j-1}^{k+1} \tag{5-16}$$

This equation, applied to all i at constant j (Figure 5-9), results in a system of linear equations which, in a compact form, has a tridiagonal matrix coefficient.

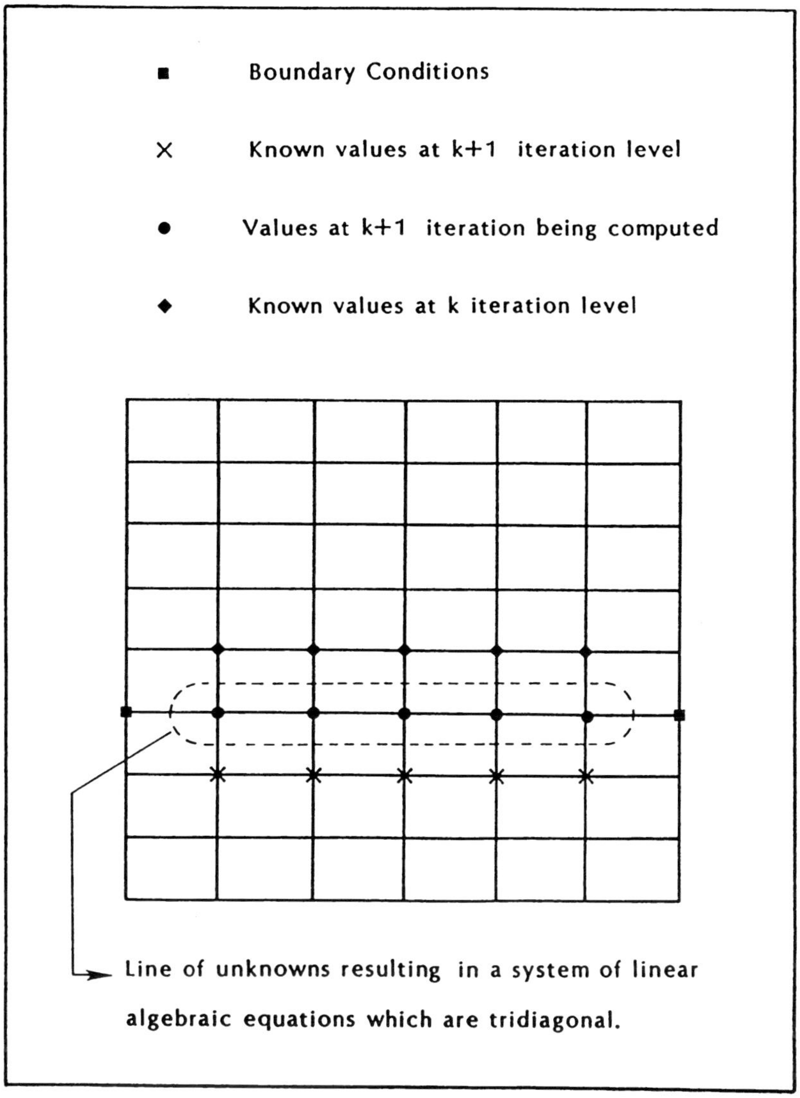

Figure 5-9. Grid points employed in the line Gauss-Seidel iteration method.

This method converges faster than the point Gauss-Seidel method (by about a factor of 1/2), but it requires more computation time per iteration, since a system of simultaneous equations is being solved. Note that, in this formulation, the boundary conditions at a line immediately affect the solution, as was the case with implicit formulation of parabolic equations. For problems in which the value of the dependent variable changes more rapidly in one direction, it is advantageous to use the line Gauss-Seidel method in that direction since the solution converges with the least number of iterations.

It is seen that, by using the updated values, the convergence rate is improved.

Still, more improvement can be incorporated to reduce the number of iterations for a converged solution, which is introduced next.

5.3.4 Point Successive Over-Relaxation Method (PSOR)

Earlier, the analogy between the iterative method for solving an elliptic equation and the FTCS explicit formulation of the unsteady parabolic equation was noted. Thus, solution by iteration can be thought of as a process beginning at an initial state and approaching a steady state. If, during this solution process, a trend in the computed values of the dependent variable is noticed, then the direction of change can be used to extrapolate for the next iteration and, thereby, accelerate the solution procedure. This procedure is known as successive over-relaxation (SOR).

First, consider the point Gauss-Seidel iteration method, given by

$$u_{i,j}^{k+1} = \frac{1}{2(1+\beta^2)}\left[u_{i+1,j}^{k} + u_{i-1,j}^{k+1} + \beta^2(u_{i,j+1}^{k} + u_{i,j-1}^{k+1})\right]$$

Adding $u_{i,j}^{k} - u_{i,j}^{k}$ to the right-hand side , and collecting terms, one obtains:

$$u_{i,j}^{k+1} = u_{i,j}^{k} + \frac{1}{2(1+\beta^2)}\left[u_{i+1,j}^{k} + u_{i-1,j}^{k+1} + \beta^2(u_{i,j+1}^{k} + u_{i,j-1}^{k+1}) - 2(1+\beta^2)u_{i,j}^{k}\right]$$

As the solution proceeds, $u_{i,j}^{k}$ must approach $u_{i,j}^{k+1}$. To accelerate the solution, the bracket term is multiplied by ω, the relaxation parameter, so that

$$u_{i,j}^{k+1} = u_{i,j}^{k} + \frac{\omega}{2(1+\beta^2)}\left[u_{i+1,j}^{k} + u_{i-1,j}^{k+1} + \beta^2(u_{i,j+1}^{k} + u_{i,j-1}^{k+1}) - 2(1+\beta^2)u_{i,j}^{k}\right] \quad (5\text{-}17)$$

For the solution to converge, it is necessary that $0 < \omega < 2$. If $0 < \omega < 1$, it is called under-relaxation. Note that for $\omega = 1$, the Gauss-Seidel iteration method is recovered. The formulation (5-17) is rearranged as

$$u_{i,j}^{k+1} = (1-\omega)u_{i,j}^{k} + \frac{\omega}{2(1+\beta^2)}\left[u_{i+1,j}^{k} + u_{i-1,j}^{k+1} + \beta^2(u_{i,j+1}^{k} + u_{i,j-1}^{k+1})\right] \quad (5\text{-}18)$$

An obvious question is: What is the optimum value of the relaxation parameter ω? Well, no general guideline exists for computing the optimum value of ω. For limited applications, some relations for which an optimum ω can be calculated have been introduced by various investigators. One such relation, for the solution of elliptic equations in a rectangular domain subject to Dirichlet boundary conditions with constant step sizes, is

$$\omega_{\text{opt}} = \frac{2 - 2\sqrt{1-a}}{a}, \quad (5\text{-}19)$$

where

$$a = \left[\frac{\cos\left(\frac{\pi}{IM-1}\right) + \beta^2\cos\left(\frac{\pi}{JM-1}\right)}{1+\beta^2}\right]^2 \quad (5\text{-}20)$$

In general, ω_{opt} cannot be determined easily. Therefore, for most cases, numerical experimentation is performed.

5.3.5 Line Successive Over-Relaxation Method (LSOR)

The line Gauss-Seidel iteration method can be accelerated by introducing a relaxation parameter similar to the one introduced into the point Gauss-Seidel method to provide the point SOR method. The line Gauss-Seidel method for the model equation is given by Equation (5-16) as

$$u_{i-1,j}^{k+1} - 2(1+\beta^2)u_{i,j}^{k+1} + u_{i+1,j}^{k+1} = -\beta^2 u_{i,j+1}^{k} - \beta^2 u_{i,j-1}^{k+1}$$

Introduction of the relaxation parameter and rearranging terms results in:

$$\omega u_{i-1,j}^{k+1} - 2(1+\beta^2)u_{i,j}^{k+1} + \omega u_{i+1,j}^{k+1} =$$
$$-(1-\omega)\left[2(1+\beta^2)\right]u_{i,j}^{k} - \omega\beta^2\left(u_{i,j+1}^{k} + u_{i,j-1}^{k+1}\right) \quad (5\text{-}21)$$

There is no simple way to determine the value of optimum ω. In practice, trial and error is used to compute ω_{opt} for a particular problem.

5.3.6 The Alternating Direction Implicit (ADI) Method

The similarity between the iterative methods and time-dependent parabolic equations suggests that some of the methods discussed earlier for the solution of parabolic equations should be investigated. Of particular interest is the application of the ADI method to the elliptic equations. An iteration cycle is considered complete once the resulting tridiagonal system is solved for all the rows and then followed by columns, or vice versa. For model Equation (5-1), the FDE takes the form

$$u_{i-1,j}^{k+\frac{1}{2}} - 2(1+\beta^2)u_{i,j}^{k+\frac{1}{2}} + u_{i+1,j}^{k+\frac{1}{2}} = -\beta^2(u_{i,j+1}^{k} + u_{i,j-1}^{k+\frac{1}{2}}) \quad (5\text{-}22)$$

and

$$\beta^2 u_{i,j-1}^{k+1} - 2(1+\beta^2)u_{i,j}^{k+1} + \beta^2 u_{i,j+1}^{k+1} = -u_{i+1,j}^{k+\frac{1}{2}} - u_{i-1,j}^{k+1} \quad (5\text{-}23)$$

In these equations, (5-22) is solved implicitly for the unknown in the x-direction and (5-23) is solved implicitly in the y-direction. The solution procedure and the grid points in Equations (5-22) and (5-23) are shown graphically in Figure 5-10.

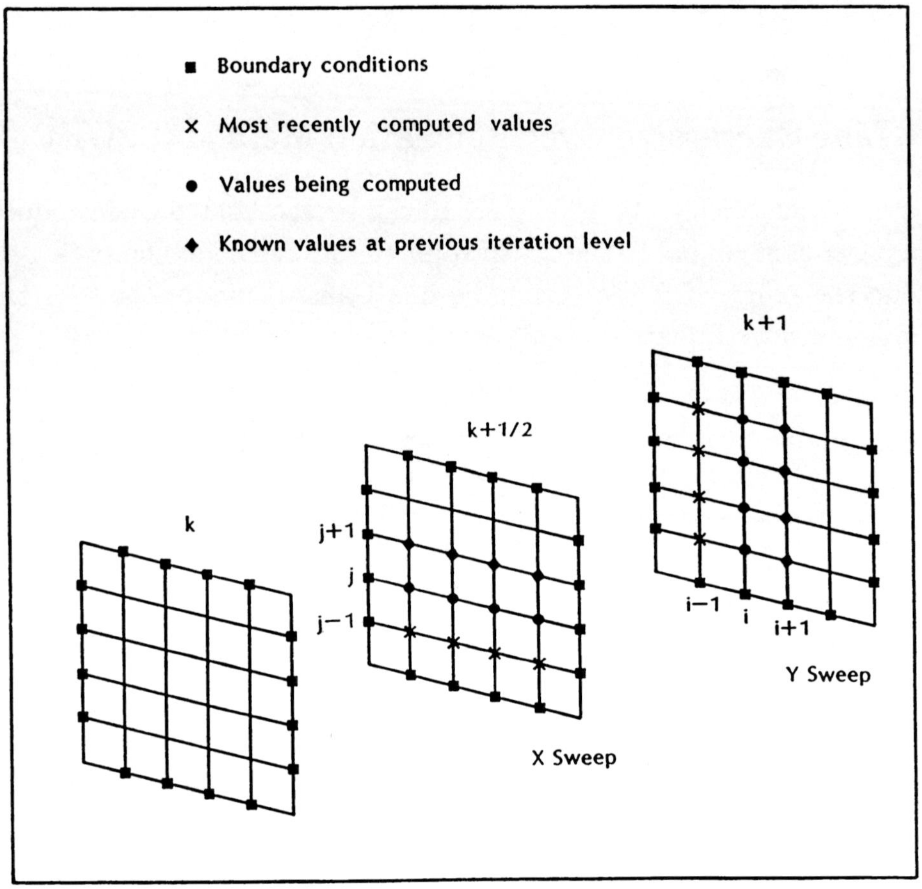

Figure 5-10. Grid points used in Equations (5-22) and (5-23).

The solution procedure can be accelerated by introducing a relaxation parameter ω into the ADI equations. The resulting formulations are:

$$\omega u_{i-1,j}^{k+\frac{1}{2}} - 2(1+\beta^2)u_{i,j}^{k+\frac{1}{2}} + \omega u_{i+1,j}^{k+\frac{1}{2}} = -(1-\omega)\left[2(1+\beta^2)\right]u_{i,j}^{k} - \omega\beta^2\left(u_{i,j+1}^{k} + u_{i,j-1}^{k+\frac{1}{2}}\right) \quad (5\text{-}24)$$

and

$$\omega\beta^2 u_{i,j-1}^{k+1} - 2(1+\beta^2)u_{i,j}^{k+1} + \omega\beta^2 u_{i,j+1}^{k+1} = -(1-\omega)\left[2(1+\beta^2)\right]u_{i,j}^{k+\frac{1}{2}} - \omega\left(u_{i+1,j}^{k+\frac{1}{2}} + u_{i-1,j}^{k+1}\right) \quad (5\text{-}25)$$

Again, because it is difficult to compute ω_{opt}, numerical experimentation is often performed to obtain the best value of ω for faster convergence. Other formulations of the ADI method and various methods of solution are described in various publications. Since it is not intended to explore every method of solution in this text, additional methods (which are usually much more complicated) are not presented. The methods which have been investigated are the basic methods of solution for elliptic equations and are sufficient for our purposes. In the next section, these methods are used to solve a steady-state heat conduction equation in two dimensions, and the results are compared to analytical solutions.

5.4 Applications

Some of the basic solution procedures for elliptic partial differential equations were just explored. In this section various methods will be used to illustrate their applications. Suppose that it is required to obtain the steady-state temperature distribution on a two-dimensional rectangular plate as shown in Figure 5-11.

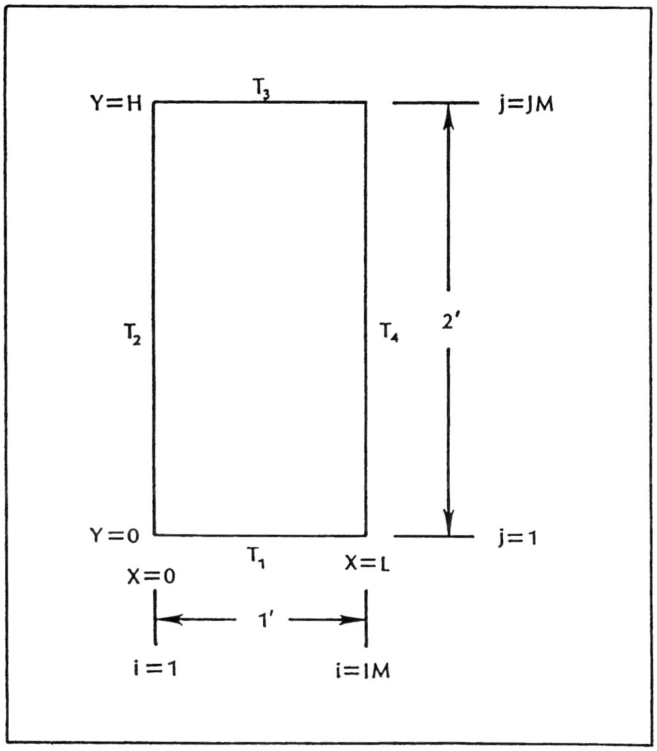

Figure 5-11. Rectangular plate subject to constant temperature distribution at the boundaries.

The imposed boundary conditions are:

$$y = 0 \qquad T = T_1$$

$$x = 0 \qquad T = T_2$$

$$y = H \qquad T = T_3$$

$$x = L \qquad T = T_4$$

The rectangular plate has dimensions of 1 ft. by 2 ft. The temperatures at the boundaries are specified as $T_1 = 100.0$ °R and $T_2 = T_3 = T_4 = 0.0$ °R. Since the values of the temperature at the boundaries of domain are specified, the boundary conditions are Dirichlet type. The governing PDE for steady, two-dimensional heat conduction is:

$$\frac{\partial^2 T}{\partial x^2} + \frac{\partial^2 T}{\partial y^2} = 0 \tag{5-26}$$

It is intended to solve this elliptic PDE, subject to the imposed boundary conditions, using the various methods discussed earlier. Select constant step sizes of $\Delta x = \Delta y =$ 0.05 ft., in which case $IM = 21$ and $JM = 41$. The temperature distribution is to be computed for a total of $(IM - 2) \times (JM - 2) = 741$ grid points.

The first method of solution to explore is the point Gauss-Seidel iteration method, represented by Equation (5-15), where $\beta = \Delta x/\Delta y = 1$. This method is simple, and coding the equation is straightforward. For this example, an initial guess of $T = 0.0$ at all interior points is selected. The point Gauss-Seidel method reaches a converged solution within 574 iterations. The imposed convergence criterion is that if $ERROR < ERRORMAX$, the solution has converged where

$$ERROR = \sum_{\substack{i=2 \\ j=2}}^{\substack{j=JMM1 \\ i=IMM1}} \left[ABS\left(T_{i,j}^{k+1} - T_{i,j}^{k}\right)\right] \tag{5-27}$$

and $ERRORMAX$ is specified to be 0.01. This condition for convergence is used in all the methods presented. The temperature distribution is given in Table 5.1 and the contours of constant temperature are shown in Figure 5-12. When the line Gauss-Seidel iteration method is used, the number of iterations are reduced as expected; in this example, the number of iterations was 308. However, since a system of tridiagonal linear equations (Equation 5-16) is being solved, the total computation time is not reduced; indeed, it increases somewhat (Table 5.2). A point to explore here is the direction to which Equation (5-16) was applied. As noted earlier, if the equation is applied in the direction of expected rapid change (in temperature, for this problem), the convergence of the solution is accelerated. The present example bears out this point. Because of the given domain and boundary conditions, larger changes in temperature in the x-direction are expected compared to the changes in the y-direction. That is the direction which Equation (5-16) was applied. The solution converged after 308 iterations. When Equation (5-16) is applied in the y-direction, the formulation (for $\beta = 1$) becomes

$$T_{i,j-1}^{k+1} - 4T_{i,j}^{k+1} + T_{i,j+1}^{k+1} = -T_{i-1,j}^{k+1} - T_{i+1,j}^{k}$$

The solution obtained from this equation converged in 315 iterations. This example illustrates that the application of a line iterative method in the direction of rapid change in the value of dependent variable increases the convergence rate.

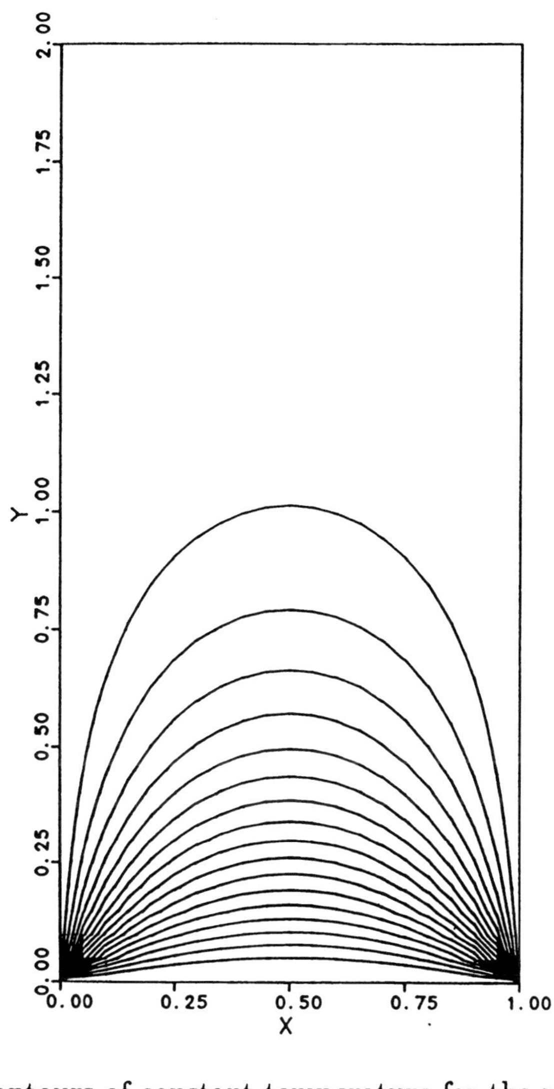

Figure 5-12. Contours of constant temperature for the rectangular plate.

The efficiency of the point Gauss-Seidel and line Gauss-Seidel iteration methods increases dramatically when a relaxation parameter is introduced to accelerate the solution. The PSOR method given by Equation (5-18) is the third method used to obtain a converged solution. For the problem at hand, where the domain is rectangular and subject to Dirichlet boundary conditions, Equation (5-19) can be used to provide the optimum value of the relaxation parameter. Thus, with $\beta = 1$, $IM = 21$, and $JM = 41$, from Equation (5-20),

$$a = \left[\frac{\cos\left(\frac{\pi}{20}\right) + \cos\left(\frac{\pi}{40}\right)}{2} \right]^2 = 0.9847$$

and, from Equation (5-19),

$$\omega_{\text{opt}} = \frac{2 - 2\sqrt{1-a}}{a} = \frac{2 - 2\sqrt{1 - 0.9847}}{0.9847} = 1.78$$

At this value of ω, the solution converges after only 52 iterations (recall that the point Gauss-Seidel method converged with 574 iterations). In general though, simple relations such as (5-19) are not available for the determination of ω_{opt}. Instead, numerical experimentation is used to determine ω_{opt}. When numerical experimentation was performed on this example (even though we know ω_{opt}), it verified ω_{opt} obtained from (5-19). The result is given in Table 5.3 and plotted in Figure 5-13. If a solution is sought only once, it is not so advantageous to compute ω_{opt} by numerical experimentation. However, if many solutions are sought, determining ω_{opt} is highly recommended.

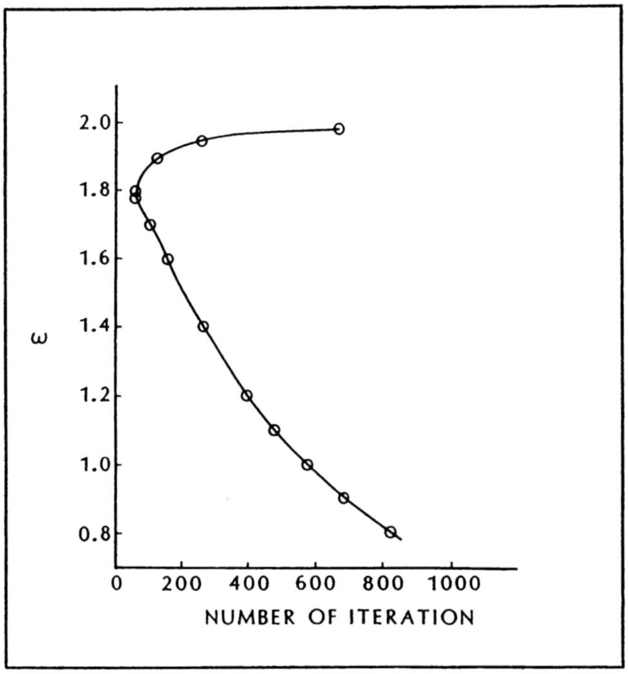

Figure 5-13. Relaxation parameter and the corresponding number of iterations for the converged solution of the rectangular plate by PSOR iteration method.

A note follows on the initial guess. Usually the imposed boundary conditions suggest a reasonable guess on the initial condition to start the solution. Of course, the best guess results in a solution with the least number of iterations. In this example, an initial guess of $T = 0.0$ °R produces a converged solution with 52 iterations. The solution was also obtained using an initial guess of $T = 100.00$ °R (which is not a good guess for this problem). The number of iterations required for the converged solution increased to 76. As this example illustrates, selecting a good guess for the initial condition is certainly desirable. For all the methods investigated in this example, an initial guess of $T = 0.0$ °R was imposed.

The number of iterations necessary for a solution is further reduced by the LSOR method, given by Equation (5-21). Numerical experimentation provided a value of $\omega_{\text{opt}} = 1.265$ for this problem. Values of ω and the corresponding number of iterations required for a converged solution appear in Table 5-4 and are shown in Figure 5-14. The number of iterations can be reduced to 36 with an optimum value of the relaxation parameter. Since the application of any line iterative method in the direction of rapid change reduces the number of iterations, the LSOR method was applied in the x-direction.

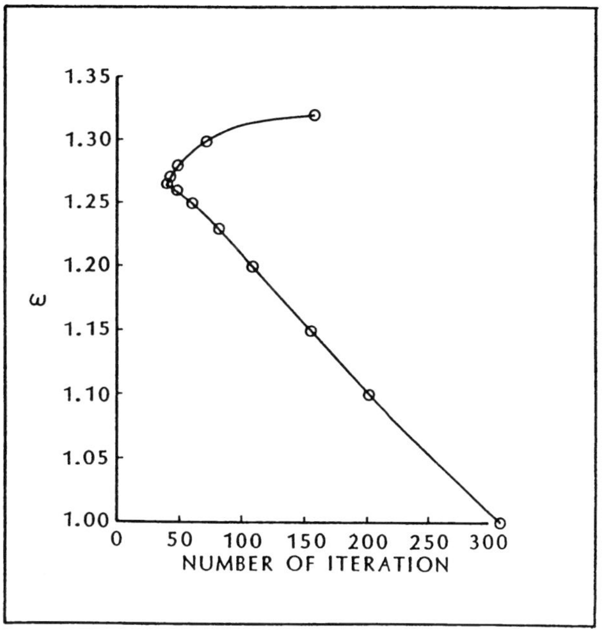

Figure 5-14. Relaxation parameter and the corresponding number of iterations for the converged solution of the rectangular plate by LSOR iteration method.

Application of the ADI method and the accelerated ADI procedure with a relaxation parameter indicated that a converged solution could be obtained with 157 iterations for ADI and only 23 iterations for the accelerated ADI with optimum ω. Table 5-5 and Figure 5-15 present the relaxation parameter ω and the corresponding number of iterations which yielded a converged solution.

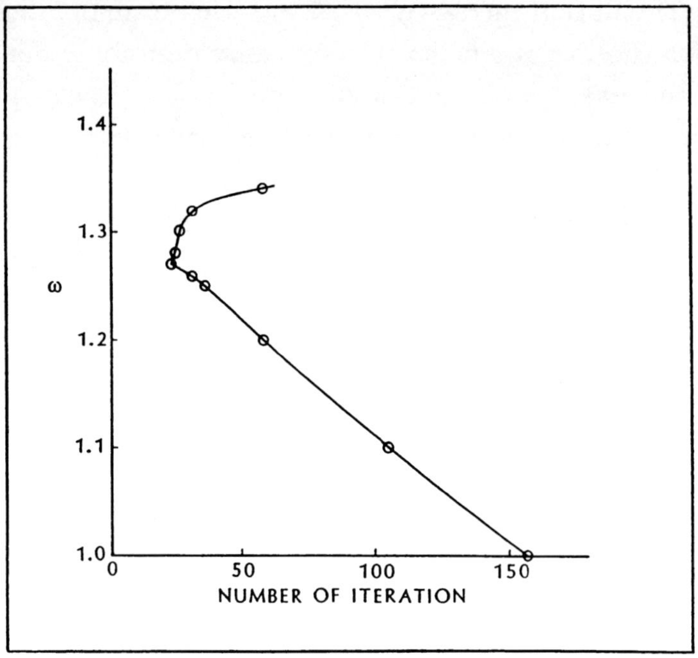

Figure 5-15. Relaxation parameter and the corresponding number of iterations for the converged solution of the rectangular plate by ADI iteration method.

Typical convergence histories are illustrated in Figure 5-16 where the convergence histories of point Gauss-Seidel, line Gauss-Seidel, and point SOR are shown. The error or residual is computed according to Equation (5-27).

Finally, Table 5-2 shows various methods used to obtain a solution, along with the number of iterations required for convergence. Also presented in the same table are the computation times. A small fraction of the computer time was used to write the computed values onto tapes, which were used for plotting the results.

The analytical solution of the two-dimensional steady-state heat conduction given by the partial differential equation (5-26) and the imposed boundary conditions specified earlier is

$$T = T_1 \left[2 \sum_{n=1}^{\infty} \frac{1-(-1)^n}{n\pi} \frac{\text{Sinh}\left(\frac{n\pi(H-y)}{L}\right)}{\text{Sinh}\left(\frac{n\pi H}{L}\right)} \text{Sin}\left(\frac{n\pi x}{L}\right) \right] \tag{5-28}$$

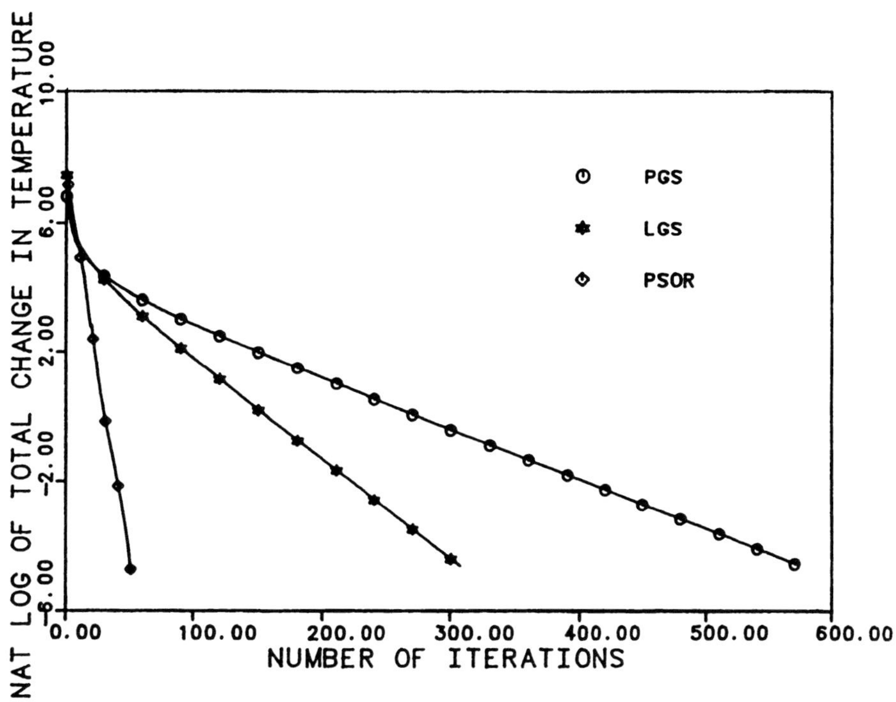

Figure 5-16. Convergence histories.

The temperature values computed from (5-28) at the locations corresponding to the grid points used in the previous computational methods are presented in Table 5-6. Comparison of the numerical solutions with the analytical solution is considered good. The computed values of temperature using Equation (5-28) were performed for $n = 1$ to $n = 20$.

In closing this chapter, the following "general" conclusions are adopted:

1. Iterative methods for solution of elliptic PDEs are preferred over direct methods.

2. Accelerated methods clearly are superior, since they drastically reduce the number of iterations and the computation time, especially if the optimum relaxation parameter is known.

3. A "good" guess on the initial data required to start the solution reduces the number of iterations.

4. For line iterative methods, application of the finite difference equation in the direction of largest change increases the convergence rate.

5.5 Summary Objectives

After studying this chapter, you should be able to do the following:

1. Describe:
   (a) The five-point formula
   (b) The Jacobi iteration method
   (c) The point Gauss-Seidel iteration method
   (d) The line Gauss-Seidel iteration method
   (e) The point successive over-relaxation (PSOR) method
   (f) The line successive over-relaxation (LSOR) method
   (g) the ADI method
2. Solve the problems for Chapter Five.

5.6 Problems

5.1 A two-dimensional inviscid, incompressible fluid is flowing steadily through a chamber between the inlet and the outlet, as shown in Figure P5-1. It is required to determine the streamline pattern within the chamber.

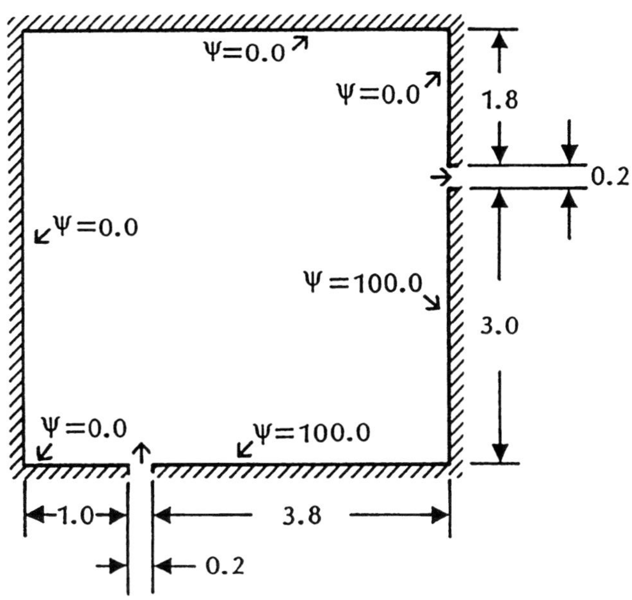

Figure P5-1. Sketch illustrating the chamber and the proposed boundary conditions .

For a two-dimensional, incompressible flow, the continuity equation is expressed as

$$\frac{\partial u}{\partial x} + \frac{\partial v}{\partial y} = 0 \tag{P5-1.1}$$

A stream function Ψ may be defined such that

$$u = \frac{\partial \Psi}{\partial y} \tag{P5-1.2}$$

and

$$v = -\frac{\partial \Psi}{\partial x} \tag{P5-1.3}$$

Recall that a streamline is a line of constant stream function. Furthermore, vorticity is defined as

$$\vec{\Omega} = \nabla \times \vec{V}$$

from which

$$\Omega_z = \frac{\partial v}{\partial x} - \frac{\partial u}{\partial y}$$

For an irrotational flow, the vorticity is zero. Therefore,

$$\frac{\partial v}{\partial x} - \frac{\partial u}{\partial y} = 0 \tag{P5-1.4}$$

Substituting (P5-1.2) and (P5-1.3) into (P5-1.4) yields

$$\frac{\partial}{\partial x}\left(-\frac{\partial \Psi}{\partial x}\right) - \frac{\partial}{\partial y}\left(\frac{\partial \Psi}{\partial y}\right) = 0$$

or

$$\frac{\partial^2 \Psi}{\partial x^2} + \frac{\partial^2 \Psi}{\partial y^2} = 0 \tag{P5-1.5}$$

The goal in this problem is to obtain the solution of this elliptic partial differential equation using the various numerical techniques discussed earlier. The solution will provide the streamline pattern within the chamber.

Since the chamber walls are streamlines, i.e., lines of constant Ψ, we will assign values for these streamlines (see Figure P5-1). The assignment of these values is totally arbitrary as long as continuity is satisfied. Note that, for this application, the boundary conditions are the Dirichlet type, i.e., the values of the dependent variable are specified.

Assume that the inlet and the outlet are 0.2 ft. (per unit depth), and the chamber is 5 ft. by 5 ft. The locations of the inlet and the outlet are shown in Figure P5-1.

The following methods are to be used:

(a) Point Gauss-Seidel

(b) Line Gauss-Seidel

(c) Point SOR

(d) Line SOR

(e) ADI

For all methods, the step sizes are specified as:

$$\Delta x = 0.2 \ , \qquad \Delta y = 0.2 \ , \quad \text{and} \qquad ERRORMAX = 0.01$$

Print the converged solution for each scheme for all y locations at $x = 0.0$, 1.0, 2.0, 3.0, 4.0, and 5.0. Use initial data distribution of $\Psi = 0.0$.

Plot: (a) The streamline pattern, i.e., lines of constant Ψs. Only one plot is sufficient since, as you will notice, solutions by various schemes are very similar; (b) The relaxation parameter versus the number of iterations for PSOR and LSOR schemes.

In addition, the following tasks are to be investigated,

I. The effect of the direction for which the finite difference formulation (i.e., LGS and LSOR) is applied on the convergence.

II. The effect of the initial data on convergence. For this investigation use the PSOR scheme with optimum value of the relaxation parameter. Suggested values of initial data are (a) 0.0, (b) 25.0, (c) 50.0, and (d) 100.0. In addition you may consider a non-uniform initial data distribution.

5.2 Flow enters at the 0.2 ft. inlet and leaves at the open-ended outlet as shown in Figure P5-2. We are interested in computing the streamline pattern within the chamber.

The assumptions stated in problem 5.1 are imposed and, therefore, the governing partial differential equation is

$$\frac{\partial^2 \Psi}{\partial x^2} + \frac{\partial^2 \Psi}{\partial y^2} = 0$$

Obtain the solution using the PSOR technique and step sizes of $\Delta x = 0.2$ and $\Delta y = 0.2$. Since the chamber walls are streamlines, we may specify the values of the stream function at these boundaries. These values are indicated in Figure P5-2. At the outlet, we will assume that the flow is parallel and, therefore, $v = 0$. This assumption implies that $\partial\Psi/\partial x = 0$ along the right boundary.

The imposed boundary conditions are summarized below.

$$\Psi(I,1) = 0.0 \qquad \text{for} \qquad I = 1,6$$

$$\Psi(I,1) = 100.0 \qquad \text{for} \qquad I = 7,31$$

$$\Psi(1,J) = 0.0 \qquad \text{for} \qquad J = 2,20$$

$$\Psi(I,JM) = 0.0 \qquad \text{for} \qquad I = 1,31$$

$$\frac{\partial\Psi}{\partial x}(IM,J) = 0.0 \qquad \text{for} \qquad J = 2,20$$

Print one solution for all the y locations at each increment of one from $x = 0.0$ to $x = 6.0$.

Plot (a) the streamline pattern, and (b) the relaxation parameter versus the number of iterations.

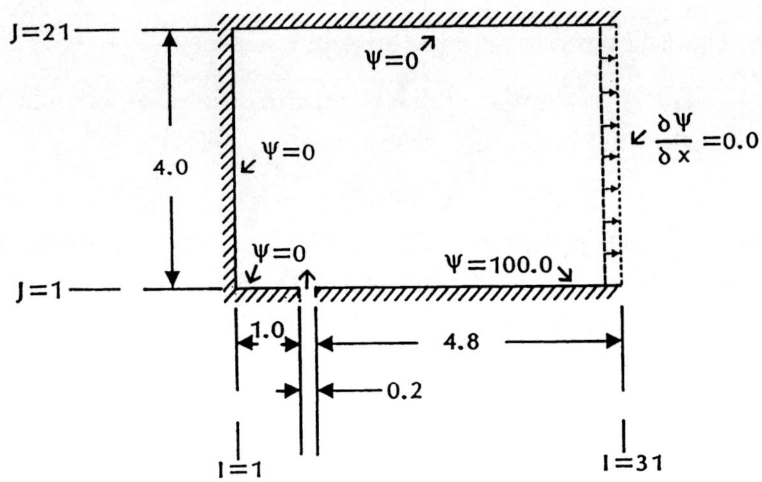

Figure P5-2. Sketch illustrating the open-ended chamber and the proposed boundary conditions.

5.3 Consider a fully developed flow in a rectangular duct with a constant streamwise pressure gradient. The cross-section of the duct, along with its dimensions and the coordinate system used, is illustrated in Figure P5-3. The x-momentum equation is reduced to

$$\mu\left(\frac{\partial^2 u}{\partial y^2}+\frac{\partial^2 u}{\partial z^2}\right)-\frac{\partial p}{\partial x}=0 \tag{P5-3.1}$$

Before attempting to numerically solve this equation, the following nondimensional variables are introduced:

$$y^* = \frac{y}{L}$$

$$z^* = \frac{z}{L}$$

$$u^* = \frac{\mu u}{L^2\left(-\dfrac{dp}{dx}\right)}$$

Upon nondimensionalization, Equation (P5-3.1) becomes

$$\frac{\partial^2 u^*}{\partial y^{*2}} + \frac{\partial^2 u^*}{\partial z^{*2}} + 1 = 0 \qquad \text{(P5-3.2)}$$

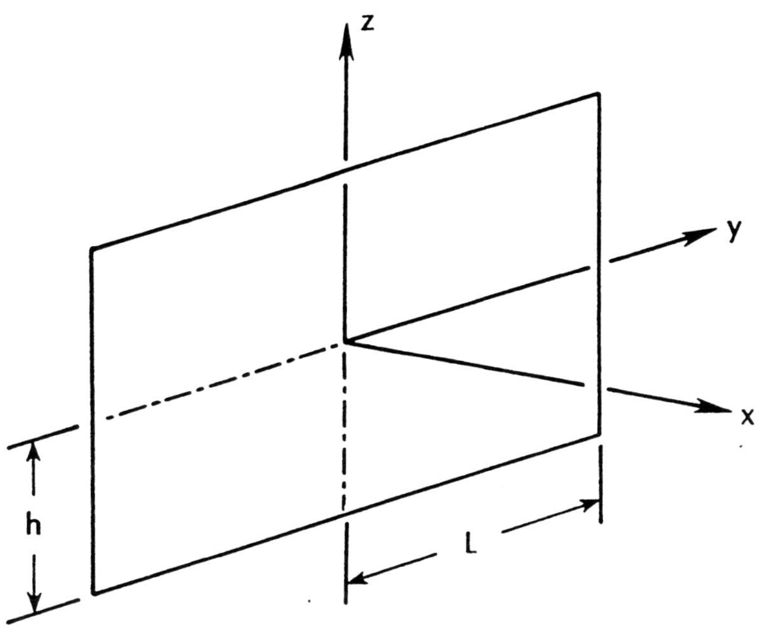

Figure P5-3. Nomenclature for problem 5.3.

The asterisk denoting the nondimensional quantities is dropped from this point on and, therefore, quantities are nondimensional unless otherwise specified. It is required to numerically solve Equation (P5-3.2) by the point Gauss-Seidel scheme subject to the following set of data:

$JM = 61$ (Maximum number of grid points in y)
$KM = 41$ (Maximum number of grid points in z)
$L = 1.5$ m and $h = 1.0$ m

The fluid is oil with a viscosity of 0.4 N.s/m^2. Two different values of pressure gradient are to be considered:

(a) $\frac{dp}{dx} = -10$ N/m^2/m, (b) $\frac{dp}{dx} = -25$ N/m^2/m

Use a convergence criterion of 0.00001 over the entire domain.

The analytical solution is given by

$$ua(y,z) = \frac{16L^2}{\mu\pi^3}\left(-\frac{\mathrm{d}p}{\mathrm{d}x}\right)\sum_{m=1,3,5,\ldots}^{\infty}(-1)^{(m-1)/2}\left[1-\frac{\cosh(m\pi z/2L)}{\cosh(m\pi h/2L}\right]\frac{\cos(m\pi y/2L)}{m^3}$$

where ua represents the velocity obtained by the analytical solution. Compute error distribution within the domain determined by the following relation

$$\mathrm{ER} = \mathrm{ABS}\left[(u(j,k) - ua(j,k))/ua(j,k)\right] * 100.0$$

Print the numerical solution, analytical solution, and the error within the domain for all the z locations and at each y location at increments of 0.5 m.

5.4 Repeat Problem 5.3, except use the PSOR scheme with the optimum relaxation parameter. Compare the number of iterations required for a converged solution and the computation time to that of Problem 5.3

Y	X = 0.0	X = 0.2	X = 0.4	X = 0.6	X = 0.8	X = 1.0
2.000	0.000	0.000	0.000	0.000	0.000	0.000
1.950	0.000	0.044	0.072	0.072	0.044	0.000
1.900	0.000	0.090	0.145	0.145	0.090	0.000
1.850	0.000	0.138	0.223	0.223	0.138	0.000
1.800	0.000	0.189	0.305	0.305	0.189	0.000
1.750	0.000	0.244	0.396	0.396	0.244	0.000
1.700	0.000	0.306	0.496	0.496	0.306	0.000
1.650	0.000	0.376	0.608	0.608	0.376	0.000
1.600	0.000	0.454	0.735	0.735	0.454	0.000
1.550	0.000	0.544	0.880	0.880	0.544	0.000
1.500	0.000	0.647	1.047	1.047	0.647	0.000
1.450	0.000	0.766	1.240	1.240	0.767	0.000
1.400	0.000	0.905	1.463	1.463	0.905	0.000
1.350	0.000	1.065	1.723	1.723	1.065	0.000
1.300	0.000	1.252	2.025	2.025	1.252	0.000
1.250	0.000	1.469	2.377	2.377	1.469	0.000
1.200	0.000	1.723	2.787	2.787	1.723	0.000
1.150	0.000	2.020	3.266	3.266	2.020	0.000
1.100	0.000	2.366	3.825	3.825	2.366	0.000
1.050	0.000	2.771	4.478	4.478	2.771	0.000
1.000	0.000	3.245	5.242	5.242	3.245	0.000
0.950	0.000	3.799	6.134	6.134	3.799	0.000
0.900	0.000	4.448	7.176	7.176	4.448	0.000
0.850	0.000	5.209	8.394	8.394	5.209	0.000
0.800	0.000	6.101	9.817	9.817	6.101	0.000
0.750	0.000	7.147	11.479	11.479	7.147	0.000
0.700	0.000	8.377	13.419	13.419	8.378	0.000
0.650	0.000	9.826	15.682	15.682	9.826	0.000
0.600	0.000	11.534	18.320	18.320	11.535	0.000
0.550	0.000	13.557	21.391	21.391	13.557	0.000
0.500	0.000	15.960	24.959	24.959	15.960	0.000
0.450	0.000	18.830	29.097	29.097	18.831	0.000
0.400	0.000	22.281	33.878	33.878	22.281	0.000
0.350	0.000	26.464	39.379	39.379	26.464	0.000
0.300	0.000	31.582	45.670	45.670	31.582	0.000
0.250	0.000	37.911	52.808	52.808	37.911	0.000
0.200	0.000	45.811	60.817	60.817	45.811	0.000
0.150	0.000	55.722	69.676	69.676	55.722	0.000
0.100	0.000	68.079	79.291	79.291	68.079	0.000
0.050	0.000	83.053	89.487	89.487	83.053	0.000
0.000	100.000	100.000	100.000	100.000	100.000	100.000

Table 5-1. Temperature distribution for the rectangular plate by the point Gauss-Seidel iteration method.

Method	Number of iterations	TM seconds
PGS	574	5.524
LGS	308	7.196
PSOR	52	1.082
LSOR	36	1.41
ADI	157	6.693
AADI	23	1.535

Table 5-2. The number of iterations and computation time required for converged solution by various methods.

Relaxation parameter	Number of iterations	TM seconds
1.00	574	6.215
1.10	478	5.282
1.20	396	4.354
1.40	261	3.071
1.60	152	2.027
1.70	102	1.550
1.78	52	1.082
1.80	55	1.100
1.90	120	1.727
1.95	256	3.023
1.98	668	6.948

Table 5-3. The relaxation parameter and the corresponding number of iterations and computation time for the PSOR method.

Relaxation parameter	Number of iterations	TM seconds
1.000	308	7.726
1.100	201	4.057
1.200	106	2.967
1.230	78	2.338
1.250	57	1.876
1.260	44	1.573
1.265	36	1.410
1.270	39	1.480
1.280	45	1.608
1.300	67	2.114
1.320	153	4.054

Table 5-4. The relaxation parameter and the corresponding number of iterations and computation time for the LSOR method.

Relaxation parameter	Number of iterations	TM seconds
1.00	157	7.081
1.10	105	4.884
1.20	58	2.940
1.25	36	2.076
1.26	31	1.867
1.27	23	1.535
1.28	25	1.620
1.30	26	1.625
1.32	31	1.862
1.34	58	2.985

Table 5-5. The relaxation parameter and the corresponding number of iterations and computation time for the accelerated ADI method.

Y	X = 0.0	X = 0.2	X = 0.4	X = 0.6	X = 0.8	X = 1.0
2.000	0.000	0.000	0.000	0.000	0.000	0.000
1.950	0.000	0.044	0.071	0.071	0.044	0.000
1.900	0.000	0.089	0.144	0.144	0.089	0.000
1.850	0.000	0.137	0.221	0.221	0.137	0.000
1.800	0.000	0.187	0.303	0.303	0.187	0.000
1.750	0.000	0.243	0.393	0.393	0.243	0.000
1.700	0.000	0.304	0.492	0.492	0.304	0.000
1.650	0.000	0.373	0.604	0.604	0.373	0.000
1.600	0.000	0.451	0.730	0.730	0.451	0.000
1.550	0.000	0.541	0.875	0.875	0.541	0.000
1.500	0.000	0.643	1.041	1.041	0.643	0.000
1.450	0.000	0.762	1.233	1.233	0.762	0.000
1.400	0.000	0.899	1.455	1.455	0.899	0.000
1.350	0.000	1.059	1.713	1.713	1.059	0.000
1.300	0.000	1.245	2.014	2.014	1.245	0.000
1.250	0.000	1.462	2.364	2.364	1.462	0.000
1.200	0.000	1.715	2.773	2.773	1.715	0.000
1.150	0.000	2.010	3.250	3.250	2.010	0.000
1.100	0.000	2.355	3.808	3.808	2.355	0.000
1.050	0.000	2.759	4.460	4.460	2.759	0.000
1.000	0.000	3.231	5.221	5.221	3.231	0.000
0.950	0.000	3.784	6.111	6.111	3.784	0.000
0.900	0.000	4.432	7.152	7.152	4.432	0.000
0.850	0.000	5.191	8.369	8.369	5.191	0.000
0.800	0.000	6.080	9.790	9.790	6.080	0.000
0.750	0.000	7.125	11.451	11.451	7.125	0.000
0.700	0.000	8.352	13.392	13.392	8.352	0.000
0.650	0.000	9.798	15.656	15.656	9.798	0.000
0.600	0.000	11.503	18.296	18.296	11.503	0.000
0.550	0.000	13.521	21.371	21.371	13.521	0.000
0.500	0.000	15.919	24.947	24.947	15.919	0.000
0.450	0.000	18.783	29.094	29.094	18.783	0.000
0.400	0.000	22.227	33.889	33.889	22.227	0.000
0.350	0.000	26.405	39.408	39.408	26.405	0.000
0.300	0.000	31.524	45.720	45.720	31.524	0.000
0.250	0.000	37.871	52.880	52.880	37.871	0.000
0.200	0.000	45.822	60.907	60.907	45.822	0.000
0.150	0.000	55.832	69.772	69.772	55.832	0.000
0.100	0.000	68.316	79.376	79.376	68.316	0.000
0.050	0.000	83.310	89.537	89.537	83.310	0.000
0.000	100.000	100.000	100.000	100.000	100.000	100.000

Table 5-6. The temperature distribution for the rectangular plate by analytical method.

Chapter 6
Hyperbolic Equations

6.1 Introductory Remarks

Hyperbolic equations and methods of solution are investigated in this chapter by considering simple model equations. A commonly used technique to solve hyperbolic equations in fluid mechanics is the method of characteristics. This method is well known and is used for the solution of inviscid supersonic flow fields. However, it is often difficult to use this method for three-dimensional problems and problems which involve nonlinear terms. Some discussion of the method of characteristics is considered in Appendices A and G. This chapter presents various methods of solution by finite difference approximations of the hyperbolic PDEs. Both linear and nonlinear PDEs are investigated by applying numerical methods to the model equations. In addition, various types of error, i.e., dispersion error and dissipation error, are explored in the example problems.

6.2 Finite Difference Formulations

The first model equation to consider is the first-order wave equation,

$$\frac{\partial u}{\partial t} = -a\frac{\partial u}{\partial x} \qquad a > 0 \tag{6-1}$$

which is a linear equation for constant speed a. For this simple equation, the characteristic lines are straight lines given by the equation $x - at = constant$. The quantity u is convected along these lines with constant speed a. A number of finite difference approximations for the first-order wave equation have been studied by various investigators. These approximations have been formulated in explicit or implicit forms. Some of these methods and their applications are presented in this section.

6.2.1 Explicit Formulations

6.2.1.1. Euler's FTFS method. In this explicit method, forward time and forward space approximations of the first-order are used, resulting in the FDE:

$$\frac{u_i^{n+1} - u_i^n}{\Delta t} = -a\frac{u_{i+1}^n - u_i^n}{\Delta x} \tag{6-2}$$

von Neumann stability analysis indicates that this method is unconditionally unstable.

6.2.1.2. Euler's FTCS method. In this formulation, central differencing of $O(\Delta x)^2$ is used for the spatial derivative, resulting in

$$\frac{u_i^{n+1} - u_i^n}{\Delta t} = -a\frac{u_{i+1}^n - u_{i-1}^n}{2\Delta x} \tag{6-3}$$

This explicit formulation is also unconditionally unstable.

6.2.1.3. The first upwind differencing method. Backward differencing of the spatial derivative produces a finite difference equation of the form

$$\frac{u_i^{n+1} - u_i^n}{\Delta t} = -a\frac{u_i^n - u_{i-1}^n}{\Delta x} \tag{6-4}$$

with $O(\Delta t,\ \Delta x)$. Von Neumann stability analysis indicates that this method is stable when $c \leq 1$, where $c = a\Delta t/\Delta x$ is the Courant number. For the model Equation (6-1), a forward differencing for the spatial derivative must be used if $a < 0$. Therefore, for $\partial u/\partial t = a(\partial u/\partial x)$, where $a < 0$, the FDE for a conditionally stable solution is

$$\frac{u_i^{n+1} - u_i^n}{\Delta t} = -a\frac{u_{i+1}^n - u_i^n}{\Delta x} \tag{6-5}$$

6.2.1.4. The Lax method. If an average value of u_i^n in the Euler's FTCS method is used, the FDE takes the form

$$u_i^{n+1} = \frac{1}{2}(u_{i+1}^n + u_{i-1}^n) - \frac{a\Delta t}{2\Delta x}(u_{i+1}^n - u_{i-1}^n) \tag{6-6}$$

Stability analysis shows that the method is stable when $c \leq 1$.

6.2.1.5. Midpoint leapfrog method. In this method, central differencing of the second order is used for both the time and space derivatives, resulting in the FDE

$$\frac{u_i^{n+1} - u_i^{n-1}}{2\Delta t} = -a\frac{u_{i+1}^n - u_{i-1}^n}{2\Delta x} \tag{6-7}$$

which is of order $[(\Delta t)^2,\ (\Delta x)^2]$. The method is stable when $c \leq 1$. As the formulation indicates, two sets of initial values are required to start the solution. The dependent variable at the advanced time level $n+1$ requires the values at time level $n-1$ and n. To provide two sets of initial data, a starter solution that requires only one set of initial data, say at $n-1$, is used. The use of a starter solution will affect the order of accuracy of the method.

It has been shown that two independent solutions can be developed as the solution proceeds forward. In summary, the midpoint leapfrog method has a higher (second) order of accuracy; however, this formulation may include some difficulties such as the starting procedure, the development of two independent solutions, and, if programming is not done carefully, a large increase in computer storage.

6.2.1.6. The Lax-Wendroff method. This finite difference representation of the PDE is derived from Taylor series expansion of the dependent variable as follows:

$$u(x,\ t+\Delta t) = u(x,t) + \frac{\partial u}{\partial t}\Delta t + \frac{\partial^2 u}{\partial t^2}\frac{(\Delta t)^2}{2!} + O(\Delta t)^3$$

or, in terms of indices,

$$u_i^{n+1} = u_i^n + \frac{\partial u}{\partial t}\Delta t + \frac{\partial^2 u}{\partial t^2}\frac{(\Delta t)^2}{2!} + O(\Delta t)^3 \tag{6-8}$$

Now consider the model equation

$$\frac{\partial u}{\partial t} = -a\frac{\partial u}{\partial x} \tag{6-9}$$

By taking the time derivative, one obtains

$$\frac{\partial^2 u}{\partial t^2} = -a\frac{\partial}{\partial t}\left(\frac{\partial u}{\partial x}\right) = -a\frac{\partial}{\partial x}\left(\frac{\partial u}{\partial t}\right) = a^2\frac{\partial^2 u}{\partial x^2} \tag{6-10}$$

Substituting (6-9) and (6-10) into (6-8) produces

$$u_i^{n+1} = u_i^n + \left(-a\frac{\partial u}{\partial x}\right)\Delta t + \frac{(\Delta t)^2}{2}\left(a^2\frac{\partial^2 u}{\partial x^2}\right)$$

When central differencing of the second order for the spatial derivatives is used, it follows that

$$u_i^{n+1} = u_i^n - a\Delta t\left[\frac{u_{i+1}^n - u_{i-1}^n}{2\Delta x}\right] + \frac{1}{2}a^2(\Delta t)^2\left[\frac{u_{i+1}^n - 2u_i^n + u_{i-1}^n}{(\Delta x)^2}\right] \tag{6-11}$$

This formulation is known as the Lax-Wendroff method and is of order $[(\Delta t)^2,\ (\Delta x)^2]$. Stability analysis shows that this explicit method is stable for $c \leq 1$.

6.2.2 Implicit Formulations

6.2.2.1. Euler's BTCS method. Implicit formulation of Euler's backward time and central space approximation is unconditionally stable and is of order $[(\Delta t), (\Delta x)^2]$. This approximation applied to model equation (6-1) yields:

$$\frac{u_i^{n+1} - u_i^n}{\Delta t} = -\frac{a}{2\Delta x}\left[u_{i+1}^{n+1} - u_{i-1}^{n+1}\right] \quad \text{or}$$

$$\frac{1}{2}c\,u_{i-1}^{n+1} - u_i^{n+1} - \frac{1}{2}c\,u_{i+1}^{n+1} = -u_i^n \tag{6-12}$$

Once this equation is applied to all grid points at the unknown time level, a set of linear algebraic equations will result. Again, these equations can be represented in a matrix form, where the coefficient matrix is tridiagonal.

6.2.2.2. Implicit first upwind differencing method. A first-order backward difference approximation in time and space results in the following implicit scheme which is $O(\Delta t, \Delta x)$

$$\frac{u_i^{n+1} - u_i^n}{\Delta t} = -a\frac{u_i^{n+1} - u_{i-1}^{n+1}}{\Delta x} \tag{6-13}$$

This equation can be rearranged as

$$c\,u_{i-1}^{n+1} - (1+c)u_i^{n+1} = -u_i^n \tag{6-14}$$

Once Equation (6-14) is applied to all the grid points, it will result in a bidiagonal system. The solution of a bidiagonal system written in a general form of

$$A_i u_{i-1}^{n+1} + B_i u_i^{n+1} = D_i$$

is given by

$$u_i^{n+1} = \frac{D_i - A_i u_{i-1}^{n+1}}{B_i}$$

6.2.2.3. Crank-Nicolson method. This is a widely used implicit method for which the model equation takes the form

$$\frac{u_i^{n+1} - u_i^n}{\Delta t} = -a\frac{1}{2}\left[\frac{u_{i+1}^{n+1} - u_{i-1}^{n+1}}{2\Delta x} + \frac{u_{i+1}^n - u_{i-1}^n}{2\Delta x}\right] \tag{6-15}$$

The order of accuracy is $[(\Delta t)^2, (\Delta x)^2]$. This formulation also results in a tridiagonal system of equations as follow.

$$\frac{1}{4}c\,u_{i-1}^{n+1} - u_i^{n+1} - \frac{1}{4}c\,u_{i+1}^{n+1} = -u_i^n + \frac{1}{4}c\left(u_{i+1}^n - u_{i-1}^n\right) \tag{6-16}$$

6.3 Splitting Methods

The methods introduced thus far are simple and their application to linear one-dimensional hyperbolic equations is straightforward. Extending these methods to multidimensional and/or nonlinear problems may create some difficulties. The implicit formulation of a multidimensional problem results in a large number of algebraic equations with a coefficient matrix that is pentadiagonal or larger. Solving such a system requires an excessive amount of computer time. Therefore, methods such as ADI, which split the multidimensional problem into a series of equations with tridiagonal coefficients, are preferred. Various formulations of ADI or, more generally, of approximate factorization discussed previously can be applied to multidimensional hyperbolic equations.

6.4 Multi-Step Methods

The second problem to address is nonlinearity. Some of the methods presented for linear equations are not well suited to the solution of nonlinear problems. Multi-step methods, which use the finite difference equations at split time levels, work well when applied to nonlinear hyperbolic equations. Some of these methods are introduced in this section and their applications to linear model equations are illustrated. Later, these methods are extended to nonlinear problems.

These methods may be referred to as predictor-corrector methods as well. In the first step, a temporary value for the dependent variable is predicted; and, subsequently in the second step, a corrected value is computed to provide the final value of the dependent variable.

6.4.1 Richtmyer/Lax-Wendroff Multi-Step Method

The Lax-Wendroff method discussed earlier is split into two time levels. There are two variations of this method in the literature. The one referred to as the Richtmyer method applies the equation at grid point "i", while the so-called Lax-Wendroff multi-step method applies the first step at the midpoint $i + \frac{1}{2}$. In both formulations, Lax's method is used at the first time level $n + \frac{1}{2}$, followed by the midpoint leapfrog method at time level $n+1$. The method is second-order accurate both in time and space.

The Richtmyer formulation is

$$\frac{u_i^{n+\frac{1}{2}} - \frac{1}{2}(u_{i+1}^n + u_{i-1}^n)}{\frac{\Delta t}{2}} = -a\frac{u_{i+1}^n - u_{i-1}^n}{2\Delta x}$$

and

$$\frac{u_i^{n+1} - u_i^n}{\Delta t} = -a\frac{u_{i+1}^{n+\frac{1}{2}} - u_{i-1}^{n+\frac{1}{2}}}{2\Delta x}$$

Rearranging terms yields

$$u_i^{n+\frac{1}{2}} = \frac{1}{2}\left(u_{i+1}^n + u_{i-1}^n\right) - \frac{a\Delta t}{4\Delta x}\left(u_{i+1}^n - u_{i-1}^n\right) \tag{6-17}$$

and

$$u_i^{n+1} = u_i^n - \frac{a\Delta t}{2\Delta x}\left(u_{i+1}^{n+\frac{1}{2}} - u_{i-1}^{n+\frac{1}{2}}\right) \tag{6-18}$$

This method is stable for $a\Delta t/\Delta x \leq 2$.

The Lax-Wendroff multi-step formulation is

$$u_{i+\frac{1}{2}}^{n+\frac{1}{2}} = \frac{1}{2}\left(u_{i+1}^n + u_i^n\right) - \frac{a\Delta t}{2\Delta x}\left(u_{i+1}^n - u_i^n\right) \tag{6-19}$$

and

$$u_i^{n+1} = u_i^n - \frac{a\Delta t}{\Delta x}\left(u_{i+\frac{1}{2}}^{n+\frac{1}{2}} - u_{i-\frac{1}{2}}^{n+\frac{1}{2}}\right) \tag{6-20}$$

The stability condition is $a\Delta t/\Delta x \leq 1$. Note that if Equation (6-19) is substituted into Equation (6-20), the original Lax-Wendroff equation given by (6-11) results. The same result can be obtained with Equations (6-17) and (6-18). Be careful about the step sizes!

6.4.2 The MacCormack Method

In this multi-level method, the first equation uses forward differencing resulting in the FDE

$$\frac{u_i^* - u_i^n}{\Delta t} = -a\frac{u_{i+1}^n - u_i^n}{\Delta x} \tag{6-21}$$

where * represents a temporary value of the dependent variable at the advanced level. The second equation uses backward differencing; thus,

$$\frac{u_i^{n+1} - u_i^{n+\frac{1}{2}}}{\frac{1}{2}\Delta t} = -a\frac{u_i^* - u_{i-1}^*}{\Delta x}$$

The value of $u_i^{n+\frac{1}{2}}$ is replaced by an average value, as follows.

$$u_i^{n+\frac{1}{2}} = \frac{1}{2}\left(u_i^n + u_i^*\right)$$

The two-level MacCormack method is organized as:

predictor step

$$u_i^* = u_i^n - \frac{a\Delta t}{\Delta x}\left(u_{i+1}^n - u_i^n\right) \tag{6-22}$$

and

corrector step

$$u_i^{n+1} = \frac{1}{2}\left[(u_i^n + u_i^*) - \frac{a\Delta t}{\Delta x}(u_i^* - u_{i-1}^*)\right] \tag{6-23}$$

This method is also second-order accurate with the stability requirement of $a\Delta t/\Delta x \leq 1$. It is well suited for nonlinear equations and is consequently a popular method in fluid mechanics. The order of differencing can be reversed for each time step, (i.e., forward/backward followed by backward/forward); and, for nonlinear problems, this procedure provides the best result. For linear problems, this method is equivalent to the Lax-Wendroff method.

A note of caution. Our classification of various formulations into splitting methods is not universal. A multi-step method might be called a splitting method by some, or a predictor-corrector method by others. The important point to recognize is the distinct procedures of the methods. What was called the splitting method applies to multidimensional problems and reduces the finite difference equation to a set of equations that are tridiagonal. This procedure is used to increase efficiency; the resulting tridiagonal system requires less computer time. The second category of methods, which was called multi-step methods, solves problems in a sequence of time steps, using various finite difference approximations. These methods are desirable for solving nonlinear hyperbolic equations.

Before proceeding to nonlinear hyperbolic equations, the applications of the methods just presented to a simple linear hyperbolic problem are considered.

6.5 Applications to a Linear Problem

As a first example, consider the first-order wave equation,

$$\frac{\partial u}{\partial t} = -a\frac{\partial u}{\partial x} \qquad a > 0 \tag{6-24}$$

where a, the speed of sound, is selected to be 250 m/s. Assume that at time $t = 0$, a disturbance of half sinusoidal shape has been generated. The initial condition is specified as

$$\left.\begin{aligned} u(x,0) &= 0 && 0 \leq x \leq 50 \\ u(x,0) &= 100\left\{\sin\left[\pi\left(\frac{x-50}{60}\right)\right]\right\} && 50 \leq x \leq 110 \\ u(x,0) &= 0 && 110 \leq x \leq 400 \end{aligned}\right\} \tag{6-25}$$

and is illustrated in Figure 6-1.

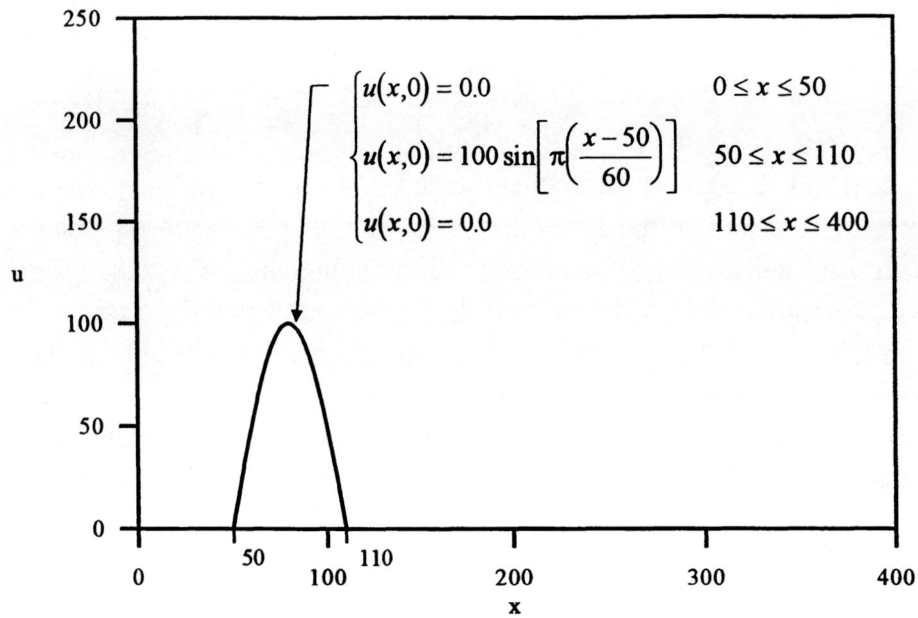

Figure 6-1. Initial distribution at $t = 0.0$ sec.

Assuming that the disturbance is introduced in a one-dimensional long tube with both ends closed, the imposed boundary conditions are

$$x = 0 \qquad u(0, t) = 0$$

and

$$x = L \qquad u(L, t) = 0$$

Due to the simplicity of the problem, only some of the methods discussed earlier will be used to illustrate the solution procedures. However, the selected methods will illustrate the various types of errors associated with each.

Start by considering an explicit first-order accurate method, such as the first upwind differencing technique. In this formulation, the finite difference equation applied to (6-24) is

$$u_i^{n+1} = u_i^n - c\,(u_i^n - u_{i-1}^n) \tag{6-26}$$

where $c = a\Delta t/\Delta x$. For a stable solution, $c \le 1$. From the finite difference formulation (6-26), it can be seen that if $c = 1$, the solution is exact, i.e.,

$$u_i^{n+1} = u_{i-1}^n$$

To investigate the effect of various step sizes on the solution, three values of the Courant number are used. The cases under study are:

(1) $\Delta x = 5.0$, $\Delta t = 0.02$ resulting in $c = 1$

(2) $\Delta x = 5.0$, $\Delta t = 0.01$ $c = 0.5$

(3) $\Delta x = 5.0$, $\Delta t = 0.005$ $c = 0.25$

Before analyzing the solutions, recall that first-order accurate methods (where second and higher order derivatives in the Taylor series expansion have been dropped in the approximation process) produce errors that are dissipative. These errors depend on the step sizes; hence, they affect the accuracy of the solution.

The resulting solutions for $c = 1.0$ and 0.25 are presented in Tables 6.1a and 6.1b, and Figures 6-2 and 6-3. Figure 6-2 clearly illustrates the propagation of solution along the characteristic line. Note that in Figure 6-2, the solution is propagated with minimum error. That is not surprising, since for $c = 1$ the solution is exact. However, as the step size (Δt) is reduced so that the Courant number is smaller than one, errors appear in the solution. Due to the dissipation error of the method, the amplitude of the original (sinusoidal) function is decreased and the solution is dissipated to neighboring points. The phenomenon is clearly evident in Figure 6-3.

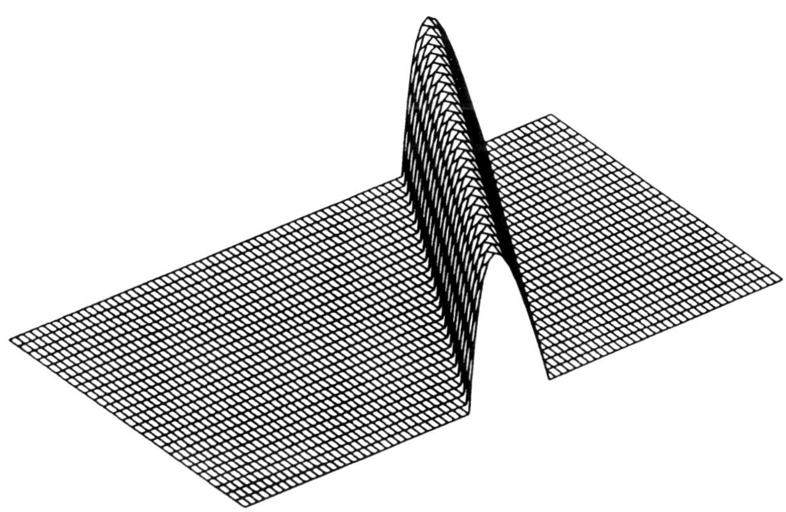

$c = 1.0$
Figure 6-2. Solution of the first-order wave equation by the explicit first upwind differencing technique.

$c = 0.25$

Figure 6-3. Solution of the first-order wave equation by the explicit first upwind differencing technique.

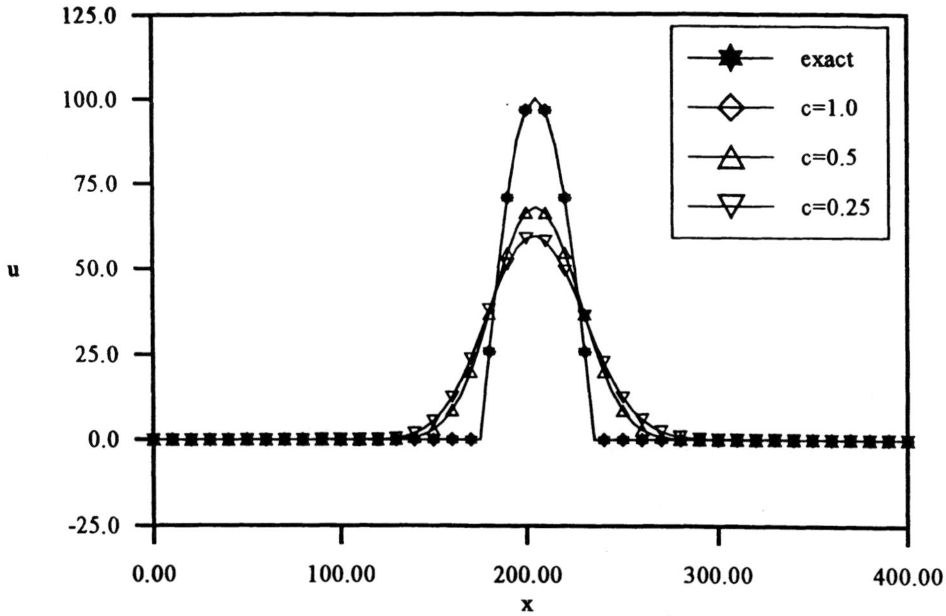

Figure 6-4. Comparison of the solutions of the wave equation by the explicit first upwind differencing method at $t = 0.5$ sec for various Courant numbers.

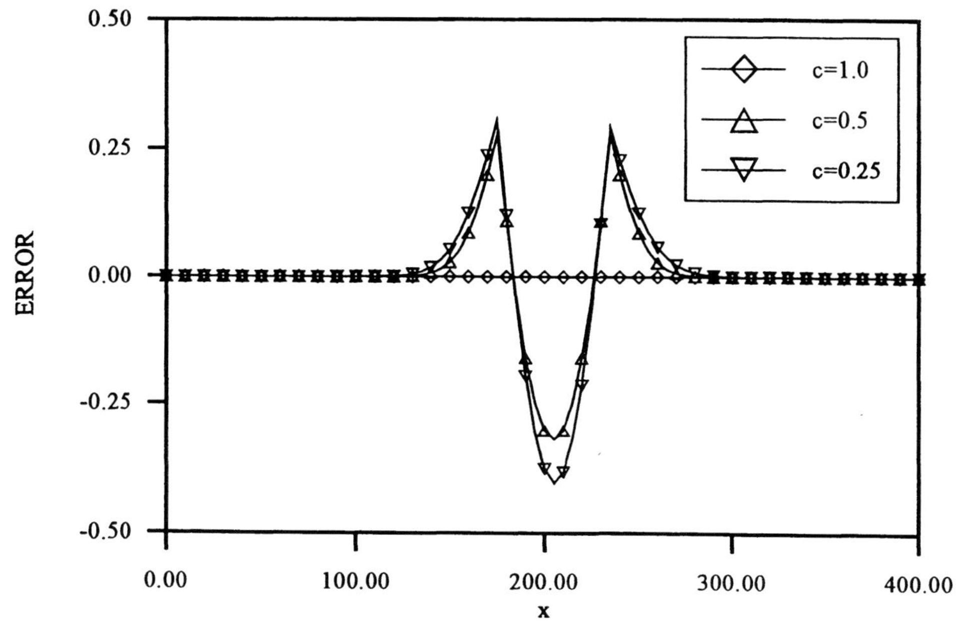

Figure 6-5. Error distributions of the explicit first upwind scheme at $t = 0.5$ sec.

The effect of the step sizes (the Courant numbers) on the solution of the wave equation is shown in Figure 6-4. These solutions are at $t = 0.5$ sec. It is clear that for $c = 1.0$, the solution is very close to the exact solution (with minimum error). As the Courant number decreases the errors increase, as is evident in the decrease in the amplitude and in the dissipation of the wave to the neighboring points. In conclusion, then, the best result is obtained when the Courant number is at the upper limit of the stability condition, i.e., when $c = 1$. A comparison of errors at $t = 0.5$ sec for the three different Courant numbers is illustrated in Figure 6-5. The error is defined as the difference between the numerical and analytical solutions.

It is interesting at this point to investigate the implicit equivalence of Equation (6-26) given by (6-14). It is important to recognize that both equations have the same order of accuracy, that is, $O(\Delta t, \Delta x)$. However, when the modified equations for the two formulations are considered, the differences between the two become apparent. In fact, the implicit formulation will have more of a dissipation error than the explicit formulation. To identify this difference, consider also the solution by the implicit formulation for Courant numbers of 1.0, 0.5, and 0.25 shown in Figure 6-6. A comparison with Figure 6-4 clearly indicates the larger dissipation error associated with the implicit formulation of (6-14). The solutions for Courant numbers of 1.0 and 0.25 are given in Tables 6-2a and 6-2b, respectively. The wave propagation for Courant number of one is shown in Figure 6-7. Note that, even at $c = 1$, the solution has considerable dissipation error. Finally, the error distributions for the three Courant numbers at $t = 0.5$ sec are shown in Figure 6-8.

Next, a second-order accurate method to the model Equation (6-24) is applied. One such method introduced earlier is the Lax-Wendroff method. The finite difference formulation applied to the model equation is

$$u_i^{n+1} = u_i^n - \frac{1}{2}c\left(u_{i+1}^n - u_{i-1}^n\right) + \frac{1}{2}c^2\left(u_{i+1}^n - 2u_i^n + u_{i-1}^n\right) \tag{6-27}$$

This method is stable when $c \leq 1$. Again, three cases with various step sizes will be investigated. A typical solution ($c = 1.0$) is presented in Table 6.3. The generated solutions for Δt step sizes of 0.02 and 0.005 are shown in Figures 6-9 and 6-10, and the solutions at $t = 0.5$ sec for the three cases are compared to the analytical solution in Figure 6-11. Furthermore, the error distributions at $t = 0.5$ sec for the three different Courant numbers are illustrated in Figure 6-12. Note that the solution behaves differently when this method is used. Since the algorithm is second-order accurate (third and higher order terms have been dropped in the approximation process of the finite difference relations), some dispersion error is expected. Indeed, the oscillatory behavior of the solution for the smaller Courant number clearly indicates the errors developed in the solution. Note that the amplitude of the solution remains the same (within small errors).

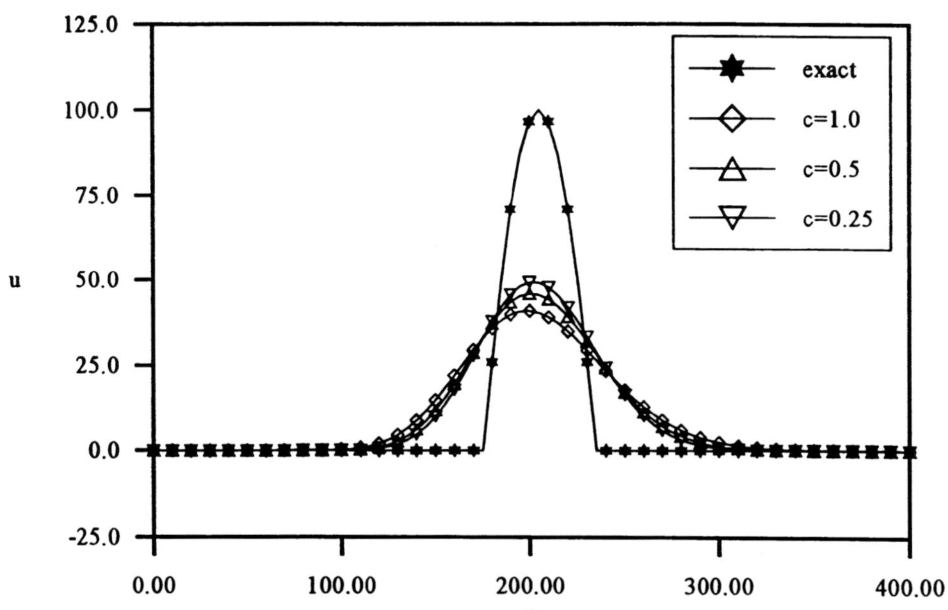

Figure 6-6. Comparison of the solutions of the wave equation by the implicit first upwind scheme at $t = 0.5$ sec for several Courant numbers.

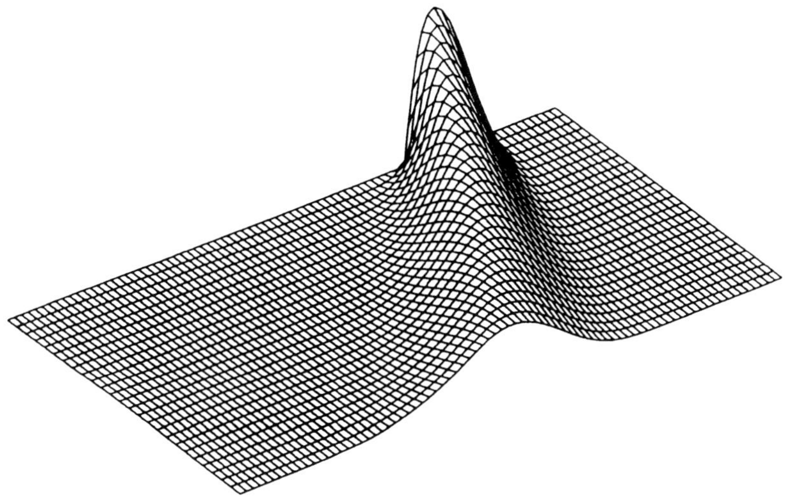

Figure 6-7. Solution of the first-order wave equation by the implicit first upwind scheme, $c = 1$.

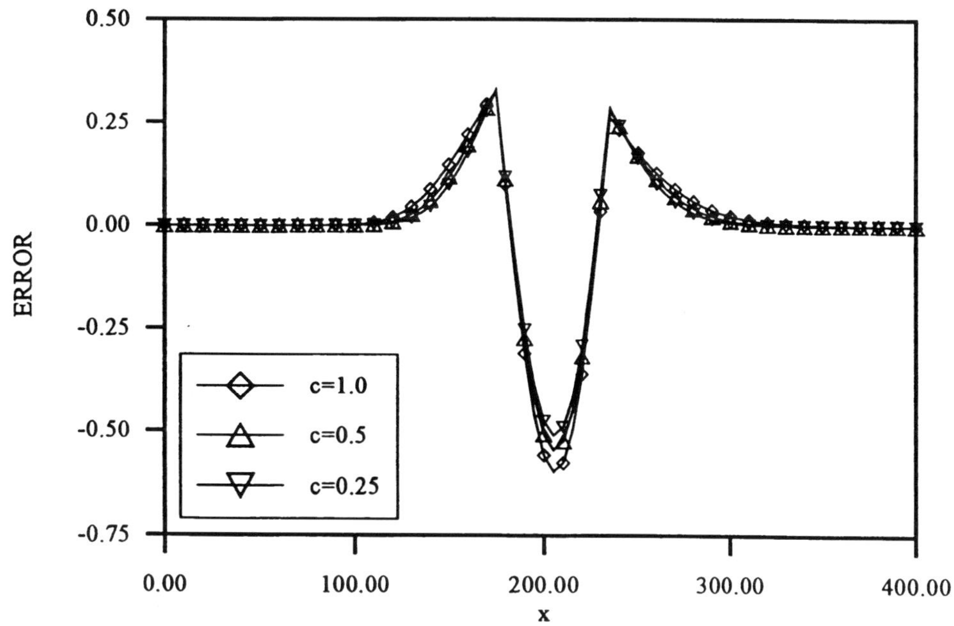

Figure 6-8. Error distributions of the implicit first upwind scheme at $t = 0.5$ sec.

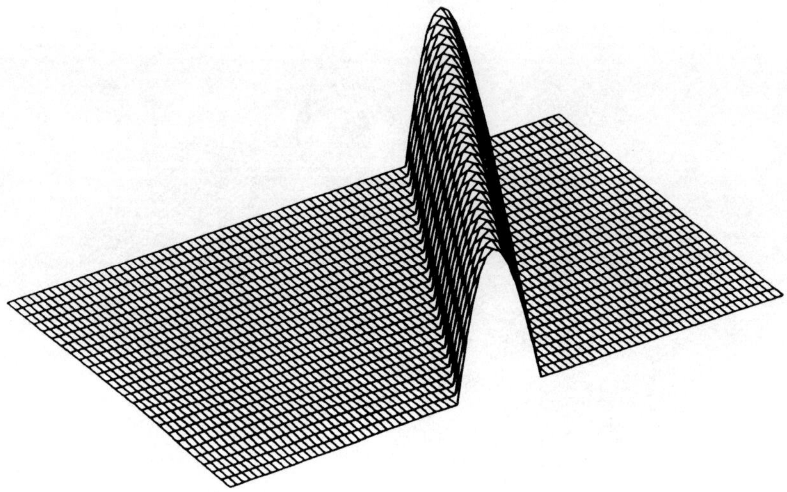

Figure 6-9. Solution of the first-order wave equation by the Lax-Wendroff method, $c = 1.0$.

Figure 6-10. Solution of the first-order wave equation by the Lax-Wendroff method, $c = 0.25$.

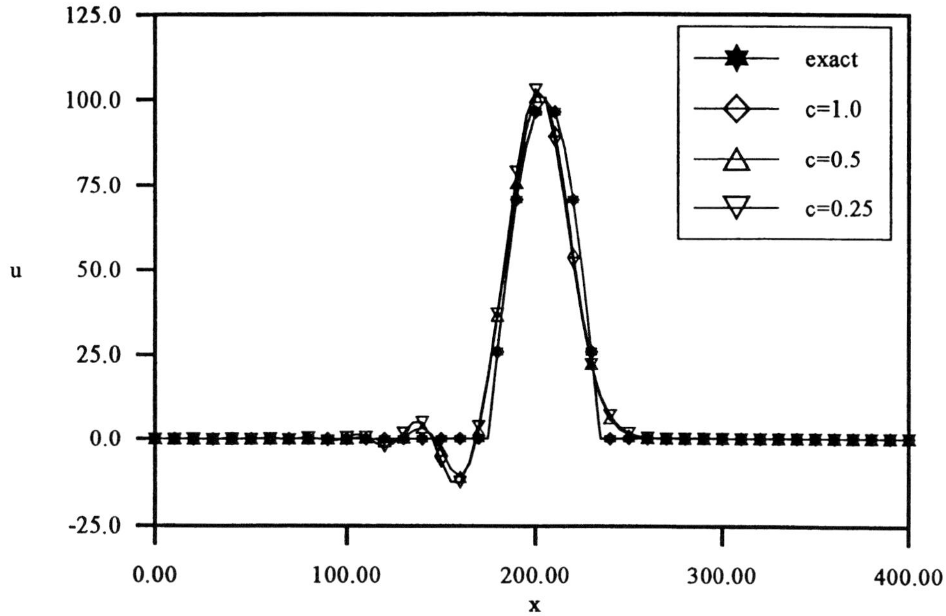

Figure 6-11. Comparison of the solutions of the wave equation by Lax-Wendroff method at $t = 0.5$ sec for various Courant numbers.

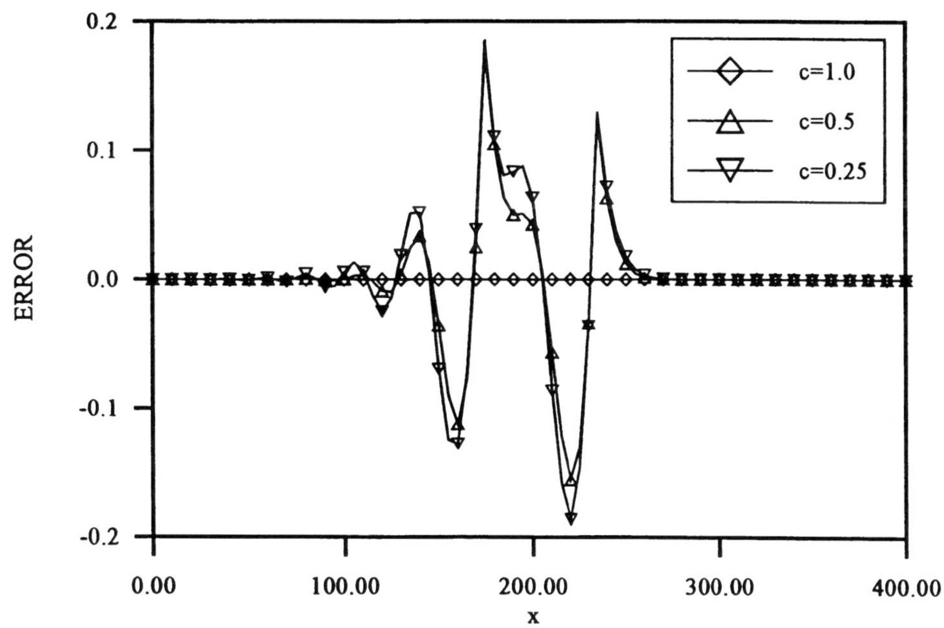

Figure 6-12. Error distributions of the Lax-Wendroff scheme at $t = 0.5$ sec.

Again, the error is smallest at the upper limit value of the Courant number, i.e., when the Courant number approaches one. Therefore, the best solution is obtained by selecting step sizes which yield a Courant number of one, or close to it. These simple applications clearly illustrated the dissipation and dispersion errors discussed in Chapter Four.

Now consider the implicit BTCS scheme. The finite difference formulation given by (6-12) is

$$\frac{1}{2}c\,u_{i-1}^{n+1} - u_i^{n+1} - \frac{1}{2}c\,u_{i+1}^{n+1} = -u_i^n$$

This equation, applied to all i at each time level, results in a system of tridiagonal equations. Solutions were obtained with step sizes (Δt) of 0.02 sec and 0.05 sec. The results are shown in Figures 6-13 and 6-14.

The solutions for Courant numbers of 1.0 and 2.5 are presented in Tables 6-4a and 6-4b, respectively. Furthermore, the solutions at $t = 0.5$ sec for several Courant numbers are compared to the analytical solution in Figure 6-15, and the error distributions are illustrated in Figure 6-16. Note that the solution at higher Courant number (larger time step) has produced a poor solution. Therefore, the advantage of an implicit scheme due to its less restrictive stability requirement has to be viewed with skepticism for some applications.

A multi-step method popular in aeronautics is the MacCormack method. But for linear problems, the method is equivalent to the Lax-Wendroff method. The solution is identical to the solution just obtained from the Lax-Wendroff method and, therefore, is not repeated here. The two methods are not identical for non-linear problems, though, and the MacCormack method will be investigated in the application of method to the inviscid Burgers equation.

As a second example, consider the second-order, one-dimensional wave equation

$$\frac{\partial^2 u}{\partial t^2} = a^2 \frac{\partial^2 u}{\partial x^2} \tag{6-28}$$

where a, the speed of sound, is assumed constant.

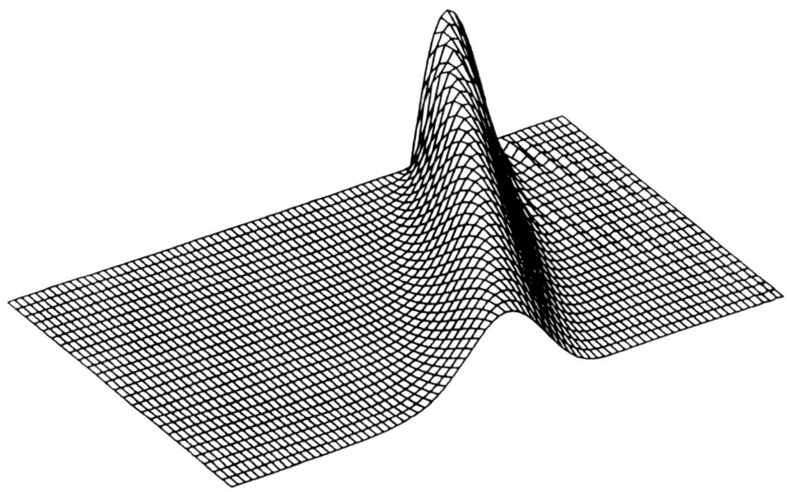

Figure 6-13. Solution of the first-order wave equation by the BTCS implicit method, $c = 1.0$.

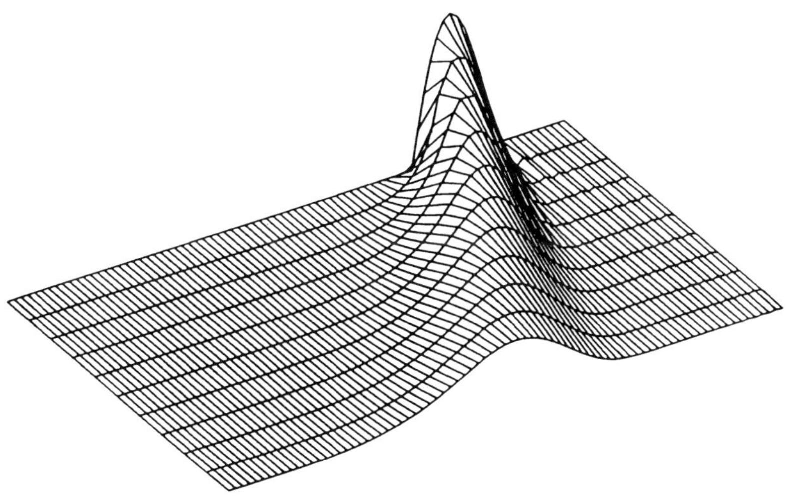

Figure 6-14. Solution of the first-order wave equation by the BTCS implicit method, $c = 2.5$.

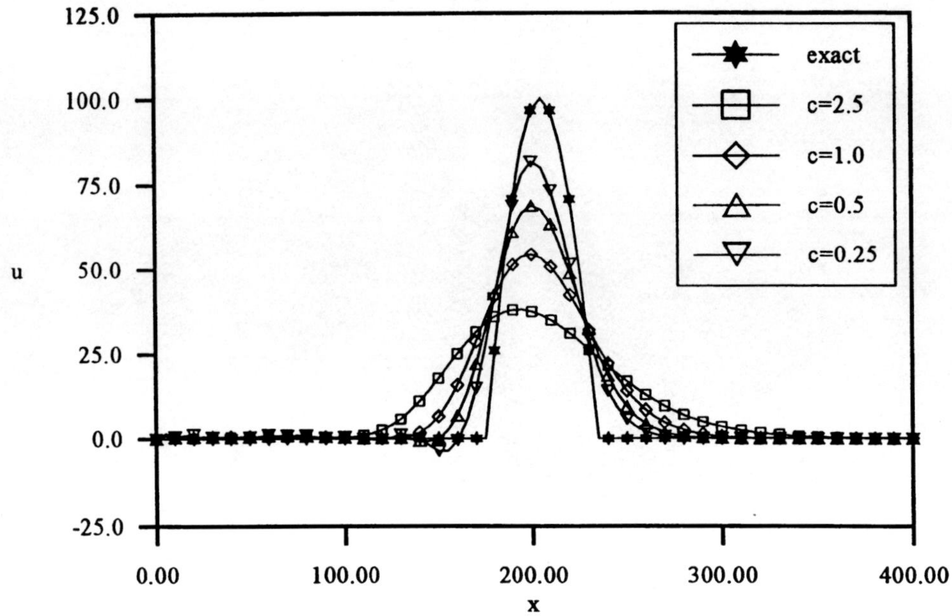

Figure 6-15. Comparison of the solutions of the wave equation by BTCS implicit method at $t = 0.5$ sec for various Courant numbers.

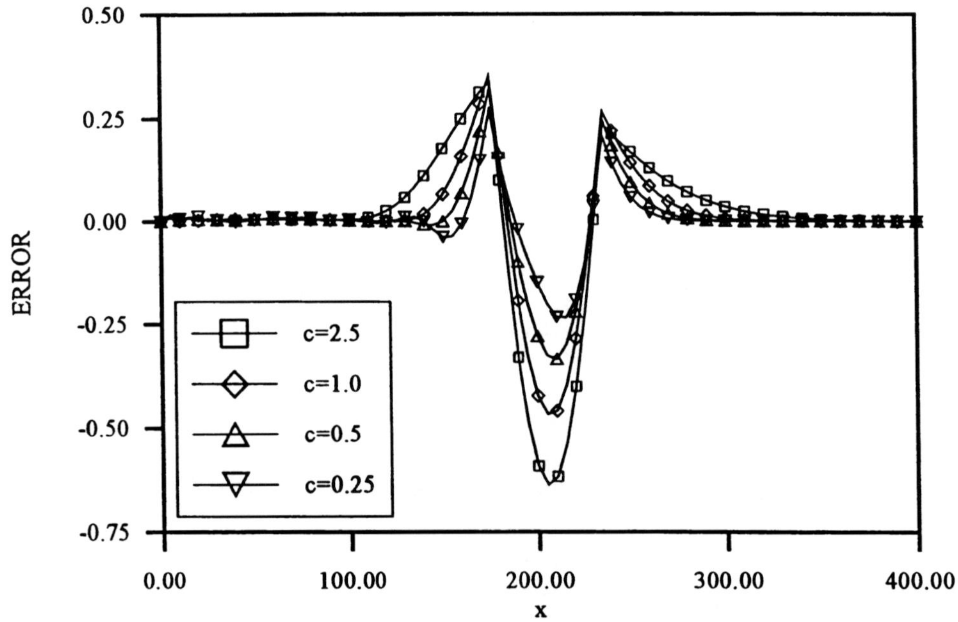

Figure 6-16. Error distributions of the implicit BTCS scheme at $t = 0.5$ sec for several Courant numbers.

For the second-order hyperbolic equation, two sets of initial conditions are required. These conditions may be expressed as

$$u(x,0) = f(x)$$

and

$$\frac{\partial u(x,0)}{\partial t} = g(x)$$

where $f(x)$ and $g(x)$ are specified for a particular problem. For this application, a u distribution similar to the one imposed on the first problem is used. In addition, $g(x) = 0$ is selected. The initial conditions are stated as

$$u(x,0) = \begin{cases} 0.0 & 0 \le x \le 100 \\ 100\left[\sin\pi\left(\frac{x-100}{120}\right)\right] & 100 \le x \le 220 \\ 0.0 & 220 \le x \le 300 \end{cases}$$

and

$$\frac{\partial u(x,0)}{\partial t} = 0.0$$

and the boundary conditions, which represent solid boundaries, are:

$$x = 0 \qquad u(0,t) = 0.0$$

$$x = L \qquad u(L,t) = 0.0$$

Since the application of the midpoint leapfrog method in the first example was not illustrated, it would be beneficial to consider its application to the problem at hand. This method, applied to (6-28), results in the FDE

$$u_i^{n+1} = 2u_i^n - u_i^{n-1} + c^2\left(u_{i-1}^n - 2u_i^n + u_{i+1}^n\right) \tag{6-29}$$

The method is three-step, i.e., the dependent variable appears at three time levels, $n-1$, n, and $n+1$. Therefore, a starter solution is required. To generate a second set of data, consider the second initial condition, namely,

$$\frac{\partial u(x,0)}{\partial t} = 0$$

Using a central differencing one obtains

$$\frac{u_i^{n+1} - u_i^{n-1}}{2\Delta t} = 0 \qquad \text{or}$$

$$u_i^{n+1} = u_i^{n-1} \tag{6-30}$$

Substituting (6-30) into (6-29) yields

$$u_i^{n+1} = 2u_i^n - u_i^{n+1} + c^2\left(u_{i-1}^n - 2u_i^n + u_{i+1}^n\right) \quad \text{or}$$

$$u_i^{n+1} = u_i^n + \frac{1}{2}c^2\left(u_{i-1}^n - 2u_i^n + u_{i+1}^n\right)$$

which can be used as a starter solution. With the first initial condition provided at $t = 0$ ($n = 1$), it follows that

$$u_i^2 = u_i^1 + \frac{1}{2}c^2\left(u_{i-1}^1 - 2u_i^1 + u_{i+1}^1\right)$$

where superscript 2 indicates time level 2. With the computed values of u_i^2 for all i, two sets of data at $n = 1$ and $n = 2$ are available for the solution of (6-29). The stability requirement of the method is $c \leq 1$.

The analytical solution of (6-28) is well known. The solution has the functional form

$$u(x,t) = f(x - at) + g(x + at) \ ,$$

where the solution propagates with constant speed a along lines $x - at = c_1$ and $x + at = c_2$ with slopes $dx/dt = \pm a$. These are the characteristic lines.

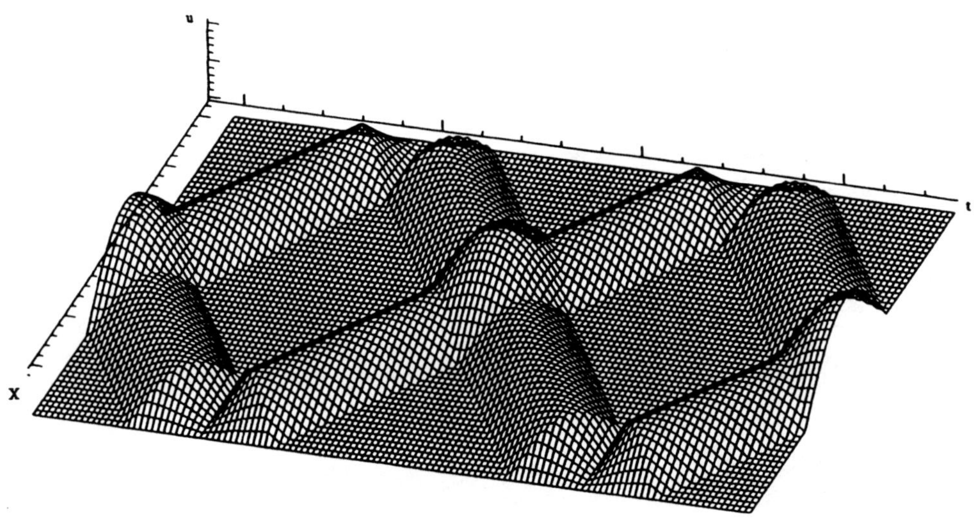

Figure 6-17. Solution of the second-order wave equation by the mid-point leapfrog method.

The numerical solution is presented in Figure 6-17. The initial wave is split into two waves (each with one-half the amplitude of the original wave) with the same wavelength and are propagated in the opposite directions, i.e., one to the right and the other to the left. As the waves reflect from the boundary, the sign of u changes. Note that to clarify the plot in Figure 6-17, all negative u were plotted as positive. The effect of various step sizes expressed by Courant numbers is shown in Figure 6-18. The oscillations (dispersion error) increase as the Courant number decreases. Again, as in the first example, the best solution is obtained for $c = 1$, which is the upper limit imposed by the stability consideration.

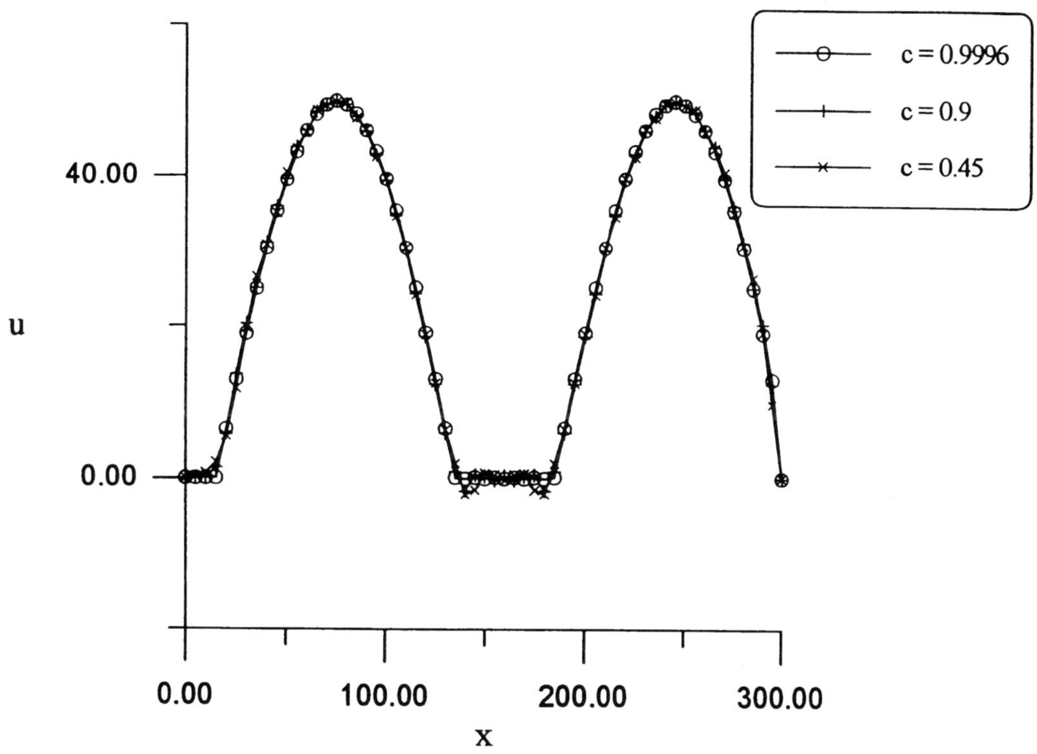

Figure 6-18. Comparison of the solutions of the wave equation by the midpoint leapfrog method at $t = 0.28$ sec. for various Courant numbers.

One further comment. Equation (6-28) is equivalent to the coupled first-order wave equations given by

$$\frac{\partial u}{\partial t} = a\frac{\partial v}{\partial x} \tag{6-31a}$$

$$\frac{\partial v}{\partial t} = a\frac{\partial u}{\partial x} \tag{6-31b}$$

Therefore a solution of the original model equation (6-28) may be obtained by solving the first-order equations (6-31a) and (6-31b).

In conclusion, when one broadly compares the implicit and explicit methods just explored, it is clear that, for linear hyperbolic equations, the explicit formulations provide better solutions than implicit methods. The advantages of implicit methods (which are usually unconditionally stable) are lost, since large step sizes produce poor results.

6.6 Nonlinear Problem

The majority of partial differential equations in fluid mechanics and heat transfer are nonlinear. The simple linear hyperbolic equation just investigated should provide some foundation to approach the nonlinear hyperbolic equations. A classical nonlinear first-order hyperbolic equation is the inviscid Burgers equation, which will be used as a model equation to investigate various solution procedures.

In this section, the numerical techniques presented earlier for the linear problem will be applied to the nonlinear model equation. The inviscid Burgers equation is

$$\frac{\partial u}{\partial t} = -u\frac{\partial u}{\partial x} \tag{6-32}$$

which, in a conservative form, may be expressed as

$$\frac{\partial u}{\partial t} = -\frac{\partial}{\partial x}\left(\frac{u^2}{2}\right) \qquad \text{or}$$

$$\frac{\partial u}{\partial t} = -\frac{\partial E}{\partial x} \tag{6-33}$$

where $E = u^2/2$. Equation (6-32) can be interpreted as the propagation of a wave with each point having a different velocity and eventually forming a discontinuity in the domain. This is similar to the formation of shock waves by a series of weak compression waves.

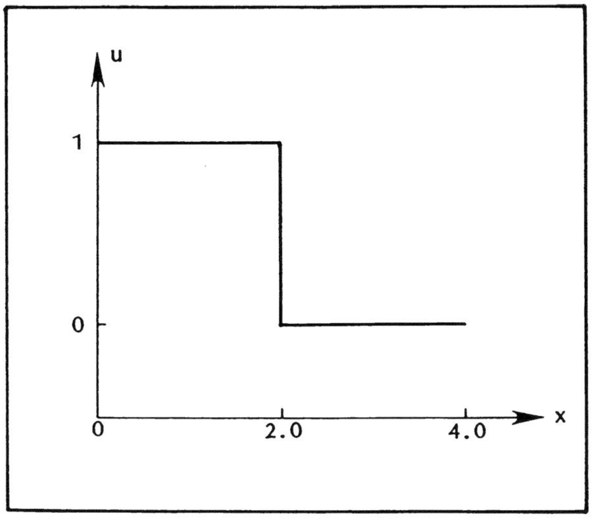

Figure 6-19. Discontinuity used as initial condition.

The discussion of various numerical schemes will be accompanied by their applications to the example problem posed as follows. A discontinuity described by the function

$$u(x,0) = 1 \qquad 0 \le x \le 2.0$$

$$u(x,0) = 0 \qquad 2.0 \le x \le 4.0$$

shown in Figure 6-19 is to be used as initial data to investigate its propagation throughout the domain.

6.6.1 The Lax Method

This explicit method uses forward time differencing of $O(\Delta t)$ and central space differencing of $O(\Delta x)^2$. The corresponding FDE for model equation (6-33) is

$$\frac{u_i^{n+1} - u_i^n}{\Delta t} = -\frac{E_{i+1}^n - E_{i-1}^n}{2\Delta x}$$

For stability consideration, u_i^n is replaced by its average at the neighboring points. Thus,

$$u_i^{n+1} = \frac{1}{2}\left(u_{i+1}^n + u_{i-1}^n\right) - \frac{\Delta t}{2\Delta x}\left(E_{i+1}^n - E_{i-1}^n\right) \quad \text{or}$$

$$u_i^{n+1} = \frac{1}{2}\left(u_{i+1}^n + u_{i-1}^n\right) - \frac{\Delta t}{4\Delta x}\left[\left(u_{i+1}^n\right)^2 - \left(u_{i-1}^n\right)^2\right] \tag{6-34}$$

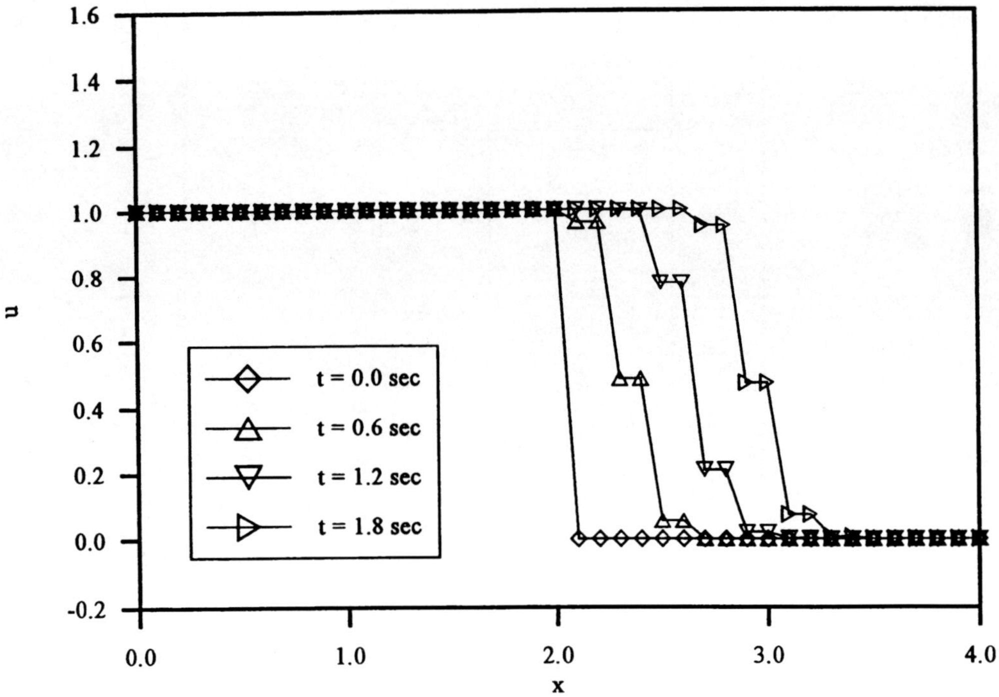

Figure 6-20. Solution of the inviscid Burgers equation by the Lax explicit method, $\Delta x = 0.1$ and $\Delta t = 0.1$.

The solution will be stable when

$$\left| \frac{\Delta t}{\Delta x} u_{\max} \right| \leq 1 \tag{6-35}$$

Since the method is first-order, it is expected that the errors will be dissipative. The solution for $\Delta x = 0.1$ and $\Delta t = 0.1$ at several time intervals is presented in Table 6.5 and Figure 6-20, which clearly reflects the dissipative nature of the solution. Note that the discontinuity is smeared over several grid points. The effect of the step size (and the corresponding Courant number) is shown in Figure 6-21, which indicates the solutions at $t = 1.8$ sec for values of $\Delta t/\Delta x = 1.0$ and $\Delta t/\Delta x = 0.5$. As in the linear problem, the best result is obtained for a Courant number of one.

6.6.2 The Lax-Wendroff Method

Here the finite difference formulation of the method is derived from a Taylor series expansion, as in the linear case. Consider the expansion

$$u_i^{n+1} = u_i^n + \frac{\partial u}{\partial t}\Delta t + \frac{\partial^2 u}{\partial t^2}\frac{(\Delta t)^2}{2!} + \cdots \tag{6-36}$$

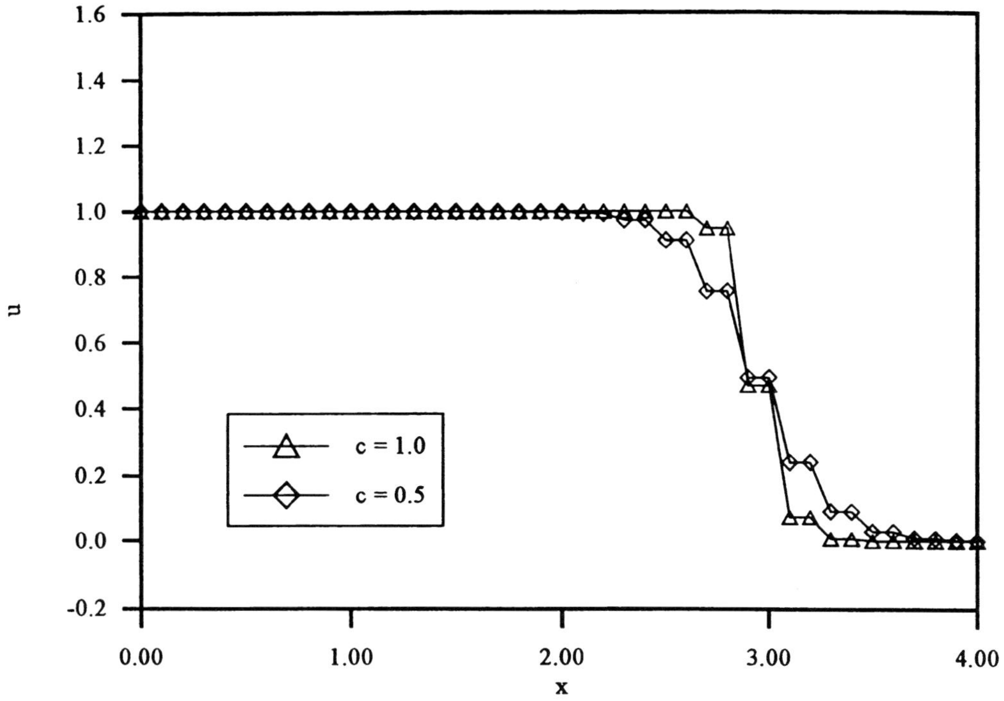

Figure 6-21. Effect of step sizes on the solution of Burgers equation by the Lax method.

The model equation is

$$\frac{\partial u}{\partial t} = -\frac{\partial E}{\partial x} \tag{6-37}$$

Therefore,

$$\frac{\partial^2 u}{\partial t^2} = -\frac{\partial}{\partial t}\left(\frac{\partial E}{\partial x}\right) = -\frac{\partial}{\partial x}\left(\frac{\partial E}{\partial t}\right)$$

But

$$\frac{\partial E}{\partial t} = \frac{\partial E}{\partial u}\frac{\partial u}{\partial t} = \frac{\partial E}{\partial u}\left(-\frac{\partial E}{\partial x}\right) = -A\left(\frac{\partial E}{\partial x}\right)$$

where $A = \partial E/\partial u$ is known as the Jacobian. Therefore,

$$\frac{\partial^2 u}{\partial t^2} = -\frac{\partial}{\partial x}\left(-A\frac{\partial E}{\partial x}\right) = \frac{\partial}{\partial x}\left(A\frac{\partial E}{\partial x}\right) \tag{6-38}$$

For the model equation, where

$$E = \frac{1}{2}u^2$$

then

$$A = \frac{\partial E}{\partial u} = \frac{\partial}{\partial u}\left(\frac{1}{2}u^2\right) = u$$

After the substitution of (6-37) and (6-38) into Taylor series expansion (6-36), one obtains

$$u_i^{n+1} = u_i^n + \left(-\frac{\partial E}{\partial x}\right)\Delta t + \frac{\partial}{\partial x}\left(A\frac{\partial E}{\partial x}\right)\frac{(\Delta t)^2}{2!} + O(\Delta t)^3 \qquad \text{or}$$

$$\frac{u_i^{n+1} - u_i^n}{\Delta t} = -\left(\frac{\partial E}{\partial x}\right) + \frac{\partial}{\partial x}\left(A\frac{\partial E}{\partial x}\right)\frac{\Delta t}{2} + O(\Delta t)^2$$

Now, the spatial derivatives are approximated by central differencing of order two, resulting in the FDE

$$\frac{u_i^{n+1} - u_i^n}{\Delta t} = -\frac{E_{i+1}^n - E_{i-1}^n}{2\Delta x} + \left[\frac{\left(A\frac{\partial E}{\partial x}\right)_{i+\frac{1}{2}}^n - \left(A\frac{\partial E}{\partial x}\right)_{i-\frac{1}{2}}^n}{\Delta x}\right]\frac{\Delta t}{2}$$

At this point, the approximation

$$\frac{\left(A\frac{\partial E}{\partial x}\right)_{i+\frac{1}{2}}^n - \left(A\frac{\partial E}{\partial x}\right)_{i-\frac{1}{2}}^n}{\Delta x} = \frac{A_{i+\frac{1}{2}}^n \frac{E_{i+1}^n - E_i^n}{\Delta x} - A_{i-\frac{1}{2}}^n \frac{E_i^n - E_{i-1}^n}{\Delta x}}{\Delta x}$$

for the last term is incorporated and the Jacobians are evaluated at the midpoints, which results in

$$\frac{\frac{1}{2\Delta x}\left(A_{i+1}^n + A_i^n\right)\left(E_{i+1}^n - E_i^n\right) - \frac{1}{2\Delta x}\left(A_i^n + A_{i-1}^n\right)\left(E_i^n - E_{i-1}^n\right)}{\Delta x}$$

Note that for the problem at hand, $A = u$. Finally, the FDE is arranged as follows:

$$u_i^{n+1} = u_i^n - \frac{\Delta t}{2\Delta x}\left(E_{i+1}^n - E_{i-1}^n\right)$$

$$+ \frac{(\Delta t)^2}{4(\Delta x)^2}\left[\left(u_{i+1}^n + u_i^n\right)\left(E_{i+1}^n - E_i^n\right) - \left(u_i^n + u_{i-1}^n\right)\left(E_i^n - E_{i-1}^n\right)\right] \qquad (6\text{-}39)$$

The method is second-order, with a stability requirement of $|u_{\max}\Delta t/\Delta x| \leq 1$. Application of the method to the sample problem yields the solution shown in Figure 6-22 and Table 6.6, where the step sizes were $\Delta x = 0.1$ and $\Delta t = 0.1$. These step sizes correspond to a Courant number of one. The dispersion error is evident by the presence of oscillations in the neighborhood of the discontinuity. Two solutions obtained with various step sizes (and, therefore, Courant numbers) at $t = 1.8$ sec are compared in Figure 6-23. When $\Delta t/\Delta x = 0.5$, the oscillations are larger and propagate further from the discontinuity. In general, as the Courant number decreases, the solution degenerates; the best solution is obtained at a Courant number of one.

6.6.3 The MacCormack Method

This multi-level method applied to the model equation yields the finite difference equations

$$u_i^* = u_i^n - \frac{\Delta t}{\Delta x}\left(E_{i+1}^n - E_i^n\right) \tag{6-40}$$

and

$$u_i^{n+1} = \frac{1}{2}\left[u_i^n + u_i^* - \frac{\Delta t}{\Delta x}\left(E_i^* - E_{i-1}^*\right)\right] \tag{6-41}$$

The stability requirement of the method is $|u_{\max}\Delta t/\Delta x| \leq 1$. The solution obtained at several time intervals with $\Delta x = 0.1$ and $\Delta t = 0.1$ is shown in Figure 6-24 and presented in Table 6-7. This solution, unlike the solution of the Lax method and the Lax-Wendroff method, is well behaved. This is due to the splitting procedure

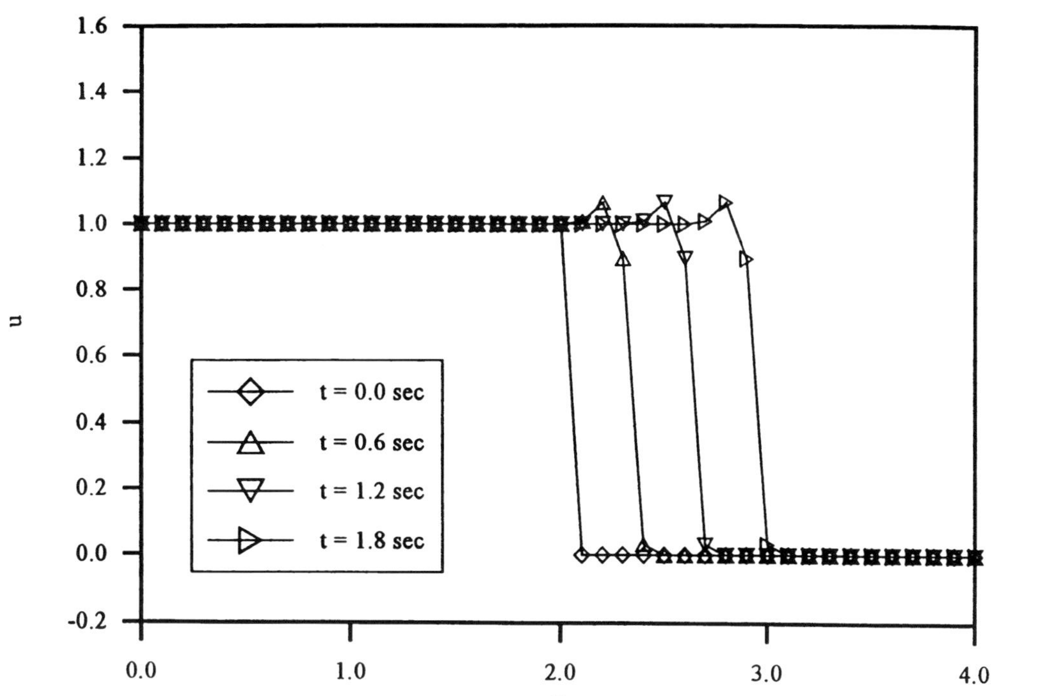

Figure 6-22. Solution of the inviscid Burgers equation by the Lax-Wendroff explicit method, $\Delta x = 0.1$ and $\Delta t = 0.1$.

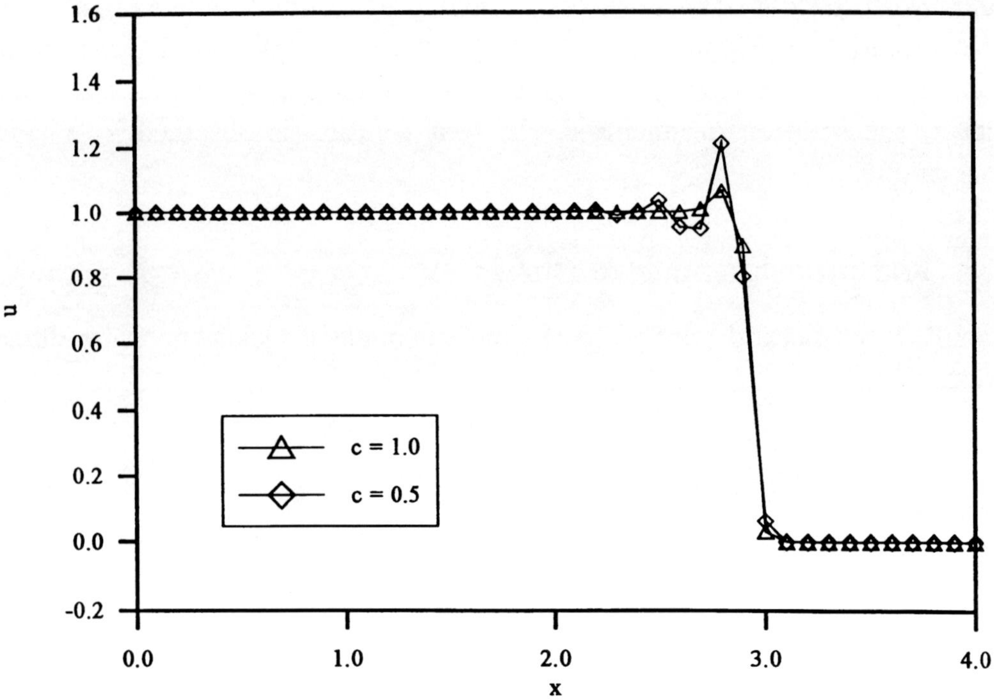

Figure 6-23. Effect of step sizes on the solution of Burgers equation by the Lax-Wendroff method.

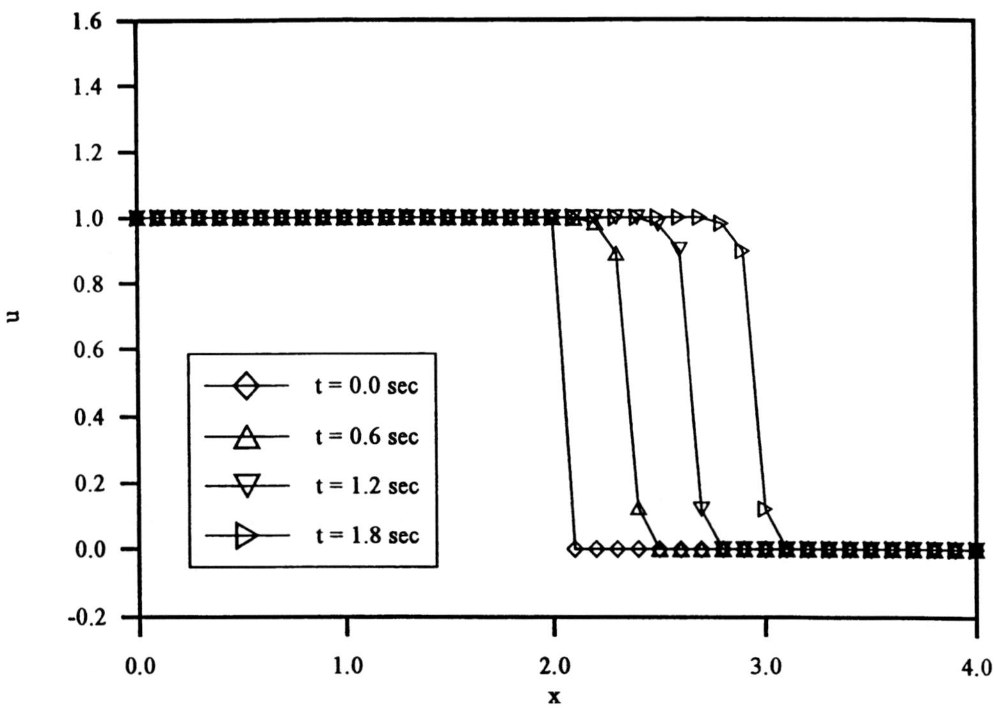

Figure 6-24. Solution of the inviscid Burgers equation by the MacCormack explicit method, $\Delta x = 0.1$ and $\Delta t = 0.1$.

and corresponding forward, backward differencing used to approximate the spatial derivative. Note that the solution is not identical to that of the Lax-Wendroff method, as it was in the linear problem (see Figures 6-22 and 6-24). As with the previous methods, the solution degrades as the Courant number decreases from the maximum allowable value of one (see Figure 6-25). As expected, the best solution is obtained with $\Delta t/\Delta x = 1.0$, i.e., when the Courant number is one.

6.6.4 The Beam and Warming Implicit Method

To understand the development of this method, start with the Taylor series expansions

$$u(x,\, t+\Delta t) = u(x,t) + \left.\frac{\partial u}{\partial t}\right|_{x,t} \Delta t + \left.\frac{\partial^2 u}{\partial t^2}\right|_{x,t} \frac{(\Delta t)^2}{2!} + O(\Delta t)^3 \qquad \text{(6-42)}$$

and

$$u(x,t) = u(x,\, t+\Delta t) - \left.\frac{\partial u}{\partial t}\right|_{x,t+\Delta t} \Delta t + \left.\frac{\partial^2 u}{\partial t^2}\right|_{x,t+\Delta t} \frac{(\Delta t)^2}{2!} + O(\Delta t)^3 \qquad \text{(6-43)}$$

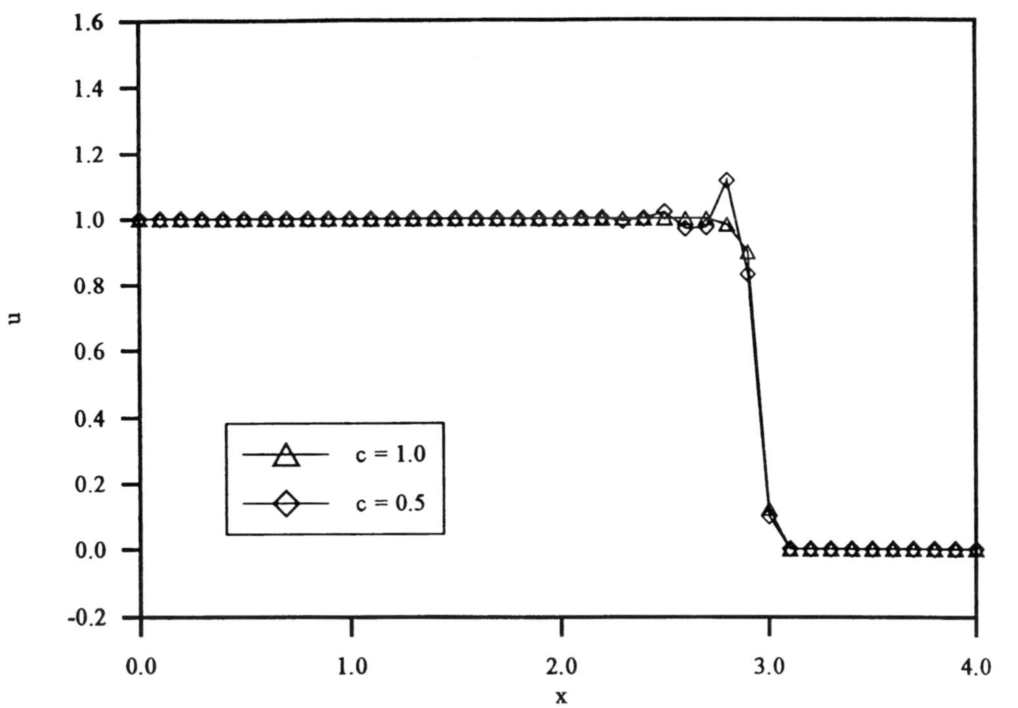

Figure 6-25. Effect of step sizes on the solution of Burgers equation by MacCormack method.

Subtracting (6-43) from (6-42), one obtains

$$2u(x,\, t+\Delta t) = 2u(x,t) + \left.\frac{\partial u}{\partial t}\right|_{x,t} \Delta t + \left.\frac{\partial u}{\partial t}\right|_{x,t+\Delta t} \Delta t$$

$$+ \left.\frac{\partial^2 u}{\partial t^2}\right|_{x,t} \frac{(\Delta t)^2}{2!} - \left.\frac{\partial^2 u}{\partial t^2}\right|_{x,t+\Delta t} \frac{(\Delta t)^2}{2!} + O(\Delta t)^3$$

or, in terms of indices,

$$u_i^{n+1} = u_i^n + \frac{1}{2}\left[\left(\frac{\partial u}{\partial t}\right)_i^n + \left(\frac{\partial u}{\partial t}\right)_i^{n+1}\right]\Delta t + \left[\left(\frac{\partial^2 u}{\partial t^2}\right)_i^n - \left(\frac{\partial^2 u}{\partial t^2}\right)_i^{n+1}\right]\frac{(\Delta t)^2}{2!} + O(\Delta t)^3$$

Additional substitution is considered for $(\partial^2 u/\partial t^2)_i^{n+1}$ using the equation

$$\left(\frac{\partial^2 u}{\partial t^2}\right)_i^{n+1} = \left(\frac{\partial^2 u}{\partial t^2}\right)_i^n + \frac{\partial}{\partial t}\left(\frac{\partial^2 u}{\partial t^2}\right)_i^n \Delta t + O(\Delta t)^2$$

Thus,

$$u_i^{n+1} = u_i^n + \frac{1}{2}\left[\left(\frac{\partial u}{\partial t}\right)_i^n + \left(\frac{\partial u}{\partial t}\right)_i^{n+1}\right]\Delta t + O(\Delta t)^3 \tag{6-44}$$

For the model equation, recall that

$$\frac{\partial u}{\partial t} = -\frac{\partial E}{\partial x}$$

which, after substituting into (6-44) and rearranging the terms, becomes

$$\frac{u_i^{n+1} - u_i^n}{\Delta t} = -\frac{1}{2}\left[\left(\frac{\partial E}{\partial x}\right)_i^n + \left(\frac{\partial E}{\partial x}\right)_i^{n+1}\right] + O(\Delta t)^2 \tag{6-45}$$

which is written in this form to show the method's second-order accuracy in time.

At this point, a distinct difference in the implicit and explicit formulation of the finite difference equations for the nonlinear model equation is recognized. Since the nonlinear term $E = u^2/2$ was applied at the known time level n, the resulting FDE in explicit formulation was linear. On the other hand, the resulting FDE in implicit formulation is nonlinear and, therefore, a procedure is used to linearize the FDE.

From Taylor series expansion,

$$E(t+\Delta t) = E(t) + \frac{\partial E}{\partial t}\Delta t + O(\Delta t)^2$$

$$= E(t) + \frac{\partial E}{\partial u}\frac{\partial u}{\partial t}\Delta t + O(\Delta t)^2$$

or, in terms of indices,

$$E^{n+1} = E^n + \frac{\partial E}{\partial u}\left(\frac{u^{n+1}-u^n}{\Delta t}\right)\Delta t + O(\Delta t)^2$$

Recall that $\partial E/\partial u = A$ is the Jacobian (for this model equation, $A = u$). Therefore, (taking the partial derivative of the equation above)

$$\left(\frac{\partial E}{\partial x}\right)^{n+1} = \left(\frac{\partial E}{\partial x}\right)^{n} + \frac{\partial}{\partial x}\left[A\left(u^{n+1}-u^n\right)\right]$$

Substitution into (6-45) yields

$$\frac{u_i^{n+1}-u_i^n}{\Delta t} = -\frac{1}{2}\left\{\left(\frac{\partial E}{\partial x}\right)_i^n + \left(\frac{\partial E}{\partial x}\right)_i^n + \frac{\partial}{\partial x}\left[A\left(u_i^{n+1}-u_i^n\right)\right]\right\}$$

or

$$u_i^{n+1} = u_i^n - \frac{1}{2}\Delta t\left\{2\left(\frac{\partial E}{\partial x}\right)_i^n + \frac{\partial}{\partial x}\left[A\left(u_i^{n+1}-u_i^n\right)\right]\right\} \tag{6-46}$$

For the term $\frac{\partial}{\partial x}\left[A\left(u_i^{n+1}-u_i^n\right)\right]$, a second-order central differencing is used so that

$$\frac{\partial}{\partial x}\left[A\left(u_i^{n+1}-u_i^n\right)\right] = \frac{A_{i+1}^n u_{i+1}^{n+1} - A_{i-1}^n u_{i-1}^{n+1}}{2\Delta x} - \frac{A_{i+1}^n u_{i+1}^n - A_{i-1}^n u_{i-1}^n}{2\Delta x}$$

Note that lagging of the Jacobian is employed, i.e., the value of A at the known time level n is used to produce a linear equation. Finally, (6-46) is written by the following FDE:

$$u_i^{n+1} = u_i^n - \frac{1}{2}\Delta t\left[2\frac{E_{i+1}^n - E_{i-1}^n}{2\Delta x} + \frac{A_{i+1}^n u_{i+1}^{n+1} - A_{i-1}^n u_{i-1}^{n+1}}{2\Delta x} - \frac{A_{i+1}^n u_{i+1}^n - A_{i-1}^n u_{i-1}^n}{2\Delta x}\right]$$

This equation may be rearranged and represented as a tridiagonal system; thus,

$$-\frac{\Delta t}{4\Delta x}A^n_{i-1}u^{n+1}_{i-1} + u^{n+1}_i + \frac{\Delta t}{4\Delta x}A^n_{i+1}u^{n+1}_{i+1} =$$

$$u^n_i - \frac{1}{2}\frac{\Delta t}{\Delta x}\left(E^n_{i+1} - E^n_{i-1}\right) + \frac{\Delta t}{4\Delta x}A^n_{i+1}u^n_{i+1} - \frac{\Delta t}{4\Delta x}A^n_{i-1}u^n_{i-1} \tag{6-47}$$

The resulting finite difference equation is second-order accurate and is unconditionally stable.

The solution of the proposed problem at several time intervals is shown in Figure 6-26 and Table 6-8. Because the method is second-order accurate, the solution has a dispersion error that is indicated by oscillations within the domain. Indeed, the oscillations are so large that the solution is clearly unacceptable. To reduce the oscillations, a fourth-order smoothing (damping) term is explicitly added to the FDE. Since the added damping term is fourth-order, it does not affect the second-order accuracy of the method; however, the solution will be stable only if $0 < \varepsilon_e \leq 0.125$.

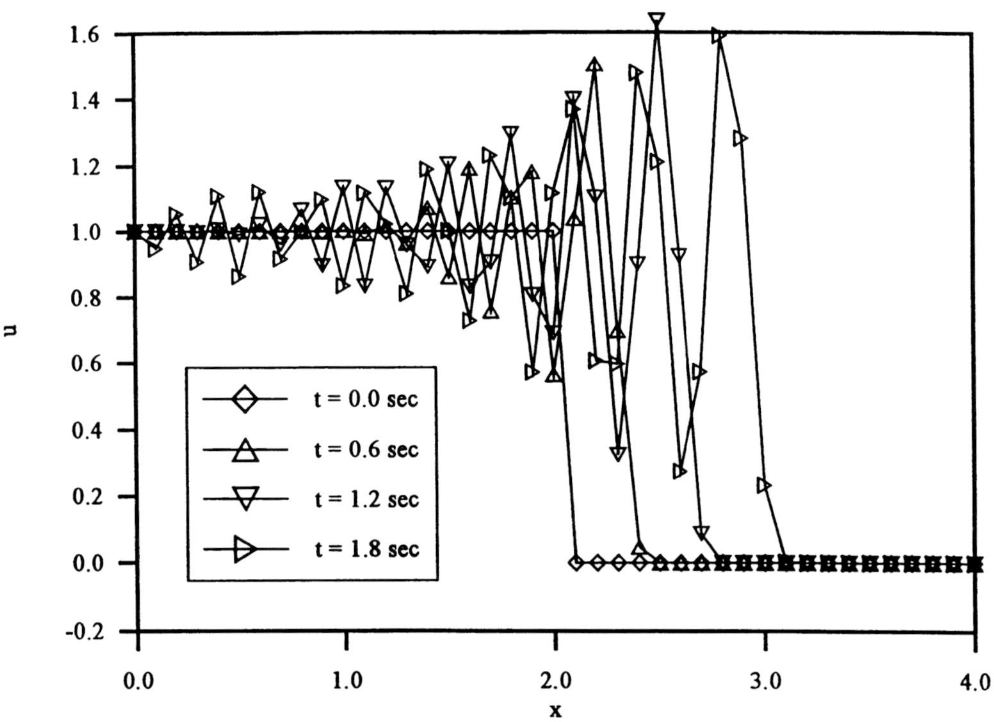

Figure 6-26. Solution of the inviscid Burgers equation by the Beam and Warming implicit method, $\Delta x = 0.1$ and $\Delta t = 0.1$.

The fourth-order damping term has the form

$$D = -\varepsilon_e \left(u_{i+2}^n - 4u_{i+1}^n + 6u_i^n - 4u_{i-1}^n + u_{i-2}^n \right) \tag{6-48}$$

Further discussion on various damping terms is provided in Section 6.7. The resulting solution with the damping term and $\varepsilon_e = 0.1$ is shown in Figure 6-27 and Table 6-9. The oscillations near the discontinuity still persist; however, the overall solution is reasonable. The effect of various step sizes on the solution is shown in Figure 6-28. The solutions are at $t = 1.8$ sec and were generated with the damping factor of $\varepsilon_e = 0.1$. For a large time step, i.e., when $\Delta t/\Delta x = 2.0$, the solution has too great a dispersion error, as the large oscillations near the discontinuity suggest. Here, as with the linear hyperbolic equation, the drawbacks in using the implicit schemes to solve nonlinear hyperbolic equations are illustrated. The advantages of implicit methods in terms of stability and the large step sizes they allow are lost, since the solution for large step sizes becomes unacceptable.

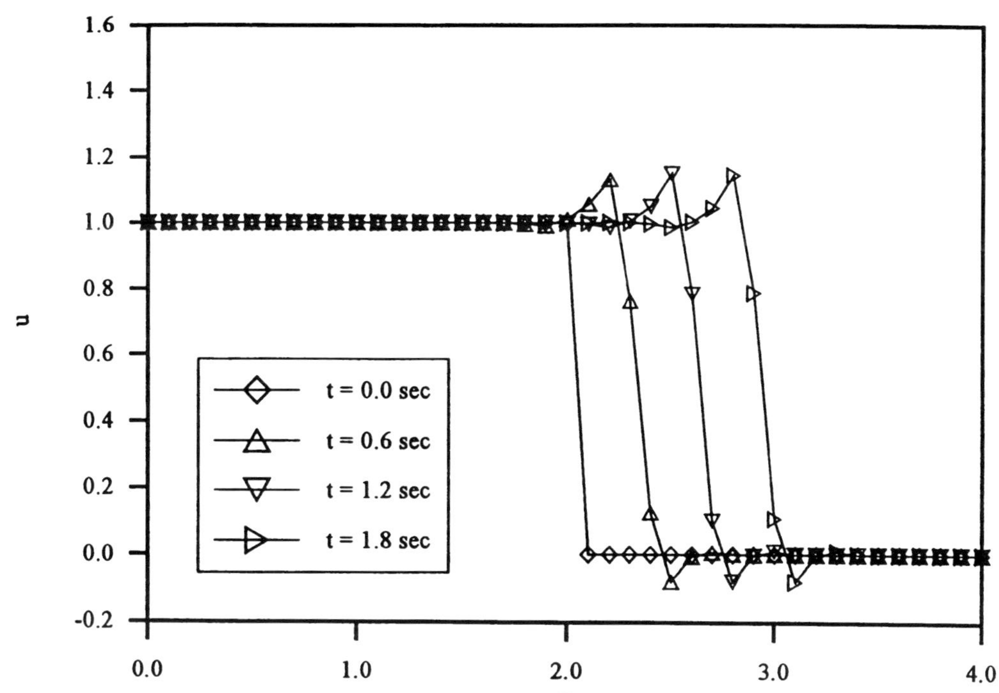

Figure 6-27. Solution of the inviscid Burgers equation by the Beam and Warming implicit method with damping of $\varepsilon_e = 0.1$, $\Delta x = 0.1$ and $\Delta t = 0.1$.

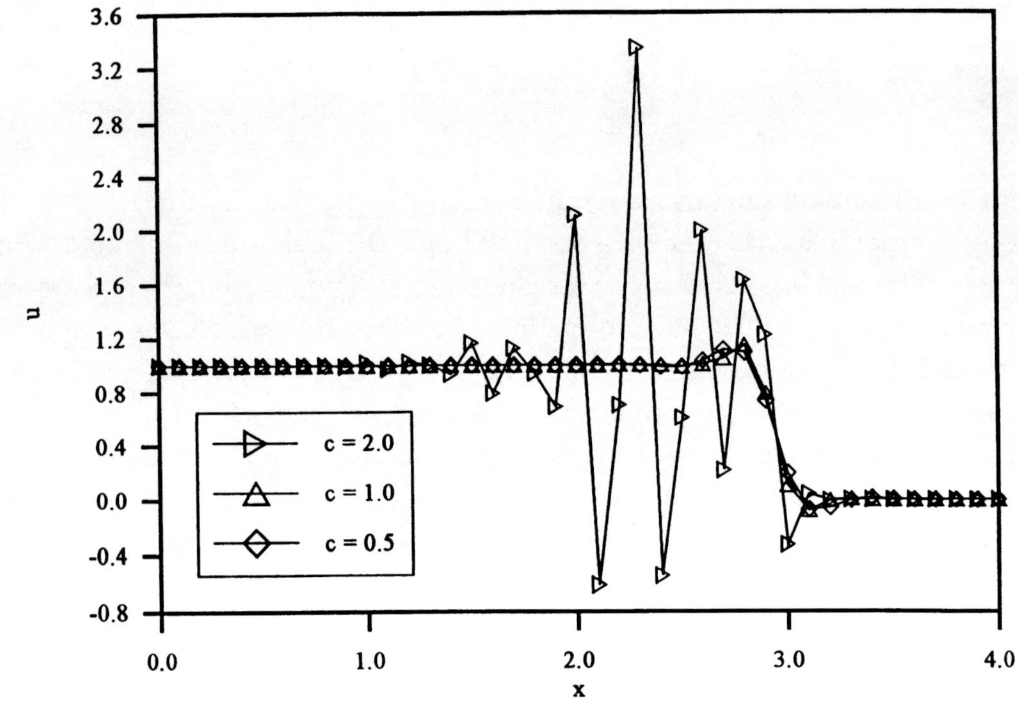

Figure 6-28. Effect of step sizes on the solution of Burgers equation by the Beam and Warming method with $\varepsilon_e = 0.1$.

6.6.5 Explicit First-Order Upwind Scheme

A first-order forward difference approximation in time and a first-order backward difference approximation in space yield the following equation.

$$\frac{u_i^{n+1} - u_i^n}{\Delta t} = -\frac{E_i^n - E_{i-1}^n}{\Delta x}$$

or

$$u_i^{n+1} = u_i^n - \frac{\Delta t}{\Delta x}\left(E_i^n - E_{i-1}^n\right) \tag{6-49}$$

The formulation is $O(\Delta t, \Delta x)$, and it is stable for $c \leq 1$. A typical solution is shown in Figure 6-29.

6.6.6 Implicit First-Order Upwind Scheme

An implicit first-order upwind scheme can be written as

$$\frac{u_i^{n+1} - u_i^n}{\Delta t} = -\frac{1}{\Delta x}\left[\frac{\left(u_i^{n+1}\right)^2}{2} - \frac{\left(u_{i-1}^{n+1}\right)^2}{2}\right] \tag{6-50}$$

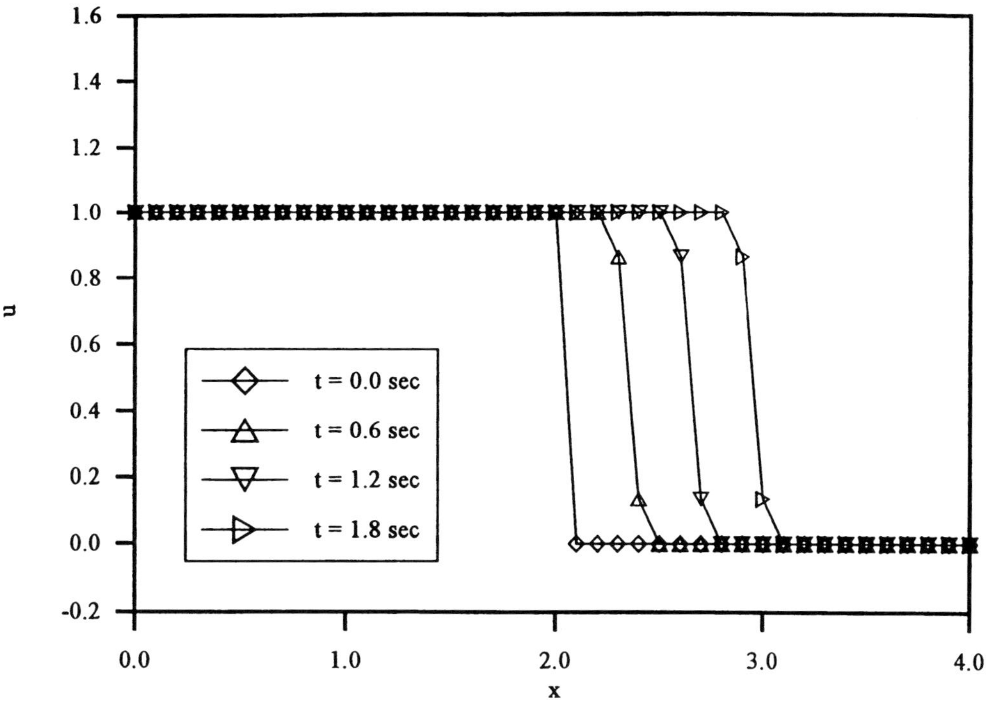

Figure 6-29. Solution of the inviscid Burgers equation by the explicit first-order upwind scheme, $\Delta x = 0.1$ and $\Delta t = 0.1$.

Now a simple lagging scheme is used to linearize Equation (6-50) to obtain

$$\frac{u_i^{n+1} - u_i^n}{\Delta t} = -\frac{1}{2\Delta x}\left(u_i^n u_i^{n+1} - u_{i-1}^n u_{i-1}^{n+1}\right)$$

which can be rearranged in a bidiagonal form as

$$\left(\frac{\Delta t}{2\Delta x}u_{i-1}^n\right)u_{i-1}^{n+1} - \left(1 + \frac{\Delta t}{2\Delta x}u_i^n\right)u_i^{n+1} = -u_i^n \tag{6-51}$$

The solution at several time levels is shown in Figure 6-30. Observe that the error is larger for the solution obtained by the implicit scheme compared to the equivalent explicit scheme. Recall that a similar conclusion was reached earlier in the analysis of linear equations in Section 6.5.

6.6.7 Runge-Kutta Method

A numerical scheme commonly used to solve initial value problems for ODE's is the Runge-Kutta (RK) method. This scheme essentially utilizes the weighted average of several solutions over the interval Δt in order to improve accuracy of solution. To clarify the statement made above, consider the model equation (6-33),

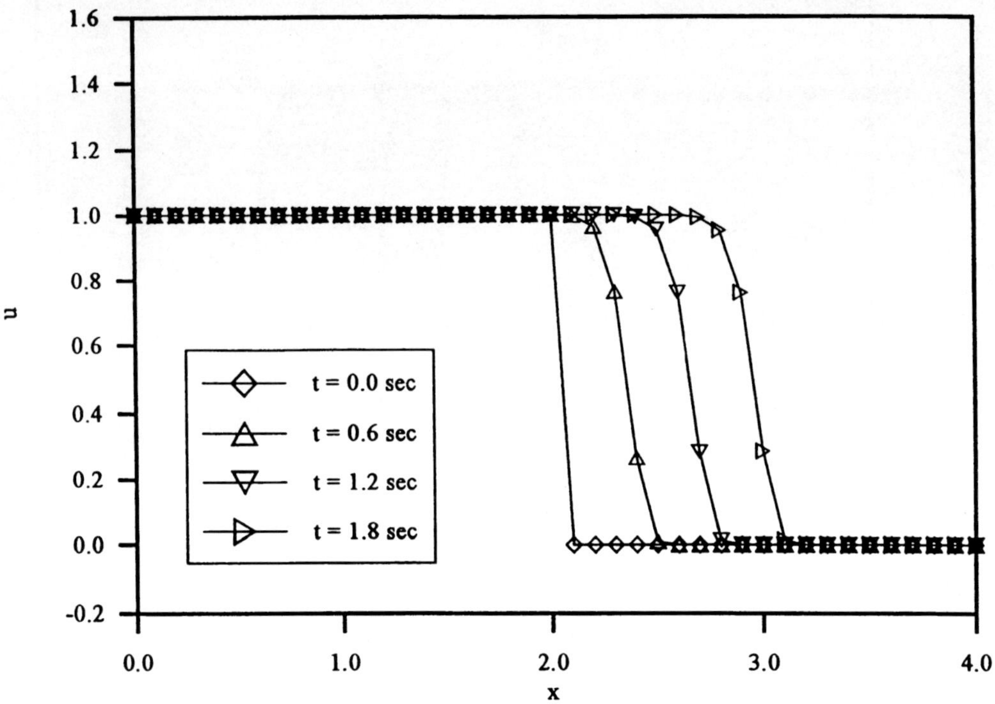

Figure 6-30. Solution of the inviscid Burgers equation by the implicit first-order upwind scheme $\Delta x = 0.1$ and $\Delta t = 0.1$.

$$\frac{\partial u}{\partial t} = -\frac{\partial E}{\partial x}$$

A first-order forward differencing provides

$$u_i^{n+1} = u_i^n - \Delta t \left(\frac{\partial E}{\partial x}\right)^n \tag{6-52}$$

Several approximations are available to evaluate the convective term $(\partial E/\partial x)^n$. For example, one may use central differencing of second order to obtain Euler's FTCS explicit scheme. Recall that, due to stability consideration, a modification was introduced and the resulting FDE is known as *the Lax method.*

Now consider evaluating the convective term $(\partial E/\partial x)$ at several time subintervals within a time interval Δt and subsequently obtain the final solution by averaging of these values. For example, first evaluate a value of u_i using a time step of $\Delta t/2$ designated by $u_i^{(2)}$. The superscripts with parentheses will be used to designate values at a time level of $\alpha\Delta t$ within the specified time step where $0 < \alpha < 1$. It is common to set the value of variable at time level of "n" to the time level designated by (1).

Now, the formulation is written as

$$u_i^{(1)} = u_i^n \tag{6-53}$$

$$u_i^{(2)} = u_i^n - \frac{\Delta t}{2}\left(\frac{\partial E}{\partial x}\right)_i^n \tag{6-54}$$

Note that Equation (6-53) is written in order to designate it as the first stage of Runge-Kutta method and to be consistent with the order of the method as it will be seen shortly. In fact, it is not required to define it in the programming process.

Again, as discussed previously, any spatial finite difference approximation can be used for the convective term $(\partial E/\partial x)$. Once $u_i^{(2)}$ is determined, a final solution for u_i^{n+1} is computed from

$$u_i^{n+1} = u_i^n - \Delta t\left\{\frac{1}{2}\left[\left(\frac{\partial E}{\partial x}\right)^{(1)} + \left(\frac{\partial E}{\partial x}\right)^{(2)}\right]\right\} \tag{6-55}$$

The scheme given by Equations (6-53) through (6-55) is known as a *two-stage Runge-Kutta method* and is second-order accurate; it will be referred to as a *second-order Runge-Kutta method.* The order of accuracy is determined by comparison of the scheme with Taylor series expansion. Note that, in Equation (6-55), equal weight is given in the averaging process of the convective term. In fact, Equation (6-55) can be expressed in a general form as

$$u_i^{n+1} = u_i^n - \Delta t\left[a\left(\frac{\partial E}{\partial x}\right)^{(1)} + b\left(\frac{\partial E}{\partial x}\right)^{(2)}\right] \tag{6-56}$$

where a and b are weighted factors and, in Equation (6-55), were set to 0.5. Note that the sum of weighted factors in the averaging process must be equal to one. That is, the sum of a and b in Equation (6-56) must be one. Furthermore, note that in evaluation of $u_i^{(2)}$ in Equation (6-54) the time interval was set at midpoint, i.e., $\Delta t/2$. In fact, the coefficient of Δt can be set to any value between zero and one. Therefore, it is seen that several (indeed infinite) number of second-order RK schemes can be developed. The second-order scheme given by Equations (6-53) through (6-55) is the most common.

A general Q-stage Runge-Kutta scheme can be written as follows:

$$u_i^{(1)} = u_i^n \tag{6-57}$$

$$u_i^{(2)} = u_i^n - \alpha_2 \Delta t \left(\frac{\partial E}{\partial x}\right)_i^{(1)} \tag{6-58}$$

$$u_i^{(3)} = u_i^n - \alpha_3 \Delta t \left(\frac{\partial E}{\partial x}\right)_i^{(2)} \tag{6-59}$$

$$u_i^{(4)} = u_i^n - \alpha_4 \Delta t \left(\frac{\partial E}{\partial x}\right)_i^{(3)} \tag{6-60}$$

$$\vdots$$

$$u_i^{(Q)} = u_i^n - \alpha_Q \Delta t \left(\frac{\partial E}{\partial x}\right)_i^{(Q-1)} \tag{6-61}$$

and

$$u_i^{n+1} = u_i^n - \Delta t \left[\sum_{q=1}^{q=Q} \beta_q \left(\frac{\partial E}{\partial x}\right)^q\right] \tag{6-62}$$

Note that β_q represent the weight factors such that $\sum_{q=1}^{q=Q} \beta_q = 1$ and the coefficient α's are specified within zero and one.

Among a variety of RK schemes, the fourth-order RK scheme is perhaps the most commonly used. A fourth-order scheme for Equation (6-33) can be written as

$$u_i^{(1)} = u_i^n \tag{6-63}$$

$$u_i^{(2)} = u_i^n - \frac{1}{2}\Delta t \left(\frac{\partial E}{\partial x}\right)_i^{(1)} \tag{6-64}$$

$$u_i^{(3)} = u_i^n - \frac{1}{2}\Delta t \left(\frac{\partial E}{\partial x}\right)_i^{(2)} \tag{6-65}$$

$$u_i^{(4)} = u_i^n - \Delta t \left(\frac{\partial E}{\partial x}\right)_i^{(3)} \tag{6-66}$$

and

$$u_i^{n+1} = u_i^n - \Delta t \left[\frac{1}{6}\left(\frac{\partial E}{\partial x}\right)^{(1)} + \frac{1}{3}\left(\frac{\partial E}{\partial x}\right)^{(2)} + \frac{1}{3}\left(\frac{\partial E}{\partial x}\right)^{(3)} + \frac{1}{6}\left(\frac{\partial E}{\partial x}\right)^{(4)}\right] \tag{6-67}$$

Runge-Kutta schemes are typically expressed in explicit form. Implicit RK schemes are computationally expensive and are rarely used.

Some of the advantages of Runge-Kutta schemes are:

1. Runge-Kutta schemes are usually expressed as explicit formulations, and, therefore, the schemes are easy to program.

2. Runge-Kutta schemes possess better stability criteria than comparable explicit schemes. Recall that, for most linear hyperbolic equations, the stability requirement of explicit formulations is $c \leq 1$. It can be shown that, for a fourth-order RK scheme with central differencing of convective term, the stability requirement is $c \leq 2\sqrt{2}$. It is emphasized, however, that the scheme may be unstable for nonlinear hyperbolic equations when central differencing of convective term is used. Therefore, damping terms are typically included to stabilize the solution. To reduce computational cost, these damping terms may be evaluated only once at time level of n and augmented to the solution after the final stage.

The primary disadvantages are:

1. Since several computations are performed for each interval, the scheme requires significantly more computation time per step.

2. Error estimates are typically difficult to obtain.

In the following applications, a second-order central difference approximation is used to compute the convective terms, that is,

$$\frac{\partial E}{\partial x} = \frac{E_{i+1} - E_{i-1}}{2\Delta x}$$

The solution of the proposed problem in Section 6.6 with a fourth-order RK governed by Equations (6-63) through (6-67) is shown in Figure 6-31 and provided in Table 6.10. The spatial and temporal steps are 0.1 and 0.1, respectively. The solution has developed large and unacceptable oscillations. That is not surprising since the central difference approximation used in the formulation possesses a large dispersion error. Similar behavior was observed in the solution by the Beam and Warming scheme. To reduce the oscillations to an acceptable level, a fourth-order damping term must be added. For this purpose, Equation (6-48) is evaluated at time "n" and augmented after the last stage. That is, after Equation (6-67), u_i^{n+1} is augmented according to $u_i^{n+1} = u_i^{n+1} + D$. Now, the solution with the added damping term and ε_e of 0.1 is shown in Figure 6-32 and presented in Table 6-11.

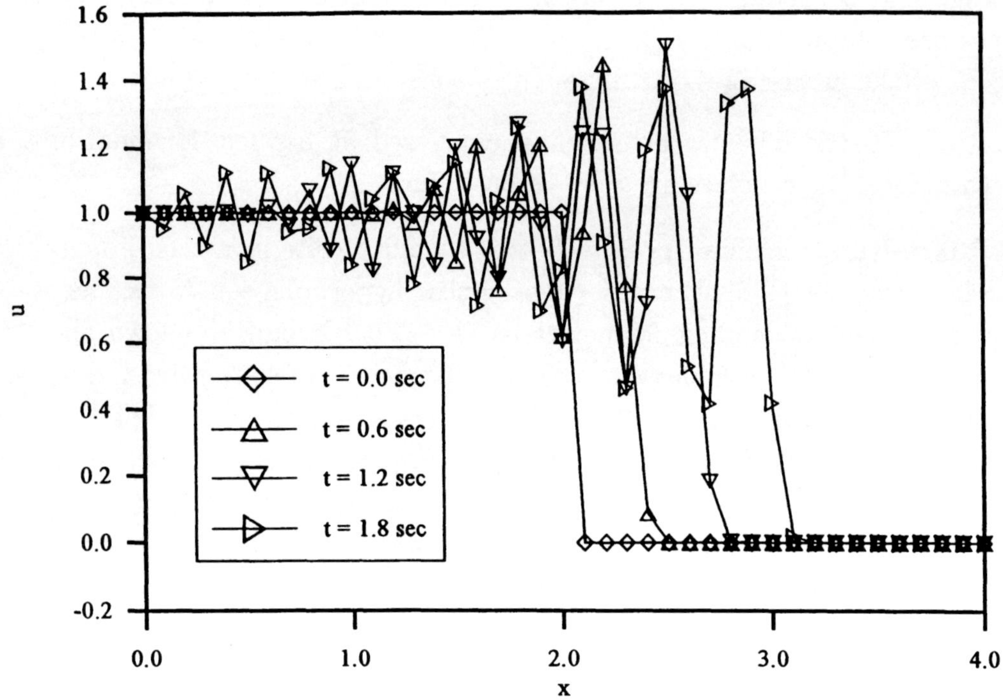

Figure 6-31. Solution of the inviscid Burgers equation by the fourth-order Runge-Kutta method, $\Delta x = 0.1$ and $\Delta t = 0.1$.

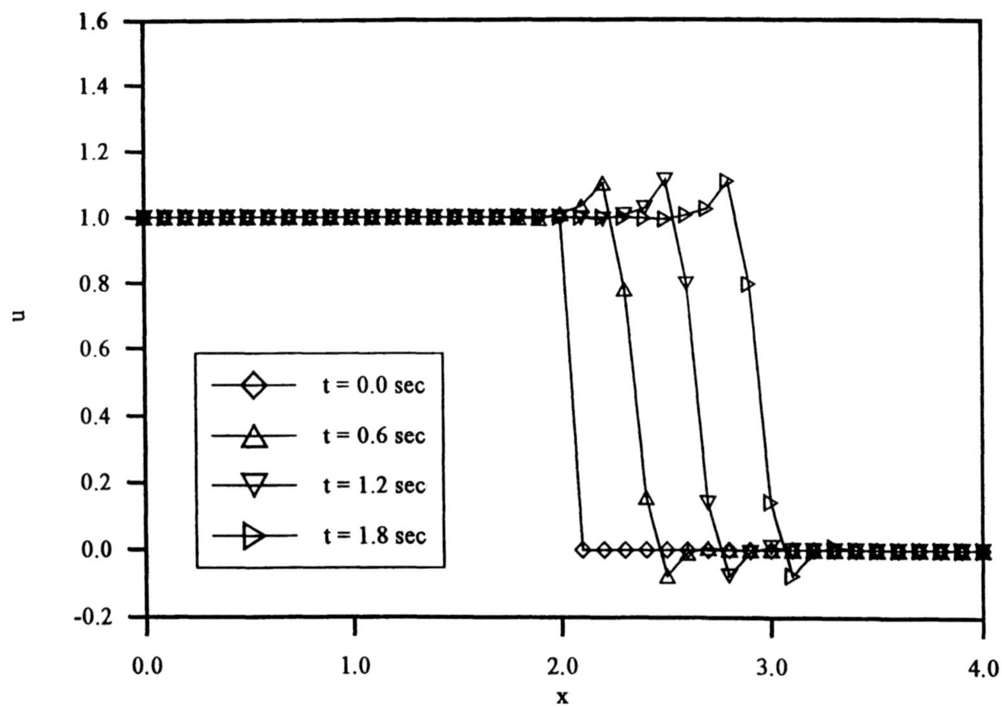

Figure 6-32. Solution of the inviscid Burgers equation by the fourth-order Runge-Kutta method with $\varepsilon_e = 0.1$, $\Delta x = 0.1$, and $\Delta t = 0.1$.

Observe that the large oscillations in the solution have been eliminated and that the solution is reasonable. In fact, the solution is very similar to that of Beam and Warming implicit scheme, shown in Figure 6-27. The effect of step size on the solution is shown in Figure 6-33 for two different Courant numbers.

6.6.8 Modified Runge-Kutta method.

In order to reduce the storage requirement of the RK scheme, a modification is introduced to eliminate the averaging step. Therefore, the equivalent formulation for a second-order RK scheme is written as

$$u_i^{(1)} = u_i^n \tag{6-68}$$

$$u_i^{(2)} = u_i^n - \frac{\Delta t}{2}\left(\frac{\partial E}{\partial x}\right)_i^{(1)} \tag{6-69}$$

$$u_i^{n+1} = u_i^n - \Delta t\left(\frac{\partial E}{\partial x}\right)_i^{(2)} \tag{6-70}$$

and a fourth-order RK scheme is

$$u_i^{(1)} = u_i^n \tag{6-71}$$

$$u_i^{(2)} = u_i^n - \frac{\Delta t}{4}\left(\frac{\partial E}{\partial x}\right)_i^{(1)} \tag{6-72}$$

$$u_i^{(3)} = u_i^n - \frac{\Delta t}{3}\left(\frac{\partial E}{\partial x}\right)_i^{(2)} \tag{6-73}$$

$$u_i^{(4)} = u_i^n - \frac{\Delta t}{2}\left(\frac{\partial E}{\partial x}\right)_i^{(3)} \tag{6-74}$$

$$u_i^{n+1} = u_i^n - \Delta t\left(\frac{\partial E}{\partial x}\right)_i^{(4)} \tag{6-75}$$

Again, a central difference approximation of second-order accuracy is used for the convective term $\partial E/\partial x$. The solution by the modified fourth-order RK is shown in Figure 6-34 and is provided in Table 6-12. The solution is similar to that of the fourth-order RK shown in Figure 6-31. Again, the addition of a damping term is necessary to produce acceptable solution. The solution with the damping term of (6-48) and a specified coefficient of $\varepsilon_e = 0.1$ is shown in Figure 6-35 and given in Table 6-13. Note that the addition of a damping term is similar to that of the previous section, that is, $u_i^{n+1} = u_i^{n+1} + D$, which is applied after (6-75).

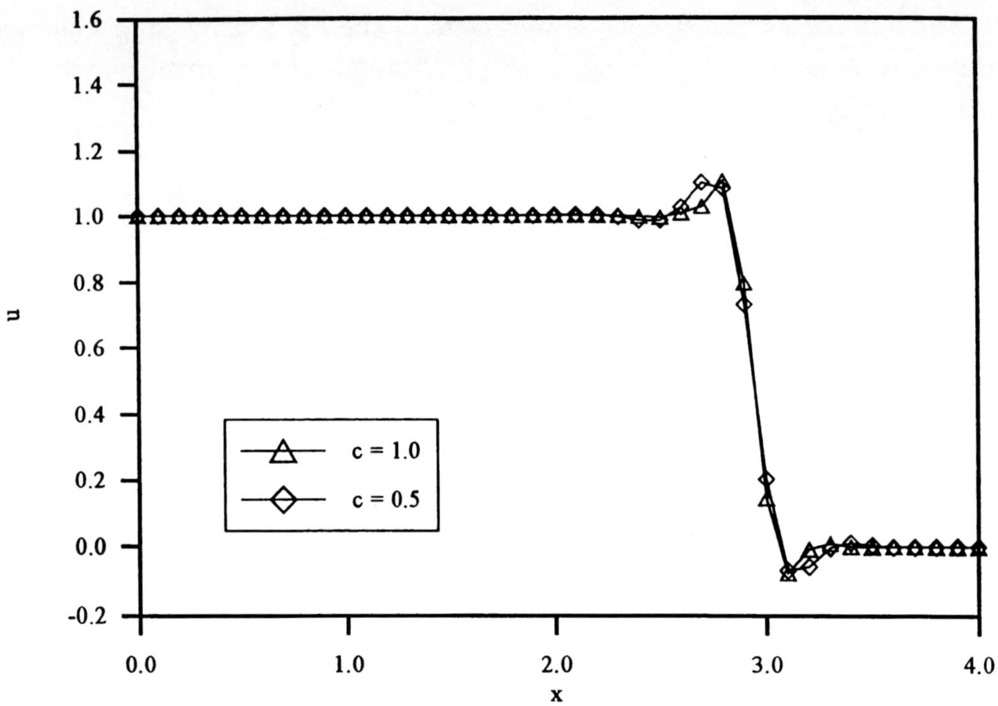

Figure 6-33. Effect of step sizes on the solution of the inviscid Burgers equation by the fourth-order Runge-Kutta method with $\varepsilon_e = 0.1$.

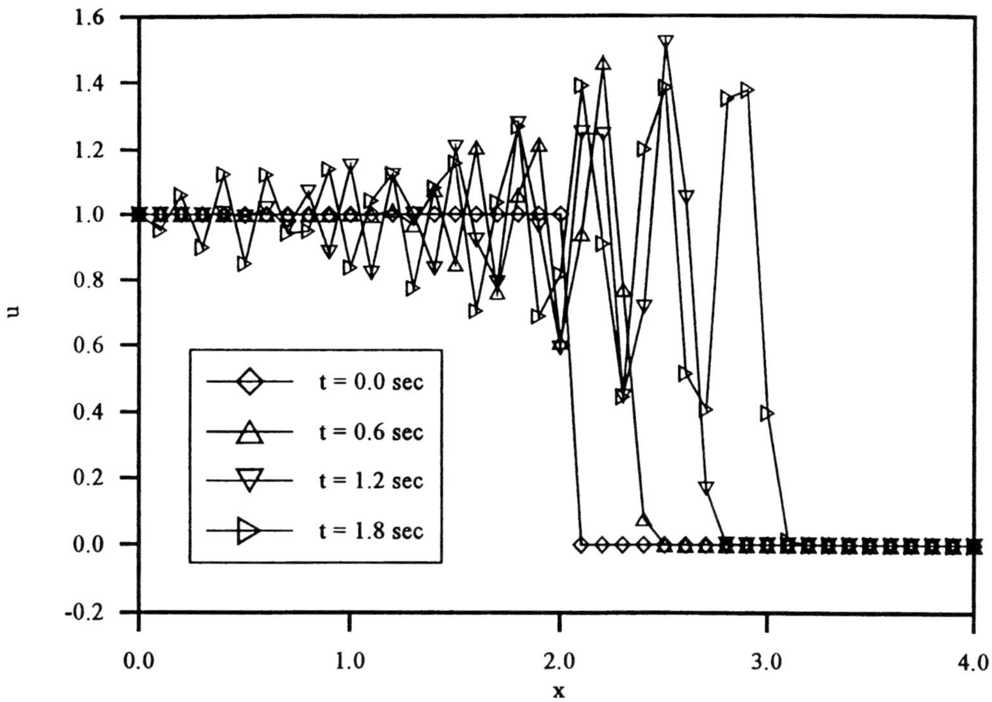

Figure 6-34. Solution of the inviscid Burgers equation by the modified fourth-order Runge-Kutta method, $\Delta x = 0.1$ and $\Delta t = 0.1$.

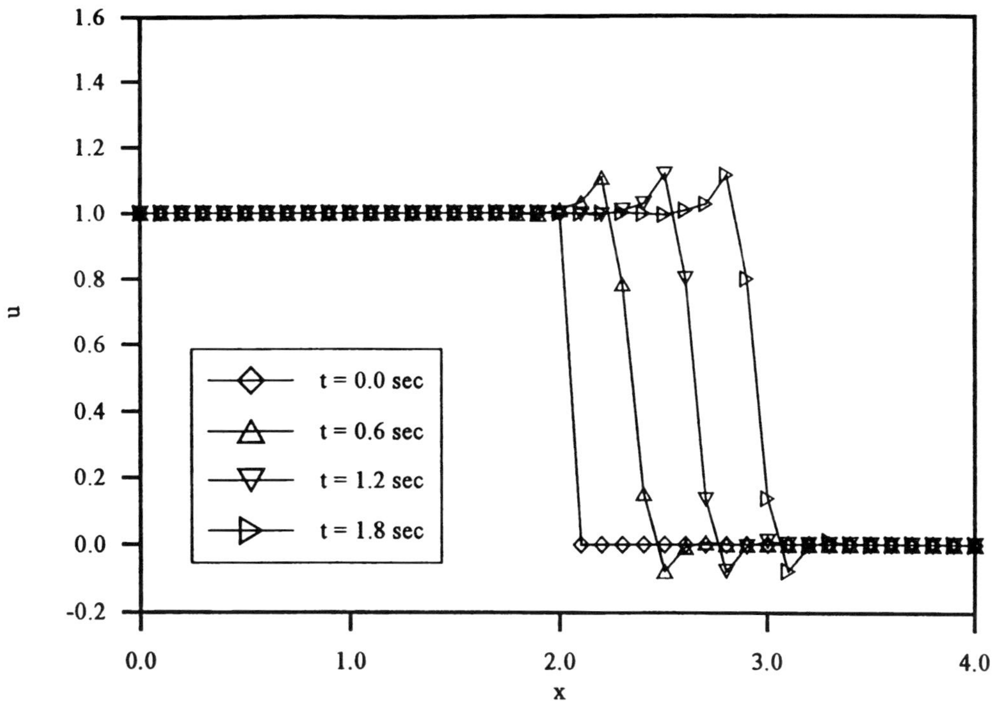

Figure 6-35. Solution of the inviscid Burgers equation by the modified fourth-order Runge-Kutta method with damping of $\varepsilon_e = 0.1$, $\Delta x = 0.1$, and $\Delta t = 0.1$.

6.7 Linear Damping

At this point further elaboration on the appearance of oscillations within the solution domain and methods to alleviate them is appropriate. The root cause of such oscillations will be reviewed by considering the simple wave equation. Recall the approximation of the wave equation by a first-order scheme. For example, consider the explicit first upwind scheme given by

$$u_i^{n+1} = u_i^n - c(u_i^n - u_{i-1}^n) \qquad a > 0$$

To identify the dominant error term, the modified equation must be investigated, which, for the explicit first upwind scheme was derived previously and is given by

$$\frac{\partial u}{\partial t} + a\frac{\partial u}{\partial x} = \frac{1}{2}a\Delta x(1-c)\frac{\partial^2 u}{\partial x^2} - \frac{1}{6}a(\Delta x)^2(2c^2 - 3c + 1)\frac{\partial^3 u}{\partial x^3}$$

The dominant error term is the even derivative term on the right-hand side of the equation. Recall that even derivative error terms are associated with dissipation error. Thus, sharp gradients within the solution are smeared, resulting in an inaccurate solution. In order to increase accuracy, i.e., reduce smearing of sharp gradients, grid refinement is required. However, that would lead to a computationally intensive and expensive operation. Indeed, in most practical applications, it would be impossible due to limited capacity of available hardware. Thus, it is desirable to increase the accuracy by incorporating higher order finite difference equations, such as second-order methods. Unfortunately, as seen previously, this approach has its own disadvantage in that second-order schemes possess a dominant odd-order derivative in their modified equations. Such error terms are associated with oscillations, i.e., dispersion error.

A simple scheme to reduce oscillations within the solution is to add second-order or fourth-order damping terms, as was shown in the previous section. Generally, the coefficients of the damping terms are specified by the user and remain constant throughout the domain as well as in the computational process. Therefore, such damping terms are referred to as linear damping or equivalently linear dissipation terms. The addition of such linear damping terms to the finite difference equation is relatively simple. The difficulty, however, is associated with the specification of the damping coefficient. It should be noted that, in general, a bound on the value of damping coefficients can be determined or estimated by stability analysis. However, the optimum value is not known a priori and is problem dependent. One must be extremely careful in the specification of the damping term(s), in particular for problems where physical dissipation (i.e., viscosity for example) is present. Addition of too much damping into the solution (unintentionally, of course) will clearly pollute the viscous region and may smear the sharp gradients in the inviscid region as well.

Obviously, inaccuracies in the solution of the viscous region would result in inaccurate velocity and temperature gradients. Recall that these gradients are used to determine skin friction and heat transfer coefficients. The inaccuracies could result in predictions of skin friction and heat transfer which may be off by as much as 100%. Since these quantities are important parameters in the design of machinery and vehicles, accurate computation of these parameters is extremely important. Therefore, it is strongly suggested that the damping coefficient be kept to a minimum value, just sufficient enough to damp out the oscillations within the solution.

Before proceeding further, some essential conclusions from numerous investigations with regard to linear damping terms are summarized.

1. Addition of a fourth-order damping term to a second-order scheme does not affect the formal order of accuracy of the algorithm.

2. Fourth-order damping terms are generally added explicitly, that is, they are added to the right-hand side of the equations. Therefore, as mentioned previously, stability analysis will impose an upper limit on the value of the damping coefficient. It is then clear that when performing computations by implicit schemes where larger step sizes (or correspondingly larger Courant numbers) can be used, the amount of added damping may be insufficient to damp out the oscillations. To overcome this limitation, a second-order damping term is added to the left-hand side of the equation (in addition to the fourth-order damping term on the right-hand side). The selection of a second-order implicit damping term will preserve the tri-diagonal nature of the implicit formulation.

3. The addition of damping terms not only eliminates or reduces oscillation in the solution, it helps stabilize the solution. That is, a solution at larger step sizes may be obtained which otherwise would have been unstable. In applications where density is being computed, oscillations in the solution may result in negative values of density. Addition of damping terms could help to alleviate this difficulty.

4. A typical fourth-order damping term for a one-dimensional problem may be expressed as

   $$D_e = -\varepsilon_e(\Delta x)^4 \frac{\partial^4 u}{\partial x^4} \tag{6-76}$$

   This term is added explicitly to the right-hand side of the finite difference equation. A negative sign is required to produce positive dissipation. The fourth-order derivative in (6-76) is approximated by a central difference expression as follows:

$$(\Delta x)^4 \frac{\partial^4 u}{\partial x^4} = u_{i-2} - 4u_{i-1} + 6u_i - 4u_{i+1} + u_{i+2} \tag{6-77}$$

The damping term given by (6-76) is easily extended to multidimensional problems. For a two-dimensional problem, for example, it may be written

$$D_e = -\varepsilon_e[(\Delta x)^4 \frac{\partial^4 u}{\partial x^4} + (\Delta y)^4 \frac{\partial^4 u}{\partial y^4}] \tag{6-78}$$

A second-order implicit damping term may be added to the left-hand side of an implicit formulation, generally defined by

$$D_i = \varepsilon_i (\Delta x)^2 \frac{\partial^2 u}{\partial x^2} \tag{6-79}$$

The second-order derivative may be approximated by

$$(\Delta x)^2 \frac{\partial^2 u}{\partial x^2} = u_{i-1} - 2u_i + u_{i+1} \tag{6-80}$$

5. Stability analysis performed on simple model equations provides a bound on the values of explicit damping coefficients ε_e and provides a relation between the explicit and implicit damping coefficients. The value/relation given in this section is provided as a general benchmark value to provide an estimate on the values of the coefficients. Clearly, these values will vary from one scheme to another. To obtain a more accurate value for a particular scheme, a stability analysis will be required. For example, an upper limit value of $\varepsilon_e = \frac{1}{8}$ for the Beam and Warming scheme is obtained. It should be recalled that stability analysis is performed on linear equations. Therefore, the actual values for nonlinear problems may be more restrictive.

   If a formulation includes both a fourth-order explicit damping term as well as a second-order implicit damping term, a typical relation is given by $\varepsilon_i = 2\varepsilon_e$. Of course, the difficult task of specification of ε_e still remains. The typical values given above can be used as a starting point. Obviously one would like to select the damping coefficients ε_e and ε_i as small as possible to damp out oscillations with minimum infusion of numerical viscosity into the solution. User experience with numerical schemes, specific application and physical insight is probably the best guide in the selection of values for the damping coefficients.

6. In applications where viscous effects are confined to a particular region (for example, the boundary layer region) physical viscosity may be sufficient enough to prevent oscillations in that region. Therefore, the damping terms are turned on only in the inviscid regions where formation of shock waves, i.e., large gradients could cause oscillations in the solution.

A logical procedure by which one may increase the accuracy of linear damping terms is by introduction of techniques in which the amount of damping is selectively added in regions where required. For this purpose, one would introduce a sensor which detects the gradients in the flowfield, e.g., computes the pressure gradient. Now, in regions where the gradient is large, the damping term is activated. Furthermore, the amount of added damping can be varied according to information provided by the sensor, i.e., in regions with severe gradients, a large amount of damping is added whereas in regions with moderate gradient, a smaller amount of damping is introduced.

Permitting such a dependency is logical from the standpoint of accuracy. Damping is increased in the vicinity of a discontinuity to maintain monotonicity, whereas damping is decreased in smooth regions, which otherwise would increase the truncation error. A procedure may also be devised whereby a second-order scheme switches to a first-order scheme by the use of a sensor. Recall that a typical first-order scheme is oscillation free; therefore, in the vicinity of large gradients, oscillations will not develop. The sensor in this case, which essentially plays the roll of limiter parameter, is defined as a flux limiter. These are used extensively in Total Variation Diminishing (TVD) schemes. Further discussion of flux limiters is postponed to Section 6.10.

6.7.1 Application

Consider the sinusoidal disturbance of Section 6.5 with $a = 250$ m/sec. The Lax-Wendroff scheme applied to the wave equation is given by

$$u_i^{n+1} = u_i^n - \frac{1}{2}c(u_{i+1}^n - u_{i-1}^n) + \frac{1}{2}c^2(u_{i+1}^n - 2u_i^n + u_{i-1}^n) \tag{6-81}$$

The solution as illustrated in Figure 6-10 includes oscillations within the domain. The solution obtained with a spatial step of 5.0 and temporal step of 0.01 seconds at time levels of 0.15, 0.3 and 0.45 seconds is shown in Figure 6-36.

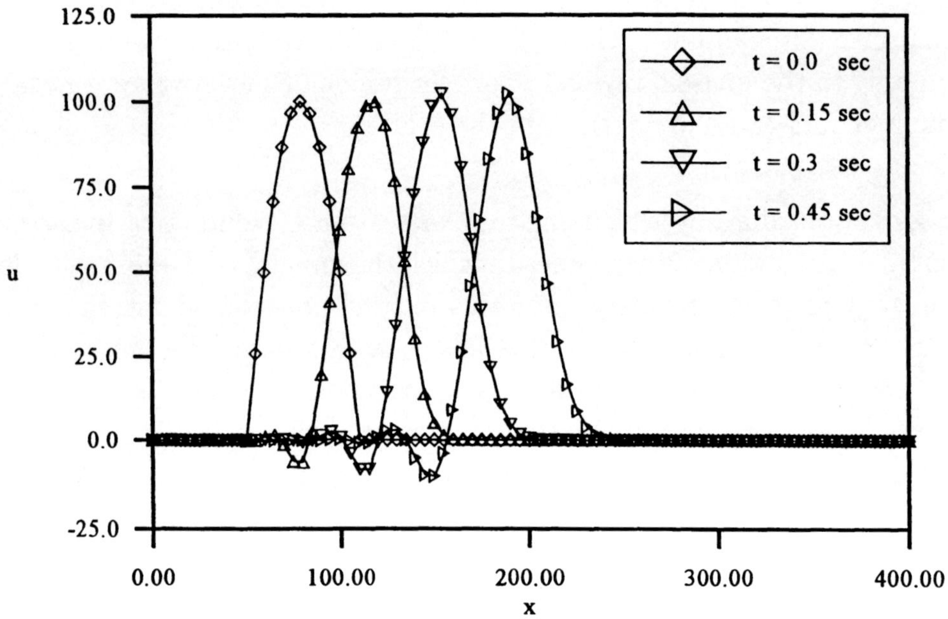

Figure 6-36. The solution of the wave equation by the Lax-Wendroff scheme.

Note that the initial profile is given by Equation (6-25). Now consider the addition of an explicit second-order damping term given by (6-80). Thus, the Lax-Wendroff scheme is expressed as

$$u_i^{n+1} = u_i^n - \frac{1}{2}c(u_{i+1}^n - u_{i-1}^n) + \frac{1}{2}c^2(u_{i+1}^n - 2u_i^n + u_{i-1}^n) + D_e$$

where

$$D_e = \varepsilon_e(u_{i+1}^n - 2u_i^n + u_{i-1}^n)$$

The solutions with damping factors ε_e of 0.03, 0.06 and 0.12 are shown in Figure 6-37. The step sizes specified previously for the solution shown in Figure 6-36 are used in this solution as well. Solution is also attempted with $\varepsilon_e = 0.40$; however, the damping coefficient of 0.40 exceeds the requirement imposed by the stability condition and an unstable solution is developed. As a comparison, the solution is shown in Figure 6-37(d).

As discussed previously, note that the addition of a damping (dissipation) term will reduce the amplitude of the wave. That is clearly evident in Figure 6-37. The larger the amount of damping, the larger the decrease in the amplitude. Thus, the amount of damping must be kept to a minimum, just sufficient enough to damp out the oscillations.

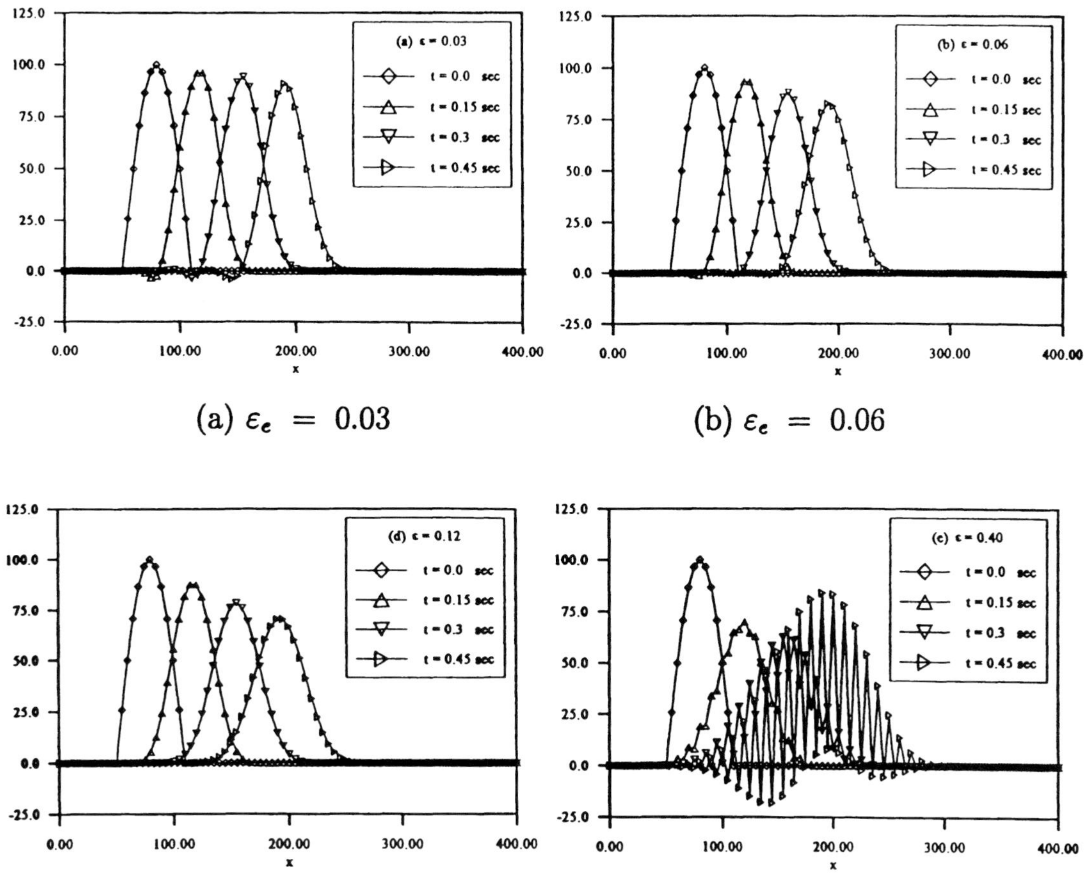

(a) $\varepsilon_e = 0.03$ (b) $\varepsilon_e = 0.06$

(c) $\varepsilon_e = 0.12$ (d) $\varepsilon_e = 0.40$

Figure 6-37. The solution of the wave equation by the Lax-Wendroff scheme with a second-order damping term.

6.8 Flux Corrected Transport

The addition of damping terms was suggested in the previous section in order to eliminate or reduce oscillations within the domain in the neighborhood of sharp gradients. Consider now an extension of the procedure whereby a second equation is added. In essence, the finite difference equation is modified to a predictor/corrector type, where in the predictor step a damping term is added and in the corrector step a certain amount of damping which may have been excessive is removed. Such a scheme is known as the Flux Corrected Transport (FCT) and was develped in Ref. [6-1]. The term which is introduced in the corrector step to remove the excessive damping is known as an anti-diffusive term.

To illustrate the procedure, consider the Lax-Wendroff scheme applied to the wave equation given by

$$u_i^{n+1} = u_i^n - \frac{1}{2}c(u_{i+1}^n - u_{i-1}^n) + \frac{1}{2}c^2(u_{i+1}^n - 2u_i^n + u_{i-1}^n)$$

A second-order damping term of the form

$$D = \varepsilon_1(u_{i+1}^n - 2u_i^n + u_{i-1}^n) \tag{6-82}$$

may be added such that the finite difference equation is written as

$$u_i^* = u_i^n - \frac{1}{2}c(u_{i+1}^n - u_{i-1}^n) + (\varepsilon_1 + \frac{1}{2}c^2)(u_{i+1}^n - 2u_i^n + u_{i-1}^n) \tag{6-83}$$

Equation (6-83) is the predictor step, which provides an intermediate value for u. Now, a corrector step where an anti-damping term is included is used to provide the value of u at time level $n+1$. The equation is expressed as

$$u_i^{n+1} = u_i^* - \varepsilon_2(u_{i+1}^* - 2u_i^* + u_{i-1}^*) \tag{6-84}$$

Expressions for the damping coefficients ε_1 and ε_2 may be developed for a specific scheme applied to a model equation. Such expressions for the Lax-Wendroff scheme applied to the continuity equation are recommended in Ref. [6-2], where

$$\varepsilon_1 = \frac{1}{6}(1 + 2c^2) \tag{6-85}$$

and

$$\varepsilon_2 = \frac{1}{6}(1 - c^2) \tag{6-86}$$

In order to preserve the conservative form of the equation for the general case of a nonlinear problem, the anti-diffusive term may be applied at $\frac{1}{2}$ points, i.e.,

$$u_i^{n+1} = u_i^* - (\bar{u}_{i+\frac{1}{2}} - \bar{u}_{i-\frac{1}{2}}) \tag{6-87}$$

where

$$\bar{u}_{i+\frac{1}{2}} = \varepsilon_2(u_{i+1}^* - u_i^*) \tag{6-88}$$

and

$$\bar{u}_{i-\frac{1}{2}} = \varepsilon_2(u_i^* - u_{i-1}^*) \tag{6-89}$$

Further investigations, however, suggest the use of alternate expressions for (6-88) and (6-89). These expressions are developed in order to prevent the injection of maxima or minima into the solution at the anti-diffusive step. The expressions to be used in (6-87) are

$$\bar{u}_{i+\frac{1}{2}} = Sgn(\Delta u_{i+\frac{1}{2}})\text{Max}\{0.0, \min[\Delta u_{i-\frac{1}{2}} Sgn(\Delta u_{i-\frac{1}{2}}), \varepsilon_2 \mid \Delta u_{i+\frac{1}{2}} \mid,$$
$$Sgn(\Delta u_{i+\frac{1}{2}})\Delta u_{i+\frac{3}{2}}]\} \qquad (6\text{-}90)$$

where

$$\Delta u_{i+\frac{1}{2}} = u^*_{i+1} - u^*_i$$

$$\Delta u_{i+\frac{3}{2}} = u^*_{i+2} - u^*_{i+1}$$

$$Sgn(A) = A/ABS(A)$$

and

$$\bar{u}_{i-\frac{1}{2}} = Sgn(\Delta u_{i-\frac{1}{2}})\text{Max}\{0.0, \min[\Delta u_{i+\frac{1}{2}} Sgn(\Delta u_{i+\frac{1}{2}}), \varepsilon_2 \mid \Delta u_{i-\frac{1}{2}} \mid,$$
$$Sgn(\Delta u_{i-\frac{1}{2}})\Delta u_{i-\frac{3}{2}}]\} \qquad (6\text{-}91)$$

where

$$\Delta u_{i-\frac{1}{2}} = u^*_i - u^*_{i-1}$$

and

$$\Delta u_{i-\frac{3}{2}} = u^*_{i-1} - u^*_{i-2}$$

6.8.1 Application

The FCT formulation of (6-83) and (6-84) is applied to the previous example as described in Section 6.5. The solution with the damping coefficients of 0.026 and 0.002 for ε_1 and ε_2, respectively, is shown in Figure 6-38.

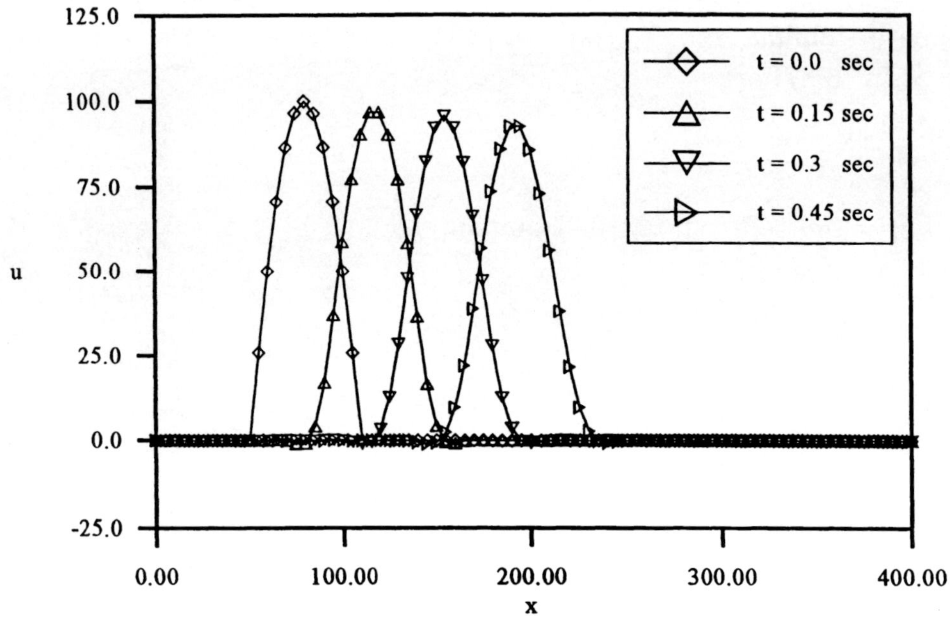

Figure 6-38. The solution of the wave equation by the Lax-Wendroff scheme with FCT.

6.9 Classification of Numerical Schemes

It is appropriate at this point to review/introduce some additional terminologies and to further classify equations utilized in conjunction with numerical approximations of hyperbolic equations. Initially, the concepts are introduced with the least amount of reference to mathematical details and equations. Subsequently, various formulations are explored.

6.9.1 Monotone Schemes

It was previously shown that certain first-order schemes, such as the upwind differencing scheme, produce oscillation-free solutions even in the presence of large gradients. Oscillation-free schemes are generally known as monotone schemes. It can be proven that monotone schemes are only first-order accurate. Therefore, as seen previously, monotone schemes are dissipative and, in general, a discontinuity would be smeared over several grid points, even though grid refinement will reduce the smearing somewhat. A formal requirement of monotonic scheme is stated by: (1) as the solution proceeds in time, no new local extremes are developed; and (2) the local minimum is non-decreasing whereas the local maximum is non-increasing.

6.9.2 Total Variation Diminishing Schemes

Define the total variation of a variable such as u by

$$TV(u) = \int \mid \frac{\partial u}{\partial x} \mid dx$$

where in a discrete form, the total variation of numerical solution may be expressed by

$$TV(u^n) = \sum_{-\infty}^{+\infty} \mid u_{i+1}^n - u_i^n \mid$$

A numerical scheme is said to be Total Variation Diminishing (TVD) in time if

$$TV(u^{n+1}) \leq TV(u^n)$$

TVD schemes can be classified as either a first-order TVD scheme or a second-order TVD scheme.

It is beneficial at this point to introduce some essential remarks about the TVD schemes, some of which will be explored in the upcoming section.

1. It can be shown that all monotone schemes are first-order TVD schemes, i.e., monotone schemes are a subset of TVD schemes.

2. A general explicit formulation for the wave equation may be expressed as

$$u_i^{n+1} = u_i^n + A_{i+\frac{1}{2}} \Delta u_{i+\frac{1}{2}}^n - B_{i-\frac{1}{2}} \Delta u_{i-\frac{1}{2}}^n \tag{6-92}$$

   where

$$\Delta u_{i+\frac{1}{2}}^n = u_{i+1} - u_i \qquad \text{and} \qquad \Delta u_{i-\frac{1}{2}}^n = u_i - u_{i-1}$$

   and the coefficients A and B are to be determined for a specific scheme. The scheme is said to be TVD if the following sufficient conditions are satisfied:

$$A_{i+\frac{1}{2}} \geq 0 \tag{6-93}$$

$$B_{i-\frac{1}{2}} \geq 0 \tag{6-94}$$

   and

$$0 \leq A_{i+\frac{1}{2}} + B_{i+\frac{1}{2}} \leq 1 \tag{6-95}$$

3. TVD property is valid only for homogenous scalar hyperbolic conservation equations. Extension to non-homogeneous hyperbolic equations is limited to special cases.

4. Second-order TVD schemes can be developed by the use of flux limiters. In such schemes, the formulation is second-order within the smooth region of the domain whereas it switches to first-order in regions of high gradients to prevent any oscillations.

5. As in any other scheme, implementation of boundary conditions may create some difficulty. Generally it is difficult to prove the TVD property for the combined interior and boundary formulation.

6. The formal extension of TVD property to systems of nonlinear equations and multi-dimensional systems has not been established. However, numerical experiments with these systems have shown that the TVD property may be preserved. Therefore, TVD schemes for multi-dimensional systems are commonly used.

7. TVD schemes possess desirable properties for the computation of domains with discontinuities. However, they are more expensive in comparison to schemes employing simple linear dissipation terms.

8. TVD schemes eliminate oscillations within the domain while shock smearing is reduced.

9. TVD schemes which are based on central difference approximations of the convective terms as well as the flux limiters are usually referred to as symmetric TVD schemes. Similarly, TVD schemes which employ one-sided differences are referred to as upwind TVD schemes.

6.9.3 Essentially Non-Oscillatory Schemes

Generally speaking, the so-called Essentially Non-Oscillating (ENO) schemes use similar principles as those of TVD schemes. The main difference between the two schemes is that some ENO schemes can retain the same order of accuracy for the entire domain. These schemes are not used as extensively as the TVD schemes; therefore, no further discussion will be presented.

6.10 TVD Formulations

Numerous TVD schemes have been developed by various investigators over the last several years and more are being introduced at present. The various TVD schemes can be broadly categorized and subcategorized. First, TVD schemes may be classified as first-order TVD schemes, second-order TVD schemes which are usually referred to as high resolution schemes, and predictor-corrector type TVD schemes. Furthermore, in each category, the formulation may be explicit or implicit. In terms of the finite difference approximation, the resulting formulation may be classified as symmetric or upwind. Furthermore, for each formulation, different functions may be available for the flux limiters. Thus, within each category, numerous formulations can be written. Since the objective at this point is to familiarize the reader with typical TVD schemes, only a selected number of schemes are introduced. Interested readers should consult Refs. [6-3] through [6-10].

6.10.1 First-Order TVD Schemes

Consider the inviscid Burgers equation given by Equation (6-33) as

$$\frac{\partial u}{\partial t} + \frac{\partial E}{\partial x} = 0 \tag{6-96}$$

An explicit, first-order upwind algorithm can be written by the following:

$$u_i^{n+1} = u_i^n - \frac{\Delta t}{\Delta x} \begin{cases} E_{i+1}^n - E_i^n & \text{for} \quad \alpha_{i+\frac{1}{2}} < 0 \\ E_i^n - E_{i-1}^n & \text{for} \quad \alpha_{i+\frac{1}{2}} > 0 \end{cases} \tag{6-97}$$

where α may be defined as in Ref. [6-10] by

$$\alpha_{i+\frac{1}{2}} = \begin{cases} u_i^n & \text{if} \quad \Delta u_{i+\frac{1}{2}}^n = 0 \\ \dfrac{E_{i+1}^n - E_i^n}{u_{i+1}^n - u_i^n} & \text{if} \quad \Delta u_{i+\frac{1}{2}}^n \neq 0 \end{cases} \tag{6-98}$$

Equation (6-97) can be expressed in a combined form by the following:

$$\begin{aligned} u_i^{n+1} = u_i^n &- \frac{1}{2}\frac{\Delta t}{\Delta x}[1 - Sgn(\alpha_{i+\frac{1}{2}})](E_{i+1}^n - E_i^n) \\ &- \frac{1}{2}\frac{\Delta t}{\Delta x}[1 + Sgn(\alpha_{i-\frac{1}{2}})](E_i^n - E_{i-1}^n) \end{aligned} \tag{6-99}$$

where $\alpha_{i-\frac{1}{2}}$ is defined similar to that of (6-98) by

$$\alpha_{i-\frac{1}{2}} = \begin{cases} u_i^n & \text{if } \Delta u_{i-\frac{1}{2}}^n = 0 \\ \dfrac{E_i^n - E_{i-1}^n}{u_i^n - u_{i-1}^n} & \text{if } \Delta u_{i-\frac{1}{2}}^n \neq 0 \end{cases} \tag{6-100}$$

Relations (6-98) and (6-100) can be used to recast Equation (6-99) in different forms. Commonly used forms of the equations are as follows:

$$u_i^{n+1} = u_i^n - \frac{1}{2}\frac{\Delta t}{\Delta x}[\alpha_{i+\frac{1}{2}}\Delta u_{i+\frac{1}{2}}^n - |\,\alpha_{i+\frac{1}{2}}\,|\,\Delta u_{i+\frac{1}{2}}^n + \alpha_{i-\frac{1}{2}}\Delta u_{i-\frac{1}{2}}^n + |\,\alpha_{i-\frac{1}{2}}\,|\,\Delta u_{i-\frac{1}{2}}^n] \tag{6-101}$$

$$u_i^{n+1} = u_i^n - \frac{1}{2}\frac{\Delta t}{\Delta x}[(E_{i+1}^n - E_i^n) + (E_i^n - E_{i-1}^n) - |\,\alpha_{i+\frac{1}{2}}\,|\,\Delta u_{i+\frac{1}{2}}^n + |\,\alpha_{i-\frac{1}{2}}\,|\,\Delta u_{i-\frac{1}{2}}^n] \tag{6-102}$$

$$u_i^{n+1} = u_i^n - \frac{1}{2}\frac{\Delta t}{\Delta x}[(E_{i+1}^n + E_i^n) - |\,\alpha_{i+\frac{1}{2}}\,|\,\Delta u_{i+\frac{1}{2}}^n - (E_i^n + E_{i-1}^n) + |\,\alpha_{i-\frac{1}{2}}\,|\,\Delta u_{i-\frac{1}{2}}^n] \tag{6-103}$$

$$u_i^{n+1} = u_i^n - \frac{1}{2}\frac{\Delta t}{\Delta x}[(E_{i+1}^n - E_{i-1}^n) - |\,\alpha_{i+\frac{1}{2}}\,|\,\Delta u_{i+\frac{1}{2}}^n + |\,\alpha_{i-\frac{1}{2}}\,|\,\Delta u_{i-\frac{1}{2}}^n] \tag{6-104}$$

Equation (6-103) is often used in conjunction with flux limiters. It is commonly rewritten as

$$u_i^{n+1} = u_i^n - \frac{\Delta t}{\Delta x}[h_{i+\frac{1}{2}}^n - h_{i-\frac{1}{2}}^n] \tag{6-105}$$

where

$$h^n_{i+\frac{1}{2}} = \frac{1}{2}[(E^n_{i+1} + E^n_i) - \mid \alpha_{i+\frac{1}{2}} \mid \Delta u^n_{i+\frac{1}{2}}] \tag{6-106}$$

and

$$h^n_{i-\frac{1}{2}} = \frac{1}{2}[(E^n_i + E^n_{i-1}) - \mid \alpha_{i-\frac{1}{2}} \mid \Delta u^n_{i-\frac{1}{2}}] \tag{6-107}$$

The expressions defined above by "h" are known as the numerical flux functions.

Recall that, in order to establish the TVD nature of the scheme, the requirements specified by (6-93) through (6-95) need to be checked. Thus, Equation (6-101) should be rewritten in the general explicit form of Equation (6-92). For this purpose, Equation (6-101) is rearranged as

$$u^{n+1}_i = u^n_i + \frac{1}{2}\frac{\Delta t}{\Delta x}(\mid \alpha_{i+\frac{1}{2}} \mid - \alpha_{i+\frac{1}{2}})\Delta u^n_{i+\frac{1}{2}} - \frac{1}{2}\frac{\Delta t}{\Delta x}(\mid \alpha_{i-\frac{1}{2}} \mid + \alpha_{i-\frac{1}{2}})\Delta u^n_{i-\frac{1}{2}}$$

or

$$u^{n+1}_i = u^n_i + A_{i+\frac{1}{2}}\Delta u^n_{i+\frac{1}{2}} - B_{i-\frac{1}{2}}\Delta u^n_{i-\frac{1}{2}} \tag{6-108}$$

where

$$A_{i+\frac{1}{2}} = \frac{1}{2}\frac{\Delta t}{\Delta x}(\mid \alpha_{i+\frac{1}{2}} \mid - \alpha_{i+\frac{1}{2}}) \tag{6-109}$$

and

$$B_{i-\frac{1}{2}} = \frac{1}{2}\frac{\Delta t}{\Delta x}(\mid \alpha_{i-\frac{1}{2}} \mid + \alpha_{i-\frac{1}{2}}) \tag{6-110}$$

Furthermore, define

$$B_{i+\frac{1}{2}} = \frac{1}{2}\frac{\Delta t}{\Delta x}(\mid \alpha_{i+\frac{1}{2}} \mid + \alpha_{i+\frac{1}{2}}) \tag{6-111}$$

Recall that a numerical scheme is said to possess TVD property if the following conditions are satisfied:

$$A_{i+\frac{1}{2}} \geq 0 \tag{6-112}$$

$$B_{i-\frac{1}{2}} \geq 0 \tag{6-113}$$

and

$$0 \leq A_{i+\frac{1}{2}} + B_{i+\frac{1}{2}} \leq 1 \tag{6-114}$$

Requirements (6-112) and (6-113) are always satisfied as seen by (6-109) and (6-110). Requirement (6-114) is rearranged with the use of (6-109) and (6-111) to provide

$$0 \leq \mid \alpha_{i+\frac{1}{2}} \frac{\Delta t}{\Delta x} \mid \leq 1 \tag{6-115}$$

which is a Courant number type requirement. Thus, it is seen that the first-order upwind scheme has TVD property and must satisfy the Courant number restriction imposed by (6-115).

Now consider formulation (6-104) given by

$$u_i^{n+1} = u_i^n - \frac{1}{2}\frac{\Delta t}{\Delta x}(E_{i+1}^n - E_{i-1}^n) + \frac{1}{2}\frac{\Delta t}{\Delta x}(\mid \alpha_{i+\frac{1}{2}} \mid \Delta u_{i+\frac{1}{2}}^n - \mid \alpha_{i-\frac{1}{2}} \mid \Delta u_{i-\frac{1}{2}}^n) \tag{6-116}$$

Observe that Equation (6-116) is composed of a central difference approximation of the convective term plus a correction term which is dissipative. Equation (6-116) may be split and solved in a multi-step fashion by the following:

$$u_i^* = u_i^n - \frac{1}{2}\frac{\Delta t}{\Delta x}(E_{i+1}^n - E_{i-1}^n) \tag{6-117}$$

and

$$u_i^{n+1} = u_i^* + \frac{1}{2}\frac{\Delta t}{\Delta x}(\phi_{i+\frac{1}{2}}^n - \phi_{i-\frac{1}{2}}^n) \tag{6-118}$$

where

$$\phi_{i+\frac{1}{2}}^n = \mid \alpha_{i+\frac{1}{2}} \mid \Delta u_{i+\frac{1}{2}}^n \tag{6-119}$$

$$\phi_{i-\frac{1}{2}}^n = \mid \alpha_{i-\frac{1}{2}} \mid \Delta u_{i-\frac{1}{2}}^n \tag{6-120}$$

The expressions defined by (6-119) and (6-120) are known as the flux limiter functions. These quantities introduce dissipation which eliminate oscillations within the solution. The scheme given by (6-117) and (6-118) suggests a simple procedure by which TVD dissipation can be incorporated to a numerical scheme. That is, an equivalent step similar to (6-118) may be added to an existing code in order to include TVD dissipation without major modifications.

6.10.2 Entropy Condition

Weak solutions (solutions with discontinuities such as shocks) of the conservation laws, e.g., inviscid Burgers equation, subject to an imposed initial data may not be unique. For example, the solution may produce a nonphysical expansion shock accompanied by a decrease in entropy. To exclude nonphysical solutions and to seek a physically relevant solution, additional conditions must be imposed. Such a requirement in fluid mechanics is provided by the second law of thermodynamics, which simply states that the entropy cannot decrease. Recall that the increase of entropy is due to irreversible processes such as viscosity or shock waves among others. In order to obtain a unique solution of the governing equation, an additional condition is imposed, known as the entropy condition.

It should be noted that some schemes provide unique solutions where the entropy condition is automatically satisfied. For example, it has been shown that [6-11, 6-12], monotone schemes applied to a single conservation law always corresponds to a physically relevant condition. On the other hand, not all TVD schemes automatically satisfy the entropy condition. In order to enforce the entropy requirement, a dissipative mechanism must be present in the numerical scheme. In the previous scheme given by Equation (6-105), the term $|\alpha|$ provided such a mechanism. However, a difficulty may appear when α becomes zero. To overcome this problem, α is replaced by an entropy correction term denoted by ψ, where

$$\psi = \begin{cases} |\alpha| & \text{for } |\alpha| \geq \varepsilon \\ \dfrac{\alpha^2 + \varepsilon^2}{2\varepsilon} & \text{for } |\alpha| < \varepsilon \end{cases} \tag{6-121}$$

The positive constant ε is selected within the range of $0.0 \leq \varepsilon \leq 0.125$.

6.10.3 Application

The nonlinear problem proposed in Section 6.6 will be used to illustrate the application of a first-order TVD scheme. Recall that the objective of the proposed problem is to solve the Burgers equation subject to an initial condition which includes a discontinuity as shown in Figure 6-19. The TVD formulation given by (6-117) and (6-118) is utilized to obtain the solution. The spatial and temporal step sizes are 0.1 and 0.1, respectively, which result in a Courant number of one. The solution at time increment of 0.6 seconds is shown in Figure 6-39 and is presented in Table 6-14. The lack of any oscillation within the domain of solution is clearly evident.

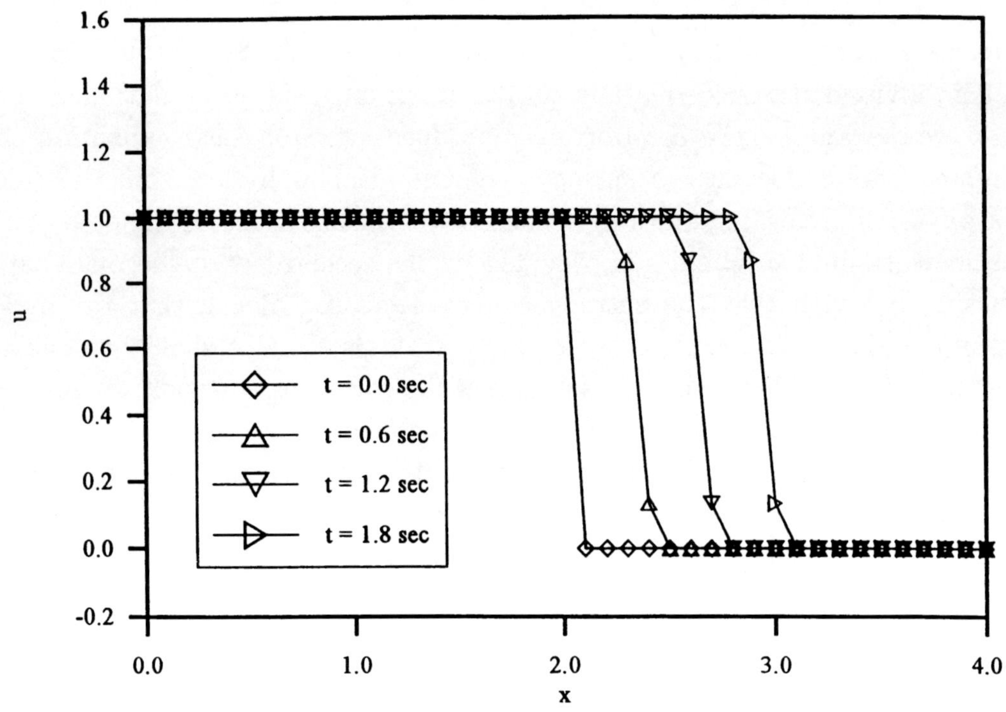

Figure 6-39. The solution of inviscid Burgers equation by the first-order TVD scheme, $\Delta x = 0.1$ and $\Delta t = 0.1$.

6.10.4 Second-Order TVD Schemes

As in the case of the first-order TVD schemes, there are numerous second-order TVD formulations. Only a selected number of schemes are introduced in this section. Additional schemes are discussed in Ref. [6-6].

One way to increase the order of accuracy from first-order to second-order is to introduce a modified flux $\bar{E}$ as proposed by Harten, Ref. [6-7] where

$$\bar{E} = E + G \tag{6-122}$$

The function G is defined so as to provide a second-order TVD, referred to as *limiter.*

The governing equations are written in a similar form as Equation (6-105). Therefore,

$$u_i^{n+1} = u_i^n - \frac{\Delta t}{\Delta x}\left[h_{i+\frac{1}{2}}^n - h_{i-\frac{1}{2}}^n\right] \tag{6-123}$$

where

$$h^n_{i+\frac{1}{2}} = \frac{1}{2}\left[E^n_{i+1} + E^n_i + \phi^n_{i+\frac{1}{2}}\right] \tag{6-124}$$

and

$$h^n_{i-\frac{1}{2}} = \frac{1}{2}\left[E^n_i + E^n_{i-1} + \phi^n_{i-\frac{1}{2}}\right] \tag{6-125}$$

where ϕ is the flux limiter function. The introduction of various limiters will be accompanied by the application of the second-order TVD scheme to the example problem of Sec. 6.6.

6.10.4.1 Harten-Yee Upwind TVD Limiters. In this scheme, the flux limiter function is defined as

$$\phi_{i+\frac{1}{2}} = (G_{i+1} + G_i) - \psi(\alpha_{i+\frac{1}{2}} + \beta_{i+\frac{1}{2}})\Delta u^n_{i+\frac{1}{2}} \tag{6-126}$$

with

$$\psi(y) = \begin{cases} \mid y \mid & \text{for } \mid y \mid \geq \varepsilon \\ \dfrac{(y)^2 + \varepsilon^2}{2\varepsilon} & \text{for } \mid y \mid < \varepsilon \end{cases} \tag{6-127}$$

where $0 \leq \varepsilon \leq 0.125$, and

$$\alpha_{i+\frac{1}{2}} = \begin{cases} \dfrac{E_{i+1} - E_i}{\Delta u_{i+\frac{1}{2}}} & \text{for } \Delta u_{i+\frac{1}{2}} \neq 0 \\ \dfrac{u_{i+1} + u_i}{2} & \text{for } \Delta u_{i+\frac{1}{2}} = 0 \end{cases} \tag{6-128}$$

$$\beta_{i+\frac{1}{2}} = \begin{cases} \dfrac{(G_{i+1} - G_i)}{\Delta u_{i+\frac{1}{2}}} & \text{for } \Delta u_{i+\frac{1}{2}} \neq 0 \\ 0 & \text{for } \Delta u_{i+\frac{1}{2}} = 0 \end{cases} \tag{6-129}$$

The limiter G may be specified as follows:

$$G_i = S * \max\{0, \min[\sigma_{i+\frac{1}{2}} \mid \Delta u_{i+\frac{1}{2}} \mid, S * \sigma_{i-\frac{1}{2}}\Delta u_{i-\frac{1}{2}}]\} \tag{6-130}$$

where

$$S = Sgn(\Delta u_{i+\frac{1}{2}}) = \frac{\Delta u_{i+\frac{1}{2}}}{\mid \Delta u_{i+\frac{1}{2}} \mid}$$

and

$$\sigma_{i+\frac{1}{2}} = \frac{1}{2}[\psi(\alpha_{i+\frac{1}{2}}) - \frac{\Delta t}{\Delta x}(\alpha_{i+\frac{1}{2}})^2]$$

The flux limiter given by (6-126) has also been modified [6-3] as follows

$$\phi_{i+\frac{1}{2}} = \sigma\left(\alpha_{i+\frac{1}{2}}\right)\left(G_{i+1} + G_i\right) - \psi\left(\alpha_{i+\frac{1}{2}} + \beta_{i+\frac{1}{2}}\right)\Delta u^n_{i+\frac{1}{2}} \quad \text{(6-131a)}$$

$$\phi_{i-\frac{1}{2}} = \sigma\left(\alpha_{i-\frac{1}{2}}\right)\left(G_i + G_{i-1}\right) - \psi\left(\alpha_{i-\frac{1}{2}} + \beta_{i-\frac{1}{2}}\right)\Delta u^n_{i-\frac{1}{2}} \quad \text{(6-131b)}$$

where

$$\sigma\left(\alpha_{i+\frac{1}{2}}\right) = \frac{1}{2}\psi\left(\alpha_{i+\frac{1}{2}}\right) + \frac{\Delta t}{\Delta x}\left(\alpha_{i+\frac{1}{2}}\right)^2$$

and

$$\beta_{i+\frac{1}{2}} = \sigma\left(\alpha_{i+\frac{1}{2}}\right)\begin{cases} \dfrac{G_{i+1} - G_i}{\Delta u_{i+\frac{1}{2}}} & , \quad \Delta u_{i+\frac{1}{2}} \neq 0 \\ 0 & , \quad \Delta u_{i+\frac{1}{2}} = 0 \end{cases}$$

A variety of limiters have been introduced to evaluate G. Several choices are as follows

$$G_i = \text{minmod}\left(\Delta u_{i-\frac{1}{2}}\ ,\ \Delta u_{i+\frac{1}{2}}\right) \quad \text{(6-132)}$$

$$G_i = \frac{\Delta u_{i+\frac{1}{2}}\,\Delta u_{i-\frac{1}{2}} + \left|\Delta u_{i+\frac{1}{2}}\,\Delta u_{i-\frac{1}{2}}\right|}{\Delta u_{i+\frac{1}{2}} + \Delta u_{i-\frac{1}{2}}} \quad \text{(6-133)}$$

If $\Delta u_{i+\frac{1}{2}} + \Delta u_{i-\frac{1}{2}} = 0$, then $G_i = 0$

$$G_i = \frac{\Delta u_{i-\frac{1}{2}}\left[\left(\Delta u_{i+\frac{1}{2}}\right)^2 + \omega\right] + \Delta u_{i+\frac{1}{2}}\left[\left(\Delta u_{i-\frac{1}{2}}\right)^2 + \omega\right]}{\left(\Delta u_{i+\frac{1}{2}}\right)^2 + \left(\Delta u_{i-\frac{1}{2}}\right)^2 + 2\omega}\ , \quad 10^{-7} \leq \omega \leq 10^{-5} \quad \text{(6-134)}$$

$$G_i = \text{minmod}\left[2\Delta u_{i-\frac{1}{2}}\ ,\ 2\Delta u_{i+\frac{1}{2}}\ ,\ \frac{1}{2}\left(\Delta u_{i+\frac{1}{2}} + \Delta u_{i-\frac{1}{2}}\right)\right] \quad \text{(6-135)}$$

$$G_i = S * \max\left[0, \min\left(2\left|\Delta u_{i+\frac{1}{2}}\right|, S * \Delta u_{i-\frac{1}{2}}\right), \min\left(\left|\Delta u_{i+\frac{1}{2}}\right|, 2S * \Delta\Delta u_{i-\frac{1}{2}}\right)\right] \quad \text{(6-136)}$$

Recall that

$$\text{minmod}\,(a, b, c, \ldots, n) = S * \max\left[0\ ,\ \min\left(|a|\ ,\ S*b\ ,\ S*c\ ,\ \ldots\ ,\ S*n\right)\right]$$

where $S = sgn(a)$.

The solution for the second-order TVD scheme given by Equation (6-123) and the flux limiter function (6-131) and limiter G given by (6-132) is shown in Figure 6-40 for several time levels and provided in Table 6.15. The solution at time level of 1.8 sec is also shown in Figure 6-41 where the effect of step size on the solution is illustrated. The solutions using limiters (6-132) through (6-136) are compared in Figure 6-42. Note that the solution by limiter (6-134) is less than desirable! The solution by limiter (6-132), which is the simplest choice, appears to be the best choice as well.

6.10.4.2 Roe-Sweby Upwind TVD Limiters. The flux limiter function is defined as

$$\phi_{i+\frac{1}{2}} = [\frac{G_i}{2}(\mid \alpha_{i+\frac{1}{2}} \mid + \frac{\Delta t}{\Delta x}\alpha^2_{i+\frac{1}{2}}) - \mid \alpha_{i+\frac{1}{2}} \mid]\Delta u_{i+\frac{1}{2}} \tag{6-137a}$$

$$\phi_{i-\frac{1}{2}} = \left[\frac{G_{i-1}}{2}(\mid \alpha_{i-\frac{1}{2}} \mid + \frac{\Delta t}{\Delta x}\alpha^2_{i-\frac{1}{2}}) - \mid \alpha_{i-\frac{1}{2}} \mid\right] \Delta u_{i-\frac{1}{2}} \tag{6-137b}$$

Several choices have been proposed for the function G, among which are:

$$G_i = \text{minmod } (1, r) \tag{6-138}$$

$$G_i = \frac{r + \mid r \mid}{1 + r} \tag{6-139}$$

$$G_i = \max[0, \min(2r, 1), \min(r, 2)] \tag{6-140}$$

where

$$r = \frac{u_{i+1+\sigma} - u_{i+\sigma}}{\Delta u_{i+\frac{1}{2}}}$$

$$\sigma = \text{Sgn } (\alpha_{i+\frac{1}{2}})$$

If at a point $\Delta u_{i+\frac{1}{2}}$ is zero, then r is set equal to zero in order to prevent a division by zero.

The solution by limiter (6-138) is shown in Figure 6-43, and solutions by limiters (6-138) through (6-140) are shown in Figure 6-44. It is observed that the solution with limiter (6-138) provides the better solution of the three limiters examined. The solution with limiter (6-138) is also presented in Table 6-16 for several time levels.

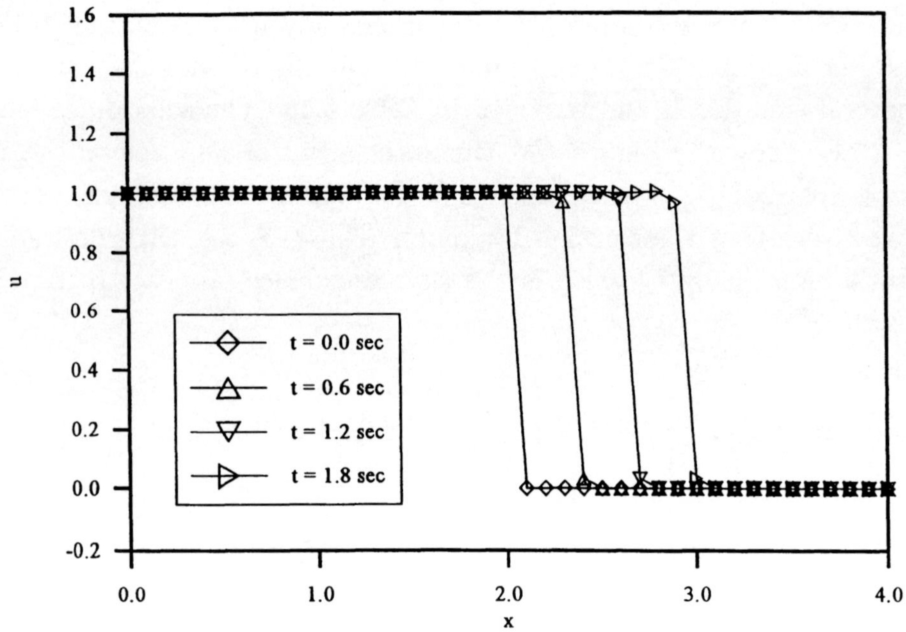

Figure 6-40. Solution of the inviscid Burgers equation by the second-order TVD method with the Harten-Yee upwind limiter (6-132), $\Delta x = 0.1$ and $\Delta t = 0.1$.

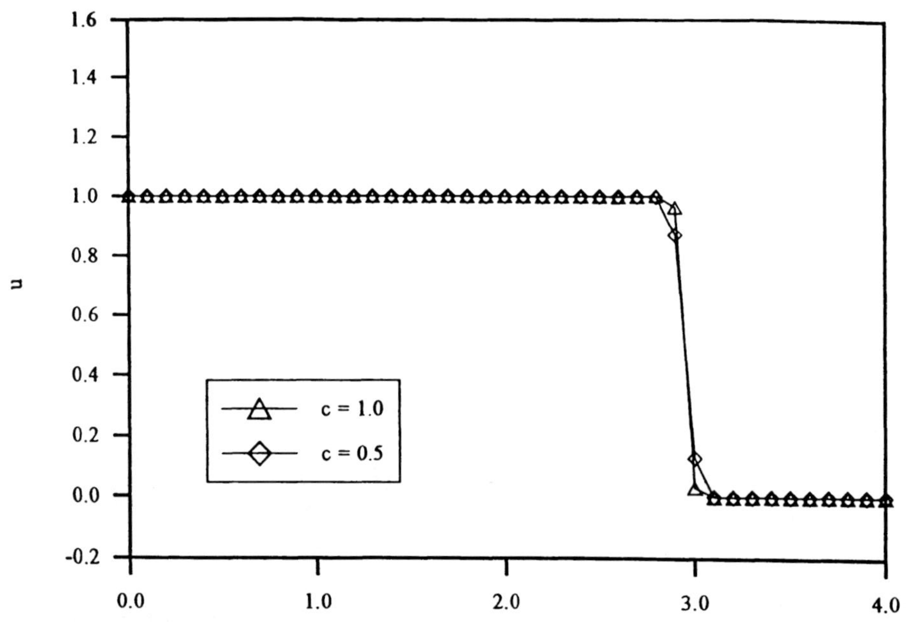

Figure 6-41. Effect of step sizes on the solution of the inviscid Burgers equation by the second-order TVD method with the Harten-Yee upwind limiter (6-132).

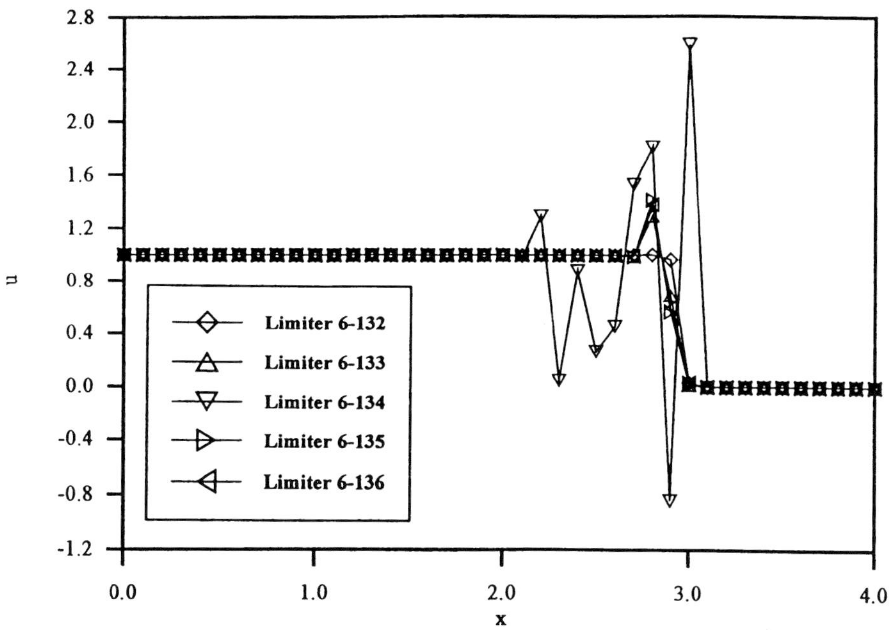

Figure 6-42. Solution of the inviscid Burgers equation by the second-order TVD method with the Harten-Yee upwind limiters at $t = 1.8$ sec, $\Delta x = 0.1$ and $\Delta t = 0.1$.

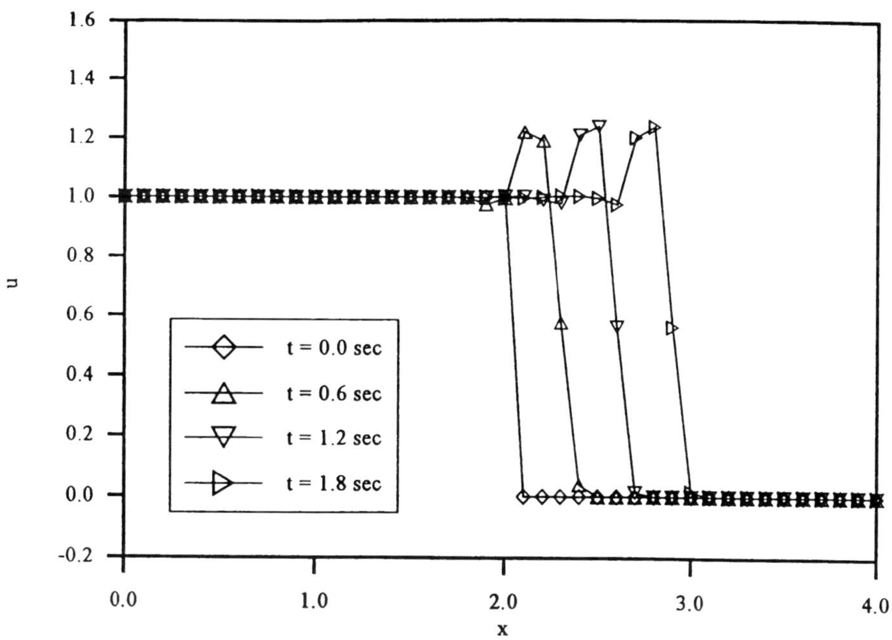

Figure 6-43. Solution of the inviscid Burgers equation by the second-order TVD method with the Roe-Sweby upwind limiter (6-138), $\Delta x = 0.1$ and $\Delta t = 0.1$.

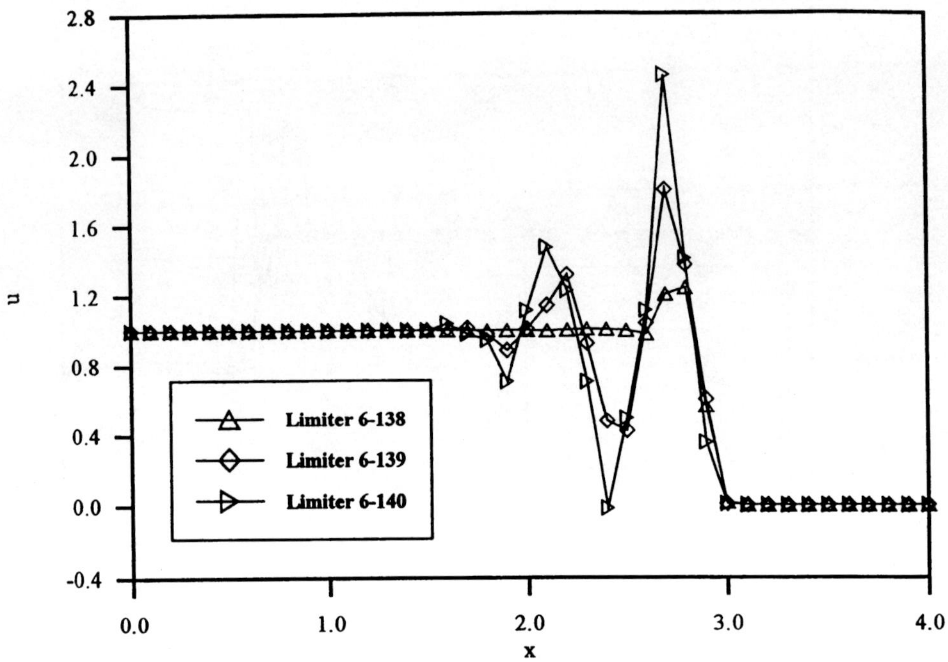

Figure 6-44. Solution of the inviscid Burgers equation by the second-order TVD method with the Roe-Sweby upwind limiters at $t = 1.8$ sec, $\Delta x = 0.1$ and $\Delta t = 0.1$.

6.10.4.3 Davis-Yee Symmetric TVD Limiters. The flux limiter function is defined as

$$\phi_{i+\frac{1}{2}} = -\left[\frac{\Delta t}{\Delta x}\left(\alpha_{i+\frac{1}{2}}\right)^2 G_{i+\frac{1}{2}} + \psi\left(\alpha_{i+\frac{1}{2}}\right)\left(\Delta u_{i+\frac{1}{2}} - G_{i+\frac{1}{2}}\right)\right] \tag{6-141a}$$

$$\phi_{i-\frac{1}{2}} = -\left[\frac{\Delta t}{\Delta x}\left(\alpha_{i-\frac{1}{2}}\right)^2 G_{i-\frac{1}{2}} + \psi\left(\alpha_{i-\frac{1}{2}}\right)\left(\Delta u_{i-\frac{1}{2}} - G_{i-\frac{1}{2}}\right)\right] \tag{6-141b}$$

Again, several choices are available for the limiters as follow

$$G_{i+\frac{1}{2}} = \text{minmod}\left[2\Delta u_{i-\frac{1}{2}}\,,\ 2\Delta u_{i+\frac{1}{2}}\,,\ 2\Delta u_{i+\frac{3}{2}}\,,\ \frac{1}{2}\left(\Delta u_{i-\frac{1}{2}} + \Delta u_{i+\frac{3}{2}}\right)\right] \tag{6-142}$$

$$G_{i+\frac{1}{2}} = \text{minmod}\left[\Delta u_{i-\frac{1}{2}}\,,\ \Delta u_{i+\frac{1}{2}}\,,\ \Delta u_{i+\frac{3}{2}}\right] \tag{6-143}$$

$$G_{i+\frac{1}{2}} = \text{minmod}\left[\Delta u_{i+\frac{1}{2}}\,,\ \Delta u_{i-\frac{1}{2}}\right] + \text{minmod}\left[\Delta u_{i+\frac{1}{2}}\,,\ \Delta u_{i+\frac{3}{2}}\right] - \Delta u_{i+\frac{1}{2}} \tag{6-144}$$

The solution by the second-order TVD scheme given by Equation (6-123) with flux limiter function defined by (6-141) and limiter (6-142) is shown in Figure 6-45 for several time levels and is presented in Table 6.17. The solutions by limiter (6-142) through (6-144) are shown in Figure 6-46. The solutions with all three limiters are similar and well behaved.

6.11 Modified Runge-Kutta Method with TVD

In the previous sections, it was observed that schemes which utilize central difference approximation of second-order will develop oscillations in the vicinity of large gradients. That is, of course, due to the dispersion error in these schemes. The Beam and Warming implicit scheme and the Runge-Kutta scheme or the modified Runge-Kutta scheme are examples of such schemes. To eliminate or reduce the oscillations and thereby improve the solution, damping terms are added, as discussed previously. Since TVD essentially performs in a similar fashion, that is, it eliminates or reduces oscillations in the solution, it can be added to the finite difference equations to provide a mechanism for reducing oscillations. An example of this procedure is the addition of a fourth-order damping term to the modified Runge-Kutta scheme discussed in Section 6.6.8. Typically, TVD is added at the final stage. Thus, the formulation (6-71) through (6-75) repeated here

$$u_i^{(1)} = u_i^n \tag{6-145}$$

$$u_i^{(2)} = u_i^n - \frac{\Delta t}{4}\left(\frac{\partial E}{\partial x}\right)_i^{(1)} \tag{6-146}$$

$$u_i^{(3)} = u_i^n - \frac{\Delta t}{3}\left(\frac{\partial E}{\partial x}\right)_i^{(2)} \tag{6-147}$$

$$u_i^{(4)} = u_i^n - \frac{\Delta t}{2}\left(\frac{\partial E}{\partial x}\right)_i^{(3)} \tag{6-148}$$

$$u_i^{n+1} = u_i^n - \Delta t\left(\frac{\partial E}{\partial x}\right)_i^{(4)} \tag{6-149}$$

is augmented by the following step.

$$u_i^{n+1} = u_i^{n+1} - \frac{\Delta t}{2\Delta x}\left(\phi_{i+\frac{1}{2}}^n - \phi_{i-\frac{1}{2}}^n\right) \tag{6-150}$$

where any one of the flux limiter functions given by (6-131), (6-137), or (6-141), along with appropriate limiters, can be used. The solution with Davis-Yee limiter (6-142) is provided in Table 6-18.

Note that this simple approach can be easily implemented into any existing code which requires the addition of a numerical damping.

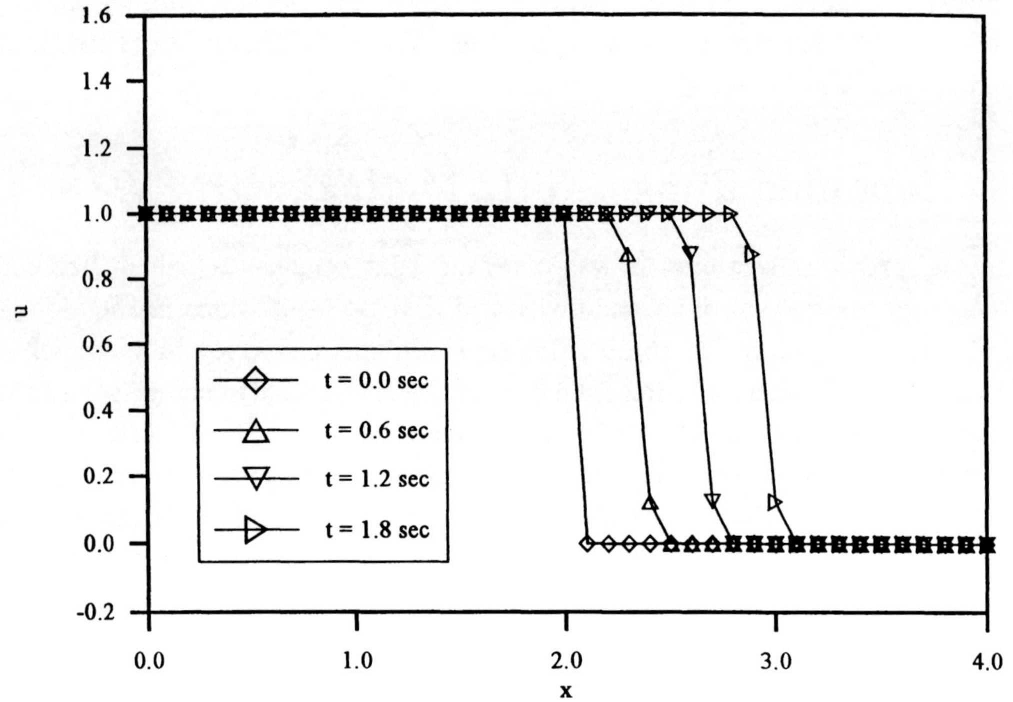

Figure 6-45. Solution of the inviscid Burgers equation by the second-order TVD method with the Davis-Yee symmetric limiter (6-142), $\Delta x = 0.1$ and $\Delta t = 0.1$.

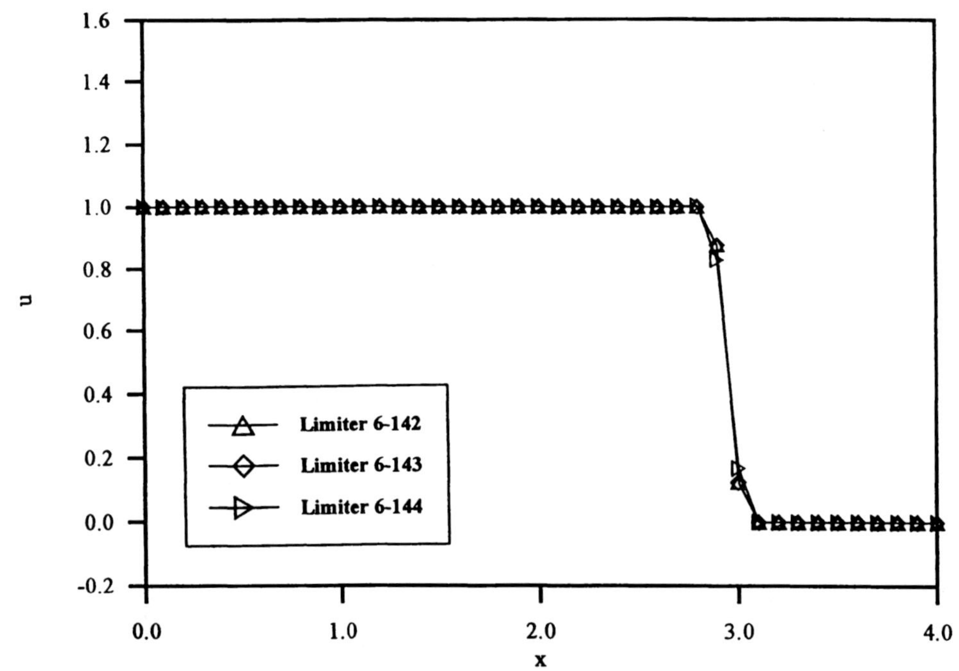

Figure 6-46. Solution of the inviscid Burgers equation by the second-order TVD method with the Davis-Yee symmetric limiters at $t = 1.8$ sec, $\Delta x = 0.1$ and $\Delta t = 0.1$.

6.12 Summary Objectives

After studying this chapter you should be able to do the following:

1. Describe:

    a. Euler's FTFS explicit method

    b. Euler's FTCS explicit method

    c. The explicit first upwind differencing method

    d. The Lax method

    e. The midpoint leapfrog method

    f. The Lax-Wendroff method

    g. Euler's BTCS implicit method

    h. The implicit first upwind differencing method

    i. The Crank-Nicolson method

    j. Splitting methods

    k. Multi-step methods

    l. The Richtmyer/Lax-Wendroff multi-step method

    m. The MacCormack method

    n. A Jacobian

    o. The Beam and Warming implicit method

    p. The Runge-Kutta method

    q. The modified Runge-Kutta method

    r. A smoothing (damping) term

    s. Flux corrected transport scheme

    t. Monotone schemes

    u. TVD schemes

    v. Essentially non-oscillating schemes

    w. Entropy condition

2. Solve the problems for Chapter Six.

6.13 Problems

6.1 A wave is propagating in a closed-end tube. Compute the wave propagation up to $t = 0.15$ sec by solving the first-order wave equation. Assume the speed of sound to be 200 m/sec. The wave has a triangular shape (see Figure P6-1) which is to be used as the initial condition at $t = 0.0$. Solve the problem by the following methods.

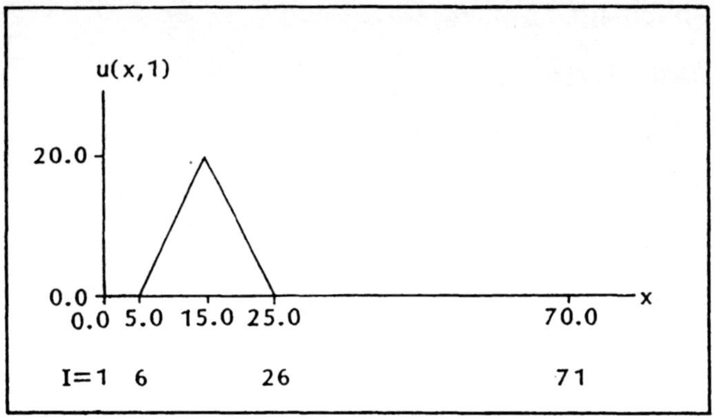

Figure P6-1. Initial wave distribution for problem 6.1.

(a) First upwind differencing

(b) Lax-Wendroff

(c) Euler's BTCS implicit

Three sets of step sizes are specified as follows:

(I) $\Delta x = 1.0 \quad (IM = 71)\ , \quad \Delta t = 0.005 \quad (NM = 31)$

(II) $\Delta x = 1.0 \quad (IM = 71)\ , \quad \Delta t = 0.0025 \quad (NM = 61)$

(III) $\Delta x = 1.0 \quad (IM = 71)\ , \quad \Delta t = 0.00125 \quad (NM = 121)$

Print the solution at intervals of 0.025 sec up to $t = 0.15$ sec.

6.2 Use each of the following methods to solve the Burgers equation:

(a) Lax

(b) Lax-Wendroff

(c) MacCormack

(d) Beam and Warming

The initial condition is specified as:

$$u(x,0) = 5.0 \qquad 0.0 \le x \le 20.0$$

$$u(x,0) = 0.0 \qquad 20.0 < x \le 40.0$$

Print the solution at intervals of 0.4 sec up to $t = 2.4$ sec. The following step-sizes are suggested:

(I) $\Delta x = 1.0$, $\Delta t = 0.1$

(II) $\Delta x = 1.0$, $\Delta t = 0.2$

6.3 An initial velocity distribution representing a compression wave is given by the following:

$$u(x,0) = \begin{cases} 1 & , \quad x < 0.25 \\ 1.25 - x & , \quad 0.25 \le x \le 1.25 \\ 0 & , \quad x > 1.25 \end{cases}$$

The inviscid Burgers equation is used to solve for the wave propagation within a domain of $0.0 \le x \le 4.0$. The solution is sought up to 6.0 seconds. Note that the initial wave is compressed (steepens) with time and subsequently forms a shock wave. Within the specified time and space intervals, no shock reflection occurs, and therefore, the boundary conditions are simply specified as

$$u(0.0, t) = 1.0$$

$$u(4.0, t) = 0.0$$

Use the Lax-Wendroff scheme with a spatial step of 0.05 m to obtain the solutions for the following cases:

(I) $\Delta t = 0.01$
(II) $\Delta t = 0.025$
(III) $\Delta t = 0.05$
(IV) $\Delta t = 0.1$

(a) Print the solution for Case II at time levels of 0.0, 2.0, 4.0, and 6.0 seconds.

(b) Plot the solutions for all cases at time levels of 0.0, 2.0, 4.0, and 6.0 seconds.

(c) Discuss the effect of the time step on stability, accuracy, and efficiency.

6.4 Use the Lax-Wendroff scheme with a second-order damping term to solve the wave propagation described in Problem 6.3. Use spatial and temporal steps of 0.05 m and 0.025 seconds, respectively. Investigate the numerical solutions for the following damping coefficients, ε.

$$\begin{aligned} &\text{(I)} \quad \varepsilon = 0.1 \\ &\text{(II)} \quad \varepsilon = 0.2 \\ &\text{(III)} \quad \varepsilon = 0.3 \end{aligned}$$

(a) Print the solution for Case I at time levels of 0.0, 2.0, 4.0, and 6.0 seconds.

(b) Plot the solutions for all three cases at time levels of 0.0, 2.0, 4.0, and 6.0 seconds.

(c) Discuss the effect of the damping coefficient on the solution.

6.5 Solve the wave propagation described in Problem 6.3 by the Lax-Wendroff scheme with added flux-corrected transport. Use spatial and temporal steps of 0.05 m and 0.025 seconds, respectively. Print and plot the solution at time levels of 0.0, 2.0, 4.0, and 6.0 seconds.

6.6 An initial velocity distribution is given by

$$u(x,0) = \begin{cases} 1.0 & x < 0.75 \\ 1.75 - x & 0.75 \le x \le 1.75 \\ 0.0 & x > 1.75 \end{cases}$$

Compute the wave propagation up to 6.0 seconds within a domain of $0.0 \le x \le 5.0$. For the specified domain and time intervals, we will enforce the following boundary conditions:

$$u(0.0, t) = 1.0$$

$$u(5.0, t) = 0.0$$

The governing equation is the inviscid Burgers equation and the numerical scheme to be used is the Beam and Warming implicit method. The spatial and temporal steps are 0.05 m and 0.025 seconds, respectively. The following tasks are to be investigated.

(I) Beam and Warming scheme with no damping term.

(II) Beam and Warming scheme with a fourth-order damping term and $\varepsilon = 0.1$.

(III) Beam and Warming scheme with added flux-corrected transport. (Use $\varepsilon_1 = -0.1$ and $\varepsilon_2 = -0.2$.)

x	t=0.00	t=0.10	t=0.20	t=0.30	t=0.40	t=0.50
0.0	0.000	0.000	0.000	0.000	0.000	0.000
10.0	0.000	0.000	0.000	0.000	0.000	0.000
20.0	0.000	0.000	0.000	0.000	0.000	0.000
30.0	0.000	0.000	0.000	0.000	0.000	0.000
40.0	0.000	0.000	0.000	0.000	0.000	0.000
50.0	0.000	0.000	0.000	0.000	0.000	0.000
60.0	50.000	0.000	0.000	0.000	0.000	0.000
70.0	86.603	0.000	0.000	0.000	0.000	0.000
80.0	100.000	25.882	0.000	0.000	0.000	0.000
90.0	86.603	70.711	0.000	0.000	0.000	0.000
100.0	50.000	96.593	0.000	0.000	0.000	0.000
110.0	0.000	96.593	50.000	0.000	0.000	0.000
120.0	0.000	70.711	86.603	0.000	0.000	0.000
130.0	0.000	25.882	100.000	25.882	0.000	0.000
140.0	0.000	0.000	86.603	70.711	0.000	0.000
150.0	0.000	0.000	50.000	96.593	0.000	0.000
160.0	0.000	0.000	0.000	96.593	50.000	0.000
170.0	0.000	0.000	0.000	70.711	86.603	0.000
180.0	0.000	0.000	0.000	25.882	100.000	25.882
190.0	0.000	0.000	0.000	0.000	86.603	70.711
200.0	0.000	0.000	0.000	0.000	50.000	96.593
210.0	0.000	0.000	0.000	0.000	0.000	96.593
220.0	0.000	0.000	0.000	0.000	0.000	70.711
230.0	0.000	0.000	0.000	0.000	0.000	25.882
240.0	0.000	0.000	0.000	0.000	0.000	0.000
250.0	0.000	0.000	0.000	0.000	0.000	0.000
260.0	0.000	0.000	0.000	0.000	0.000	0.000
270.0	0.000	0.000	0.000	0.000	0.000	0.000
280.0	0.000	0.000	0.000	0.000	0.000	0.000
290.0	0.000	0.000	0.000	0.000	0.000	0.000
300.0	0.000	0.000	0.000	0.000	0.000	0.000
310.0	0.000	0.000	0.000	0.000	0.000	0.000
320.0	0.000	0.000	0.000	0.000	0.000	0.000
330.0	0.000	0.000	0.000	0.000	0.000	0.000
340.0	0.000	0.000	0.000	0.000	0.000	0.000
350.0	0.000	0.000	0.000	0.000	0.000	0.000
360.0	0.000	0.000	0.000	0.000	0.000	0.000
370.0	0.000	0.000	0.000	0.000	0.000	0.000
380.0	0.000	0.000	0.000	0.000	0.000	0.000
390.0	0.000	0.000	0.000	0.000	0.000	0.000
400.0	0.000	0.000	0.000	0.000	0.000	0.000

Table 6-1a. Solution of the first-order wave equation by the explicit first upwind differencing scheme, $\Delta x=5.0$, $\Delta t=0.02$, $c=1.0$.

x	t=0.00	t=0.10	t=0.20	t=0.30	t=0.40	t=0.50
0.0	0.000	0.000	0.000	0.000	0.000	0.000
10.0	0.000	0.000	0.000	0.000	0.000	0.000
20.0	0.000	0.000	0.000	0.000	0.000	0.000
30.0	0.000	0.000	0.000	0.000	0.000	0.000
40.0	0.000	0.000	0.000	0.000	0.000	0.000
50.0	0.000	0.000	0.000	0.000	0.000	0.000
60.0	50.000	0.706	0.004	0.000	0.000	0.000
70.0	86.603	8.582	0.149	0.001	0.000	0.000
80.0	100.000	32.346	1.622	0.032	0.000	0.000
90.0	86.603	64.019	8.231	0.327	0.007	0.000
100.0	50.000	85.192	24.363	1.923	0.068	0.001
110.0	0.000	84.812	48.199	7.256	0.438	0.015
120.0	0.000	62.532	69.770	19.115	1.936	0.099
130.0	0.000	30.862	77.923	37.372	6.238	0.489
140.0	0.000	9.109	68.207	56.721	15.303	1.823
150.0	0.000	1.492	46.141	68.880	29.577	5.311
160.0	0.000	0.129	23.493	67.970	46.249	12.407
170.0	0.000	0.006	8.786	54.654	59.647	23.751
180.0	0.000	0.000	2.373	35.623	64.235	37.904
190.0	0.000	0.000	0.458	18.669	58.113	51.109
200.0	0.000	0.000	0.063	7.808	44.225	58.787
210.0	0.000	0.000	0.006	2.591	28.268	58.035
220.0	0.000	0.000	0.000	0.680	15.137	49.332
230.0	0.000	0.000	0.000	0.141	6.774	36.151
240.0	0.000	0.000	0.000	0.023	2.528	22.836
250.0	0.000	0.000	0.000	0.003	0.786	12.429
260.0	0.000	0.000	0.000	0.000	0.203	5.824
270.0	0.000	0.000	0.000	0.000	0.044	2.349
280.0	0.000	0.000	0.000	0.000	0.008	0.815
290.0	0.000	0.000	0.000	0.000	0.001	0.243
300.0	0.000	0.000	0.000	0.000	0.000	0.062
310.0	0.000	0.000	0.000	0.000	0.000	0.014
320.0	0.000	0.000	0.000	0.000	0.000	0.003
330.0	0.000	0.000	0.000	0.000	0.000	0.000
340.0	0.000	0.000	0.000	0.000	0.000	0.000
350.0	0.000	0.000	0.000	0.000	0.000	0.000
360.0	0.000	0.000	0.000	0.000	0.000	0.000
370.0	0.000	0.000	0.000	0.000	0.000	0.000
380.0	0.000	0.000	0.000	0.000	0.000	0.000
390.0	0.000	0.000	0.000	0.000	0.000	0.000
400.0	0.000	0.000	0.000	0.000	0.000	0.000

Table 6-1b. Solution of the first-order wave equation by the explicit first upwind differencing scheme, $\Delta x=5.0$, $\Delta t=0.005$, $c=0.25$.

x	t=0.00	t=0.10	t=0.20	t=0.30	t=0.40	t=0.50
0.0	0.000	0.000	0.000	0.000	0.000	0.000
10.0	0.000	0.000	0.000	0.000	0.000	0.000
20.0	0.000	0.000	0.000	0.000	0.000	0.000
30.0	0.000	0.000	0.000	0.000	0.000	0.000
40.0	0.000	0.000	0.000	0.000	0.000	0.000
50.0	0.000	0.000	0.000	0.000	0.000	0.000
60.0	50.000	3.585	0.175	0.007	0.000	0.000
70.0	86.603	17.629	1.796	0.132	0.008	0.000
80.0	100.000	40.508	7.394	0.867	0.078	0.006
90.0	86.603	62.930	18.670	3.298	0.418	0.042
100.0	50.000	74.727	34.239	8.730	1.520	0.202
110.0	0.000	69.696	49.511	17.725	4.138	0.714
120.0	0.000	51.043	58.804	29.229	8.996	1.983
130.0	0.000	30.755	58.812	40.559	16.279	4.527
140.0	0.000	16.034	50.653	48.521	25.235	8.777
150.0	0.000	7.508	38.388	50.986	34.218	14.787
160.0	0.000	3.240	26.105	47.807	41.250	22.039
170.0	0.000	1.312	16.197	40.550	44.794	29.465
180.0	0.000	0.505	9.297	31.488	44.309	35.744
190.0	0.000	0.187	4.993	22.618	40.313	39.723
200.0	0.000	0.067	2.532	15.165	34.019	40.777
210.0	0.000	0.023	1.222	9.564	26.825	38.946
220.0	0.000	0.008	0.565	5.711	19.893	34.827
230.0	0.000	0.003	0.251	3.248	13.954	29.325
240.0	0.000	0.001	0.108	1.767	9.305	23.365
250.0	0.000	0.000	0.045	0.924	5.925	17.695
260.0	0.000	0.000	0.018	0.466	3.617	12.789
270.0	0.000	0.000	0.007	0.228	2.124	8.852
280.0	0.000	0.000	0.003	0.108	1.203	5.887
290.0	0.000	0.000	0.001	0.050	0.660	3.772
300.0	0.000	0.000	0.000	0.022	0.351	2.335
310.0	0.000	0.000	0.000	0.010	0.181	1.400
320.0	0.000	0.000	0.000	0.004	0.091	0.814
330.0	0.000	0.000	0.000	0.002	0.045	0.460
340.0	0.000	0.000	0.000	0.001	0.022	0.254
350.0	0.000	0.000	0.000	0.000	0.010	0.136
360.0	0.000	0.000	0.000	0.000	0.005	0.072
370.0	0.000	0.000	0.000	0.000	0.002	0.037
380.0	0.000	0.000	0.000	0.000	0.001	0.018
390.0	0.000	0.000	0.000	0.000	0.000	0.009
400.0	0.000	0.000	0.000	0.000	0.000	0.000

Table 6-2a. Solution of the first-order wave equation by the implicit first upwind differencing scheme, $\Delta x=5.0$, $\Delta t=0.02$, $c=1.0$.

x	t=0.00	t=0.10	t=0.20	t=0.30	t=0.40	t=0.50
0.0	0.000	0.000	0.000	0.000	0.000	0.000
10.0	0.000	0.000	0.000	0.000	0.000	0.000
20.0	0.000	0.000	0.000	0.000	0.000	0.000
30.0	0.000	0.000	0.000	0.000	0.000	0.000
40.0	0.000	0.000	0.000	0.000	0.000	0.000
50.0	0.000	0.000	0.000	0.000	0.000	0.000
60.0	50.000	1.770	0.034	0.001	0.000	0.000
70.0	86.603	12.778	0.621	0.019	0.000	0.000
80.0	100.000	36.264	3.865	0.214	0.008	0.000
90.0	86.603	63.138	13.090	1.249	0.074	0.003
100.0	50.000	79.939	29.470	4.625	0.418	0.026
110.0	0.000	77.891	49.131	12.188	1.633	0.143
120.0	0.000	57.494	64.406	24.566	4.778	0.579
130.0	0.000	31.585	68.367	39.696	11.020	1.824
140.0	0.000	13.151	59.547	53.022	20.772	4.647
150.0	0.000	4.305	42.941	59.706	32.853	9.849
160.0	0.000	1.151	25.914	57.434	44.435	17.746
170.0	0.000	0.259	13.259	47.676	52.121	27.647
180.0	0.000	0.051	5.833	34.458	53.587	37.733
190.0	0.000	0.009	2.238	21.873	48.708	45.584
200.0	0.000	0.001	0.759	12.299	39.433	49.156
210.0	0.000	0.000	0.230	6.177	28.627	47.653
220.0	0.000	0.000	0.063	2.793	18.754	41.783
230.0	0.000	0.000	0.016	1.146	11.152	33.319
240.0	0.000	0.000	0.004	0.429	6.055	24.286
250.0	0.000	0.000	0.001	0.148	3.017	16.256
260.0	0.000	0.000	0.000	0.047	1.387	10.036
270.0	0.000	0.000	0.000	0.014	0.591	5.739
280.0	0.000	0.000	0.000	0.004	0.234	3.051
290.0	0.000	0.000	0.000	0.001	0.087	1.514
300.0	0.000	0.000	0.000	0.000	0.030	0.703
310.0	0.000	0.000	0.000	0.000	0.010	0.307
320.0	0.000	0.000	0.000	0.000	0.003	0.126
330.0	0.000	0.000	0.000	0.000	0.001	0.049
340.0	0.000	0.000	0.000	0.000	0.000	0.018
350.0	0.000	0.000	0.000	0.000	0.000	0.006
360.0	0.000	0.000	0.000	0.000	0.000	0.002
370.0	0.000	0.000	0.000	0.000	0.000	0.001
380.0	0.000	0.000	0.000	0.000	0.000	0.000
390.0	0.000	0.000	0.000	0.000	0.000	0.000
400.0	0.000	0.000	0.000	0.000	0.000	0.000

Table 6-2b. Solution of the first-order wave equation by the implicit first upwind differencing scheme, $\Delta x=5.0$, $\Delta t=0.005$, $c=0.25$.

x	t=0.00	t=0.10	t=0.20	t=0.30	t=0.40	t=0.50
0.0	0.000	0.000	0.000	0.000	0.000	0.000
10.0	0.000	0.000	0.000	0.000	0.000	0.000
20.0	0.000	0.000	0.000	0.000	0.000	0.000
30.0	0.000	0.000	0.000	0.000	0.000	0.000
40.0	0.000	0.000	0.000	0.000	0.000	0.000
50.0	0.000	0.000	0.000	0.000	0.000	0.000
60.0	50.000	0.000	0.000	0.000	0.000	0.000
70.0	86.603	0.000	0.000	0.000	0.000	0.000
80.0	100.000	25.882	0.000	0.000	0.000	0.000
90.0	86.603	70.711	0.000	0.000	0.000	0.000
100.0	50.000	96.593	0.000	0.000	0.000	0.000
110.0	0.000	96.593	50.000	0.000	0.000	0.000
120.0	0.000	70.711	86.603	0.000	0.000	0.000
130.0	0.000	25.882	100.000	25.882	0.000	0.000
140.0	0.000	0.000	86.603	70.711	0.000	0.000
150.0	0.000	0.000	50.000	96.593	0.000	0.000
160.0	0.000	0.000	0.000	96.593	50.000	0.000
170.0	0.000	0.000	0.000	70.711	86.603	0.000
180.0	0.000	0.000	0.000	25.882	100.000	25.882
190.0	0.000	0.000	0.000	0.000	86.603	70.711
200.0	0.000	0.000	0.000	0.000	50.000	96.593
210.0	0.000	0.000	0.000	0.000	0.000	96.593
220.0	0.000	0.000	0.000	0.000	0.000	70.711
230.0	0.000	0.000	0.000	0.000	0.000	25.882
240.0	0.000	0.000	0.000	0.000	0.000	0.000
250.0	0.000	0.000	0.000	0.000	0.000	0.000
260.0	0.000	0.000	0.000	0.000	0.000	0.000
270.0	0.000	0.000	0.000	0.000	0.000	0.000
280.0	0.000	0.000	0.000	0.000	0.000	0.000
290.0	0.000	0.000	0.000	0.000	0.000	0.000
300.0	0.000	0.000	0.000	0.000	0.000	0.000
310.0	0.000	0.000	0.000	0.000	0.000	0.000
320.0	0.000	0.000	0.000	0.000	0.000	0.000
330.0	0.000	0.000	0.000	0.000	0.000	0.000
340.0	0.000	0.000	0.000	0.000	0.000	0.000
350.0	0.000	0.000	0.000	0.000	0.000	0.000
360.0	0.000	0.000	0.000	0.000	0.000	0.000
370.0	0.000	0.000	0.000	0.000	0.000	0.000
380.0	0.000	0.000	0.000	0.000	0.000	0.000
390.0	0.000	0.000	0.000	0.000	0.000	0.000
400.0	0.000	0.000	0.000	0.000	0.000	0.000

Table 6-3. Solution of the first-order wave equation by the explicit Lax-Wendroff differencing scheme, Δx=5.0, Δt=0.02, c=1.0.

x	t=0.00	t=0.10	t=0.20	t=0.30	t=0.40	t=0.50
0.0	0.000	0.000	0.000	0.000	0.000	0.000
10.0	0.000	-0.302	-0.380	0.390	-0.152	0.128
20.0	0.000	-0.777	-0.345	0.374	-0.108	0.250
30.0	0.000	-1.323	-0.019	0.004	0.118	0.376
40.0	0.000	-1.041	-0.137	-0.303	0.364	0.497
50.0	0.000	0.684	-0.618	-0.408	0.513	0.581
60.0	50.000	-1.430	-0.628	-0.284	0.482	0.618
70.0	86.603	9.296	-1.215	-0.017	0.338	0.594
80.0	100.000	36.074	-0.083	0.087	0.192	0.511
90.0	86.603	65.288	9.559	-0.239	0.141	0.389
100.0	50.000	83.220	29.137	1.330	0.006	0.276
110.0	0.000	82.211	52.608	8.663	-0.007	0.179
120.0	0.000	56.301	70.691	23.405	1.673	0.048
130.0	0.000	28.147	73.776	42.499	7.599	0.183
140.0	0.000	11.374	60.581	59.121	19.129	1.771
150.0	0.000	3.971	40.536	65.906	34.698	6.588
160.0	0.000	1.249	22.969	60.617	49.610	15.791
170.0	0.000	0.363	11.383	47.271	58.383	28.571
180.0	0.000	0.099	5.061	32.042	58.201	41.801
190.0	0.000	0.026	2.058	19.284	50.289	51.387
200.0	0.000	0.006	0.777	10.489	38.395	54.474
210.0	0.000	0.002	0.275	5.230	26.326	50.792
220.0	0.000	0.000	0.092	2.419	16.434	42.333
230.0	0.000	0.000	0.030	1.048	9.448	31.964
240.0	0.000	0.000	0.009	0.429	5.051	22.112
250.0	0.000	0.000	0.003	0.167	2.531	14.150
260.0	0.000	0.000	0.001	0.062	1.197	8.445
270.0	0.000	0.000	0.000	0.022	0.538	4.733
280.0	0.000	0.000	0.000	0.008	0.230	2.506
290.0	0.000	0.000	0.000	0.003	0.095	1.260
300.0	0.000	0.000	0.000	0.001	0.037	0.605
310.0	0.000	0.000	0.000	0.000	0.014	0.278
320.0	0.000	0.000	0.000	0.000	0.005	0.123
330.0	0.000	0.000	0.000	0.000	0.002	0.052
340.0	0.000	0.000	0.000	0.000	0.001	0.022
350.0	0.000	0.000	0.000	0.000	0.000	0.009
360.0	0.000	0.000	0.000	0.000	0.000	0.003
370.0	0.000	0.000	0.000	0.000	0.000	0.001
380.0	0.000	0.000	0.000	0.000	0.000	0.000
390.0	0.000	0.000	0.000	0.000	0.000	0.000
400.0	0.000	0.000	0.000	0.000	0.000	0.000

Table 6-4a. Solution of the first-order wave equation by the implicit BTCS differencing scheme, Δx=5.0, Δt=0.02, c=1.0.

x	t=0.00	t=0.10	t=0.20	t=0.30	t=0.40	t=0.50
0.0	0.000	0.000	0.000	0.000	0.000	0.000
10.0	0.000	-0.234	-0.250	0.034	0.089	0.082
20.0	0.000	-0.512	-0.404	0.052	0.144	0.173
30.0	0.000	-0.838	-0.392	0.021	0.158	0.272
40.0	0.000	-1.068	-0.246	-0.088	0.162	0.367
50.0	0.000	-0.590	-0.227	-0.219	0.185	0.436
60.0	50.000	2.548	-0.627	-0.231	0.213	0.461
70.0	86.603	19.443	0.739	-0.227	0.231	0.442
80.0	100.000	43.760	7.790	0.511	0.212	0.391
90.0	86.603	64.816	21.376	3.652	0.439	0.329
100.0	50.000	73.640	38.235	10.552	1.725	0.385
110.0	0.000	66.183	52.507	21.037	5.107	0.925
120.0	0.000	45.201	58.362	33.043	11.168	2.545
130.0	0.000	27.532	54.601	42.971	19.504	5.845
140.0	0.000	15.743	45.075	48.014	28.552	11.062
150.0	0.000	8.648	33.979	47.586	36.257	17.775
160.0	0.000	4.620	23.934	42.896	41.012	24.954
170.0	0.000	2.419	16.001	35.847	42.200	31.322
180.0	0.000	1.246	10.266	28.173	40.152	35.802
190.0	0.000	0.635	6.371	21.050	35.784	37.814
200.0	0.000	0.320	3.847	15.078	30.177	37.342
210.0	0.000	0.160	2.271	10.422	24.277	34.806
220.0	0.000	0.079	1.314	6.987	18.753	30.859
230.0	0.000	0.039	0.748	4.562	13.983	26.189
240.0	0.000	0.019	0.420	2.911	10.108	21.387
250.0	0.000	0.009	0.233	1.820	7.110	16.881
260.0	0.000	0.005	0.127	1.118	4.882	12.926
270.0	0.000	0.002	0.069	0.676	3.279	9.632
280.0	0.000	0.001	0.037	0.403	2.160	7.003
290.0	0.000	0.001	0.020	0.237	1.398	4.980
300.0	0.000	0.000	0.010	0.138	0.891	3.471
310.0	0.000	0.000	0.006	0.079	0.559	2.374
320.0	0.000	0.000	0.003	0.045	0.346	1.597
330.0	0.000	0.000	0.002	0.025	0.212	1.057
340.0	0.000	0.000	0.001	0.014	0.128	0.689
350.0	0.000	0.000	0.000	0.008	0.076	0.442
360.0	0.000	0.000	0.000	0.004	0.045	0.278
370.0	0.000	0.000	0.000	0.002	0.026	0.169
380.0	0.000	0.000	0.000	0.001	0.014	0.096
390.0	0.000	0.000	0.000	0.001	0.006	0.043
400.0	0.000	0.000	0.000	0.000	0.000	0.000

Table 6-4b. Solution of the first-order wave equation by the implicit BTCS differencing scheme, Δx=5.0, Δt=0.05, c=2.5.

x	t=0.0	t=0.3	t=0.6	t=0.9	t=1.2	t=1.5	t=1.8
0.00	1.00000	1.00000	1.00000	1.00000	1.00000	1.00000	1.00000
0.20	1.00000	1.00000	1.00000	1.00000	1.00000	1.00000	1.00000
0.40	1.00000	1.00000	1.00000	1.00000	1.00000	1.00000	1.00000
0.60	1.00000	1.00000	1.00000	1.00000	1.00000	1.00000	1.00000
0.80	1.00000	1.00000	1.00000	1.00000	1.00000	1.00000	1.00000
1.00	1.00000	1.00000	1.00000	1.00000	1.00000	1.00000	1.00000
1.20	1.00000	1.00000	1.00000	1.00000	1.00000	1.00000	1.00000
1.40	1.00000	1.00000	1.00000	1.00000	1.00000	1.00000	1.00000
1.60	1.00000	1.00000	1.00000	1.00000	1.00000	1.00000	1.00000
1.80	1.00000	0.99994	1.00000	1.00000	1.00000	1.00000	1.00000
2.00	1.00000	0.92578	0.99997	1.00000	1.00000	1.00000	1.00000
2.20	0.00000	0.32428	0.96045	0.98520	1.00000	1.00000	1.00000
2.40	0.00000	0.00000	0.48537	0.63793	0.99605	0.99929	1.00000
2.60	0.00000	0.00000	0.05420	0.11976	0.77501	0.88082	0.99992
2.80	0.00000	0.00000	0.00000	0.00711	0.20861	0.32732	0.94819
3.00	0.00000	0.00000	0.00000	0.00000	0.01943	0.03965	0.47212
3.20	0.00000	0.00000	0.00000	0.00000	0.00089	0.00281	0.07313
3.40	0.00000	0.00000	0.00000	0.00000	0.00000	0.00011	0.00624
3.60	0.00000	0.00000	0.00000	0.00000	0.00000	0.00000	0.00039
3.80	0.00000	0.00000	0.00000	0.00000	0.00000	0.00000	0.00001
4.00	0.00000	0.00000	0.00000	0.00000	0.00000	0.00000	0.00000

Table 6-5. Solution of inviscid Burgers equation by the Lax scheme, $\Delta x = 0.1$, $\Delta t = 0.1$.

x	t=0.0	t=0.3	t=0.6	t=0.9	t=1.2	t=1.5	t=1.8
0.00	1.00000	1.00000	1.00000	1.00000	1.00000	1.00000	1.00000
0.20	1.00000	1.00000	1.00000	1.00000	1.00000	1.00000	1.00000
0.40	1.00000	1.00000	1.00000	1.00000	1.00000	1.00000	1.00000
0.60	1.00000	1.00000	1.00000	1.00000	1.00000	1.00000	1.00000
0.80	1.00000	1.00000	1.00000	1.00000	1.00000	1.00000	1.00000
1.00	1.00000	1.00000	1.00000	1.00000	1.00000	1.00000	1.00000
1.20	1.00000	1.00000	1.00000	1.00000	1.00000	1.00000	1.00000
1.40	1.00000	1.00000	1.00000	1.00000	1.00000	1.00000	1.00000
1.60	1.00000	1.00000	1.00000	1.00000	1.00000	1.00000	1.00000
1.80	1.00000	1.00000	1.00000	1.00000	1.00000	1.00000	1.00000
2.00	1.00000	1.00341	1.00002	1.00000	1.00000	1.00000	1.00000
2.20	0.00000	0.33616	1.06496	1.00105	1.00000	1.00000	1.00000
2.40	0.00000	0.00000	0.03165	1.16745	1.00868	1.00002	1.00000
2.60	0.00000	0.00000	0.00000	0.00024	0.89524	1.00712	1.00003
2.80	0.00000	0.00000	0.00000	0.00000	0.00000	0.32363	1.06544
3.00	0.00000	0.00000	0.00000	0.00000	0.00000	0.00000	0.03067
3.20	0.00000	0.00000	0.00000	0.00000	0.00000	0.00000	0.00000
3.40	0.00000	0.00000	0.00000	0.00000	0.00000	0.00000	0.00000
3.60	0.00000	0.00000	0.00000	0.00000	0.00000	0.00000	0.00000
3.80	0.00000	0.00000	0.00000	0.00000	0.00000	0.00000	0.00000
4.00	0.00000	0.00000	0.00000	0.00000	0.00000	0.00000	0.00000

Table 6-6. Solution of inviscid Burgers equation by the Lax-Wendroff scheme, $\Delta x = 0.1$, $\Delta t = 0.1$.

x	t=0.0	t=0.3	t=0.6	t=0.9	t=1.2	t=1.5	t=1.8
0.00	1.00000	1.00000	1.00000	1.00000	1.00000	1.00000	1.00000
0.20	1.00000	1.00000	1.00000	1.00000	1.00000	1.00000	1.00000
0.40	1.00000	1.00000	1.00000	1.00000	1.00000	1.00000	1.00000
0.60	1.00000	1.00000	1.00000	1.00000	1.00000	1.00000	1.00000
0.80	1.00000	1.00000	1.00000	1.00000	1.00000	1.00000	1.00000
1.00	1.00000	1.00000	1.00000	1.00000	1.00000	1.00000	1.00000
1.20	1.00000	1.00000	1.00000	1.00000	1.00000	1.00000	1.00000
1.40	1.00000	1.00000	1.00000	1.00000	1.00000	1.00000	1.00000
1.60	1.00000	1.00000	1.00000	1.00000	1.00000	1.00000	1.00000
1.80	1.00000	1.00000	1.00000	1.00000	1.00000	1.00000	1.00000
2.00	1.00000	0.99722	1.00000	1.00000	1.00000	1.00000	1.00000
2.20	0.00000	0.52908	0.98402	0.99989	1.00000	1.00000	1.00000
2.40	0.00000	0.00000	0.12585	0.95309	0.99961	1.00000	1.00000
2.60	0.00000	0.00000	0.00000	0.00400	0.90062	0.99853	1.00000
2.80	0.00000	0.00000	0.00000	0.00000	0.00000	0.53880	0.98088
3.00	0.00000	0.00000	0.00000	0.00000	0.00000	0.00000	0.12221
3.20	0.00000	0.00000	0.00000	0.00000	0.00000	0.00000	0.00000
3.40	0.00000	0.00000	0.00000	0.00000	0.00000	0.00000	0.00000
3.60	0.00000	0.00000	0.00000	0.00000	0.00000	0.00000	0.00000
3.80	0.00000	0.00000	0.00000	0.00000	0.00000	0.00000	0.00000
4.00	0.00000	0.00000	0.00000	0.00000	0.00000	0.00000	0.00000

Table 6-7. Solution of inviscid Burgers equation by the MacCormack scheme, $\Delta x = 0.1$, $\Delta t = 0.1$.

x	t=0.0	t=0.3	t=0.6	t=0.9	t=1.2	t=1.5	t=1.8
0.00	1.00000	1.00000	1.00000	1.00000	1.00000	1.00000	1.00000
0.20	1.00000	1.00000	1.00000	1.00004	1.00107	1.01163	1.05227
0.40	1.00000	1.00000	1.00000	1.00027	1.00530	1.03858	1.10826
0.60	1.00000	1.00000	1.00004	1.00170	1.02115	1.09095	1.12052
0.80	1.00000	1.00000	1.00033	1.00898	1.06480	1.13787	0.99854
1.00	1.00000	1.00002	1.00251	1.03761	1.13405	1.06811	0.83563
1.20	1.00000	1.00029	1.01558	1.11041	1.13168	0.84434	1.02225
1.40	1.00000	1.00308	1.07161	1.17653	0.89280	0.93520	1.18901
1.60	1.00000	1.02757	1.19184	0.98016	0.83288	1.24605	0.73017
1.80	1.00000	1.16399	1.10403	0.71707	1.29572	0.72217	1.10264
2.00	1.00000	1.28523	0.56385	1.35769	0.69121	1.07562	1.11638
2.20	0.00000	0.26982	1.50247	0.54340	1.10400	1.10630	0.60736
2.40	0.00000	0.00000	0.04774	1.31217	0.90055	0.63963	1.47633
2.60	0.00000	0.00000	0.00000	0.00396	0.92439	1.30304	0.27484
2.80	0.00000	0.00000	0.00000	0.00000	0.00013	0.52559	1.59021
3.00	0.00000	0.00000	0.00000	0.00000	0.00000	0.00000	0.23271
3.20	0.00000	0.00000	0.00000	0.00000	0.00000	0.00000	0.00000
3.40	0.00000	0.00000	0.00000	0.00000	0.00000	0.00000	0.00000
3.60	0.00000	0.00000	0.00000	0.00000	0.00000	0.00000	0.00000
3.80	0.00000	0.00000	0.00000	0.00000	0.00000	0.00000	0.00000
4.00	0.00000	0.00000	0.00000	0.00000	0.00000	0.00000	0.00000

Table 6-8. Solution of inviscid Burgers equation by the Beam-Warming scheme, $\Delta x = 0.1$, $\Delta t = 0.1$.

x	t=0.0	t=0.3	t=0.6	t=0.9	t=1.2	t=1.5	t=1.8
0.00	1.00000	1.00000	1.00000	1.00000	1.00000	1.00000	1.00000
0.20	1.00000	1.00000	1.00000	1.00000	1.00000	1.00000	1.00000
0.40	1.00000	1.00000	1.00000	1.00000	1.00000	1.00000	1.00000
0.60	1.00000	1.00000	1.00000	1.00000	1.00000	1.00000	1.00000
0.80	1.00000	1.00000	1.00000	1.00000	1.00000	1.00000	1.00000
1.00	1.00000	1.00000	1.00000	1.00000	1.00000	1.00000	1.00000
1.20	1.00000	1.00000	0.99999	1.00000	1.00000	1.00000	1.00000
1.40	1.00000	1.00001	0.99997	0.99997	1.00001	1.00000	1.00000
1.60	1.00000	0.99966	1.00022	1.00008	0.99995	1.00000	1.00001
1.80	1.00000	1.00486	0.99621	1.00033	1.00041	0.99988	0.99999
2.00	1.00000	1.08229	1.01119	0.98931	1.00077	1.00080	0.99979
2.20	0.00000	0.41905	1.13208	1.01986	0.98603	1.00042	1.00093
2.40	0.00000	-0.05886	0.12621	1.04333	1.04934	0.99240	0.99702
2.60	0.00000	-0.00100	-0.00594	-0.06268	0.78578	1.11332	1.00320
2.80	0.00000	0.00000	0.00147	0.00843	-0.07988	0.42860	1.14567
3.00	0.00000	0.00000	0.00027	-0.00119	0.00893	-0.03864	0.10703
3.20	0.00000	0.00000	0.00000	0.00005	-0.00087	0.00336	-0.00336
3.40	0.00000	0.00000	0.00000	-0.00002	0.00010	-0.00028	-0.00034
3.60	0.00000	0.00000	0.00000	0.00000	0.00000	0.00002	0.00011
3.80	0.00000	0.00000	0.00000	0.00000	0.00000	0.00000	-0.00002
4.00	0.00000	0.00000	0.00000	0.00000	0.00000	0.00000	0.00000

Table 6-9. Solution of inviscid Burgers equation by the Beam-Warming scheme with a damping of $\varepsilon_e = 0.1$, $\Delta x = 0.1$, $\Delta t = 0.1$.

x	t=0.0	t=0.3	t=0.6	t=0.9	t=1.2	t=1.5	t=1.8
0.00	1.00000	1.00000	1.00000	1.00000	1.00000	1.00000	1.00000
0.20	1.00000	1.00000	1.00000	1.00000	1.00040	1.00922	1.05810
0.40	1.00000	1.00000	1.00000	1.00005	1.00320	1.03779	1.12298
0.60	1.00000	1.00000	1.00000	1.00069	1.01809	1.10047	1.12165
0.80	1.00000	1.00000	1.00006	1.00612	1.06796	1.15003	0.95036
1.00	1.00000	1.00000	1.00109	1.03563	1.14897	1.03556	0.84019
1.20	1.00000	1.00003	1.01201	1.12073	1.12066	0.81144	1.12040
1.40	1.00000	1.00130	1.07384	1.18303	0.83816	1.03705	1.08015
1.60	1.00000	1.02431	1.20528	0.92345	0.91948	1.18274	0.71304
1.80	1.00000	1.17251	1.05811	0.78059	1.27003	0.65292	1.26079
2.00	1.00000	1.23969	0.61833	1.34386	0.60444	1.24083	0.82269
2.20	0.00000	0.31216	1.44482	0.50268	1.23605	0.83760	0.90897
2.40	0.00000	0.00001	0.08645	1.33564	0.71803	0.89331	1.19113
2.60	0.00000	0.00000	0.00000	0.01752	1.05190	1.04211	0.52733
2.80	0.00000	0.00000	0.00000	0.00000	0.00259	0.71646	1.33155
3.00	0.00000	0.00000	0.00000	0.00000	0.00000	0.00026	0.41765
3.20	0.00000	0.00000	0.00000	0.00000	0.00000	0.00000	0.00001
3.40	0.00000	0.00000	0.00000	0.00000	0.00000	0.00000	0.00000
3.60	0.00000	0.00000	0.00000	0.00000	0.00000	0.00000	0.00000
3.80	0.00000	0.00000	0.00000	0.00000	0.00000	0.00000	0.00000
4.00	0.00000	0.00000	0.00000	0.00000	0.00000	0.00000	0.00000

Table 6-10. Solution of inviscid Burgers equation by the 4th-order Runge-Kutta scheme, $\Delta x = 0.1$, $\Delta t = 0.1$.

x	t=0.0	t=0.3	t=0.6	t=0.9	t=1.2	t=1.5	t=1.8
0.00	1.00000	1.00000	1.00000	1.00000	1.00000	1.00000	1.00000
0.20	1.00000	1.00000	1.00000	1.00000	1.00000	1.00000	1.00000
0.40	1.00000	1.00000	1.00000	1.00000	1.00000	1.00000	1.00000
0.60	1.00000	1.00000	1.00000	1.00000	1.00000	1.00000	1.00000
0.80	1.00000	1.00000	1.00000	1.00000	1.00000	1.00000	1.00000
1.00	1.00000	1.00000	1.00000	1.00000	1.00000	1.00000	1.00000
1.20	1.00000	1.00000	1.00000	1.00000	1.00000	1.00000	1.00000
1.40	1.00000	1.00000	0.99999	1.00000	1.00000	1.00000	1.00000
1.60	1.00000	0.99944	0.99993	1.00007	0.99999	0.99999	1.00000
1.80	1.00000	1.00576	0.99742	0.99999	1.00024	0.99996	0.99999
2.00	1.00000	1.06444	1.01050	0.99377	1.00026	1.00032	0.99995
2.20	0.00000	0.43611	1.10320	1.01209	0.99394	0.99990	1.00021
2.40	0.00000	-0.05882	0.15709	1.02820	1.02810	1.00041	0.99754
2.60	0.00000	-0.00092	-0.00831	-0.04253	0.79695	1.07728	1.00726
2.80	0.00000	0.00000	0.00159	0.00697	-0.07710	0.46160	1.10930
3.00	0.00000	0.00000	0.00025	-0.00110	0.00898	-0.04324	0.14324
3.20	0.00000	0.00000	0.00000	0.00005	-0.00091	0.00399	-0.00764
3.40	0.00000	0.00000	0.00000	-0.00002	0.00010	-0.00035	0.00013
3.60	0.00000	0.00000	0.00000	0.00000	0.00000	0.00002	0.00006
3.80	0.00000	0.00000	0.00000	0.00000	0.00000	0.00000	-0.00001
4.00	0.00000	0.00000	0.00000	0.00000	0.00000	0.00000	0.00000

Table 6-11. Solution of inviscid Burgers equation by the 4th-order Runge-Kutta scheme with a damping of $\varepsilon_e = 0.1$, $\Delta x = 0.1$, $\Delta t = 0.1$.

x	t=0.0	t=0.3	t=0.6	t=0.9	t=1.2	t=1.5	t=1.8
0.00	1.00000	1.00000	1.00000	1.00000	1.00000	1.00000	1.00000
0.20	1.00000	1.00000	1.00000	1.00000	1.00040	1.00928	1.05858
0.40	1.00000	1.00000	1.00000	1.00005	1.00322	1.03810	1.12386
0.60	1.00000	1.00000	1.00000	1.00069	1.01823	1.10129	1.12209
0.80	1.00000	1.00000	1.00006	1.00616	1.06855	1.15102	0.94802
1.00	1.00000	1.00000	1.00109	1.03592	1.15018	1.03457	0.83768
1.20	1.00000	1.00003	1.01208	1.12181	1.12118	0.80721	1.12496
1.40	1.00000	1.00129	1.07449	1.18466	0.83359	1.04035	1.08148
1.60	1.00000	1.02443	1.20731	0.92044	0.92000	1.18761	0.70484
1.80	1.00000	1.17415	1.05852	0.77694	1.27816	0.64191	1.27060
2.00	1.00000	1.24551	0.60973	1.35491	0.59200	1.25058	0.81721
2.20	0.00000	0.30458	1.45922	0.48984	1.24544	0.83496	0.90978
2.40	0.00000	0.00000	0.07750	1.34527	0.71734	0.89268	1.20179
2.60	0.00000	0.00000	0.00000	0.01331	1.05015	1.05521	0.51539
2.80	0.00000	0.00000	0.00000	0.00000	0.00150	0.70288	1.35335
3.00	0.00000	0.00000	0.00000	0.00000	0.00000	0.00010	0.39710
3.20	0.00000	0.00000	0.00000	0.00000	0.00000	0.00000	0.00000
3.40	0.00000	0.00000	0.00000	0.00000	0.00000	0.00000	0.00000
3.60	0.00000	0.00000	0.00000	0.00000	0.00000	0.00000	0.00000
3.80	0.00000	0.00000	0.00000	0.00000	0.00000	0.00000	0.00000
4.00	0.00000	0.00000	0.00000	0.00000	0.00000	0.00000	0.00000

Table 6-12. Solution of inviscid Burgers equation by the modified 4th-order Runge-Kutta scheme, $\Delta x = 0.1$, $\Delta t = 0.1$.

x	t=0.0	t=0.3	t=0.6	t=0.9	t=1.2	t=1.5	t=1.8
0.00	1.00000	1.00000	1.00000	1.00000	1.00000	1.00000	1.00000
0.20	1.00000	1.00000	1.00000	1.00000	1.00000	1.00000	1.00000
0.40	1.00000	1.00000	1.00000	1.00000	1.00000	1.00000	1.00000
0.60	1.00000	1.00000	1.00000	1.00000	1.00000	1.00000	1.00000
0.80	1.00000	1.00000	1.00000	1.00000	1.00000	1.00000	1.00000
1.00	1.00000	1.00000	1.00000	1.00000	1.00000	1.00000	1.00000
1.20	1.00000	1.00000	1.00000	1.00000	1.00000	1.00000	1.00000
1.40	1.00000	1.00000	0.99999	1.00000	1.00000	1.00000	1.00000
1.60	1.00000	0.99944	0.99992	1.00007	0.99999	0.99999	1.00000
1.80	1.00000	1.00589	0.99736	0.99998	1.00024	0.99996	0.99999
2.00	1.00000	1.06655	1.01063	0.99355	1.00029	1.00032	0.99994
2.20	0.00000	0.43394	1.10830	1.01148	0.99374	0.99996	1.00021
2.40	0.00000	-0.05888	0.15137	1.03343	1.02767	1.00052	0.99750
2.60	0.00000	-0.00095	-0.00771	-0.04733	0.79890	1.07963	1.00741
2.80	0.00000	0.00000	0.00155	0.00741	-0.07855	0.45866	1.11457
3.00	0.00000	0.00000	0.00026	-0.00114	0.00907	-0.04276	0.13713
3.20	0.00000	0.00000	0.00000	0.00005	-0.00091	0.00391	-0.00682
3.40	0.00000	0.00000	0.00000	-0.00002	0.00010	-0.00034	0.00003
3.60	0.00000	0.00000	0.00000	0.00000	0.00000	0.00002	0.00007
3.80	0.00000	0.00000	0.00000	0.00000	0.00000	0.00000	-0.00001
4.00	0.00000	0.00000	0.00000	0.00000	0.00000	0.00000	0.00000

Table 6-13. Solution of inviscid Burgers equation by the modified 4th-order Runge-Kutta scheme with a damping of $\varepsilon_e = 0.1$, $\Delta x = 0.1$, $\Delta t = 0.1$.

x	t=0.0	t=0.3	t=0.6	t=0.9	t=1.2	t=1.5	t=1.8
0.00	1.00000	1.00000	1.00000	1.00000	1.00000	1.00000	1.00000
0.20	1.00000	1.00000	1.00000	1.00000	1.00000	1.00000	1.00000
0.40	1.00000	1.00000	1.00000	1.00000	1.00000	1.00000	1.00000
0.60	1.00000	1.00000	1.00000	1.00000	1.00000	1.00000	1.00000
0.80	1.00000	1.00000	1.00000	1.00000	1.00000	1.00000	1.00000
1.00	1.00000	1.00000	1.00000	1.00000	1.00000	1.00000	1.00000
1.20	1.00000	1.00000	1.00000	1.00000	1.00000	1.00000	1.00000
1.40	1.00000	1.00000	1.00000	1.00000	1.00000	1.00000	1.00000
1.60	1.00000	1.00000	1.00000	1.00000	1.00000	1.00000	1.00000
1.80	1.00000	1.00000	1.00000	1.00000	1.00000	1.00000	1.00000
2.00	1.00000	1.00000	1.00000	1.00000	1.00000	1.00000	1.00000
2.20	0.00000	0.50000	0.99996	1.00000	1.00000	1.00000	1.00000
2.40	0.00000	0.00000	0.13381	0.99099	1.00000	1.00000	1.00000
2.60	0.00000	0.00000	0.00000	0.00901	0.86603	1.00000	1.00000
2.80	0.00000	0.00000	0.00000	0.00000	0.00004	0.50000	0.99996
3.00	0.00000	0.00000	0.00000	0.00000	0.00000	0.00000	0.13397
3.20	0.00000	0.00000	0.00000	0.00000	0.00000	0.00000	0.00000
3.40	0.00000	0.00000	0.00000	0.00000	0.00000	0.00000	0.00000
3.60	0.00000	0.00000	0.00000	0.00000	0.00000	0.00000	0.00000
3.80	0.00000	0.00000	0.00000	0.00000	0.00000	0.00000	0.00000
4.00	0.00000	0.00000	0.00000	0.00000	0.00000	0.00000	0.00000

Table 6-14. Solution of inviscid Burgers equation by the first-order TVD scheme, $\Delta x = 0.1$, $\Delta t = 0.1$.

x	t=0.0	t=0.3	t=0.6	t=0.9	t=1.2	t=1.5	t=1.8
0.00	1.00000	1.00000	1.00000	1.00000	1.00000	1.00000	1.00000
0.20	1.00000	1.00000	1.00000	1.00000	1.00000	1.00000	1.00000
0.40	1.00000	1.00000	1.00000	1.00000	1.00000	1.00000	1.00000
0.60	1.00000	1.00000	1.00000	1.00000	1.00000	1.00000	1.00000
0.80	1.00000	1.00000	1.00000	1.00000	1.00000	1.00000	1.00000
1.00	1.00000	1.00000	1.00000	1.00000	1.00000	1.00000	1.00000
1.20	1.00000	1.00000	1.00000	1.00000	1.00000	1.00000	1.00000
1.40	1.00000	1.00000	1.00000	1.00000	1.00000	1.00000	1.00000
1.60	1.00000	1.00000	1.00000	1.00000	1.00000	1.00000	1.00000
1.80	1.00000	1.00000	1.00000	1.00000	1.00000	1.00000	1.00000
2.00	1.00000	1.00000	1.00000	1.00000	1.00000	1.00000	1.00000
2.20	0.00000	0.48462	1.00032	1.00000	1.00000	1.00000	1.00000
2.40	0.00000	0.00000	0.03052	1.01351	1.00000	1.00004	1.00000
2.60	0.00000	0.00000	0.00000	0.00023	0.97055	0.99796	0.99878
2.80	0.00000	0.00000	0.00000	0.00000	0.00000	0.48130	1.00409
3.00	0.00000	0.00000	0.00000	0.00000	0.00000	0.00000	0.03080
3.20	0.00000	0.00000	0.00000	0.00000	0.00000	0.00000	0.00000
3.40	0.00000	0.00000	0.00000	0.00000	0.00000	0.00000	0.00000
3.60	0.00000	0.00000	0.00000	0.00000	0.00000	0.00000	0.00000
3.80	0.00000	0.00000	0.00000	0.00000	0.00000	0.00000	0.00000
4.00	0.00000	0.00000	0.00000	0.00000	0.00000	0.00000	0.00000

Table 6-15. Solution of inviscid Burgers equation by the second-order TVD scheme with a Harten-Yee limiter, $\Delta x = 0.1$, $\Delta t = 0.1$.

x	t=0.0	t=0.3	t=0.6	t=0.9	t=1.2	t=1.5	t=1.8
0.00	1.00000	1.00000	1.00000	1.00000	1.00000	1.00000	1.00000
0.20	1.00000	1.00000	1.00000	1.00000	1.00000	1.00000	1.00000
0.40	1.00000	1.00000	1.00000	1.00000	1.00000	1.00000	1.00000
0.60	1.00000	1.00000	1.00000	1.00000	1.00000	1.00000	1.00000
0.80	1.00000	1.00000	1.00000	1.00000	1.00000	1.00000	1.00000
1.00	1.00000	1.00000	1.00000	1.00000	1.00000	1.00000	1.00000
1.20	1.00000	1.00000	1.00000	1.00000	1.00000	1.00000	1.00000
1.40	1.00000	1.00000	1.00000	1.00000	1.00000	1.00000	1.00000
1.60	1.00000	1.00000	1.00000	1.00000	1.00000	1.00005	0.99990
1.80	1.00000	1.00000	1.00000	1.00162	0.99915	0.99975	1.00062
2.00	1.00000	1.19231	0.99664	0.99714	1.00333	0.99968	0.99887
2.20	0.00000	0.35138	1.19100	0.96737	0.98957	1.00756	1.00074
2.40	0.00000	0.00000	0.03827	0.93191	1.20954	0.97366	1.00421
2.60	0.00000	0.00000	0.00000	0.00022	0.56346	1.42878	0.97672
2.80	0.00000	0.00000	0.00000	0.00000	0.00000	0.16949	1.23845
3.00	0.00000	0.00000	0.00000	0.00000	0.00000	0.00000	0.01495
3.20	0.00000	0.00000	0.00000	0.00000	0.00000	0.00000	0.00000
3.40	0.00000	0.00000	0.00000	0.00000	0.00000	0.00000	0.00000
3.60	0.00000	0.00000	0.00000	0.00000	0.00000	0.00000	0.00000
3.80	0.00000	0.00000	0.00000	0.00000	0.00000	0.00000	0.00000
4.00	0.00000	0.00000	0.00000	0.00000	0.00000	0.00000	0.00000

Table 6-16. Solution of inviscid Burgers equation by the second-order TVD scheme with a Roe-Sweby limiter, $\Delta x = 0.1$, $\Delta t = 0.1$.

x	t=0.0	t=0.3	t=0.6	t=0.9	t=1.2	t=1.5	t=1.8
0.00	1.00000	1.00000	1.00000	1.00000	1.00000	1.00000	1.00000
0.20	1.00000	1.00000	1.00000	1.00000	1.00000	1.00000	1.00000
0.40	1.00000	1.00000	1.00000	1.00000	1.00000	1.00000	1.00000
0.60	1.00000	1.00000	1.00000	1.00000	1.00000	1.00000	1.00000
0.80	1.00000	1.00000	1.00000	1.00000	1.00000	1.00000	1.00000
1.00	1.00000	1.00000	1.00000	1.00000	1.00000	1.00000	1.00000
1.20	1.00000	1.00000	1.00000	1.00000	1.00000	1.00000	1.00000
1.40	1.00000	1.00000	1.00000	1.00000	1.00000	1.00000	1.00000
1.60	1.00000	1.00000	1.00000	1.00000	1.00000	1.00000	1.00000
1.80	1.00000	1.00000	1.00000	1.00000	1.00000	1.00000	1.00000
2.00	1.00000	1.00000	1.00000	1.00000	1.00000	1.00000	1.00000
2.20	0.00000	0.48438	0.99997	1.00000	1.00000	1.00000	1.00000
2.40	0.00000	0.00000	0.12361	1.00778	1.00000	1.00000	1.00000
2.60	0.00000	0.00000	0.00000	0.00767	0.87639	1.00000	1.00000
2.80	0.00000	0.00000	0.00000	0.00000	0.00004	0.48455	0.99997
3.00	0.00000	0.00000	0.00000	0.00000	0.00000	0.00000	0.12361
3.20	0.00000	0.00000	0.00000	0.00000	0.00000	0.00000	0.00000
3.40	0.00000	0.00000	0.00000	0.00000	0.00000	0.00000	0.00000
3.60	0.00000	0.00000	0.00000	0.00000	0.00000	0.00000	0.00000
3.80	0.00000	0.00000	0.00000	0.00000	0.00000	0.00000	0.00000
4.00	0.00000	0.00000	0.00000	0.00000	0.00000	0.00000	0.00000

Table 6-17. Solution of inviscid Burgers equation by the second-order TVD scheme with a Davis-Yee limiter, $\Delta x = 0.1$, $\Delta t = 0.1$.

x	t=0.0	t=0.3	t=0.6	t=0.9	t=1.2	t=1.5	t=1.8
0.00	1.00000	1.00000	1.00000	1.00000	1.00000	1.00000	1.00000
0.20	1.00000	1.00000	1.00000	1.00000	1.00000	1.00000	1.00000
0.40	1.00000	1.00000	1.00000	1.00000	1.00000	1.00000	1.00000
0.60	1.00000	1.00000	1.00000	1.00000	1.00000	1.00000	1.00000
0.80	1.00000	1.00000	1.00000	1.00000	1.00000	1.00000	1.00000
1.00	1.00000	1.00000	1.00000	1.00000	1.00000	1.00000	1.00000
1.20	1.00000	1.00000	1.00000	1.00000	1.00000	1.00000	1.00000
1.40	1.00000	1.00000	1.00000	1.00000	1.00000	1.00000	1.00000
1.60	1.00000	1.00000	1.00000	1.00000	1.00000	1.00000	1.00000
1.80	1.00000	1.00000	1.00000	1.00000	1.00000	1.00000	1.00000
2.00	1.00000	0.99953	0.99997	1.00000	1.00000	1.00000	1.00000
2.20	0.00000	0.48760	0.98977	0.99987	1.00000	1.00000	1.00000
2.40	0.00000	0.00050	0.22436	0.94295	0.99951	0.99999	1.00000
2.60	0.00000	0.00000	0.00002	0.06731	0.77552	0.99797	0.99997
2.80	0.00000	0.00000	0.00000	0.00000	0.01048	0.49126	0.98949
3.00	0.00000	0.00000	0.00000	0.00000	0.00000	0.00067	0.22744
3.20	0.00000	0.00000	0.00000	0.00000	0.00000	0.00000	0.00002
3.40	0.00000	0.00000	0.00000	0.00000	0.00000	0.00000	0.00000
3.60	0.00000	0.00000	0.00000	0.00000	0.00000	0.00000	0.00000
3.80	0.00000	0.00000	0.00000	0.00000	0.00000	0.00000	0.00000
4.00	0.00000	0.00000	0.00000	0.00000	0.00000	0.00000	0.00000

Table 6-18. Solution of inviscid Burgers equation by the modified 4th-order Runge-Kutta scheme with a Davis-Yee limiter, $\Delta x = 0.1$, $\Delta t = 0.001$.

Chapter 7

Scalar Representation of the Navier-Stokes Equations

7.1 Introductory Remarks

In the previous chapters, simple scalar model equations were used to investigate solution procedures for parabolic, elliptic, and hyperbolic equations. It is beneficial at this point to study a scalar model equation which is of similar character as the Navier-Stokes equations. That is, the model equation is unsteady and includes both convection and diffusion terms. Such an equation is the viscous Burgers equation.

Before proceeding with the model equation and numerical schemes, a summary of approximation for each term in the model equation is reviewed. (1) The time derivative is typically approximated by either a forward difference or a backward difference approximation resulting in an explicit or an implicit formulation, respectively. This approximation is typically first-order, even though second-order accuracy may be used. (2) There are several methods by which the convective term can be approximated. That includes central difference approximation, forward/backward difference approximations, TVD formulation, and the Runge-Kutta scheme. Furthermore, the addition of damping terms or TVD terms may be required for some of the schemes in order to reduce or to eliminate oscillations (due to dispersion error) in the solution. (3) The diffusion term is typically approximated by a central difference (usually second-order) formulation. A limited number of numerical schemes will be investigated in this chapter.

This chapter also serves as a divider between the introductory material and intermediate/advanced topics. The introductory topics were primarily studied by scalar model equations. Furthermore, coordinate transformation and complex geometries were avoided. In the upcoming chapters these issues will be addressed and the equations of fluid motion in various forms will be investigated as well.

7.2 Model Equation

The viscous Burgers equation may be written in a general form as

$$\frac{\partial u}{\partial t} + (\alpha + \beta u)\frac{\partial u}{\partial x} = \nu\frac{\partial^2 u}{\partial x^2} \tag{7-1}$$

where α and β are some prescribed parameters. When $\alpha = a$, the speed of sound, and $\beta = 0$, the familiar linear Burgers equation is produced, i.e.,

$$\frac{\partial u}{\partial t} + a\frac{\partial u}{\partial x} = \nu\frac{\partial^2 u}{\partial x^2} \tag{7-2}$$

Note that a and ν are assumed to be constants. When $\alpha = 0$ and $\beta = 1$, the standard nonlinear Burgers equation is produced as

$$\frac{\partial u}{\partial t} + u\frac{\partial u}{\partial x} = \nu\frac{\partial^2 u}{\partial x^2} \tag{7-3}$$

This equation may be rearranged such that

$$\frac{\partial u}{\partial t} + \frac{\partial E}{\partial x} = \nu\frac{\partial^2 u}{\partial x^2} \tag{7-4}$$

where $E = (1/2)u^2$.

A two-dimensional Burgers equation may be expressed by the following

$$\frac{\partial u}{\partial t} + u\frac{\partial u}{\partial x} + v\frac{\partial u}{\partial y} = \nu\left(\frac{\partial^2 u}{\partial x^2} + \frac{\partial^2 u}{\partial y^2}\right) \tag{7-5}$$

and

$$\frac{\partial v}{\partial t} + u\frac{\partial v}{\partial x} + v\frac{\partial v}{\partial y} = \nu\left(\frac{\partial^2 v}{\partial x^2} + \frac{\partial^2 v}{\partial y^2}\right) \tag{7-6}$$

Observe that Equations (7-5) and (7-6) represent the x and y components of the momentum equation for an incompressible flow in the absence of a pressure gradient. If the incompressible continuity equation is incorporated into Equations (7-5) and (7-6), the conservative form of the equations is obtained as follows

$$\frac{\partial u}{\partial t} + \frac{\partial}{\partial x}(u^2) + \frac{\partial}{\partial y}(uv) = \nu\left(\frac{\partial^2 u}{\partial x^2} + \frac{\partial^2 u}{\partial y^2}\right) \tag{7-7}$$

and

$$\frac{\partial v}{\partial t} + \frac{\partial}{\partial x}(uv) + \frac{\partial}{\partial y}(v^2) = \nu\left(\frac{\partial^2 v}{\partial x^2} + \frac{\partial^2 v}{\partial y^2}\right) \tag{7-8}$$

Equations (7-7) and (7-8) can be written in a vector form by the following

$$\frac{\partial Q}{\partial t} + \frac{\partial E}{\partial x} + \frac{\partial F}{\partial y} = \nu \left(\frac{\partial^2 Q}{\partial x^2} + \frac{\partial^2 Q}{\partial y^2} \right) \tag{7-9}$$

where

$$Q = \begin{bmatrix} u \\ v \end{bmatrix} \quad , \quad E = \begin{bmatrix} u^2 \\ uv \end{bmatrix} \quad , \quad F = \begin{bmatrix} uv \\ v^2 \end{bmatrix}$$

Before proceeding with a review of the numerical schemes, the equations of fluid motion are briefly reviewed. A summary of the equations of fluid motion is provided in Appendix D, and the nondimensional form and the transformation of the equations from physical space to computational space are given in Chapter 11.

7.3 Equations of Fluid Motion

The equations of fluid motion include conservations of mass, momentum, and energy. These equations in a differential form are known as the Navier-Stokes equations. The nondimensionalized conservative form of the equations are:

1. Continuity:

$$\frac{\partial \rho^*}{\partial t^*} + \frac{\partial}{\partial x^*}(\rho^* u^*) + \frac{\partial}{\partial y^*}(\rho^* v^*) + \frac{\partial}{\partial z^*}(\rho^* w^*) = 0 \tag{7-10}$$

2. X-component of the momentum equation:

$$\frac{\partial}{\partial t^*}(\rho^* u^*) + \frac{\partial}{\partial x^*}(\rho^* u^{*2} + p^*) + \frac{\partial}{\partial y^*}(\rho^* u^* v^*) + \frac{\partial}{\partial z^*}(\rho^* u^* w^*) = \frac{\partial}{\partial x^*}(\tau^*_{xx}) + \frac{\partial}{\partial y^*}(\tau^*_{xy}) + \frac{\partial}{\partial z^*}(\tau^*_{xz}) \tag{7-11}$$

3. Y-component of the momentum equation:

$$\frac{\partial}{\partial t^*}(\rho^* v^*) + \frac{\partial}{\partial x^*}(\rho^* u^* v^*) + \frac{\partial}{\partial y^*}(\rho^* v^{*2} + p^*) + \frac{\partial}{\partial z^*}(\rho^* v^* w^*) = \frac{\partial}{\partial x^*}(\tau^*_{xy}) + \frac{\partial}{\partial y^*}(\tau^*_{yy}) + \frac{\partial}{\partial z^*}(\tau^*_{yz}) \tag{7-12}$$

4. Z-component of the momentum equation:

$$\frac{\partial}{\partial t^*}(\rho^* w^*) + \frac{\partial}{\partial x^*}(\rho^* u^* w^*) + \frac{\partial}{\partial y^*}(\rho^* v^* w^*) + \frac{\partial}{\partial z^*}(\rho^* w^{*2} + p^*) = \frac{\partial}{\partial x^*}(\tau^*_{xz}) + \frac{\partial}{\partial y^*}(\tau^*_{yz}) + \frac{\partial}{\partial z^*}(\tau^*_{zz}) \tag{7-13}$$

5. Energy:

$$\frac{\partial}{\partial t^*}(\rho^* e_t^*) + \frac{\partial}{\partial x^*}(\rho^* u^* e_t^* + p^* u^*) + \frac{\partial}{\partial y^*}(\rho^* v^* e_t^* + p^* v^*) +$$

$$\frac{\partial}{\partial z^*}(\rho^* w^* e_t^* + p^* w^*) = \frac{\partial}{\partial x^*}[u^* \tau_{xx}^* + v^* \tau_{xy}^* + w^* \tau_{xz}^* - q_x^*] +$$

$$\frac{\partial}{\partial y^*}[u^* \tau_{yx}^* + v^* \tau_{yy}^* + w^* \tau_{yz}^* - q_y^*] + \frac{\partial}{\partial z^*}[u^* \tau_{zx}^* + v^* \tau_{zy}^* + w^* \tau_{zz}^* - q_z^*] \quad \text{(7-14)}$$

These equations may be written in a vector form as

$$\frac{\partial Q^*}{\partial t^*} + \frac{\partial E^*}{\partial x^*} + \frac{\partial F^*}{\partial y^*} + \frac{\partial G^*}{\partial z^*} = \frac{\partial E_v^*}{\partial x^*} + \frac{\partial F_v^*}{\partial y^*} + \frac{\partial G_v^*}{\partial z^*} \quad \text{(7-15)}$$

where

$$Q^* = \begin{bmatrix} \rho^* \\ \rho^* u^* \\ \rho^* v^* \\ \rho^* w^* \\ \rho^* e_t^* \end{bmatrix}$$

$$E^* = \begin{bmatrix} \rho^* u^* \\ \rho^* u^{*^2} + p^* \\ \rho^* u^* v^* \\ \rho^* u^* w^* \\ (\rho^* e_t^* + p^*) u^* \end{bmatrix} \qquad E_v^* = \begin{bmatrix} 0 \\ \tau_{xx}^* \\ \tau_{xy}^* \\ \tau_{xz}^* \\ u^* \tau_{xx}^* + v^* \tau_{xy}^* + w^* \tau_{xz}^* - q_x^* \end{bmatrix}$$

$$F^* = \begin{bmatrix} \rho^* v^* \\ \rho^* v^* u^* \\ \rho^* v^{*^2} + p^* \\ \rho^* v^* w^* \\ (\rho^* e_t^* + p^*) v^* \end{bmatrix} \qquad F_v^* = \begin{bmatrix} 0 \\ \tau_{yx}^* \\ \tau_{yy}^* \\ \tau_{yz}^* \\ u^* \tau_{yx}^* + v^* \tau_{yy}^* + w^* \tau_{yz}^* - q_y^* \end{bmatrix}$$

$$G^* = \begin{bmatrix} \rho^* w^* \\ \rho^* w^* u^* \\ \rho^* w^* v^* \\ \rho^* w^{*^2} + p^* \\ (\rho^* e_t^* + p^*) w^* \end{bmatrix} \qquad G_v^* = \begin{bmatrix} 0 \\ \tau_{zx}^* \\ \tau_{zy}^* \\ \tau_{zz}^* \\ u^* \tau_{zx}^* + v^* \tau_{zy}^* + w^* \tau_{zz}^* - q_z^* \end{bmatrix}$$

When the right-hand side of Equation (7-15) is set equal to zero, the Euler equation is produced, i.e.,

$$\frac{\partial Q^*}{\partial t^*} + \frac{\partial E^*}{\partial x^*} + \frac{\partial F^*}{\partial y^*} + \frac{\partial G^*}{\partial z^*} = 0 \tag{7-16}$$

This vector equation is a hyperbolic equation. The Navier-Stokes equations are in general a mixed hyperbolic (in inviscid region), parabolic (in viscous region) equation. Note that both sets of equations, i.e., (7-15) and (7-16), are solved by marching in time. When the Navier-Stokes equation is reduced by imposing the steady-state assumption, then it is classified as a mixed hyperbolic (in the inviscid region), elliptic (in the viscous region) equation. It is clear that for the steady-state forms of the equations, space marching procedures are utilized.

The brief review of the Navier-Stokes equations was presented above to identify the similarity of the terms in Equation (7-15) to that of the scalar model equation given by (7-4). In the subsequent sections, several numerical schemes are presented for the solution of the viscous Burgers equation. The numerical schemes are extended to a system of equations in the following chapters.

7.4 Numerical Algorithms

In this section a limited number of numerical schemes for the solution of the model scalar equation (7-1) is investigated.

7.4.1 FTCS Explicit

In this explicit formulation, a first-order forward difference approximation and second-order central difference approximation for the time derivative and spatial derivatives are used, respectively. Hence, the FDE of the PDE given by (7-2) is

$$\frac{u_i^{n+1} - u_i^n}{\Delta t} + a\frac{u_{i+1}^n - u_{i-1}^n}{2\Delta x} = \nu\frac{u_{i+1}^n - 2u_i^n + u_{i-1}^n}{(\Delta x)^2} \tag{7-17}$$

with a truncation error of $[(\Delta t), (\Delta x)^2]$. The amplification factor for this FDE was obtained in Chapter 4 as

$$G = [1 + 2d(\cos\theta - 1)] - I\,[c\sin\theta]$$

which resulted in the following stability requirements:

$$d \leq \frac{1}{2} \qquad \text{and} \qquad Re_c \leq 2/c \tag{7-18}$$

7.4.2 FTBCS Explicit

In this explicit scheme, a first-order forward difference approximation for the time derivative and a second-order central difference approximation for the diffusion term is used. When a first-order backward difference approximation ($a > 0$) for the convective term is employed, then the FDE is expressed as

$$\frac{u_i^{n+1} - u_i^n}{\Delta t} + a\frac{u_i^n - u_{i-1}^n}{\Delta x} = \nu\frac{u_{i+1}^n - 2u_i^n + u_{i-1}^n}{(\Delta x)^2} \tag{7-19}$$

However, the first-order approximation of the convective term may introduce excessive dissipation error (artificial viscosity) such that it is of the same order as the natural viscosity. As a result, an accurate solution is not achieved. To overcome this problem, one may use a second-order approximation. However, as it was shown previously in Chapter 6, this approximation may produce dispersion error with oscillations near discontinuities. To solve this problem, damping terms must be added. The addition of damping is problem-dependent, and it should be handled with extreme care. The third alternative is to use a third-order scheme resulting in the following finite difference equation for (7-2):

$$\begin{aligned}\frac{u_i^{n+1} - u_i^n}{\Delta t} &+ a\left(\frac{u_{i+1}^n - u_{i-1}^n}{2\Delta x} - \frac{u_{i+1}^n - 3u_i^n + 3u_{i-1}^n - u_{i-2}^n}{6(\Delta x)}\right) \\ &= \nu\frac{u_{i+1}^n - 2u_i^n + u_{i-1}^n}{(\Delta x)^2}\end{aligned} \tag{7-20}$$

7.4.3 DuFort-Frankel Explicit

The DuFort-Frankel method is obtained by modification of Richardson's scheme. Second-order central difference approximations for all the derivatives are used in this method. The FDE is

$$\frac{u_i^{n+1} - u_i^{n-1}}{2\Delta t} + a\frac{u_{i+1}^n - u_{i-1}^n}{2\Delta x} = \nu\frac{u_{i+1}^n - (u_i^{n-1} + u_i^{n+1}) + u_{i-1}^n}{(\Delta x)^2}$$

and the truncation error is $O[(\Delta t)^2,\ (\Delta x)^2,\ (\Delta t/\Delta x)^2]$. This FDE may be rearranged as

$$u_i^{n+1} = \left(\frac{1-2d}{1+2d}\right)u_i^{n-1} + \left(\frac{c+2d}{1+2d}\right)u_{i-1}^n - \left(\frac{c-2d}{1+2d}\right)u_{i+1}^n \tag{7-21}$$

The stability requirement of the scheme is $c \leq 1$.

7.4.4 MacCormack Explicit

This multi-step or predictor-corrector scheme is performed in two steps as follows:

1. $$u_i^* = u_i^n - a\frac{\Delta t}{\Delta x}\left(u_{i+1}^n - u_i^n\right) + \nu\frac{\Delta t}{(\Delta x)^2}\left(u_{i+1}^n - 2u_i^n + u_{i-1}^n\right) \tag{7-22}$$

2. $$u_i^{n+1} = \frac{1}{2}\left[u_i^n + u_i^* - a\frac{\Delta t}{\Delta x}\left(u_i^* - u_{i-1}^*\right) + \nu\frac{\Delta t}{(\Delta x)^2}\left(u_{i+1}^* - 2u_i^* + u_{i-1}^*\right)\right] \tag{7-23}$$

The method is second-order accurate with the stability requirement of

$$\Delta t \leq \frac{1}{\dfrac{a}{\Delta x} + \dfrac{2\nu}{(\Delta x)^2}}$$

Equations (7-22) and (7-23) may be written in a delta form as:

1. $$\begin{cases} \Delta u_i^n = -a\dfrac{\Delta t}{\Delta x}\left(u_{i+1}^n - u_i^n\right) + \nu\dfrac{\Delta t}{(\Delta x)^2}\left(u_{i+1}^n - 2u_i^n + u_{i-1}^n\right) & \text{(7-24a)} \\ u_i^* = u_i^n + \Delta u_i^n & \text{(7-24b)} \end{cases}$$

2. $$\begin{cases} \Delta u_i^* = -a\dfrac{\Delta t}{\Delta x}\left(u_i^* - u_{i-1}^*\right) + \nu\dfrac{\Delta t}{(\Delta x)^2}\left(u_{i+1}^* - 2u_i^* + u_{i-1}^*\right) & \text{(7-25a)} \\ u_i^{n+1} = \dfrac{1}{2}\left(u_i^n + u_i^* + \Delta u_i^*\right) & \text{(7-25b)} \end{cases}$$

7.4.5 MacCormack Implicit

This scheme is an implicit analog of the explicit method just described. The formulation is as follows:

1. $$\begin{cases} \left(1 + \lambda\dfrac{\Delta t}{\Delta x}\right)\delta u_i^* = \Delta u_i^n + \lambda\dfrac{\Delta t}{\Delta x}\delta u_{i+1}^* & \text{(7-26a)} \\ u_i^* = u_i^n + \delta u_i^* & \text{(7-26b)} \end{cases}$$

where Δu_i^n is computed from Equation (7-24a).

2. $$\begin{cases} \left(1 + \lambda\dfrac{\Delta t}{\Delta x}\right)\delta u_i^{n+1} = \Delta u_i^* + \lambda\dfrac{\Delta t}{\Delta x}\delta u_{i-1}^{n+1} & \text{(7-27a)} \\ u_i^{n+1} = \dfrac{1}{2}\left(u_i^n + u_i^* + \delta u_i^{n+1}\right) & \text{(7-27b)} \end{cases}$$

In this equation, Δu_i^* is provided from Equation (7-25a). The parameter, λ, used in Equations (7-26a) and (7-27a) is selected such that

$$\lambda \geq \max\left[\frac{1}{2}\left(|a| + 2\frac{\nu}{\Delta x} - \frac{\Delta x}{\Delta t},\ 0.0\right)\right]$$

Equations (7-26a) and (7-27a) form bidiagonal systems which can be solved efficiently by various routines. The method is unconditionally stable and second-order accurate as long as the diffusion number $\nu[\Delta t/(\Delta x)^2]$ is bounded for the limiting process for which Δt, Δx approaches zero.

7.4.6 BTCS Implicit

This implicit formulation utilizes a first-order backward difference approximation for the time derivative and a second-order central difference approximation for the spatial derivatives. The resulting FDE is

$$\frac{u_i^{n+1} - u_i^n}{\Delta t} + a\frac{u_{i+1}^{n+1} - u_{i-1}^{n+1}}{2\Delta x} = \nu\frac{u_{i+1}^{n+1} - 2u_i^{n+1} + u_{i-1}^{n+1}}{(\Delta x)^2}$$

which may be rearranged as

$$-\left(\frac{1}{2}c + d\right)u_{i-1}^{n+1} + \left(1 + 2d\right)u_i^{n+1} + \left(\frac{1}{2}c - d\right)u_{i+1}^{n+1} = u_i^n \tag{7-28}$$

This formulation will result in a system of equations with a tridiagonal coefficient matrix. As seen in Chapter 3, an efficient method for solving such a system is available.

7.4.7 BTBCS Implicit

In this implicit scheme, a first-order backward difference approximation for the time derivative and a second-order central difference approximation for the diffusion term are used. As stated previously, the convective term may be approximated by a backward difference approximation which is either first-, second-, or third-order accurate ($a > 0$). Again, the use of the first-order approximation may introduce too much artificial viscosity, whereas a second-order approximation may cause some oscillations (dispersion error) in the solution. The finite difference formulations are as follows:

a. First-order approximation of the convective term

$$\frac{u_i^{n+1} - u_i^n}{\Delta t} + a\frac{u_i^{n+1} - u_{i-1}^{n+1}}{\Delta x} = \nu\frac{u_{i+1}^{n+1} - 2u_i^{n+1} + u_{i-1}^{n+1}}{(\Delta x)^2}$$

which may be rearranged as a tridiagonal system as

$$\left[-\nu\frac{\Delta t}{(\Delta x)^2} - a\frac{\Delta t}{\Delta x}\right]u_{i-1}^{n+1} + \left[1 + 2\nu\frac{\Delta t}{(\Delta x)^2} + a\frac{\Delta t}{\Delta x}\right]u_i^{n+1}$$

$$-\left[\nu\frac{\Delta t}{(\Delta x)^2}\right]u_{i+1}^{n+1} = u_i^n \tag{7-29}$$

b. Second-order approximation of the convective term

$$\frac{u_i^{n+1} - u_i^n}{\Delta t} + a\frac{u_{i-2}^{n+1} - 4u_{i-1}^{n+1} + 3u_i^{n+1}}{2(\Delta x)} = \nu\frac{u_{i+1}^{n+1} - 2u_i^{n+1} + u_{i-1}^{n+1}}{(\Delta x)^2}$$

which is rearranged as

$$\left[-\nu\frac{\Delta t}{(\Delta x)^2} - 2a\frac{\Delta t}{\Delta x}\right]u_{i-1}^{n+1} + \left[1 + 2\nu\frac{\Delta t}{(\Delta x)^2} + 3a\frac{\Delta t}{2\Delta x}\right]u_i^{n+1}$$

$$-\left[\nu\frac{\Delta t}{(\Delta x)^2}\right]u_{i+1}^{n+1} = u_i^n - a\frac{\Delta t}{2\Delta x}u_{i-2}^n \tag{7-30}$$

Note that, in order to preserve the tridiagonal nature of the system, the $i-2$ point was moved to the right-hand side and evaluated explicitly.

c. Third-order approximation of the convective term

$$\frac{u_i^{n+1} - u_i^n}{\Delta t} + a\left(\frac{u_{i+1}^{n+1} - u_{i-1}^{n+1}}{2\Delta x} - \frac{u_{i+1}^{n+1} - 3u_i^{n+1} + 3u_{i-1}^{n+1} - u_{i-2}^{n+1}}{6\Delta x}\right)$$

$$= \nu\frac{u_{i+1}^{n+1} - 2u_i^{n+1} + u_{i-1}^{n+1}}{(\Delta x)^2}$$

which is rearranged as

$$\left[-\nu\frac{\Delta t}{(\Delta x)^2} - a\frac{\Delta t}{\Delta x}\right]u_{i-1}^{n+1} + \left[1 + 2\nu\frac{\Delta t}{(\Delta x)^2} + a\frac{\Delta t}{2\Delta x}\right]u_i^{n+1}$$

$$\left[-\nu\frac{\Delta t}{(\Delta x)^2} + a\frac{\Delta t}{3\Delta x}\right]u_{i+1}^{n+1} = u_i^n - a\frac{\Delta t}{6\Delta x}u_{i-2}^n \tag{7-31}$$

7.5 Applications: Nonlinear Problem

Various algorithms described in the previous section are now used to illustrate their applications. In order to generalize the schemes just introduced, the nonlinear Burgers equation will be used. After all, the Navier-Stokes equations are nonlinear and, therefore, this application should provide some fundamental insight into the procedures and steps required for the solution of the Navier-Stokes equation.

The nonlinear Burgers equation is

$$\frac{\partial u}{\partial t} + u\frac{\partial u}{\partial x} = \nu\frac{\partial^2 u}{\partial x^2} \tag{7-32}$$

This equation is nondimensionalized by the following:

$$x^* = \frac{x}{L}\ , \qquad u^* = \frac{uL}{\nu}\ , \qquad t^* = \frac{\nu}{L^2}t$$

Hence,

$$\frac{\partial u^*}{\partial t^*} + u^*\frac{\partial u^*}{\partial x^*} = \frac{\partial^2 u^*}{\partial x^{*2}} \tag{7-33}$$

In Reference [7-1], 35 different analytical solutions of Burgers equation are given. One such stationary solution will be used to verify the numerical solutions and estimate the accuracy of the methods. The selected solution is

$$u^* = -\frac{2\sinh x^*}{\cosh x^* - e^{-t^*}} \tag{7-34}$$

which is shown in Figure 7-1 and tabulated at various time levels in Table 7-1. The notation "$*$", which is used to designate the nondimensional quantities, is dropped in the following discussions. Equation (7-33) may be expressed in a conservative form as

$$\frac{\partial u}{\partial t} + \frac{\partial E}{\partial x} = \frac{\partial^2 u}{\partial x^2} \tag{7-35}$$

where

$$E = \frac{1}{2}u^2$$

Rearranging this equation in terms of the Jacobian, one has

$$\frac{\partial u}{\partial t} + A\frac{\partial u}{\partial x} = \frac{\partial^2 u}{\partial x^2} \tag{7-36}$$

where

$$A = \frac{\partial E}{\partial u}$$

Note that the second term, i.e., convection term in either of the Equations (7-35) or (7-36), introduces nonlinearity into the equation.

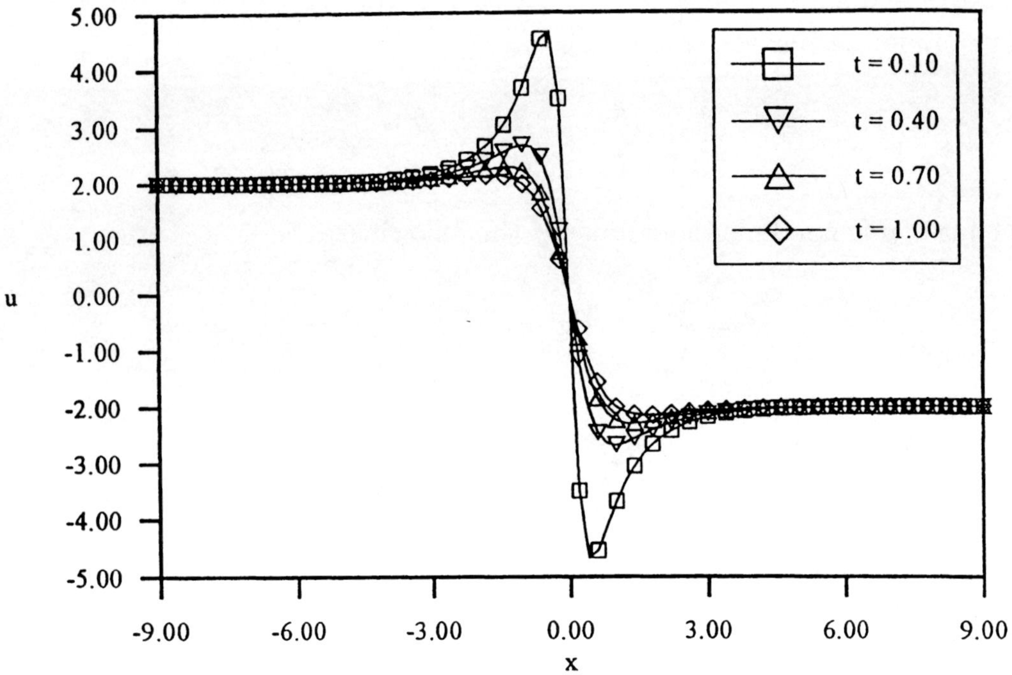

Figure 7-1. Analytical solution given by Equation (7-34) at various time levels.

7.5.1 FTCS Explicit

When Equation (7-36) is approximated by FDE (7-17) introduced previously, one obtains

$$\frac{u_i^{n+1}-u_i^n}{\Delta t}+A_i^n\frac{u_{i+1}^n-u_{i-1}^n}{2\Delta x}=\frac{u_{i+1}^n-2u_i^n+u_{i-1}^n}{(\Delta x)^2} \tag{7-37}$$

An important point about Equation (7-37) is explored as follows. When an explicit scheme is used, the nonlinear term in Equation (7-36) is evaluated at the known location and, therefore, no linearization is required. That is not the case when an implicit formulation is employed; as a result, a linearization procedure must be introduced. An alternate way to evaluate the coefficient A in Equation (7-37) is to use an average value, in which case the formulation becomes

$$\frac{u_i^{n+1}-u_i^n}{\Delta t}+\frac{1}{2}(A_{i+1}^n+A_{i-1}^n)\frac{u_{i+1}^n-u_{i-1}^n}{2\Delta x}=\frac{u_{i+1}^n-2u_i^n+u_{i-1}^n}{(\Delta x)^2} \tag{7-38}$$

The FTCS explicit formulation may be directly applied to Equation (7-35), in which case the FDE becomes

$$\frac{u_i^{n+1}-u_i^n}{\Delta t}+\frac{E_{i+1}^n-E_{i-1}^n}{2\Delta x}=\frac{u_{i+1}^n-2u_i^n+u_{i-1}^n}{(\Delta x)^2} \tag{7-39}$$

Again, since the nonlinear term E is applied at the known time level "n", no linearization is required. Either one of the FDEs, (7-37), (7-38), or (7-39) may be used to obtain a solution.

To start the solution, an initial condition must be specified. This initial condition is taken from the analytical solution at $t = 0.1$ and is shown in Figure 7-1. The spatial domain is taken to be between -9 and 9. The boundary conditions at $x = -9$ and $x = 9$ are specified as $u = 2.0$ and $u = -2.0$, respectively. This imposition is an approximation. Indeed, the value of u at the boundaries will change as a function of time. However, these changes are small (see for example Table 7-1) and, for the sake of simplicity, they will be neglected. If the exact values of u at the boundaries are desired, they may be obtained from Equation (7-34) and input to the numerical code.

The next consideration is the selection of step sizes. The FTCS explicit scheme requires the following constraints for stability:

$$d \leq \frac{1}{2} \qquad \text{and} \qquad Re_c \leq 2/c$$

For the result given in Table 7-2 obtained from the formulation of (7-39), the following step sizes which satisfy the stability requirements are used:

$$\Delta x = 0.2$$

$$\Delta t = 0.01$$

The solution starts from $t = 0.1$ and proceeds to $t = 1.0$ and is illustrated in Figure 7-2.

It is interesting to compare the solutions obtained by the three finite difference equations of (7-37), (7-38), and (7-39). For this purpose, the error distributions at time levels of 0.4 and 1.0 are shown in Figures (7-3) and (7-4), respectively. Observe that the formulation (7-38) has the least error and that the errors decrease as the solutions proceed in time. The error is simply defined as the difference between the analytical solution and the numerical solution.

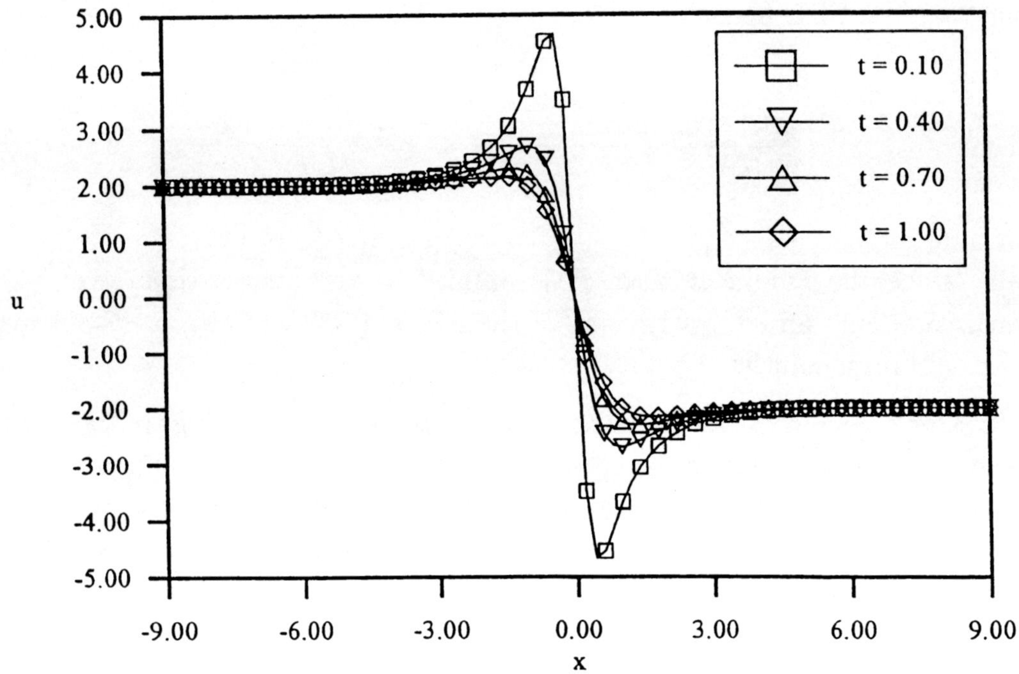

Figure 7-2. Solution of the viscous Burgers equation by FTCS explicit scheme of Equation (7-39).

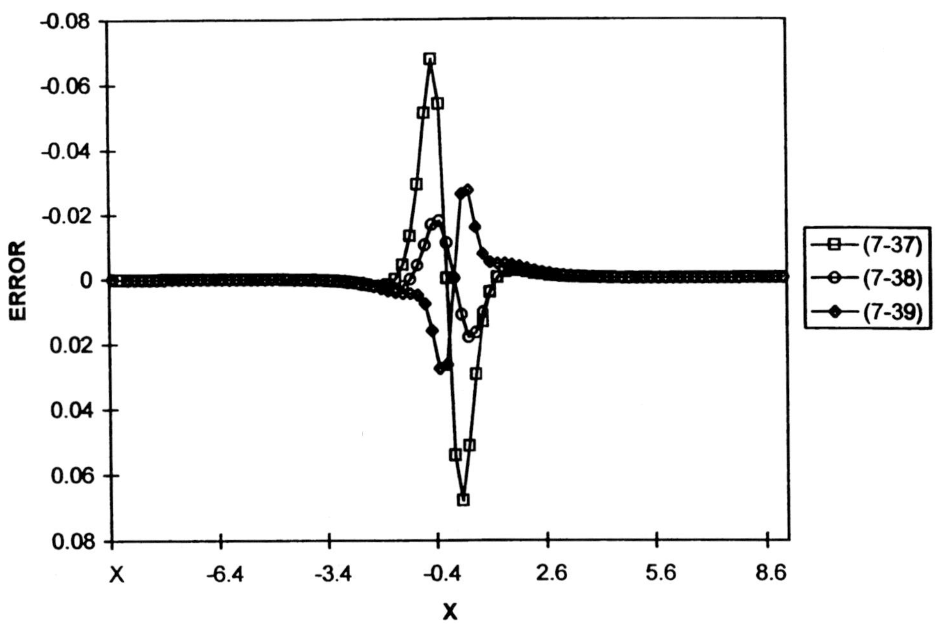

Figure 7-3. Comparison of error distributions for FTCS formulations at $t = 0.4$.

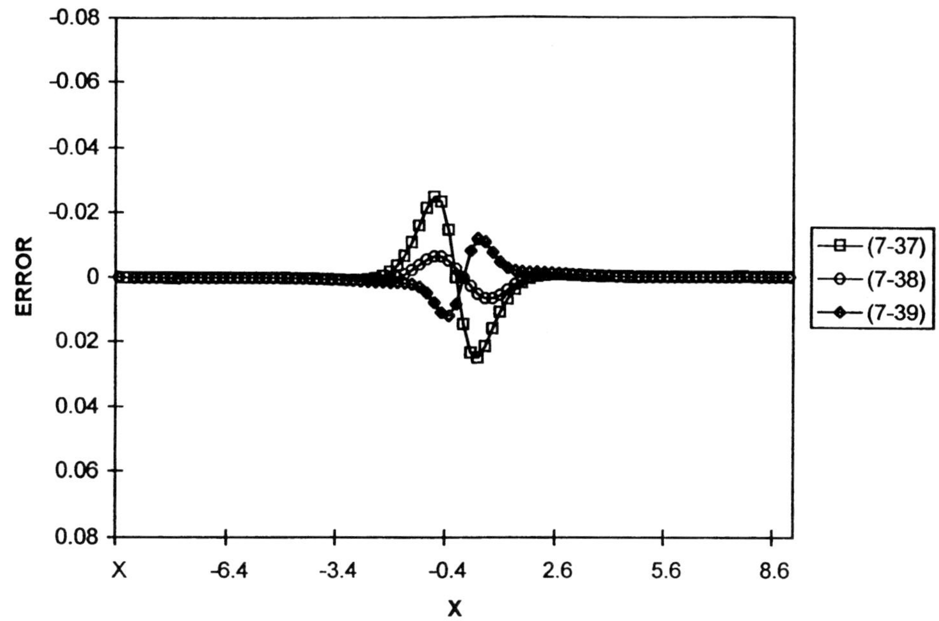

Figure 7-4. Comparison of error distributions for FTCS formulations at $t = 1.0$.

7.5.2 FTBCS Explicit

This formulation is similar to the method just used, except a first-order backward approximation is employed for the convective term. The resulting FDE for PDE (7-35) is

$$\frac{u_i^{n+1} - u_i^n}{\Delta t} + \frac{E_i^n - E_{i-1}^n}{\Delta x} = \frac{u_{i+1}^n - 2u_i^n + u_{i-1}^n}{(\Delta x)^2} \tag{7-40}$$

and for (7-36) is

$$\frac{u_i^{n+1} - u_i^n}{\Delta t} + A_i^n \frac{u_i^n - u_{i-1}^n}{\Delta x} = \frac{u_{i+1}^n - 2u_i^n + u_{i-1}^n}{(\Delta x)^2} \tag{7-41}$$

The argument on linearity, stability, and imposition of initial and boundary conditions discussed previously applies to this scheme as well.

7.5.3 DuFort-Frankel Explicit

Application of Equation (7-21) to the model equation (7-36) yields

$$\frac{u_i^{n+1} - u_i^{n-1}}{2\Delta t} + A_i^n \frac{u_{i+1}^n - u_{i-1}^n}{2\Delta x} = \frac{u_{i+1}^n - (u_i^{n-1} + u_i^{n+1}) + u_{i-1}^n}{(\Delta x)^2}$$

This equation may be rearranged as

$$u_i^{n+1} = \left(\frac{1-2d}{1+2d}\right) u_i^{n-1} + \left(\frac{c+2d}{1+2d}\right) u_{i-1}^n - \left(\frac{c-2d}{1+2d}\right) u_{i+1}^n \tag{7-42}$$

where

$$c = A_i^n \frac{\Delta t}{\Delta x} \qquad \text{and} \qquad d = \frac{\Delta t}{(\Delta x)^2}$$

When Equation (7-35) is used, the following FDE is obtained:

$$\frac{u_i^{n+1} - u_i^{n-1}}{2\Delta t} + \frac{E_{i+1}^n - E_{i-1}^n}{2\Delta x} = \frac{u_{i+1}^n - (u_i^{n-1} + u_i^{n+1}) + u_{i-1}^n}{(\Delta x)^2}$$

From which

$$u_i^{n+1} = \left(\frac{1-2d}{1+2d}\right) u_i^{n-1} + \left(\frac{2d}{1+2d}\right) (u_{i+1}^n + u_{i-1}^n) - \frac{c}{1+2d} (E_{i+1}^n - E_{i-1}^n) \tag{7-43}$$

As seen previously in Chapter 3, a minor difficulty is encountered with regard to the starting procedure. This scheme requires two sets of initial conditions. Various procedures may be used to provide the second set of necessary data. Usually, a one-step scheme such as FTCS explicit is used to provide the required second set of data. However, the stability requirement of the starter solution must be checked to ensure a stable initial set of data.

For the results shown in Table (7-3), the second set of data was also provided by the analytical solution.

7.5.4 MacCormack Explicit

Application of this multi-step scheme given by Equations (7-24) and (7-25) to the model equation (7-35) provides the following FDEs:

$$\Delta u_i^n = -\frac{\Delta t}{\Delta x}(E_{i+1}^n - E_i^n) + \frac{\Delta t}{(\Delta x)^2}(u_{i+1}^n - 2u_i^n + u_{i-1}^n) \tag{7-44a}$$

$$u_i^* = u_i^n + \Delta u_i^n \tag{7-44b}$$

and

$$\Delta u_i^* = -\frac{\Delta t}{\Delta x}(E_i^* - E_{i-1}^*) + \frac{\Delta t}{(\Delta x)^2}(u_{i+1}^* - 2u_i^* + u_{i-1}^*) \tag{7-45a}$$

$$u_i^{n+1} = \frac{1}{2}(u_i^n + u_i^* + \Delta u_i^*) \tag{7-45b}$$

7.5.5 MacCormack Implicit

In this implicit formulation, the FDEs are

$$\begin{cases} \left(1+\lambda\dfrac{\Delta t}{\Delta x}\right)\delta u_i^* - \lambda\dfrac{\Delta t}{\Delta x}\delta u_{i+1}^* = \Delta u_i^n & \text{(7-46a)} \\ \\ u_i^* = u_i^n + \delta u_i^* & \text{(7-46b)} \end{cases}$$

and

$$\begin{cases} \left(1+\lambda\dfrac{\Delta t}{\Delta x}\right)\delta u_i^{n+1} - \lambda\dfrac{\Delta t}{\Delta x}\delta u_{i-1}^{n+1} = \Delta u_i^* & \text{(7-47a)} \\ \\ u_i^{n+1} = \dfrac{1}{2}(u_i^n + u_i^* + \delta u_i^{n+1}) & \text{(7-47b)} \end{cases}$$

The right-hand sides of Equations (7-46a) and (7-47a), i.e., Δu_i^n and Δu_i^*, are evaluated by Equations (7-44a) and (7-45a). The parameter λ is selected such that

$$\lambda \geq \max\left[\frac{1}{2}\left(|u| + \frac{2}{\Delta x} - \frac{\Delta x}{\Delta t}\ ,\ 0.0\right)\right]$$

Equations (7-46a) and (7-47a) form upper and lower bidiagonal systems, respectively. To clearly identify the bidiagonal system, define the following:

$$1 + \lambda\frac{\Delta t}{\Delta x} = a$$

$$-\lambda\frac{\Delta t}{\Delta x} = b$$

Thus, Equation (7-46a) may be expressed as

$$a_i\delta u_i^* + b_i\delta u_{i+1}^* = \Delta u_i^n \tag{7-48}$$

Now apply this equation to grid points from $i = 2$ to $i = IMM1$, to obtain the following equations:

$$\begin{aligned}
i = 2 &: \quad a_2\delta u_2^* + b_2\delta u_3^* = \Delta u_2 \\
i = 3 &: \quad a_3\delta u_3^* + b_3\delta u_4^* = \Delta u_3 \\
&\ \ \vdots \\
i = IMM2 &: \quad a_{IMM2}\delta u_{IMM2}^* + b_{IMM2}\delta u_{IMM1}^* = \Delta u_{IMM2} \\
i = IMM1 &: \quad a_{IMM1}\delta u_{IMM1}^* + b_{IMM1}\delta u_{IM}^* = \Delta u_{IMM1}
\end{aligned}$$

In the last equation, δu^*_{IM} is provided by the boundary condition. Since in this example the boundary condition at IM is specified as a constant, δu^*_{IM} will be zero. Now, the equations above are combined in a matrix formulation as

$$\begin{bmatrix} a_2 & b_2 & & & \\ & a_3 & b_3 & & \\ & & & & \\ & & a_{IMM2} & b_{IMM2} & \\ & & & & a_{IMM1} \end{bmatrix} \begin{bmatrix} \delta u^*_2 \\ \delta u^*_3 \\ | \\ \delta u^*_{IMM2} \\ \delta u^*_{IMM1} \end{bmatrix} = \begin{bmatrix} \Delta u_2 \\ \Delta u_3 \\ | \\ \Delta u_{IMM2} \\ \Delta u_{IMM1} \end{bmatrix}$$

This system is an upper bidiagonal system which may be solved efficiently by the following:

$$\delta u^*_i = \frac{(\Delta u)_i - b_i(\delta u^*_{i+1})}{a_i}$$

Now, consider Equation (7-47a), which may be expressed as

$$a_i \delta u^{n+1}_i + b_i \delta u^{n+1}_{i-1} = \Delta u^*_i \tag{7-49}$$

For grid points $i = 2$ to $i = IMM1$, the following equations will result:

$$i = 2 : \quad a_2 \delta u^{n+1}_2 + b_2 \delta u^{n+1}_1 = \Delta u^*_2$$

$$i = 3 : \quad a_3 \delta u^{n+1}_3 + b_3 \delta u^{n+1}_2 = \Delta u^*_3$$

$$\vdots$$

$$i = IMM2 : \quad a_{IMM2} \delta u^{n+1}_{IMM2} + b_{IMM2} \delta u^{n+1}_{IMM3} = \Delta u^*_{IMM2}$$

$$i = IMM1 : \quad a_{IMM1} \delta u^{n+1}_{IMM1} + b_{IMM1} \delta u^{n+1}_{IMM2} = \Delta u^*_{IMM1}$$

Since at the boundary $i = 1$, the value of u is assigned a constant value of 2 for all time levels; therefore $\delta u^{n+1}_1 = 0$. The system of equations above form a lower bidiagonal system as

$$\begin{bmatrix} a_2 & & & & \\ b_3 & a_3 & & & \\ & & & & \\ & b_{IMM2} & a_{IMM2} & & \\ & & b_{IMM1} & a_{IMM1} & \end{bmatrix} \begin{bmatrix} \delta u^{n+1}_2 \\ \delta u^{n+1}_3 \\ | \\ \delta u^{n+1}_{IMM2} \\ \delta u^{n+1}_{IMM1} \end{bmatrix} = \begin{bmatrix} \Delta u^*_2 \\ \Delta u^*_3 \\ | \\ \Delta u^*_{IMM2} \\ \Delta u^*_{IMM1} \end{bmatrix}$$

The solution of this system is given by

$$\delta u^{n+1}_i = \frac{\Delta u^*_i - b_i \delta u^{n+1}_{i-1}}{a_i}$$

7.5.6 BTCS Implicit

The implicit formulation previously described is applied to the model equation (7-36) where lagging of the Jacobian is used for linearization purposes. The FDE is

$$\frac{u_i^{n+1} - u_i^n}{\Delta t} + A_i^n \frac{u_{i+1}^{n+1} - u_{i-1}^{n+1}}{2\Delta x} = \frac{u_{i+1}^{n+1} - 2u_i^{n+1} + u_{i-1}^{n+1}}{(\Delta x)^2}$$

which is rearranged as

$$-\left[\frac{\Delta t}{(\Delta x)^2} + A_i^n \frac{\Delta t}{2\Delta x}\right] u_{i-1}^{n+1} + \left[1 + 2\frac{\Delta t}{(\Delta x)^2}\right] u_i^{n+1}$$

$$+ \left[-\frac{\Delta t}{(\Delta x)^2} + A_i^n \frac{\Delta t}{2\Delta x}\right] u_{i+1}^{n+1} = u_i^n \tag{7-50}$$

Equation (7-50) may be expressed as a general tridiagonal system by the following relations:

$$a_i = -\frac{\Delta t}{(\Delta x)^2} - A_i^n \frac{\Delta t}{2\Delta x}$$

$$b_i = 1 + 2\frac{\Delta t}{(\Delta x)^2}$$

$$c_i = -\frac{\Delta t}{(\Delta x)^2} + A_i^n \frac{\Delta t}{2\Delta x}$$

and

$$D_i = u_i^n$$

Hence,

$$a_i u_{i-1}^{n+1} + b_i u_i^{n+1} + c_i u_{i+1}^{n+1} = D_i$$

This tridiagonal system is easily solved by the method introduced in Chapter 3. The solution is presented in Table 7-4. The stepsizes used are $\Delta x = 0.2$ and $\Delta t = 0.01$.

7.5.7 BTBCS Implicit

The implicit formulations, where backward difference approximations of first, second, or third order is used for the convective terms, may be expressed as

$$a_i u_{i-1}^{n+1} + b_i u_i^{n+1} + c_i u_{i+1}^{n+1} = D_i$$

where

$$a_i = -\frac{\Delta t}{(\Delta x)^2} - \theta_1 \left(A \frac{\Delta t}{\Delta x} \right)$$

$$b_i = 1 + 2\frac{\Delta t}{(\Delta x)^2} + \theta_2 \left(A \frac{\Delta t}{\Delta x} \right)$$

$$c_i = -\frac{\Delta t}{(\Delta x)^2} + \theta_3 \left(A \frac{\Delta t}{\Delta x} \right)$$

and

$$D_i = u_i^n + \theta_4 \left(A \frac{\Delta t}{\Delta x} u_{i-2}^n \right)$$

The coefficients θ_1 through θ_4 are defined as

Order of Approximation for the the Convective Term	θ_1	θ_2	θ_3	θ_4
First-Order	1	1	0	0
Second-Order	2	$\frac{3}{2}$	0	$-\frac{1}{2}$
Third-Order	1	$\frac{1}{2}$	$\frac{1}{3}$	$-\frac{1}{6}$

A comparison of errors by the three approximations described above is shown in Figure (7-5). The error is determined as the difference between the analytical and numerical solutions. As expected, the third-order accurate approximation of the convective term provides the solution with the least error.

7.5.8 Modified Runge-Kutta

The modified Runge-Kutta scheme, described in Section 6.6.8, is now applied to Equation (7-35). There are several methods by which the viscous term can be incorporated into the finite difference equation. In the first scheme, the viscous terms are evaluated at each stage, resulting in the following equations.

$$u_i^{(1)} = u_i^n \tag{7-51}$$

$$u_i^{(2)} = u_i^n - \frac{\Delta t}{4} \left(\frac{\partial E}{\partial x} \right)_i^{(1)} + \frac{\Delta t}{4} \left(\frac{\partial^2 u}{\partial x^2} \right)_i^{(1)} \tag{7-52}$$

$$u_i^{(3)} = u_i^n - \frac{\Delta t}{3} \left(\frac{\partial E}{\partial x} \right)_i^{(2)} + \frac{\Delta t}{3} \left(\frac{\partial^2 u}{\partial x^2} \right)_i^{(2)} \tag{7-53}$$

$$u_i^{(4)} = u_i^n - \frac{\Delta t}{2}\left(\frac{\partial E}{\partial x}\right)_i^{(3)} + \frac{\Delta t}{2}\left(\frac{\partial^2 u}{\partial x^2}\right)_i^{(3)} \tag{7-54}$$

$$u_i^{n+1} = u_i^n - \Delta t\left(\frac{\partial E}{\partial x}\right)_i^{(4)} + \Delta t\left(\frac{\partial^2 u}{\partial x^2}\right)_i^{(4)} \tag{7-55}$$

where

$$\left(\frac{\partial E}{\partial x}\right)_i = \frac{E_{i+1} - E_{i-1}}{2\Delta x} \tag{7-56}$$

$$\left(\frac{\partial^2 u}{\partial x^2}\right)_i = \frac{u_{i+1} - 2u_i + u_{i-1}}{(\Delta x)^2} \tag{7-57}$$

A solution by the modified Runge-Kutta scheme given by Equations (7-51) through (7-55) is provided in Table 7.5.

A second method of approximation of the viscous term is to evaluate it only at the first stage and to use it at the subsequent stages. Finally, a third method is to compute only the convective term using the modified Runge-Kutta scheme and, subsequently, add the viscous term after the final stage, that is,

$$u_i^{n+1} = u_i^{n+1} + \Delta t\left(\frac{\partial^2 u}{\partial x^2}\right)^{n+1}$$

Obviously, the second and third methods identified above require less computation time; however, accuracy may be compromised.

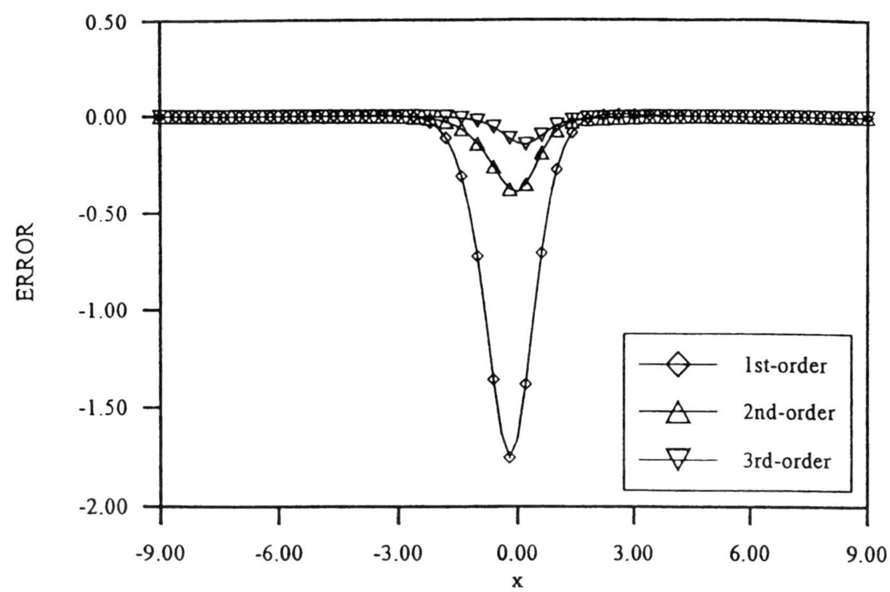

Figure 7-5. Comparison of errors for the BTBCS implicit scheme at $t = 1$, for $\Delta x = 0.2$, $\Delta t = 0.01$.

7.5.9 Second-Order TVD Schemes

Numerous TVD schemes along with their applications were presented in Chapter 6. The general formulation is reviewed in this section, and a sample application is illustrated. Recall that the concept of TVD is introduced to eliminate or reduce oscillations in the solution due to dispersion error. In fact, a TVD scheme operates similarly to that of the addition of a damping term. For equations such as (7-35), which contains both convective and diffusive terms, TVD is considered on the convective term only. Recall some of the criteria for TVD schemes presented in Section 6.9.2. The finite difference equation following (6-123) through (6-125) may be expressed as

$$u_i^{n+1} = u_i^n - \frac{\Delta t}{\Delta x}\left(h_{i+\frac{1}{2}}^n - h_{i-\frac{1}{2}}^n\right) + \frac{\Delta t}{(\Delta x)^2}\left(u_{i+1}^n - 2u_i^n + u_{i-1}^n\right) \tag{7-58}$$

where

$$h_{i+\frac{1}{2}}^n = \frac{1}{2}\left[E_{i+1}^n + E_i^n + \phi_{i+\frac{1}{2}}^n\right] \tag{7-59}$$

and

$$h_{i-\frac{1}{2}}^n = \frac{1}{2}\left[E_i^n + E_{i-1}^n + \phi_{i-\frac{1}{2}}^n\right] \tag{7-60}$$

Several flux limiter functions were introduced in Chapter 6 for Harten-Yee upwind TVD, Roe-Sweby upwind TVD, and Davis-Yee symmetric TVD. These functions are given by Equations (6-131), (6-137), and (6-141), respectively. Furthermore, several limiters are provided for each of these limiter functions.

A solution by the Davis-Yee symmetric TVD and limiter (6-143) is illustrated in Figure (7-6) and provided in Table 7.6.

Comparison of the numerical solutions obtained by various methods and analytical solution reveals the accuracy of the methods and, indeed for the stepsizes used, all methods considered provide accurate solutions.

For the specified stepsizes of $\Delta x = 0.2$ and $\Delta t = 0.01$, all schemes investigated are stable. Comparison of errors at time levels 0.4 and 1.0 are shown in Figures 7-7 and 7-8. The schemes employed are the FTCS explicit, the Dufort-Frankel explicit, the MacCormack explicit, the BTCS implicit, and the modified Runge-Kutta. Note that the error is decreased as the solution proceeds forward in time for all three schemes.

It is also interesting to note that the BTCS implicit scheme produces more error compared to the FTCS explicit scheme. Recall that a similar conclusion was reached in Section 6.6.6.

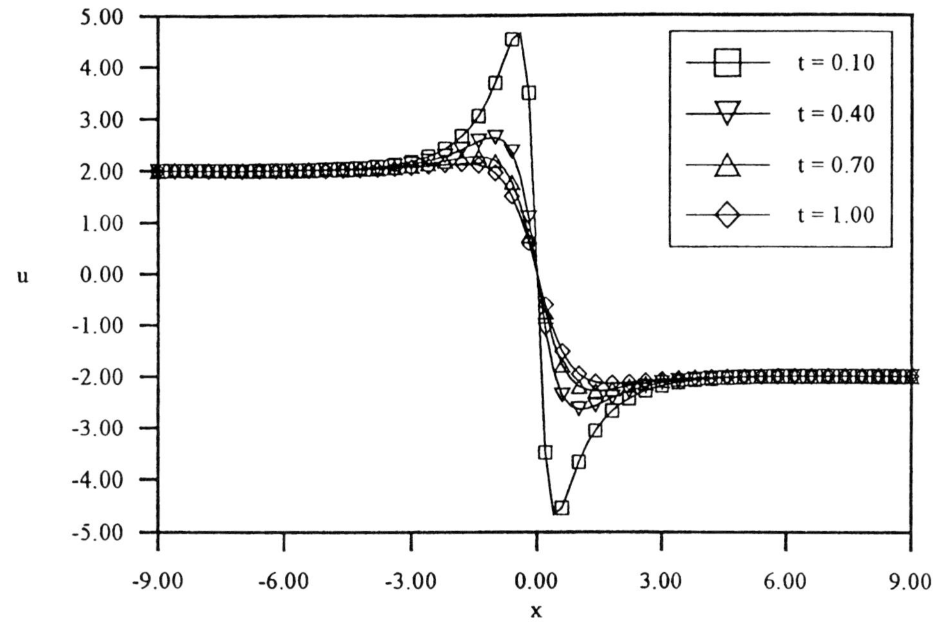

Figure 7-6. Solution of the viscous Burgers equation by the Davis-Yee symmetric TVD scheme and limiter 6-143.

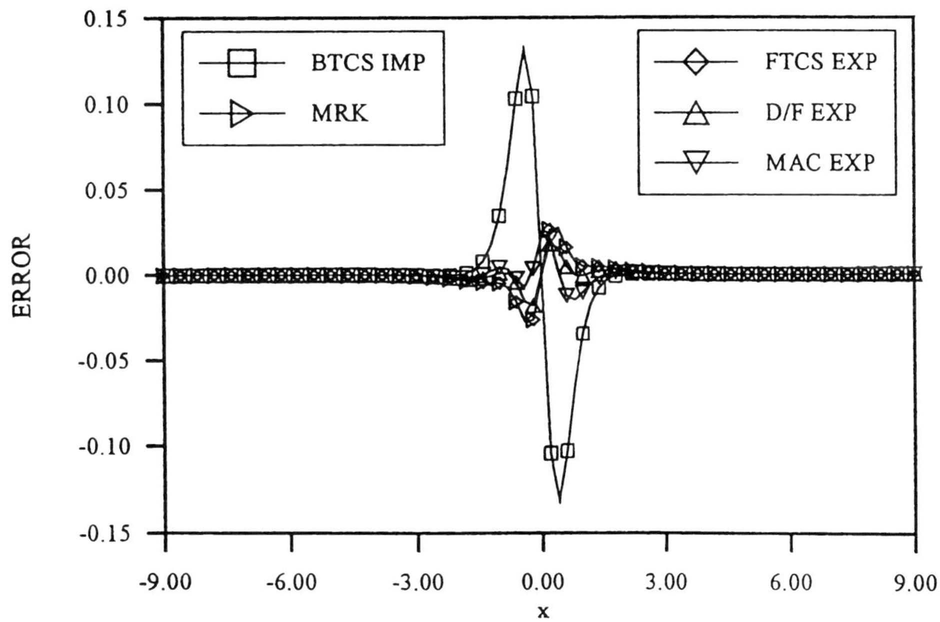

Figure 7-7. Comparison of errors at time level 0.4 for the FTCS explicit, DuFort-Frankel explicit, MacCormack explicit, modified Runge-Kutta, and BTCS implicit.

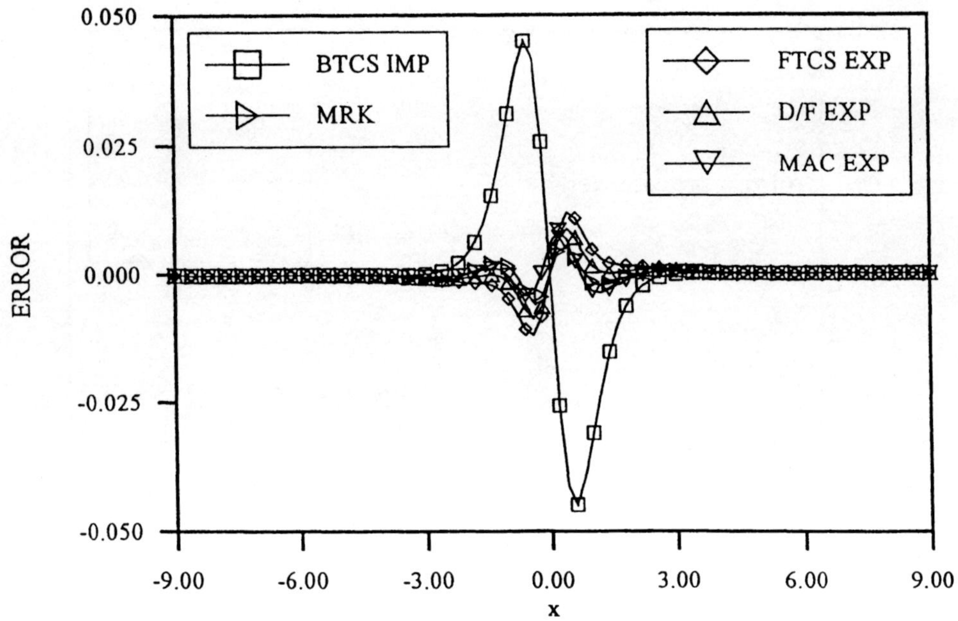

Figure 7-8. Comparison of errors at time level 1.0 for the FTCS explicit, DuFort-Frankel explicit, MacCormack explicit, modified Runge-Kutta, and BTCS implicit.

7.6 Summary Objectives

After studying the material in this chapter, you should be able to do the following:

1. Describe:
    a. Burgers equation and its relation to Navier-Stokes equation
    b. Linearization of Burgers equation
    c. FTCS explicit scheme
    d. FTBCS explicit scheme
    e. DuFort-Frankel explicit scheme
    f. MacCormack explicit scheme
    g. MacCormack implicit scheme
    h. BTCS implicit scheme
    i. BTBCS implicit scheme
    j. Modified Runge-Kutta scheme
    k. Second-order TVD scheme
2. Solve the Problems for Chapter Seven.

7.7 Problems

7.1 Solve the viscous Burgers equation by the following schemes: (a) FTCS explicit, (b) MacCormack explicit, (c) BTCS implicit. The domain of interest is $-9 \le x \le 9$, and the initial distribution of u is given as

$$u = -\frac{2\sinh x}{\cosh x - e^{-0.1}}$$

i.e., at $t = 0.1$. Compare the numerical solutions to the analytical solution given by

$$u = -\frac{2\sinh x}{\cosh x - e^{-t}}$$

Assume that the values of u at the boundaries are fixed according to

$$x = -9 \qquad u = 2.0$$

$$x = 9 \qquad u = -2.0$$

Use stepsizes of $\Delta x = 0.2$ and $\Delta t = 0.01$. Print and plot the solutions for all x locations at $t = 0.1$ (initial distribution), 0.4, 0.7, and 1.0.

To investigate the effect of stepsizes on accuracy and stability, consider the following cases:

(a) $\Delta x = 0.2$ $\quad\Delta t = 0.02$

(b) $\Delta x = 0.2$ $\quad\Delta t = 0.05$

(c) $\Delta x = 0.5$ $\quad\Delta t = 0.01$

(d) $\Delta x = 0.5$ $\quad\Delta t = 0.05$

7.2 Consider a fluid with a temperature of T_c and a constant velocity of u travelling from left to right in a channel. The temperature at the end of the channel is suddenly changed to T_h and is maintained at that constant value. It is required to compute the time-accurate solution for the temperature distribution within the channel. The governing equation is given by the one-dimensional energy equation as

$$\frac{\partial T}{\partial t} + u\frac{\partial T}{\partial x} = \alpha\frac{\partial^2 T}{\partial x^2}$$

where α is the thermal diffusivity, which is assumed constant for this problem. The initial and boundary conditions are specified as

$$t = 0 \;;\quad T(x, 0) = T_c$$

$$t \geq 0 \;, \qquad x = 0 \qquad , \; T = T_c$$

$$x = L \qquad , \; T = T_h$$

(a) Nondimensionalize the equation by

$$t^* = \frac{tu_o}{L} \;, \quad x^* = \frac{x}{L} \;, \quad u^* = \frac{u}{u_o} \;, \quad T^* = \frac{T - T_c}{T_h - T_c}$$

and define

$$\alpha^* = \frac{\alpha}{Lu_o}$$

(b) Obtain the analytical solution.

(c) Use a first-order backward differencing of the convective term and a second-order central difference approximation of the diffusion term to write an explicit formulation. The time derivative is approximated by a first-order accurate differencing.

(d) Repeat (c), but use a second-order backward differencing for the convective term.

(e) Repeat (c), but use a second-order central differencing for the convective term.

(f) Repeat (e), but write an implicit formulation.

(g) Obtain the numerical solutions with the formulations developed in (c), (d), (e), and (f). Use a time step of 0.01 sec. and 21 equally spaced grid points, i.e., IM = 21. Assume a fluid with a diffusivity of 0.04 m^2/sec., a velocity of 0.2 m/s, temperature of $T_c = 20°C$. The sudden change in temperature T_h at the boundary is $T_h = 100$ °C and the length of the channel is 2 m.

Print the solutions at time levels of 1.0, 2.0, and 3.0 seconds for all schemes. Plot a typical temperature profile at these time levels from the solution of (c). Plot the error distributions at a time level of 3.0 sec. for all schemes, where

Error = ABS (TA - TN)

TA $\Longrightarrow$ Analytical solution

TN $\Longrightarrow$ Numerical solution.

7.3 A fluid with a temperature of T_c and a constant velocity of u_o is travelling from left to right in a channel of length L. The temperature at the end of the channel is suddenly changed to T_h and is maintained at the constant value of T_h. The governing equation for this one-dimension flow is given by

$$\frac{\partial T}{\partial t} + u\frac{\partial T}{\partial x} = \alpha\frac{\partial^2 T}{\partial x^2} \tag{7.3.1}$$

where α is the thermal diffusivity assumed to be a constant. The initial and boundary conditions are specified as

$$t = 0; \qquad T(x, 0) = T_c$$

$$t \geq 0; \qquad x = 0, \qquad T = T_c$$

$$x = L, \qquad T = T_h$$

(a) Nondimensionalize the equation according to

$$t^* = \frac{tu_o}{L}, \quad x^* = \frac{x}{L}, \quad u^* = \frac{u}{u_o}, \quad T^* = \frac{T - T_c}{T_h - T_c}$$

and define

$$\alpha^* = \frac{\alpha}{Lu_o}$$

For the following, the nondimensional equation is to be used.

(b) Obtain the analytical solution.

(c) Use a second-order central differencing for the spatial derivatives to write an explicit, first-order in time finite difference equation.

(d) Repeat (c), but use nonequal grid spacing. Denote the variable spatial step as follows:

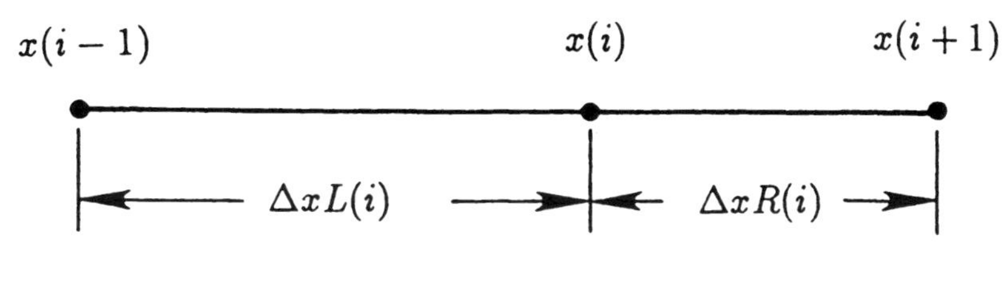

where $\Delta xL(i) = x(i) - x(i-1)$

and $\Delta xR(i) = x(i+1) - x(i)$

The ratio of stepsizes is defined as

$$\gamma_i = \frac{\Delta xR(i)}{\Delta xL(i)}$$

To simplify the formulation, use

$$\Delta xL(i) = \Delta x \quad , \qquad \Delta xR(i) = \gamma \Delta x$$

However, bear in mind that Δx and γ will change from grid point to grid point.

In this problem, the grid point clustering near the boundary at $x = L$ is desired to better resolve the temperature gradient within that region. The grid point clustering procedure suggested in problem 3.9 may be used in this problem.

(e) Obtain the numerical solutions for the following set of data:

$$T_c = 20°\text{C} \quad , \quad T_h = 100°\text{C} \quad , \quad u_o = 0.2 \text{ m/sec.}$$

$$\alpha = 0.04 \text{ m}^2/\text{sec.} \quad , \qquad L = 2 \text{ m}$$

$$\Delta t = 0.01\text{sec.} \quad , \qquad IM = 21$$

(I) Use the formulation developed in (c)

(II) Use the formulation developed in (d) with a clustering parameter β of 1.2.

(III) Repeat (II) with $\beta = 1.1$.

Print the solutions at time levels of 1, 2, and 3 seconds. Plot the temperature and error distributions at a time level of 3 seconds, where the error distribution is determined as defined in problem 7.2, part (g).

x	t=0.1	t=0.4	t=0.7	t=1.0
-9.00	2.0004	2.0003	2.0002	2.0002
-8.60	2.0007	2.0005	2.0004	2.0003
-8.20	2.0010	2.0007	2.0005	2.0004
-7.80	2.0015	2.0011	2.0008	2.0006
-7.40	2.0022	2.0016	2.0012	2.0009
-7.00	2.0033	2.0024	2.0018	2.0013
-6.60	2.0049	2.0036	2.0027	2.0020
-6.20	2.0074	2.0054	2.0040	2.0030
-5.80	2.0110	2.0081	2.0060	2.0044
-5.40	2.0164	2.0121	2.0089	2.0066
-5.00	2.0245	2.0180	2.0133	2.0098
-4.60	2.0366	2.0269	2.0198	2.0145
-4.20	2.0549	2.0401	2.0293	2.0214
-3.80	2.0823	2.0597	2.0434	2.0314
-3.40	2.1237	2.0889	2.0639	2.0457
-3.00	2.1866	2.1321	2.0934	2.0656
-2.60	2.2833	2.1955	2.1347	2.0917
-2.20	2.4335	2.2871	2.1895	2.1224
-1.80	2.6715	2.4144	2.2538	2.1479
-1.40	3.0565	2.5724	2.3022	2.1360
-1.00	3.6826	2.6931	2.2460	2.0000
-0.60	4.5374	2.4717	1.8484	1.5574
-0.20	3.4945	1.1513	0.7692	0.6174
0.20	-3.4945	-1.1513	-0.7692	-0.6174
0.60	-4.5374	-2.4717	-1.8484	-1.5574
1.00	-3.6826	-2.6931	-2.2460	-2.0000
1.40	-3.0565	-2.5724	-2.3022	-2.1360
1.80	-2.6715	-2.4144	-2.2538	-2.1479
2.20	-2.4335	-2.2871	-2.1895	-2.1224
2.60	-2.2833	-2.1955	-2.1347	-2.0917
3.00	-2.1866	-2.1321	-2.0934	-2.0656
3.40	-2.1237	-2.0889	-2.0639	-2.0457
3.80	-2.0823	-2.0597	-2.0434	-2.0314
4.20	-2.0549	-2.0401	-2.0293	-2.0214
4.60	-2.0366	-2.0269	-2.0198	-2.0145
5.00	-2.0245	-2.0180	-2.0133	-2.0098
5.40	-2.0164	-2.0121	-2.0089	-2.0066
5.80	-2.0110	-2.0081	-2.0060	-2.0044
6.20	-2.0074	-2.0054	-2.0040	-2.0030
6.60	-2.0049	-2.0036	-2.0027	-2.0020
7.00	-2.0033	-2.0024	-2.0018	-2.0013
7.40	-2.0022	-2.0016	-2.0012	-2.0009
7.80	-2.0015	-2.0011	-2.0008	-2.0006
8.20	-2.0010	-2.0007	-2.0005	-2.0004
8.60	-2.0007	-2.0005	-2.0004	-2.0003
9.00	-2.0004	-2.0003	-2.0002	-2.0002

Table 7-1. Analytical solution given by Equation (7-34).

x	t=0.1	t=0.4	t=0.7	t=1.0
-9.00	2.0000	2.0000	2.0000	2.0000
-8.60	2.0007	2.0002	2.0001	2.0001
-8.20	2.0010	2.0005	2.0003	2.0002
-7.80	2.0015	2.0010	2.0006	2.0004
-7.40	2.0022	2.0016	2.0010	2.0007
-7.00	2.0033	2.0024	2.0017	2.0011
-6.60	2.0049	2.0036	2.0026	2.0018
-6.20	2.0074	2.0054	2.0039	2.0028
-5.80	2.0110	2.0081	2.0059	2.0043
-5.40	2.0164	2.0120	2.0088	2.0064
-5.00	2.0245	2.0180	2.0132	2.0096
-4.60	2.0366	2.0268	2.0196	2.0143
-4.20	2.0549	2.0399	2.0290	2.0211
-3.80	2.0823	2.0594	2.0429	2.0310
-3.40	2.1237	2.0884	2.0632	2.0450
-3.00	2.1866	2.1312	2.0923	2.0646
-2.60	2.2833	2.1940	2.1331	2.0905
-2.20	2.4335	2.2846	2.1874	2.1209
-1.80	2.6715	2.4107	2.2512	2.1462
-1.40	3.0565	2.5678	2.2993	2.1339
-1.00	3.6826	2.6881	2.2408	1.9952
-0.60	4.5374	2.4558	1.8347	1.5465
-0.20	3.4945	1.1250	0.7564	0.6091
0.20	-3.4945	-1.1250	-0.7564	-0.6091
0.60	-4.5374	-2.4558	-1.8347	-1.5465
1.00	-3.6826	-2.6881	-2.2408	-1.9952
1.40	-3.0565	-2.5678	-2.2993	-2.1339
1.80	-2.6715	-2.4107	-2.2512	-2.1462
2.20	-2.4335	-2.2846	-2.1874	-2.1209
2.60	-2.2833	-2.1940	-2.1331	-2.0905
3.00	-2.1866	-2.1312	-2.0923	-2.0646
3.40	-2.1237	-2.0884	-2.0632	-2.0450
3.80	-2.0823	-2.0594	-2.0429	-2.0310
4.20	-2.0549	-2.0399	-2.0290	-2.0211
4.60	-2.0366	-2.0268	-2.0196	-2.0143
5.00	-2.0245	-2.0180	-2.0132	-2.0096
5.40	-2.0164	-2.0120	-2.0088	-2.0064
5.80	-2.0110	-2.0081	-2.0059	-2.0043
6.20	-2.0074	-2.0054	-2.0039	-2.0028
6.60	-2.0049	-2.0036	-2.0026	-2.0018
7.00	-2.0033	-2.0024	-2.0017	-2.0011
7.40	-2.0022	-2.0016	-2.0010	-2.0007
7.80	-2.0015	-2.0010	-2.0006	-2.0004
8.20	-2.0010	-2.0005	-2.0003	-2.0002
8.60	-2.0007	-2.0002	-2.0001	-2.0001
9.00	-2.0000	-2.0000	-2.0000	-2.0000

Table 7-2. Solution of the viscous Burgers equation by the FTCS scheme.

x	t=0.1	t=0.4	t=0.7	t=1.0
-9.00	2.0000	2.0000	2.0000	2.0000
-8.60	2.0007	2.0002	2.0001	2.0001
-8.20	2.0010	2.0005	2.0003	2.0002
-7.80	2.0015	2.0010	2.0006	2.0004
-7.40	2.0022	2.0016	2.0010	2.0007
-7.00	2.0033	2.0024	2.0017	2.0011
-6.60	2.0049	2.0036	2.0026	2.0018
-6.20	2.0074	2.0054	2.0039	2.0028
-5.80	2.0110	2.0081	2.0059	2.0043
-5.40	2.0164	2.0121	2.0089	2.0065
-5.00	2.0245	2.0180	2.0132	2.0096
-4.60	2.0366	2.0268	2.0196	2.0143
-4.20	2.0549	2.0399	2.0291	2.0211
-3.80	2.0823	2.0595	2.0430	2.0310
-3.40	2.1237	2.0885	2.0633	2.0452
-3.00	2.1866	2.1313	2.0925	2.0649
-2.60	2.2833	2.1942	2.1334	2.0909
-2.20	2.4335	2.2851	2.1880	2.1216
-1.80	2.6715	2.4117	2.2525	2.1474
-1.40	3.0565	2.5702	2.3018	2.1360
-1.00	3.6826	2.6939	2.2453	1.9984
-0.60	4.5374	2.4672	1.8410	1.5504
-0.20	3.4945	1.1334	0.7599	0.6111
0.20	-3.4945	-1.1334	-0.7599	-0.6111
0.60	-4.5374	-2.4672	-1.8410	-1.5504
1.00	-3.6826	-2.6939	-2.2453	-1.9984
1.40	-3.0565	-2.5702	-2.3018	-2.1360
1.80	-2.6715	-2.4117	-2.2525	-2.1474
2.20	-2.4335	-2.2851	-2.1880	-2.1216
2.60	-2.2833	-2.1942	-2.1334	-2.0909
3.00	-2.1866	-2.1313	-2.0925	-2.0649
3.40	-2.1237	-2.0885	-2.0633	-2.0452
3.80	-2.0823	-2.0595	-2.0430	-2.0310
4.20	-2.0549	-2.0399	-2.0291	-2.0211
4.60	-2.0366	-2.0268	-2.0196	-2.0143
5.00	-2.0245	-2.0180	-2.0132	-2.0096
5.40	-2.0164	-2.0121	-2.0089	-2.0065
5.80	-2.0110	-2.0081	-2.0059	-2.0043
6.20	-2.0074	-2.0054	-2.0039	-2.0028
6.60	-2.0049	-2.0036	-2.0026	-2.0018
7.00	-2.0033	-2.0024	-2.0017	-2.0011
7.40	-2.0022	-2.0016	-2.0010	-2.0007
7.80	-2.0015	-2.0010	-2.0006	-2.0004
8.20	-2.0010	-2.0005	-2.0003	-2.0002
8.60	-2.0007	-2.0002	-2.0001	-2.0001
9.00	-2.0000	-2.0000	-2.0000	-2.0000

Table 7-3. Solution of the viscous Burgers equation by the DuFort-Frankel scheme.

x	t=0.1	t=0.4	t=0.7	t=1.0
-9.00	2.0000	2.0000	2.0000	2.0000
-8.60	2.0007	2.0002	2.0001	2.0001
-8.20	2.0010	2.0005	2.0003	2.0002
-7.80	2.0015	2.0010	2.0006	2.0004
-7.40	2.0022	2.0016	2.0010	2.0007
-7.00	2.0033	2.0024	2.0017	2.0012
-6.60	2.0049	2.0036	2.0026	2.0018
-6.20	2.0074	2.0054	2.0040	2.0029
-5.80	2.0110	2.0081	2.0060	2.0043
-5.40	2.0164	2.0121	2.0089	2.0065
-5.00	2.0245	2.0180	2.0132	2.0097
-4.60	2.0366	2.0269	2.0197	2.0144
-4.20	2.0549	2.0400	2.0292	2.0213
-3.80	2.0823	2.0596	2.0432	2.0313
-3.40	2.1237	2.0887	2.0637	2.0456
-3.00	2.1866	2.1318	2.0932	2.0657
-2.60	2.2833	2.1951	2.1347	2.0924
-2.20	2.4335	2.2868	2.1906	2.1246
-1.80	2.6715	2.4153	2.2583	2.1542
-1.40	3.0565	2.5800	2.3166	2.1513
-1.00	3.6826	2.7277	2.2827	2.0311
-0.60	4.5374	2.5747	1.9144	1.6023
-0.20	3.4945	1.2556	0.8136	0.6432
0.20	-3.4945	-1.2556	-0.8136	-0.6432
0.60	-4.5374	-2.5747	-1.9144	-1.6023
1.00	-3.6826	-2.7277	-2.2827	-2.0311
1.40	-3.0565	-2.5800	-2.3166	-2.1513
1.80	-2.6715	-2.4153	-2.2583	-2.1542
2.20	-2.4335	-2.2868	-2.1906	-2.1246
2.60	-2.2833	-2.1951	-2.1347	-2.0924
3.00	-2.1866	-2.1318	-2.0932	-2.0657
3.40	-2.1237	-2.0887	-2.0637	-2.0456
3.80	-2.0823	-2.0596	-2.0432	-2.0313
4.20	-2.0549	-2.0400	-2.0292	-2.0213
4.60	-2.0366	-2.0269	-2.0197	-2.0144
5.00	-2.0245	-2.0180	-2.0132	-2.0097
5.40	-2.0164	-2.0121	-2.0089	-2.0065
5.80	-2.0110	-2.0081	-2.0060	-2.0043
6.20	-2.0074	-2.0054	-2.0040	-2.0029
6.60	-2.0049	-2.0036	-2.0026	-2.0018
7.00	-2.0033	-2.0024	-2.0017	-2.0012
7.40	-2.0022	-2.0016	-2.0010	-2.0007
7.80	-2.0015	-2.0010	-2.0006	-2.0004
8.20	-2.0010	-2.0005	-2.0003	-2.0002
8.60	-2.0007	-2.0002	-2.0001	-2.0001
9.00	-2.0000	-2.0000	-2.0000	-2.0000

Table 7-4. Solution of the viscous Burgers equation by the BTCS scheme.

x	t=0.1	t=0.4	t=0.7	t=1.0
-9.00	2.0000	2.0000	2.0000	2.0000
-8.60	2.0007	2.0002	2.0001	2.0001
-8.20	2.0010	2.0005	2.0003	2.0002
-7.80	2.0015	2.0010	2.0006	2.0004
-7.40	2.0022	2.0016	2.0010	2.0007
-7.00	2.0033	2.0024	2.0017	2.0011
-6.60	2.0049	2.0036	2.0026	2.0018
-6.20	2.0074	2.0054	2.0040	2.0028
-5.80	2.0110	2.0081	2.0059	2.0043
-5.40	2.0164	2.0121	2.0089	2.0065
-5.00	2.0245	2.0180	2.0132	2.0097
-4.60	2.0366	2.0268	2.0196	2.0143
-4.20	2.0549	2.0400	2.0291	2.0212
-3.80	2.0823	2.0595	2.0431	2.0311
-3.40	2.1237	2.0885	2.0634	2.0453
-3.00	2.1866	2.1315	2.0927	2.0651
-2.60	2.2833	2.1945	2.1338	2.0912
-2.20	2.4335	2.2856	2.1887	2.1222
-1.80	2.6715	2.4128	2.2537	2.1486
-1.40	3.0565	2.5726	2.3042	2.1380
-1.00	3.6826	2.6995	2.2498	2.0016
-0.60	4.5374	2.4780	1.8471	1.5541
-0.20	3.4945	1.1414	0.7634	0.6130
0.20	-3.4945	-1.1414	-0.7634	-0.6130
0.60	-4.5374	-2.4780	-1.8471	-1.5541
1.00	-3.6826	-2.6995	-2.2498	-2.0016
1.40	-3.0565	-2.5726	-2.3042	-2.1380
1.80	-2.6715	-2.4128	-2.2537	-2.1486
2.20	-2.4335	-2.2856	-2.1887	-2.1222
2.60	-2.2833	-2.1945	-2.1338	-2.0912
3.00	-2.1866	-2.1315	-2.0927	-2.0651
3.40	-2.1237	-2.0885	-2.0634	-2.0453
3.80	-2.0823	-2.0595	-2.0431	-2.0311
4.20	-2.0549	-2.0400	-2.0291	-2.0212
4.60	-2.0366	-2.0268	-2.0196	-2.0143
5.00	-2.0245	-2.0180	-2.0132	-2.0097
5.40	-2.0164	-2.0121	-2.0089	-2.0065
5.80	-2.0110	-2.0081	-2.0059	-2.0043
6.20	-2.0074	-2.0054	-2.0040	-2.0028
6.60	-2.0049	-2.0036	-2.0026	-2.0018
7.00	-2.0033	-2.0024	-2.0017	-2.0011
7.40	-2.0022	-2.0016	-2.0010	-2.0007
7.80	-2.0015	-2.0010	-2.0006	-2.0004
8.20	-2.0010	-2.0005	-2.0003	-2.0002
8.60	-2.0007	-2.0002	-2.0001	-2.0001
9.00	-2.0000	-2.0000	-2.0000	-2.0000

Table 7-5. Solution of the viscous Burgers equation by the modified Runge-Kutta scheme.

x	t=0.1	t=0.4	t=0.7	t=1.0
-9.00	2.0000	2.0000	2.0000	2.0000
-8.60	2.0007	2.0006	2.0006	2.0006
-8.20	2.0010	2.0008	2.0007	2.0007
-7.80	2.0015	2.0011	2.0009	2.0008
-7.40	2.0022	2.0017	2.0013	2.0011
-7.00	2.0033	2.0025	2.0019	2.0015
-6.60	2.0049	2.0037	2.0028	2.0021
-6.20	2.0074	2.0055	2.0041	2.0031
-5.80	2.0110	2.0082	2.0061	2.0045
-5.40	2.0164	2.0122	2.0090	2.0067
-5.00	2.0245	2.0182	2.0135	2.0100
-4.60	2.0366	2.0271	2.0200	2.0148
-4.20	2.0549	2.0404	2.0297	2.0217
-3.80	2.0823	2.0601	2.0439	2.0319
-3.40	2.1237	2.0895	2.0646	2.0462
-3.00	2.1866	2.1329	2.0942	2.0660
-2.60	2.2833	2.1967	2.1354	2.0918
-2.20	2.4335	2.2886	2.1898	2.1211
-1.80	2.6715	2.4162	2.2512	2.1381
-1.40	3.0565	2.5710	2.2805	2.1124
-1.00	3.6826	2.6421	2.1980	1.9615
-0.60	4.5374	2.3669	1.7891	1.5183
-0.20	3.4945	1.0982	0.7461	0.6038
0.20	-3.4945	-1.0982	-0.7461	-0.6038
0.60	-4.5374	-2.3669	-1.7891	-1.5183
1.00	-3.6826	-2.6421	-2.1980	-1.9615
1.40	-3.0565	-2.5710	-2.2805	-2.1124
1.80	-2.6715	-2.4162	-2.2512	-2.1381
2.20	-2.4335	-2.2886	-2.1898	-2.1211
2.60	-2.2833	-2.1967	-2.1354	-2.0918
3.00	-2.1866	-2.1329	-2.0942	-2.0660
3.40	-2.1237	-2.0895	-2.0646	-2.0462
3.80	-2.0823	-2.0601	-2.0439	-2.0319
4.20	-2.0549	-2.0404	-2.0297	-2.0217
4.60	-2.0366	-2.0271	-2.0200	-2.0148
5.00	-2.0245	-2.0182	-2.0135	-2.0100
5.40	-2.0164	-2.0122	-2.0090	-2.0067
5.80	-2.0110	-2.0082	-2.0061	-2.0045
6.20	-2.0074	-2.0055	-2.0041	-2.0031
6.60	-2.0049	-2.0037	-2.0028	-2.0021
7.00	-2.0033	-2.0025	-2.0019	-2.0015
7.40	-2.0022	-2.0017	-2.0013	-2.0011
7.80	-2.0015	-2.0011	-2.0009	-2.0008
8.20	-2.0010	-2.0008	-2.0007	-2.0007
8.60	-2.0007	-2.0006	-2.0006	-2.0006
9.00	-2.0000	-2.0000	-2.0000	-2.0000

Table 7-6. Solution of the viscous Burgers equation by the second-order TVD scheme.

Chapter 8

Incompressible Navier-Stokes Equations

8.1 Introductory Remarks

The equations of motion for a homogeneous fluid in the absence of a finite rate chemical reaction or mass diffusion are based on three physical conservation laws. If a fluid is composed of various chemical species with mass diffusion and/or chemical reaction, additional conservation laws may be required, e.g., conservation laws for species. Since, for most engineering applications, the average measurable values of the flow properties are desired, the assumption of continuous distribution of matter is imposed. This assumption is known as continuum and is valid as long as the smallest length in a physical domain is much larger than the mean free path of molecules.

The fundamental equations of fluid motion in differential form are derived from:

1. Conservation of mass (continuity)

2. Conservation of linear momentum (Newton's second law)

3. Conservation of energy (first law of thermodynamics)

The resulting system of equations is known as the Navier-Stokes equations and was introduced in Chapter 7.

For applications for which the density remains uniform throughout the domain, the assumption of incompressible flow is invoked. Therefore, for an incompressible flow, the density is constant and no longer an unknown. Furthermore, variations in the coefficient of viscosity are essentially negligible and it is assumed constant as well. The incompressible Navier-Stokes equations can be obtained from the Navier-Stokes equations for the limiting case of $M \to 0$ ($a \to \infty$). Before the governing equations are reviewed, some essential remarks with regard to the incompressible Navier-Stokes equations are in order. First, the energy equation is decoupled from the system of equations composed of the continuity and momentum equations. Therefore, the velocity and pressure fields are computed initially, and subsequently the energy equation may be solved for the temperature distribution if required. Second, due to the reduction of unknowns within the domain, the computer storage and memory requirements are reduced. Third, depending on the formulations employed, a specific numerical scheme must be utilized which will permit coupling the velocity and the pressure. Fourth, using the Navier-Stokes equations[1] to solve for incompressible flowfields is not computationally efficient. In addition to the storage and memory requirements identified earlier, the computational time would be excessive due to the stability requirement of the numerical schemes. For example, as was seen in the previous chapter, the time step for most schemes is limited by Courant number. A typical restriction of most explicit schemes for the Navier-Stokes or Euler equations in the computational space may be expressed as

$$(CFL)_\eta = (\lambda_\eta)_{\max}\frac{\Delta\tau}{\Delta\eta} < 1$$

From which

$$\Delta\tau < \frac{\Delta\eta}{(\lambda_\eta)_{\max}}$$

where

$$(\lambda_\eta)_{\max} = \eta_x u + \eta_y v + a\sqrt{\eta_x^2 + \eta_y^2}$$

(The exact details will be shown later.) Note that as the speed of sound, "a", approaches infinity (as for incompressible flow), $\Delta\eta/(\lambda_\eta)_{\max}$ will approach zero. Thus, an extremely small time step is required. To reach a steady state solution, a tremendous amount of computations will be required. Therefore, efficient solution of incompressible flows requires utilization of appropriate incompressible Navier-Stokes equations. However, the application of the incompressible Navier-Stokes equations requires special considerations with regard to domain discretization, boundary conditions and solution procedures. The implications described above, along with the special considerations, are explored in the following sections.

8.2 Incompressible Navier-Stokes Equations

In this section, the incompressible Navier-Stokes equations are reviewed. As mentioned previously, since the energy equation is decoupled from the continuity

[1]Note that whenever one refers to Navier-Stokes equations, it normally implies unsteady compressible Navier-Stokes equations.

and momentum equations, its discussion will be postponed to the latter part of this chapter. Generally speaking, the governing equations for an incompressible flow may be expressed in two different formulations based on the dependent variables used. First is the primitive variable formulation expressed in terms of the pressure and velocity. The second form of the equations is the so-called vorticity-stream function equations which are derived from the Navier-Stokes equations by incorporating the definitions for the vorticity and the stream function. This formulation is primarily used for two-dimensional applications. Obviously the implication is due to the definition of stream function, which exists for a two-dimensional (planar or axisymmetric) flow only. It should be noted that for a three-dimensional flow, it is possible to extend the approach of stream function by the use of the so-called vector potential [8.1]. However, additional complications arise. Therefore, extension of the vorticity-stream function formulation for three dimensions is not explored further. Interested readers are encouraged to consult Refs. [8.2] through [8.4]. For three-dimensional applications, the extension of the equations in primitive variable formulation is recommended.

Either one of the formulations described above can be expressed in dimensional or nondimensional form. Furthermore, they may be either in conservative or non-conservative form. For completeness, various forms of the incompressible Navier-Stokes equations without body forces are provided.

8.2.1 Primitive Variable Formulations

(I) Dimensional, conservative form

1. Vector form

$$\nabla \cdot \vec{V} = 0 \tag{8-1}$$

$$\frac{\partial \vec{V}}{\partial t} + \nabla \cdot (\vec{V}\vec{V}) + \frac{\nabla p}{\rho} = \nu \nabla^2 \vec{V} \tag{8-2}$$

2. Two-dimensional Cartesian coordinate (Extensions to three-dimensions is straightforward)

$$\frac{\partial u}{\partial x} + \frac{\partial v}{\partial y} = 0 \tag{8-3}$$

$$\frac{\partial u}{\partial t} + \frac{\partial}{\partial x}\left(u^2 + \frac{p}{\rho}\right) + \frac{\partial}{\partial y}(uv) = \nu \left(\frac{\partial^2 u}{\partial x^2} + \frac{\partial^2 u}{\partial y^2}\right) \tag{8-4}$$

$$\frac{\partial v}{\partial t} + \frac{\partial}{\partial x}(uv) + \frac{\partial}{\partial y}\left(v^2 + \frac{p}{\rho}\right) = \nu \left(\frac{\partial^2 v}{\partial x^2} + \frac{\partial^2 v}{\partial y^2}\right) \tag{8-5}$$

(II) Dimensional, nonconservative form

1. Vector form

$$\nabla \cdot \vec{V} = 0 \tag{8-6}$$

$$\frac{\partial \vec{V}}{\partial t} + (\vec{V} \cdot \nabla)\vec{V} + \frac{\nabla p}{\rho} = \nu \nabla^2 \vec{V} \tag{8-7}$$

2. Two-dimensional Cartesian coordinate

$$\frac{\partial u}{\partial x} + \frac{\partial v}{\partial y} = 0 \tag{8-8}$$

$$\frac{\partial u}{\partial t} + u\frac{\partial u}{\partial x} + v\frac{\partial u}{\partial y} + \frac{1}{\rho}\frac{\partial p}{\partial x} = \nu\left(\frac{\partial^2 u}{\partial x^2} + \frac{\partial^2 u}{\partial y^2}\right) \tag{8-9}$$

$$\frac{\partial v}{\partial t} + u\frac{\partial v}{\partial x} + v\frac{\partial v}{\partial y} + \frac{1}{\rho}\frac{\partial p}{\partial y} = \nu\left(\frac{\partial^2 v}{\partial x^2} + \frac{\partial^2 v}{\partial y^2}\right) \tag{8-10}$$

(III) Nondimensional, conservative form

1. Vector form

$$\nabla^* \cdot \vec{V}^* = 0 \tag{8-11}$$

$$\frac{\partial \vec{V}^*}{\partial t^*} + \nabla^* \cdot (\vec{V}^*\vec{V}^*) + \nabla^* p^* = \frac{1}{Re}\nabla^{*2}\vec{V}^* \tag{8-12}$$

2. Two-dimensional Cartesian coordinate

$$\frac{\partial u^*}{\partial x^*} + \frac{\partial v^*}{\partial y^*} = 0 \tag{8-13}$$

$$\frac{\partial u^*}{\partial t^*} + \frac{\partial}{\partial x^*}(u^{*2} + p^*) + \frac{\partial}{\partial y^*}(u^* v^*) = \frac{1}{Re}\left(\frac{\partial^2 u^*}{\partial x^{*2}} + \frac{\partial^2 u^*}{\partial y^{*2}}\right) \tag{8-14}$$

$$\frac{\partial v^*}{\partial t^*} + \frac{\partial}{\partial x^*}(u^* v^*) + \frac{\partial}{\partial y^*}(v^{*2} + p^*) = \frac{1}{Re}\left(\frac{\partial^2 v^*}{\partial x^{*2}} + \frac{\partial^2 v^*}{\partial y^{*2}}\right) \tag{8-15}$$

(IV) Nondimensional, nonconservative form

1. Vector form

$$\nabla^* \cdot \vec{V}^* = 0 \tag{8-16}$$

$$\frac{\partial \vec{V}^*}{\partial t^*} + (\vec{V}^* \cdot \nabla^*)\vec{V}^* + \nabla^* p^* = \frac{1}{Re}\nabla^{*2}\vec{V}^* \tag{8-17}$$

2. Two-dimensional Cartesian coordinate

$$\frac{\partial u^*}{\partial x^*} + \frac{\partial v^*}{\partial y^*} = 0 \tag{8-18}$$

$$\frac{\partial u^*}{\partial t^*} + u^*\frac{\partial u^*}{\partial x^*} + v^*\frac{\partial u^*}{\partial y^*} + \frac{\partial p^*}{\partial x^*} = \frac{1}{Re}\left(\frac{\partial^2 u^*}{\partial x^{*2}} + \frac{\partial^2 u^*}{\partial y^{*2}}\right) \tag{8-19}$$

$$\frac{\partial v^*}{\partial t^*} + u^*\frac{\partial v^*}{\partial x^*} + v^*\frac{\partial v^*}{\partial y^*} + \frac{\partial p^*}{\partial y^*} = \frac{1}{Re}\left(\frac{\partial^2 v^*}{\partial x^{*2}} + \frac{\partial^2 v^*}{\partial y^{*2}}\right) \tag{8-20}$$

The variables in the equations above are nondimensionalized as follows,

$$t^* = \frac{tu_\infty}{L} \qquad x^* = \frac{x}{L} \qquad y^* = \frac{y}{L}$$
$$u^* = \frac{u}{u_\infty} \qquad v^* = \frac{v}{u_\infty} \qquad p^* = \frac{p}{\rho_\infty u_\infty^2} \tag{8-21}$$

where L is a characteristic length, and ρ_∞ and u_∞ are the reference (e.g., freestream) density and velocity, respectively. The nondimensional parameter Reynolds number is defined as

$$Re = \frac{\rho_\infty u_\infty L}{\mu_\infty}$$

It should be emphasized that other reference variables can be used to nondimensionalize the equations. For example, one may nondimensionalize the pressure with respect to the free stream pressure, p_∞, or nondimensionalize the velocity with respect to the free stream speed of sound (for high speed flows). Therefore, in reviewing various publications, one should pay close attention to the procedure by which the equations have been nondimensionalized. At this point, a comment with respect to nondimensionalization of time is in order. For applications where the Reynolds number is high (i.e., $Re \gg 1$), time is nondimensionalized with respect to tu_∞/L as defined previously. That is due to the fact that in these types of problems, the convective term dominates the viscous term and L/u_∞ is a natural representation of the time interval for the problem, i.e., the time by which a particle with the free stream velocity of u_∞ is convected a characteristic length of L. On the other hand, for a low Reynolds number flow (i.e., $Re \ll 1$), diffusion dominates over convection and L^2/ν is a more appropriate factor to nondimensionalize time. Thus, $t^* = t\nu/L^2$.

As an example, the nondimensional momentum equation in conservative form using the nondimensionalized time defined for diffusion-dominating problems becomes

$$\frac{\partial \vec{V}^*}{\partial t^*} + Re\left[\nabla^* \cdot (\vec{V}^*\vec{V}^*) + \nabla^* p^*\right] = \nabla^{*2}\vec{V}^* \tag{8-22}$$

8.2.2 Vorticity-Stream Function Formulations

The vorticity at a fluid point is defined as twice the angular velocity and is

$$\vec{\Omega} = 2\vec{\omega} = \nabla \times \vec{V}$$

which, for a two-dimensional flow, is reduced to

$$\Omega_z = \frac{\partial v}{\partial x} - \frac{\partial u}{\partial y} \tag{8-23}$$

Now, for a two-dimensional, incompressible flow, a function may be defined which satisfies the continuity equation. Such a function is known as the *stream function* and, in Cartesian coordinate system, is given by

$$u = \frac{\partial \psi}{\partial y} \tag{8-24}$$

$$v = -\frac{\partial \psi}{\partial x} \tag{8-25}$$

From a physical point of view, the lines of constant ψ represent stream lines, and the difference in the values of ψ between two streamlines gives the volumetric flow rate between the two.

In order to derive the vorticity transport equation, the pressure is eliminated from the momentum equations by cross-differentiation. Differentiation with respect to y of Equation (8-9) yields

$$\frac{\partial^2 u}{\partial y \partial t} + \frac{\partial u}{\partial y}\frac{\partial u}{\partial x} + u\frac{\partial^2 u}{\partial x \partial y} + \frac{\partial v}{\partial y}\frac{\partial u}{\partial y} + v\frac{\partial^2 u}{\partial y^2} = -\frac{1}{\rho}\frac{\partial^2 p}{\partial x \partial y} + \nu\left(\frac{\partial^3 u}{\partial y \partial x^2} + \frac{\partial^3 u}{\partial y^3}\right) \tag{8-26}$$

whereas the differentiation with respect to x of Equation (8-10) yields

$$\frac{\partial^2 v}{\partial x \partial t} + \frac{\partial u}{\partial x}\frac{\partial v}{\partial x} + u\frac{\partial^2 v}{\partial x^2} + \frac{\partial v}{\partial x}\frac{\partial v}{\partial y} + v\frac{\partial^2 v}{\partial x \partial y} = -\frac{1}{\rho}\frac{\partial^2 p}{\partial x \partial y} + \nu\left(\frac{\partial^3 v}{\partial x^3} + \frac{\partial^3 v}{\partial x \partial y^2}\right) \tag{8-27}$$

Subtract Equation (8-27) from Equation (8-26) to obtain

$$\frac{\partial}{\partial t}\left(\frac{\partial u}{\partial y} - \frac{\partial v}{\partial x}\right) + u\frac{\partial}{\partial x}\left(\frac{\partial u}{\partial y} - \frac{\partial v}{\partial x}\right) + v\frac{\partial}{\partial y}\left(\frac{\partial u}{\partial y} - \frac{\partial v}{\partial x}\right) + \left(\frac{\partial u}{\partial x} + \frac{\partial v}{\partial y}\right)\left(\frac{\partial u}{\partial y} - \frac{\partial v}{\partial x}\right)$$

$$= \nu\left[\frac{\partial^2}{\partial x^2}\left(\frac{\partial u}{\partial y} - \frac{\partial v}{\partial x}\right) + \frac{\partial^2}{\partial y^2}\left(\frac{\partial u}{\partial y} - \frac{\partial v}{\partial x}\right)\right]$$

Note that the fourth term on the left-hand side is zero by continuity. Now, upon substitution of the vorticity defined by (8-23), one obtains

$$\frac{\partial \Omega}{\partial t} + u\frac{\partial \Omega}{\partial x} + v\frac{\partial \Omega}{\partial y} = \nu\left(\frac{\partial^2 \Omega}{\partial x^2} + \frac{\partial^2 \Omega}{\partial y^2}\right) \tag{8-28}$$

where the subscript z is dropped from Ω_z. Thus, in the remainder of this section, Ω will designate the z-component of the vorticity, unless otherwise specified. Equation (8-28) is known as the *vorticity transport equation* and is classified as a parabolic equation with the unknown being the vorticity Ω.

Now, reconsider the definition of vorticity given by

$$\Omega = \frac{\partial v}{\partial x} - \frac{\partial u}{\partial y} \tag{8-29}$$

Substitution of relations (8-24) and (8-25) yields

$$\frac{\partial^2 \psi}{\partial x^2} + \frac{\partial^2 \psi}{\partial y^2} = -\Omega \tag{8-30}$$

This equation is known as the *stream function equation* and is classified as an elliptic PDE. The unknown is the stream function ψ, whose Ω is provided from the solution of Equation (8-28). Once the stream function has been computed, the velocity components may be determined from relations (8-24) and (8-25).

The vorticity equation may be expressed in a nondimensional form by using the nondimensional quantities defined previously and a nondimensional vorticity defined as

$$\Omega^* = \frac{\Omega L}{u_\infty}$$

The nondimensional form of the vorticity equation can be expressed as

$$\frac{\partial \Omega^*}{\partial t^*} + u^* \frac{\partial \Omega^*}{\partial x^*} + v^* \frac{\partial \Omega^*}{\partial y^*} = \frac{1}{Re_\infty} \left(\frac{\partial^2 \Omega^*}{\partial x^{*2}} + \frac{\partial^2 \Omega^*}{\partial y^{*2}} \right) \tag{8-31}$$

Similarly, the nondimensional form of the stream function equation given by Equation (8-30) is

$$\frac{\partial^2 \psi^*}{\partial x^{*2}} + \frac{\partial^2 \psi^*}{\partial y^{*2}} = -\Omega^* \tag{8-32}$$

where

$$\psi^* = \frac{\psi}{u_\infty L}$$

A summary of the vorticity-stream function formulations is provided below.

(I) Dimensional, conservative form in Cartesian coordinate

$$\frac{\partial \Omega}{\partial t} + \frac{\partial}{\partial x}(u\Omega) + \frac{\partial}{\partial y}(v\Omega) = \nu \left(\frac{\partial^2 \Omega}{\partial x^2} + \frac{\partial^2 \Omega}{\partial y^2} \right) \tag{8-33}$$

$$\frac{\partial^2 \psi}{\partial x^2} + \frac{\partial^2 \psi}{\partial y^2} = -\Omega \tag{8-34}$$

(II) Dimensional, nonconservative form in Cartesian coordinate

$$\frac{\partial \Omega}{\partial t} + u\frac{\partial \Omega}{\partial x} + v\frac{\partial \Omega}{\partial y} = \nu\left(\frac{\partial^2 \Omega}{\partial x^2} + \frac{\partial^2 \Omega}{\partial y^2}\right) \tag{8-35}$$

$$\frac{\partial^2 \psi}{\partial x^2} + \frac{\partial^2 \psi}{\partial y^2} = -\Omega \tag{8-36}$$

(III) Nondimensional, conservative form in Cartesian coordinate

$$\frac{\partial \Omega^*}{\partial t^*} + \frac{\partial}{\partial x^*}(u^*\Omega^*) + \frac{\partial}{\partial y^*}(v^*\Omega^*) = \frac{1}{Re}\left(\frac{\partial^2 \Omega^*}{\partial x^{*2}} + \frac{\partial^2 \Omega^*}{\partial y^{*2}}\right) \tag{8-37}$$

$$\frac{\partial^2 \psi^*}{\partial x^{*2}} + \frac{\partial^2 \psi^*}{\partial y^{*2}} = -\Omega^* \tag{8-38}$$

(IV) Nondimensional, nonconservative form in Cartesian coordinate

$$\frac{\partial \Omega^*}{\partial t^*} + u^*\frac{\partial \Omega^*}{\partial x^*} + v^*\frac{\partial \Omega^*}{\partial y^*} = \frac{1}{Re}\left(\frac{\partial^2 \Omega^*}{\partial x^{*2}} + \frac{\partial^2 \Omega^*}{\partial y^{*2}}\right) \tag{8-39}$$

$$\frac{\partial^2 \psi^*}{\partial x^{*2}} + \frac{\partial^2 \psi^*}{\partial y^{*2}} = -\Omega^* \tag{8-40}$$

8.2.3 Comments on Formulations

Before proceeding to the numerical algorithms, a few comments with regard to the incompressible Navier-Stokes equations expressed in either the primitive variable or the vorticity-stream function formulation are in order.

1. Primitive variable formulation

(a) The governing equations are a mixed elliptic-parabolic system of equations which are solved simultaneously. The unknowns in the equations are velocity and pressure.

(b) There is no direct link for the pressure between the continuity and momentum equations. To establish a connection between the two equations, mathematical manipulations are introduced. Generally speaking there are two procedures for this purpose. The first is that of the Poisson equation for pressure which is developed in the next section; and the second is the introduction of artificial compressibility into the continuity equation. This procedure will be addressed shortly in Section 8.5.1.1. Note that this difficulty does not exist for the compressible Navier-Stokes equations. That is because there is a linkage between the continuity and momentum equations through the density which appears in both equations.

(c) Specification of boundary conditions and in particular for pressure may be nonexistent. To overcome this difficulty, a special procedure must be introduced.

(d) Extension to three dimensions is straightforward with the least amount of complications.

2. Vorticity-stream function formulation

(a) By introduction of new variables, namely the vorticity and the stream function, the incompressible Navier-Stokes equations are decoupled into one elliptic equation and one parabolic equation which can be solved sequentially.

(b) Vorticity-stream function formulation does not include the pressure term. Therefore, the velocity field is determined initially and, subsequently, the Poisson equation for pressure (which is described in the next section) is employed to solve for the pressure field.

(c) Due to lack of a simple stream function in three dimensions, extension of the vorticity-stream function formulation to three dimensions loses its attractiveness.

8.3 Poisson Equation for Pressure: Primitive Variables

In this section an equation is developed which may be used for the computation of the pressure field. The reason for incorporating the Poisson equation for pressure, which is usually used in lieu of the continuity equation, is the lack of a direct link for pressure between continuity and momentum equations. A typical numerical scheme for the solution of the Poisson equation for pressure is investigated in the subsequent sections. For the time being, the steps required to obtain such a formulation are illustrated. The conservative form of the x- and y-components of the momentum equation obtained previously are

$$\frac{\partial u}{\partial t} + \frac{\partial}{\partial x}(u^2) + \frac{\partial p}{\partial x} + \frac{\partial}{\partial y}(uv) = \frac{1}{Re}\nabla^2 u \tag{8-41}$$

$$\frac{\partial v}{\partial t} + \frac{\partial}{\partial x}(uv) + \frac{\partial}{\partial y}(v^2) + \frac{\partial p}{\partial y} = \frac{1}{Re}\nabla^2 v \tag{8-42}$$

Equations (8-41) and (8-42) are differentiated with respect to x and y, respectively, to provide

$$\frac{\partial}{\partial t}\left(\frac{\partial u}{\partial x}\right) + \frac{\partial^2}{\partial x^2}(u^2) + \frac{\partial^2 p}{\partial x^2} + \frac{\partial^2}{\partial x \partial y}(uv) = \frac{1}{Re}\frac{\partial}{\partial x}(\nabla^2 u) \tag{8-43}$$

and

$$\frac{\partial}{\partial t}\left(\frac{\partial v}{\partial y}\right)+\frac{\partial^2}{\partial x \partial y}(uv)+\frac{\partial^2}{\partial y^2}(v^2)+\frac{\partial^2 p}{\partial y^2}=\frac{1}{Re}\frac{\partial}{\partial y}(\nabla^2 v) \tag{8-44}$$

Addition of equations (8-43) and (8-44) yields

$$\frac{\partial}{\partial t}\left(\frac{\partial u}{\partial x}+\frac{\partial v}{\partial y}\right)+\frac{\partial^2}{\partial x^2}(u^2)+2\frac{\partial^2}{\partial x \partial y}(uv)+\frac{\partial^2}{\partial y^2}(v^2)+\frac{\partial^2 p}{\partial x^2}+\frac{\partial^2 p}{\partial y^2}$$

$$=\frac{1}{Re}\left[\frac{\partial}{\partial x}(\nabla^2 u)+\frac{\partial}{\partial y}(\nabla^2 v)\right] \tag{8-45}$$

The right-hand side is rearranged as

$$\frac{\partial}{\partial x}\left(\frac{\partial^2 u}{\partial x^2}+\frac{\partial^2 u}{\partial y^2}\right)+\frac{\partial}{\partial y}\left(\frac{\partial^2 v}{\partial x^2}+\frac{\partial^2 v}{\partial y^2}\right)=\frac{\partial^2}{\partial x^2}\left(\frac{\partial u}{\partial x}+\frac{\partial v}{\partial y}\right)+\frac{\partial^2}{\partial y^2}\left(\frac{\partial u}{\partial x}+\frac{\partial v}{\partial y}\right)$$

Finally, Equation (8-45) can be rewritten in the form of Poisson equation

$$\frac{\partial^2 p}{\partial x^2}+\frac{\partial^2 p}{\partial y^2}=-\frac{\partial D}{\partial t}-\frac{\partial^2}{\partial x^2}(u^2)-2\frac{\partial^2}{\partial x \partial y}(uv)-\frac{\partial^2}{\partial y^2}(v^2)+\frac{1}{Re}\left[\frac{\partial^2}{\partial x^2}(D)+\frac{\partial^2}{\partial y^2}(D)\right] \tag{8-46}$$

where

$$D=\frac{\partial u}{\partial x}+\frac{\partial v}{\partial y}$$

is known as dilatation.

It is obvious that for an incompressible flow, the dilatation term is zero by continuity. However, due to numerical considerations, this term will not be set to zero in Equation (8-46). Indeed it must be evaluated within Equation (8-46) to prevent error accumulation in the process of iterative solution of the equation, as well as to prevent nonlinear instability.

8.4 Poisson Equation for Pressure: Vorticity-Stream Function Formulation

Consider the conservative form of the momentum equation given by Equations (8-14) and (8-15) as repeated here with the asterisk dropped,

$$\frac{\partial u}{\partial t}+\frac{\partial}{\partial x}(u^2)+\frac{\partial p}{\partial x}+\frac{\partial}{\partial y}(uv)=\frac{1}{Re}\left(\frac{\partial^2 u}{\partial x^2}+\frac{\partial^2 u}{\partial y^2}\right) \tag{8-47}$$

$$\frac{\partial v}{\partial t}+\frac{\partial}{\partial x}(uv)+\frac{\partial}{\partial y}(v^2)+\frac{\partial p}{\partial y}=\frac{1}{Re}\left(\frac{\partial^2 v}{\partial x^2}+\frac{\partial^2 v}{\partial y^2}\right) \tag{8-48}$$

Equation (8-47) is now differentiated with respect to x to provide

$$\frac{\partial}{\partial x}\left(\frac{\partial u}{\partial t}\right)+\frac{\partial}{\partial x}\left(2u\frac{\partial u}{\partial x}\right)+\frac{\partial}{\partial x}\left(\frac{\partial p}{\partial x}\right)+\frac{\partial}{\partial x}\left(u\frac{\partial v}{\partial y}+v\frac{\partial u}{\partial y}\right)=\frac{1}{Re}\frac{\partial}{\partial x}(\nabla^2 u)$$

or

$$\frac{\partial}{\partial t}\left(\frac{\partial u}{\partial x}\right)+2\frac{\partial u}{\partial x}\frac{\partial u}{\partial x}+2u\frac{\partial^2 u}{\partial x^2}+\frac{\partial^2 p}{\partial x^2}+\frac{\partial u}{\partial x}\frac{\partial v}{\partial y}+u\frac{\partial^2 v}{\partial x\partial y}+\frac{\partial v}{\partial x}\frac{\partial u}{\partial y}+v\frac{\partial^2 u}{\partial x\partial y}=\frac{1}{Re}\frac{\partial}{\partial x}(\nabla^2 u)$$

The second term (only one) and the fifth term are combined to provide

$$\frac{\partial u}{\partial x}\left(\frac{\partial u}{\partial x}+\frac{\partial v}{\partial y}\right)=0$$

due to continuity. Similarly, term three (only one) and term six are added to yield

$$u\frac{\partial^2 v}{\partial x\partial y}+u\frac{\partial^2 u}{\partial x^2}=u\frac{\partial}{\partial x}\left(\frac{\partial u}{\partial x}+\frac{\partial v}{\partial y}\right)=0$$

Thus, one has

$$\frac{\partial}{\partial t}\left(\frac{\partial u}{\partial x}\right)+\left(\frac{\partial u}{\partial x}\right)^2+u\frac{\partial^2 u}{\partial x^2}+\frac{\partial v}{\partial x}\frac{\partial u}{\partial y}+v\frac{\partial^2 u}{\partial x\partial y}+\frac{\partial^2 p}{\partial x^2}=\frac{1}{Re}\frac{\partial}{\partial x}(\nabla^2 u) \tag{8-49}$$

Similarly, the y-component of momentum becomes

$$\frac{\partial}{\partial t}\left(\frac{\partial v}{\partial y}\right)+\left(\frac{\partial v}{\partial y}\right)^2+v\frac{\partial^2 v}{\partial y^2}+\frac{\partial u}{\partial y}\frac{\partial v}{\partial x}+u\frac{\partial^2 v}{\partial x\partial y}+\frac{\partial^2 p}{\partial y^2}=\frac{1}{Re}\frac{\partial}{\partial y}(\nabla^2 v) \tag{8-50}$$

Addition of Equations (8-49) and (8-50) yields

$$\frac{\partial}{\partial t}\left(\frac{\partial u}{\partial x}+\frac{\partial v}{\partial y}\right)+\left(\frac{\partial u}{\partial x}\right)^2+\left(\frac{\partial v}{\partial y}\right)^2+2\frac{\partial u}{\partial y}\frac{\partial v}{\partial x}+u\left(\frac{\partial^2 u}{\partial x^2}+\frac{\partial^2 v}{\partial x\partial y}\right)$$

$$+v\left(\frac{\partial^2 u}{\partial x\partial y}+\frac{\partial^2 v}{\partial y^2}\right)+u\left(\frac{\partial^2 p}{\partial x^2}+\frac{\partial^2 p}{\partial y^2}\right)=\frac{1}{Re}\left[\frac{\partial}{\partial x}(\nabla^2 u)+\frac{\partial}{\partial y}(\nabla^2 v)\right] \tag{8-51}$$

Note that the first, fifth, and sixth terms each contain the continuity equation and therefore disappear

$$\frac{\partial}{\partial t}\left(\frac{\partial u}{\partial x}+\frac{\partial v}{\partial y}\right)=0$$

$$\frac{\partial^2 u}{\partial x^2}+\frac{\partial^2 v}{\partial x\partial y}=\frac{\partial}{\partial x}\left(\frac{\partial u}{\partial x}+\frac{\partial v}{\partial y}\right)=0$$

and

$$\frac{\partial^2 u}{\partial x\partial y}+\frac{\partial^2 v}{\partial y^2}=\frac{\partial}{\partial y}\left(\frac{\partial u}{\partial x}+\frac{\partial v}{\partial y}\right)=0$$

The right-hand side is now rearranged to provide

$$\frac{\partial}{\partial x}\left(\frac{\partial^2 u}{\partial x^2}+\frac{\partial^2 u}{\partial y^2}\right)+\frac{\partial}{\partial y}\left(\frac{\partial^2 v}{\partial x^2}+\frac{\partial^2 v}{\partial y^2}\right)=\frac{\partial^3 u}{\partial x^3}+\frac{\partial^3 u}{\partial x\partial y^2}+\frac{\partial^3 v}{\partial y\partial x^2}+\frac{\partial^3 v}{\partial x^3}$$

$$=\frac{\partial^2}{\partial x^2}\left(\frac{\partial u}{\partial x}+\frac{\partial v}{\partial y}\right)+\frac{\partial^2}{\partial y^2}\left(\frac{\partial u}{\partial x}+\frac{\partial v}{\partial y}\right)=0$$

Therefore Equation (8-51) is reduced to

$$\left(\frac{\partial u}{\partial x}\right)^2+\left(\frac{\partial v}{\partial y}\right)^2+2\frac{\partial u}{\partial y}\frac{\partial v}{\partial x}=-\left(\frac{\partial^2 p}{\partial x^2}+\frac{\partial^2 p}{\partial y^2}\right) \tag{8-52}$$

Now, the left-hand side can be further reduced by considering the continuity equation as follows

$$\left(\frac{\partial u}{\partial x}+\frac{\partial v}{\partial y}\right)^2=\left(\frac{\partial u}{\partial x}\right)^2+\left(\frac{\partial v}{\partial y}\right)^2+2\left(\frac{\partial u}{\partial x}\right)\left(\frac{\partial v}{\partial y}\right)=0$$

from which

$$\left(\frac{\partial u}{\partial x}\right)^2+\left(\frac{\partial v}{\partial y}\right)^2=-2\left(\frac{\partial u}{\partial x}\right)\left(\frac{\partial v}{\partial y}\right)$$

Substituting into Equation (8-52) yields

$$-\left(\frac{\partial^2 p}{\partial x^2}+\frac{\partial^2 p}{\partial y^2}\right)=2\left(\frac{\partial u}{\partial y}\frac{\partial v}{\partial x}-\frac{\partial u}{\partial x}\frac{\partial v}{\partial y}\right) \tag{8-53}$$

This equation can be written in terms of the stream function by using relations (8-24) and (8-25)

$$-\left(\frac{\partial^2 p}{\partial x^2}+\frac{\partial^2 p}{\partial y^2}\right)=2\left[\left(\frac{\partial^2\psi}{\partial y^2}\right)\left(-\frac{\partial^2\psi}{\partial x^2}\right)-\left(\frac{\partial^2\psi}{\partial x\partial y}\right)\left(-\frac{\partial^2\psi}{\partial x\partial y}\right)\right]$$

or

$$\frac{\partial^2 p}{\partial x^2}+\frac{\partial^2 p}{\partial y^2}=2\left[\left(\frac{\partial^2\psi}{\partial x^2}\right)\left(\frac{\partial^2\psi}{\partial y^2}\right)-\left(\frac{\partial^2\psi}{\partial x\partial y}\right)^2\right] \tag{8-54}$$

Observe that Equation (8-54) is in nondimensional form. However, it may be expressed in a dimensional form as

$$\frac{\partial^2 p}{\partial x^2}+\frac{\partial^2 p}{\partial y^2}=2\rho\left[\left(\frac{\partial^2\psi}{\partial x^2}\right)\left(\frac{\partial^2\psi}{\partial y^2}\right)-\left(\frac{\partial^2\psi}{\partial x\partial y}\right)^2\right] \tag{8-55}$$

8.5 Numerical Algorithms: Primitive Variables

The incompressible Navier-Stokes equations in primitive variables expressed in various forms were reviewed in Section 8.2.1. From a physical point of view, a problem may be classified as steady state or unsteady. Obviously, for a steady state problem, time dependent terms are omitted from the governing equations. However, due to numerical considerations, the solution of steady incompressible Navier-Stokes equations, in general, incorporates a pseudo-transient scheme. In this technique a nonphysical time dependent term is added to the continuity equation and, along with the momentum equation, an equivalent unsteady system of equations is constructed. Subsequently, unsteady forms of the equations are solved numerically. Note that the basic philosophy of the pseudo-transient method was explored previously in Chapter 5 where the relation between the iterative solution of an elliptic equation and the time marching solution of a parabolic equation was identified. It was observed that the solution during the iterative process leading up to the converged solution is physically meaningless and is only a vehicle to reach the desired steady-state solution. A similar philosophy is used in the pseudotransient method. Namely, a time dependent term is added, thus allowing time marching of the solution until the steady state solution is achieved. It is therefore obvious that time plays the role of iteration in this type procedure with no physical significance. Consequently it can be argued that the numerical procedure for both categories of steady and unsteady flows is similar. However, a few important points, some of which were mentioned previously, are outlined at this time. First, for unsteady problems a physically correct and reasonably accurate initial condition must be provided; whereas for steady problems, the solution may start with an assumed (i.e., arbitrary) set of initial data. For example one may impose the free stream conditions over the entire domain to initialize the numerical procedure. Second, the numerical time step for an unsteady problem must be consistent with the physical time and with regard to changes which could occur during that time period. On the other hand, as mentioned previously, the intermediate solutions of steady problems are physically meaningless and, therefore, for steady-state solutions the maximum allowable time step imposed by the stability requirement can be used.

In light of the above discussion and based on the similarity of the numerical approaches for steady and unsteady flows, selected numerical schemes commonly used for steady problems will be explored in detail. Extension to unsteady problems is rather straightforward. The concern, however, lies in the specification of initial conditions, time step, and, in some applications, the specification of boundary conditions.

Before proceeding with specific numerical schemes, it is essential to emphasize a difficulty which exists with respect to the solution of the incompressible Navier-Stokes equations. To make the point clear, revisit the governing equations given by

$$\frac{\partial u}{\partial x} + \frac{\partial v}{\partial y} = 0 \tag{8-56}$$

$$\frac{\partial u}{\partial t} + \frac{\partial}{\partial x}(u^2 + p) + \frac{\partial}{\partial y}(uv) = \frac{1}{Re}\left(\frac{\partial^2 u}{\partial x^2} + \frac{\partial^2 u}{\partial y^2}\right) \tag{8-57}$$

$$\frac{\partial v}{\partial t} + \frac{\partial}{\partial x}(uv) + \frac{\partial}{\partial y}(v^2 + p) = \frac{1}{Re}\left(\frac{\partial^2 v}{\partial x^2} + \frac{\partial^2 v}{\partial y^2}\right) \tag{8-58}$$

Obviously, the system of equations given by (8-56) through (8-58) includes three unknowns: u, v, and p. If one considers a simple explicit formulation, it would seem logical to use Equation (8-57) to solve for u (where p and v are lagged), whereas one would solve Equation (8-58) for v. Now, one is left with Equation (8-56) to solve for the pressure, but unfortunately pressure does not appear in that equation! This difficulty is overcome, however, by manipulation of the continuity equation to include the pressure term. Two procedures have been introduced for this purpose. One procedure involves the manipulation of the momentum equation along with the continuity equation. The mathematical details were described previously which resulted in the Poisson equation for pressure. A second procedure incorporates the addition of a time-dependent pressure term to the continuity equation, i.e., to Equation (8-56). This approach is generally known as the artificial compressibility method which is investigated in the following section.

8.5.1 Steady Flows

In this section various numerical schemes commonly used to solve the steady incompressible Navier-Stokes equations within the pseudotransient category are explored. The schemes include the application to the artificial compressibility formulations, as well as to the Poisson equation for pressure in the subsequent sections.

8.5.1.1 Artificial Compressibility

The application of the scheme to the steady incompressible Navier-Stokes equations was introduced by Chorin (Ref. [8-5]). The continuity equation is modified by inclusion of a time-dependent term and is given by

$$\frac{\partial p}{\partial t} + \frac{1}{\tau}\left(\frac{\partial u}{\partial x} + \frac{\partial v}{\partial y}\right) = 0 \tag{8-59}$$

where τ can be interpreted as the "artificial compressibility" of the fluid. Following the equation of state the compressibility can be related to a pseudo-speed of sound and to an artificial density by the following relations

$$\tau = \frac{1}{a^2}$$

$$a^2 = \frac{p}{\rho}$$

where all the quantities are in nondimensional form. Note that in the limit, as one approaches steady state, i.e., $t \to \infty$, Equation (8-59) is indeed reduced to the incompressible continuity equation.

Thus, the steady incompressible Navier-Stokes equations (two-dimensional Cartesian coordinates) are expressed in a pseudotransient form as

$$\frac{\partial p}{\partial t} + a^2\left(\frac{\partial u}{\partial x} + \frac{\partial v}{\partial y}\right) = 0 \tag{8-60}$$

$$\frac{\partial u}{\partial t} + \frac{\partial}{\partial x}(u^2 + p) + \frac{\partial}{\partial y}(uv) = \frac{1}{Re}\left(\frac{\partial^2 u}{\partial x^2} + \frac{\partial^2 u}{\partial y^2}\right) \tag{8-61}$$

$$\frac{\partial v}{\partial t} + \frac{\partial}{\partial x}(uv) + \frac{\partial}{\partial y}(v^2 + p) = \frac{1}{Re}\left(\frac{\partial^2 v}{\partial x^2} + \frac{\partial^2 v}{\partial y^2}\right) \tag{8-62}$$

Note that Equations (8-60) through (8-62) are in nondimensional form, where the asterisk denoting nondimensional quantities has been dropped. The nondimensional quantities were defined previously and in addition

$$a^{*2} = \frac{p^*}{\rho^*}$$

where

$$\rho^* = \frac{\rho}{\rho_\infty}$$

The numerical solutions of the system of equations composed of Equations (8-60) through (8-62) can be placed in two categories depending on the grid system employed. In the first category, the equations are solved on a regular grid whereas the second category involves the solution of the equations in a staggered grid.

8.5.1.2 Solution on Regular Grid

To facilitate the application of the finite difference formulations, the conservative form of the governing equations from Equations (8-60) through (8-62) are written in a flux vector form as

$$\frac{\partial Q}{\partial t} + \frac{\partial E}{\partial x} + \frac{\partial F}{\partial y} = \frac{1}{Re}[N]\nabla^2 Q \tag{8-63}$$

where

$$Q = \begin{bmatrix} p \\ u \\ v \end{bmatrix} \quad , \quad E = \begin{bmatrix} a^2 u \\ u^2 + p \\ uv \end{bmatrix} \quad , \quad F = \begin{bmatrix} a^2 v \\ uv \\ v^2 + p \end{bmatrix}$$

and

$$N = \begin{bmatrix} 0 & 0 & 0 \\ 0 & 1 & 0 \\ 0 & 0 & 1 \end{bmatrix}$$

Obviously, Equation (8-63) is a nonlinear system of equations. As was seen previously, explicit formulation of nonlinear equations can be formulated with no difficulty. However, when implicit formulations are utilized, a linearization procedure must be introduced. For this purpose, consider the following Taylor series expansion

$$E^{n+1} = E^n + \frac{\partial E}{\partial t}\Delta t + O(\Delta t)^2$$

which may be rewritten as

$$E^{n+1} = E^n + \frac{\partial E}{\partial Q}\frac{\partial Q}{\partial t}\Delta t + O(\Delta t)^2$$

or

$$E^{n+1} = E^n + \frac{\partial E}{\partial Q}\frac{\Delta Q}{\Delta t}\Delta t + O(\Delta t)^2$$

from which

$$E^{n+1} = E^n + \frac{\partial E}{\partial Q}\Delta Q + O(\Delta t)^2 \tag{8-64}$$

and

$$\Delta Q = Q^{n+1} - Q^n$$

Terms such as $\dfrac{\partial E}{\partial Q}$ are known as flux Jacobian matrices. The Jacobian matrices $\dfrac{\partial E}{\partial Q}$ and $\dfrac{\partial F}{\partial Q}$ will be denoted by A and B, respectively. The Jacobian matrix A is given by

$$A = \frac{\partial[E_1, E_2, E_3]}{\partial[Q_1, Q_2, Q_3]} = \begin{bmatrix} \dfrac{\partial E_1}{\partial Q_1} & \dfrac{\partial E_1}{\partial Q_2} & \dfrac{\partial E_1}{\partial Q_3} \\ \dfrac{\partial E_2}{\partial Q_1} & \dfrac{\partial E_2}{\partial Q_2} & \dfrac{\partial E_2}{\partial Q_3} \\ \dfrac{\partial E_3}{\partial Q_1} & \dfrac{\partial E_3}{\partial Q_2} & \dfrac{\partial E_3}{\partial Q_3} \end{bmatrix}$$

The elements of the Jacobian matrix are determined as follows. First, the components of vector E are written in terms of the components of Q and are subsequently differentiated. Recall that the vectors Q and E are

$$Q = \begin{bmatrix} p \\ u \\ v \end{bmatrix} = \begin{bmatrix} Q_1 \\ Q_2 \\ Q_3 \end{bmatrix} \quad \text{and} \quad E = \begin{bmatrix} a^2 u \\ u^2 + p \\ uv \end{bmatrix} = \begin{bmatrix} a^2 Q_2 \\ Q_2^2 + Q_1 \\ Q_2 Q_3 \end{bmatrix} = \begin{bmatrix} E_1 \\ E_2 \\ E_3 \end{bmatrix}$$

Thus,

$$\frac{\partial E_1}{\partial Q_1} = \frac{\partial (a^2 Q_2)}{\partial Q_1} = 0$$

$$\frac{\partial E_1}{\partial Q_2} = \frac{\partial (a^2 Q_2)}{\partial Q_2} = a^2$$

$$\frac{\partial E_1}{\partial Q_3} = \frac{\partial (a^2 Q_2)}{\partial Q_3} = 0$$

The remaining elements are determined in a similar fashion resulting in

$$A = \begin{bmatrix} 0 & a^2 & 0 \\ 1 & 2u & 0 \\ 0 & v & u \end{bmatrix}$$

Similarly, the Jacobian matrix B is obtained as

$$B = \begin{bmatrix} 0 & 0 & a^2 \\ 0 & v & u \\ 1 & 0 & 2v \end{bmatrix}$$

Now Equation (8-63) can be written as

$$\frac{\Delta Q}{\Delta t} + \frac{\partial}{\partial x}(E^n + A\Delta Q) + \frac{\partial}{\partial y}(F^n + B\Delta Q) = \frac{1}{Re}[N]\nabla^2 Q \tag{8-65}$$

Consequently, one may rearrange (8-65) to provide

$$\Delta Q + \Delta t \left[\frac{\partial}{\partial x}(A\Delta Q) + \frac{\partial}{\partial y}(B\Delta Q) - \frac{N}{Re}\left(\frac{\partial^2}{\partial x^2} + \frac{\partial^2}{\partial y^2} \right) \Delta Q \right]$$
$$= \Delta t \left[-\frac{\partial E^n}{\partial x} - \frac{\partial F^n}{\partial y} + \frac{N}{Re}\left(\frac{\partial^2 Q}{\partial x^2} + \frac{\partial^2 Q}{\partial y^2} \right) \right]$$

or

$$\left\{ I + \Delta t \left[\frac{\partial A}{\partial x} + \frac{\partial B}{\partial y} - \frac{N}{Re}\left(\frac{\partial^2}{\partial x^2} + \frac{\partial^2}{\partial y^2} \right) \right] \right\} \Delta Q = \text{RHS} \tag{8-66}$$

Among the various options available for the finite difference formulation of Equation (8-66), an implicit formulation is investigated initially and, subsequently, other schemes are explored.

Prior to writing an implicit finite difference formulation, two points must be noted. First, for the two-dimensional problems under consideration, the implicit formulation will result in a block pentadiagonal system of equations. As seen previously, the solution of such a system is expensive. To overcome this problem, approximate factorization, which was introduced previously, will be implemented. Second, to overcome any possible instability in the solution, artificial viscosity (damping terms) will be added to the equations. Thus, with the stated considerations, Equation (8-66) is formulated as

$$\left[I + \Delta t\left(\frac{\partial A}{\partial x} - \frac{N}{Re}\frac{\partial^2}{\partial x^2}\right)\right]\left[I + \Delta t\left(\frac{\partial B}{\partial y} - \frac{N}{Re}\frac{\partial^2}{\partial y^2}\right)\right]\Delta Q = \text{RHS}$$

The solution will proceed in two steps as

$$\left[I + \Delta t\left(\frac{\partial A}{\partial x} - \frac{N}{Re}\frac{\partial^2}{\partial x^2}\right) + \epsilon_i(\Delta x)^2\frac{\partial^2}{\partial x^2}\right]\Delta Q^* = \text{RHS} - \epsilon_e\left[(\Delta x)^4\frac{\partial^4}{\partial x^4} + (\Delta y)^4\frac{\partial^4}{\partial y^4}\right]Q \tag{8-67}$$

and

$$\left[I + \Delta t\left(\frac{\partial B}{\partial y} - \frac{N}{Re}\frac{\partial^2}{\partial y^2}\right) + \epsilon_i(\Delta y)^2\frac{\partial^2}{\partial y^2}\right]\Delta Q = \Delta Q^* \tag{8-68}$$

Note that implicit and explicit damping terms have been added to Equations (8-67) and (8-68), where ϵ_i and ϵ_e are damping coefficients to be specified by the user. Typical values of the damping coefficients for the Burgers equation was investigated in Chapter 6.

Now, a second-order central difference approximation is applied spatially to Equation (8-67) to provide

$$\begin{aligned}
&\Delta Q^*_{i,j} + \frac{\Delta t}{2\Delta x}(A_{i+1,j}\Delta Q^*_{i+1,j} - A_{i-1,j}\Delta Q^*_{i-1,j}) \\
&- \frac{N}{Re}\frac{\Delta t}{(\Delta x)^2}(\Delta Q^*_{i+1,j} - 2\Delta Q^*_{i,j} + \Delta Q^*_{i-1,j}) + \epsilon_i(\Delta Q^*_{i+1,j} - 2\Delta Q^*_{i,j} + \Delta Q^*_{i-1,j}) \\
&= \Delta t\left\{ - \frac{E_{i+1,j} - E_{i-1,j}}{2\Delta x} - \frac{F_{i,j+1} - F_{i,j-1}}{2\Delta y} + \frac{N}{Re}\frac{Q_{i+1,j} - 2Q_{i,j} + Q_{i-1,j}}{(\Delta x)^2}\right. \\
&\left. + \frac{N}{Re}\frac{Q_{i,j+1} - 2Q_{i,j} + Q_{i,j-1}}{(\Delta y)^2}\right\} - \epsilon_e\Big\{(Q_{i-2,j} - 4Q_{i-1,j} + 6Q_{i,j} - 4Q_{i+1,j} + Q_{i+2,j}) \\
&+ (Q_{i,j-2} - 4Q_{i,j-1} + 6Q_{i,j} - 4Q_{i,j+1} + Q_{i,j+2})\Big\}
\end{aligned} \tag{8-69}$$

Equation (8-69) is now rearranged to provide a block tridiagonal system as follows

$$\left[-\left(\frac{\Delta t}{2\Delta x}\right)A_{i-1,j}-\left(\frac{N}{Re}\frac{\Delta t}{(\Delta x)^2}\right)+\epsilon_i\right]\Delta Q^*_{i-1,j}+\left[I+\left(\frac{2N}{Re}\frac{\Delta t}{(\Delta x)^2}\right)-2\epsilon_i\right]\Delta Q^*_{i,j}$$
$$+\left[\left(\frac{\Delta t}{2\Delta x}\right)A_{i+1,j}-\left(\frac{N}{Re}\frac{\Delta t}{(\Delta x)^2}\right)+\epsilon_i\right]\Delta Q^*_{i+1,j}=\text{RHS}_{i,j} \tag{8-70}$$

Similarly, Equation (8-68) is expressed in a block tridiagonal form:

$$\left[-\left(\frac{\Delta t}{2\Delta y}\right)B_{i,j-1}-\left(\frac{N}{Re}\frac{\Delta t}{(\Delta y)^2}\right)+\epsilon_i\right]\Delta Q_{i,j-1}+\left[I+\left(\frac{2N}{Re}\frac{\Delta t}{(\Delta y)^2}\right)-2\epsilon_i\right]\Delta Q_{i,j}$$
$$+\left[\left(\frac{\Delta t}{2\Delta y}\right)B_{i,j+1}-\left(\frac{N}{Re}\frac{\Delta t}{(\Delta y)^2}\right)+\epsilon_i\right]\Delta Q_{i,j+1}=\Delta Q^*_{i,j} \tag{8-71}$$

Equations (8-70) and (8-71) are written in a compact form as

$$CAM_{i,j}\Delta Q^*_{i-1,j}+CA_{i,j}\Delta Q^*_{i,j}+CAP_{i,j}\Delta Q^*_{i+1,j}=\text{RHS}_{i,j} \tag{8-72}$$

and

$$CBM_{i,j}\Delta Q_{i,j-1}+CB_{i,j}\Delta Q_{i,j}+CAP_{i,j}\Delta Q_{i,j+1}=\Delta Q^*_{i,j} \tag{8-73}$$

where

$$CAM_{i,j}=-\left(\frac{\Delta t}{2\Delta x}\right)A_{i-1,j}-\frac{N}{Re}\frac{\Delta t}{(\Delta x)^2}+\epsilon_i$$

$$CA_{i,j}=I+\frac{2N}{Re}\frac{\Delta t}{(\Delta x)^2}-2\epsilon_i$$

$$CAP_{i,j}=\left(\frac{\Delta t}{2\Delta x}\right)A_{i+1,j}-\frac{N}{Re}\frac{\Delta t}{(\Delta x)^2}+\epsilon_i$$

$$CBM_{i,j}=-\left(\frac{\Delta t}{2\Delta y}\right)B_{i,j-1}-\frac{N}{Re}\frac{\Delta t}{(\Delta y)^2}+\epsilon_i$$

$$CB_{i,j}=I+\frac{2N}{Re}\frac{\Delta t}{(\Delta y)^2}-2\epsilon_i$$

$$CBP_{i,j}=\left(\frac{\Delta t}{2\Delta y}\right)B_{i,j+1}-\frac{N}{Re}\frac{\Delta t}{(\Delta y)^2}+\epsilon_i$$

A procedure to solve block tridiagonal systems such as (8-72) and (8-73) is provided in Appendix E.

Other numerical techniques investigated previously in Chapters 3 and 7, such as Dufort-Frankel, Beam and Warming implicit, or Crank-Nicolson schemes, may be applied to Equations (8-67) and (8-68) as well. For example, the Crank-Nicolson scheme is briefly described in the following section.

8.5.1.3 Crank-Nicolson Implicit

Recall that the Crank-Nicolson formulation requires an averaging of the terms at time levels of n and $n+1$. Thus, the formulation applied to Equation (8-63) yields

$$\frac{\Delta Q}{\Delta t} + \frac{1}{2}\left[\left(\frac{\partial E}{\partial x}\right)^n + \left(\frac{\partial E}{\partial x}\right)^{n+1}\right] + \frac{1}{2}\left[\left(\frac{\partial F}{\partial y}\right)^n + \left(\frac{\partial F}{\partial y}\right)^{n+1}\right]$$
$$= \frac{1}{2}\frac{1}{Re}[N]\left\{\frac{\partial^2}{\partial x^2} + \frac{\partial^2}{\partial y^2}\right\}(Q^n + Q^{n+1})$$

Linearization such as (8-64) is employed to provide

$$\frac{\Delta Q}{\Delta t} + \frac{1}{2}\left[\left(\frac{\partial E}{\partial x}\right)^n + \frac{\partial}{\partial x}(E^n + A\Delta Q)\right] + \frac{1}{2}\left[\left(\frac{\partial F}{\partial y}\right)^n + \frac{\partial}{\partial y}(F^n + B\Delta Q)\right]$$
$$= \frac{1}{2}\frac{1}{Re}[N]\left\{\frac{\partial^2}{\partial x^2} + \frac{\partial^2}{\partial y^2}\right\}(Q^{n+1} - Q^n + 2Q^n)$$

from which

$$\Delta Q + \frac{1}{2}\Delta t\left[\frac{\partial}{\partial x}(A\Delta Q) + \frac{\partial}{\partial y}(B\Delta Q) - \frac{N}{Re}\left(\frac{\partial^2}{\partial x^2} + \frac{\partial^2}{\partial y^2}\right)\Delta Q\right]$$
$$= \Delta t\left[-\frac{\partial E^n}{\partial x} - \frac{\partial F^n}{\partial y} + \frac{N}{Re}\left(\frac{\partial^2 Q}{\partial x^2} + \frac{\partial^2 Q}{\partial y^2}\right)\right]$$

or

$$\left\{I + \frac{\Delta t}{2}\left[\frac{\partial A}{\partial x} + \frac{\partial B}{\partial y} - \frac{N}{Re}\left(\frac{\partial^2 Q}{\partial x^2} + \frac{\partial^2 Q}{\partial y^2}\right)\right]\right\}\Delta Q = \text{RHS} \tag{8-74}$$

Application of approximate factorization and addition of damping terms results in

$$\left[I + \frac{1}{2}\Delta t\left(\frac{\partial A}{\partial x} - \frac{N}{Re}\frac{\partial^2}{\partial x^2}\right) + \epsilon_i(\Delta x)^2\frac{\partial^2}{\partial x^2}\right]\Delta Q^*$$
$$= \text{RHS} - \epsilon_e\left[(\Delta x)^4\frac{\partial^4}{\partial x^4} + (\Delta y)^4\frac{\partial^4}{\partial y^4}\right]Q \tag{8-75}$$

and

$$\left[I + \frac{1}{2}\Delta t\left(\frac{\partial B}{\partial y} - \frac{N}{Re}\frac{\partial^2}{\partial y^2}\right) + \epsilon_i(\Delta y)^2\frac{\partial^2}{\partial y^2}\right]\Delta Q = \Delta Q^* \tag{8-76}$$

Now central difference approximations are applied to Equations (8-75) and (8-76) to provide block tridiagonal systems.

8.6 Boundary Conditions

Physical boundaries of a specified domain upon which boundary conditions are generally required or where the values of the dependent variables must be determined as a part of the overall solution can be categorized into five groups. They are: Body surface, far-field, symmetry line (or surface in 3D), inflow, and outflow boundaries. Various categories of physical boundaries are illustrated in Fig. 8-1. Physical

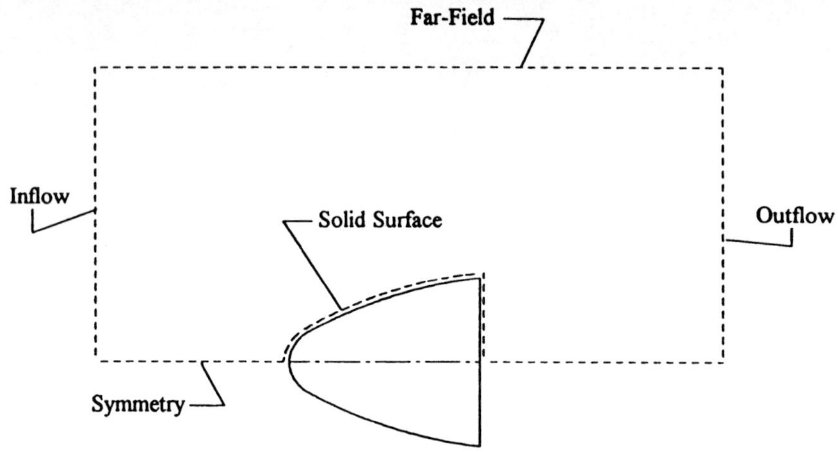

Figure 8-1. Illustration of typical boundaries.

or numerical specification and implementation of the boundary conditions along various boundaries are generally challenging. Specification of the boundary conditions for incompressible Navier-Stokes equations is no exception. Of course physical considerations usually provide some clues on boundary conditions, some of which are relatively simple to implement. For example, at a solid surface the condition of no slip is used to specify the boundary condition on the velocity. However, the specification of boundary conditions for the velocity components at the inflow, outflow, or the far-field is not usually straightforward. Obviously the manner by which any boundary condition is specified would depend largely on the physics as well as the domain of the problem. For example, if the far-field boundary is truly set far from the region where all of the "flow activity" takes place, one may indeed impose the freestream conditions along the boundary. However, if the far-field boundary is relatively close to the "action," then the far-field boundary may be treated as an inflow/outflow boundary depending on the sign of the normal (to the boundary) component of the velocity. Primarily two factors must be considered as one sets the boundary conditions on the inflow and outflow. First, the velocity and/or pressure at the outflow is usually not known a priori and must be determined as the overall solution evolves. Second, due to the influence of the interior solution on the inflow or the far-field (if treated as inflow) conditions, updating of the boundary values may be required. Of course, these factors are due to the physical phenomenon of signal propagation, i.e., for an incompressible flow, a disturbance is propagated in all directions! Therefore, it is clear that the specification of the boundary conditions is very much dependent on the specific problem of interest. That is, the location of the inflow, outflow, and far-field boundaries with respect to the location where changes in the flow properties occur. With these comments in mind, some of the options available in the specification of boundary conditions are reviewed.

Recall that the incompressible Navier-Stokes equations in primitive variable formulation in two space dimensions involve three unknowns, namely, u, v, and p. In this section the boundary conditions for the velocity and pressure are investigated. Discussion of the necessary boundary conditions for the vorticity-stream function formulation is deferred until Section 8.9.

8.6.1 Body Surface

No slip velocity boundary conditions are used at the body surface. Therefore, the surface velocity is imposed at the boundary. For most applications where there is no relative motion between the solid surface and the fluid, the velocity components are set to zero according to the no slip condition. If the surface is porous where fluid is injected or extracted at some specified velocity, the injection or extraction velocity is used. Usually the pressure at the surface is not known and must be determined as a part of the overall solution. Generally speaking, the Neumann-type boundary condition is imposed for the pressure. For this purpose a relation involving the normal pressure gradient is obtained from the appropriate momentum equation. For example, the following expression can be utilized along the solid boundary aligned parallel to the y-coordinate:

$$\frac{\partial p}{\partial x} = \frac{1}{Re}\frac{\partial^2 u}{\partial x^2} \tag{8-77}$$

This pressure boundary condition may be specified along the boundary AB of the classical steady flow over step as illustrated in Fig. 8-2. Similarly, the condition

$$\frac{\partial p}{\partial y} = \frac{1}{Re}\frac{\partial^2 v}{\partial y^2}$$

would be imposed along the boundary BC. These boundary conditions are implemented in the solution of the Poisson equation for pressure. The Neumann conditions stated above may be reduced to the simple zero normal pressure gradient for applications where the primary flow is parallel to the surface and the Reynolds number is high. Note that this requirement is consistent with the boundary layer assumption.

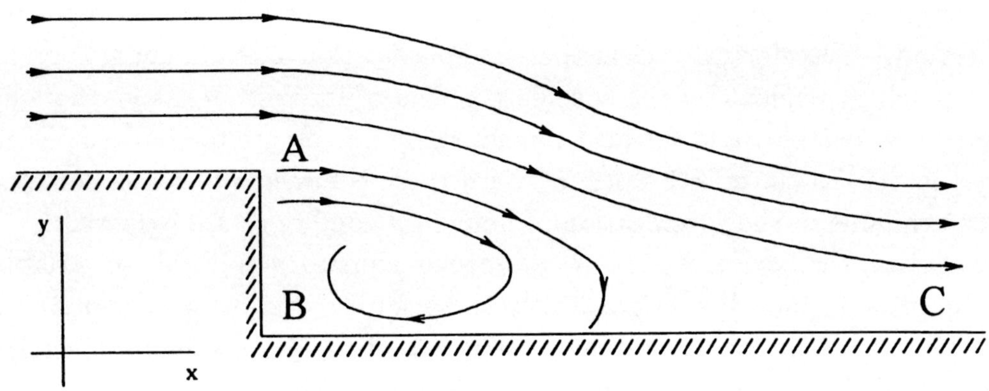

Figure 8-2. Schematic of step flow.

A few comments with regard to pressure are in order. First, specific numerical schemes have been developed which do not require specification of any boundary condition for the pressure. The Marker and Cell formulation used on a staggered grid, for example, is well adapted for this purpose. Second, computation of pressure over the entire domain may not be required, in which case the appropriate equations may be solved to provide the velocity field. Third, the value of pressure may be required only along the surface, for which there is no need to compute the pressure over the entire domain. For example one may need to compute the pressure force acting on an airfoil. For such applications, subsequent to the computation of the velocity field, the tangential derivative of pressure (which is obtained from the appropriate momentum equation) is integrated along the surface to provide the required pressure distribution along the body. Fourth, when applying Neumann-type boundary condition on pressure, e.g., $\partial p/\partial n = 0$ along the surface of internal flows for the solution of the Poisson equation, an additional global integral constraint must be satisfied. The application of the divergence theorem to the Poisson equation provides the global constraints given by

$$\iint_A \left(\frac{\partial^2 p}{\partial x^2} + \frac{\partial^2 p}{\partial y^2} \right) dx dy = \oint_C \frac{\partial p}{\partial n} \, d\ell \tag{8-78}$$

where A is the area of the computational domain enclosed by the boundary C and $d\ell$ is a differential length along the boundary. Failure to satisfy the requirement stated by (8-78) could result in either a slow convergence or, most likely, a slow divergence of the solution for the Poisson equation for pressure.

8.6.2 Far-Field

Specification of boundary conditions on the far-field boundary is very much problem dependent. If the boundary is located far away such that the flow properties on the boundary are not influenced by the interior solution, then it is indeed far-field and usually the freestream conditions are imposed. On the other hand, if the boundary is located relatively close to the "action," the boundary can no longer be considered as far-field and must be dealt with as an inflow and/or outflow boundary. Whether the boundary is considered as an outflow boundary or an inflow boundary depends on the sign of the velocity component normal to the boundary. If the velocity is into the domain, that portion is considered as inflow boundary; otherwise it is considered as an outflow and appropriate boundary conditions must be incorporated.

8.6.3 Symmetry

For applications where the configuration and the domain of solution are symmetrical, the axis of symmetry (or surface of symmetry) may be used as a boundary. The boundary location may be defined in two fashions. First, the boundary is set on the axis of symmetry as shown in Fig. 8-3.

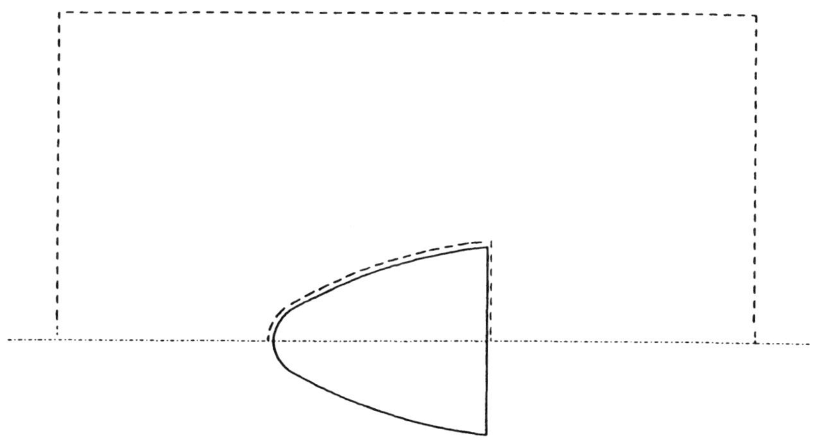

Figure 8-3. Illustration of boundary set on the axis of symmetry.

In this case the net flow across the symmetry line is zero. Therefore, the component of the velocity normal to the boundary is set to zero. Furthermore, the shear stress along the axis of symmetry may be zero in some applications. Thus, the velocity gradient is set to zero. Second, the boundary may be set below the axis of symmetry as shown in Fig. 8-4, in which case the symmetry of flow variables are used as the required boundary conditions.

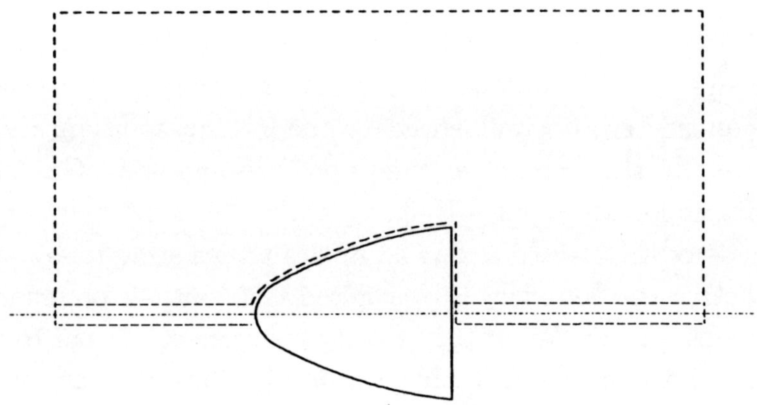

Figure 8-4. Illustration of boundary set below the axis of symmetry.

8.6.4 Inflow

Usually two boundary conditions are required at the inflow. For most applications, pressure and one component of the velocity are provided. Typically then $u(y) = u_o$ and p are provided. The y-component of the velocity may be set to zero or may be determined by setting the velocity gradient $\partial v/\partial x$ to zero. If one uses a first-order approximation, then $v_{1,j} = v_{2,j}$. Higher order approximations can be used as well; for example a second-order approximation yields

$$v_{1,j} = 2v_{2,j} - v_{3,j}$$

8.6.5 Outflow

Generally speaking the value of the velocity and/or pressure are not known at the outflow boundary. Therefore, for most applications Neumann type boundary conditions are imposed. The specification of zero velocity gradient at the outflow may be appropriate for most applications.

8.6.6 An Example

The classical step problem shown in Fig. 8-5 is used as an example to illustrate the specification of various boundary conditions. The domain of solution is contained within the region specified by $ABCDEF$, where flow enters at surface EF and exits at CD. Furthermore, surfaces AB, BC, and FA are solid surfaces. But what about surface ED? If the surface ED is located sufficiently far from the step where all the "action" takes place, one may consider it as a far-field boundary.

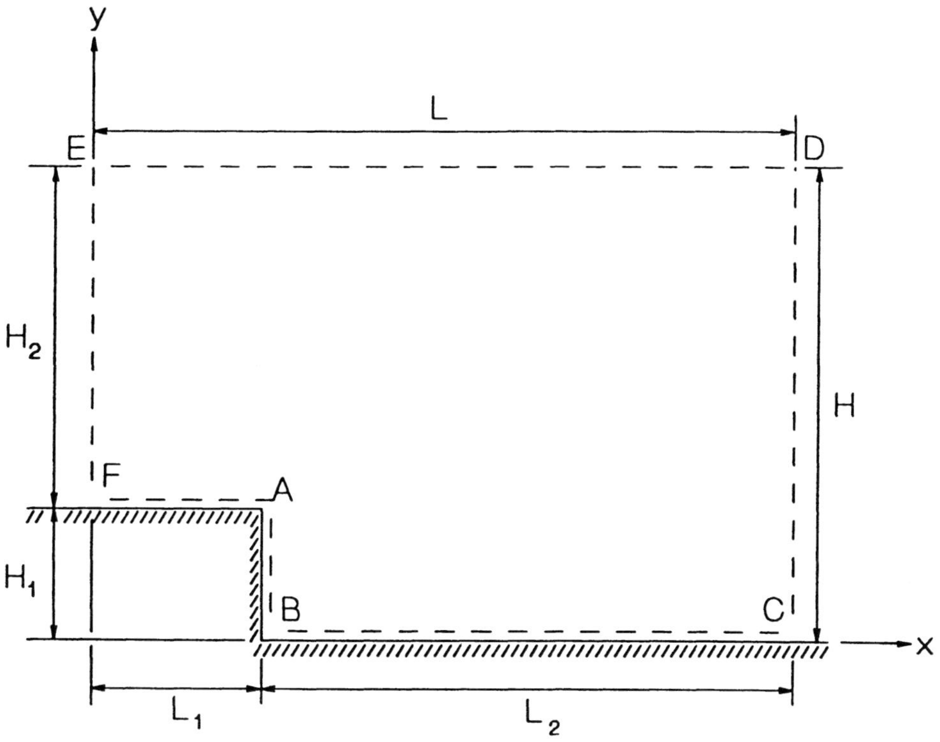

Figure 8-5. Illustration of various boundaries for the step flow.

Thus, freestream conditions are imposed. Otherwise, the boundary ED could be either an inflow or an outflow boundary depending on the sign of the y-component of the velocity. In reality, the physical domain of interest and the specified flow conditions play a dominant role in how the boundary conditions must be specified. The following boundary conditions for the step problem shown in Fig. 8-5 should not be used universally, but instead are given here to illustrate a set of typical boundary conditions. Indeed, some of the boundary values may not be available and special procedures such as extrapolation or solution of the momentum equation at the boundary must be used. Specifications of typical boundary conditions for the step problem shown in Fig. 8-5 are:

Boundary EF (Inflow):

$$H_1 < y < H \qquad \begin{cases} u = u_\infty \\ v = 0 \quad \text{or} \quad \dfrac{\partial v}{\partial x} = 0 \\ p = p_\infty \quad \text{or} \quad \dfrac{\partial p}{\partial x} = \dfrac{1}{Re}\dfrac{\partial^2 u}{\partial x^2} \end{cases}$$

Boundary FA (Solid surface):

$$0 < x < L_1 \qquad \begin{cases} u = 0 \\ v = 0 \\ \dfrac{\partial p}{\partial y} = \dfrac{1}{Re}\dfrac{\partial^2 v}{\partial y^2} \end{cases}$$

Boundary AB (Solid surface):

$$0 < y < H_1 \qquad \begin{cases} u = 0 \\ v = 0 \\ \dfrac{\partial p}{\partial x} = \dfrac{1}{Re}\dfrac{\partial^2 u}{\partial x^2} \end{cases}$$

Boundary BC (Solid surface):

$$L_1 < x < L \qquad \begin{cases} u = 0 \\ v = 0 \\ \dfrac{\partial p}{\partial y} = \dfrac{1}{Re}\dfrac{\partial^2 v}{\partial y^2} \end{cases}$$

Boundary CD (Outflow):

$$0 < y < H \qquad \begin{cases} \dfrac{\partial u}{\partial x} = 0 \\ \dfrac{\partial v}{\partial y} = 0 \\ p = p_\infty \quad \text{or} \quad \dfrac{\partial p}{\partial x} = 0 \end{cases}$$

Boundary ED (Far-Field):

$$0 < x < L \qquad \begin{cases} u = u_\infty \\ v = 0 \\ p = p_\infty \end{cases}$$

8.7 Staggered Grid

For a certain class of problems where the governing equations can be solved sequentially, it is advantageous to write the finite difference formulations suited for a so-called staggered grid. The procedure allows the coupling of variables and consequently improves stability constraints. Staggered grids may be constructed by several methods. For example, one may shift the grid along a coordinate line one-half of a grid spacing or shift the grid diagonally. A commonly used staggered grid is shown in Fig. 8-6, which will be used as a prototype to investigate numerical solutions. Since it is apparent that two superimposed grids are involved, they will be identified as primary and secondary grids. The grid points of the primary system are identified as in a standard grid by i, j, whereas the grid points of the secondary

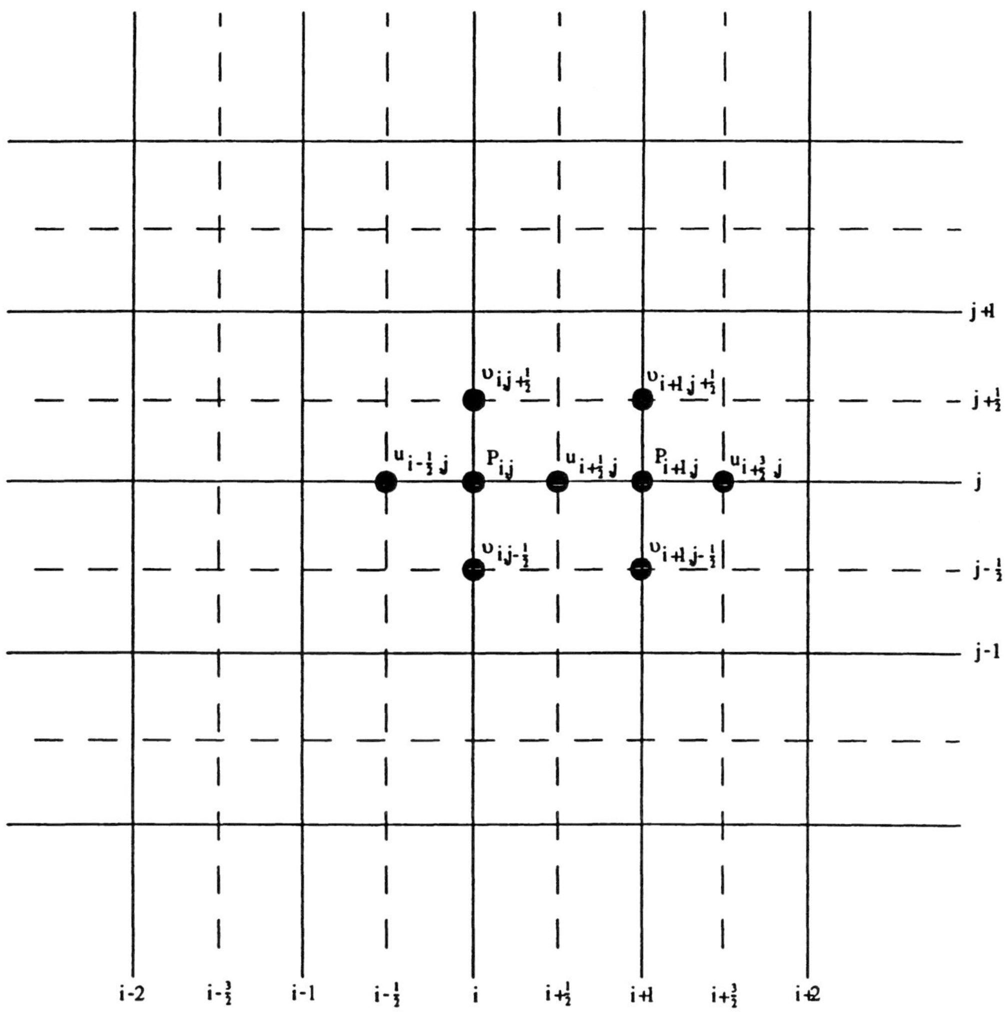

Figure 8-6. Typical staggered grid and assignment of the flow variables.

system are distinguished by increments of one-half. The primary grid is shown in Fig. 8-6 by solid lines, whereas the secondary grid is shown as dashed lines.

Different variables are assigned to specific grid points on the primary or secondary systems. To illustrate the procedure, consider the solution of the incompressible Navier-Stokes equations which, after all, is the subject of this chapter. The unknowns are the velocity components u and v (for two-dimensional problems) and the pressure p. A typical assignment of the variables at different grid points is as follows. The pressure is defined on the primary grid whereas the velocity components are defined on the cell faces of the secondary grid. More precisely, the x-component of the velocity, u, is assigned on the one-half grid line in the x-direction at the same y-location on the primary grid. The y-component of the velocity, v, is defined in a similar fashion. The specification of the flow variables is illustrated in Fig. 8-6. A typical formulation applied to the incompressible Navier-Stokes equations is provided in the following section.

8.7.1 Marker and Cell Method

A formulation well adapted for application on a staggered grid is the Marker and Cell (MAC) scheme introduced by the Harlow and Welch [8-6]. The Navier-Stokes equations (8-60) through (8-62) are used to illustrate the application of the scheme on the staggered grid shown in Fig. 8-7. To write an explicit formulation, a first-order approximation is used for temporal derivative, and second-order central difference approximations are used for space discretization. The continuity equation given by (8-60) is applied at grid point (i,j) providing the following finite difference equation:

$$\frac{p_{i,j}^{n+1}-p_{i,j}^{n}}{\Delta t}+a^2\left(\frac{u_{i+\frac{1}{2},j}^{n+1}-u_{i-\frac{1}{2},j}^{n+1}}{\Delta x}+\frac{v_{i,j+\frac{1}{2}}^{n+1}-v_{i,j-\frac{1}{2}}^{n+1}}{\Delta y}\right)=0 \tag{8-79}$$

Observe that the velocity components are specified at the $(n+1)$ time level. However, no difficulty is created because the momentum equation will be solved first to provide u and v at the $n+1$ level. Subsequently, Equation (8-79) is solved for the pressure. The x-momentum equation is applied on the secondary grid at grid point $(i+\frac{1}{2},j)$ to provide

$$\frac{u_{i+\frac{1}{2},j}^{n+1}-u_{i+\frac{1}{2},j}^{n}}{\Delta t}+\frac{p_{i+1,j}^{n}-p_{i,j}^{n}}{\Delta x}+\frac{(u^2)_{i+1,j}^{n}-(u^2)_{i,j}^{n}}{\Delta x}+\frac{(uv)_{i+\frac{1}{2},j+\frac{1}{2}}^{n}-(uv)_{i+\frac{1}{2},j-\frac{1}{2}}^{n}}{\Delta y}$$

$$=\frac{1}{Re}\frac{u_{i-\frac{1}{2},j}^{n}-2u_{i+\frac{1}{2},j}^{n}+u_{i+\frac{3}{2},j}^{n}}{(\Delta x)^2}+\frac{1}{Re}\frac{u_{i+\frac{1}{2},j-1}^{n}-2u_{i+\frac{1}{2},j}^{n}+u_{i+\frac{1}{2},j+1}^{n}}{(\Delta y)^2} \tag{8-80}$$

Similarly the y-component of momentum is applied at grid point $i,j+\frac{1}{2}$ resulting

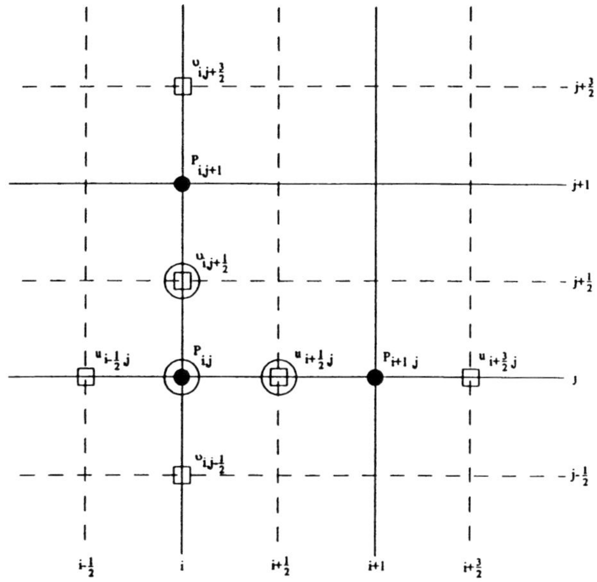

Figure 8-7. Assignment of the variables on the staggered grid.

in the following finite difference equation

$$\frac{v^{n+1}_{i,j+\frac{1}{2}} - v^n_{i,j+\frac{1}{2}}}{\Delta t} + \frac{(uv)^n_{i+\frac{1}{2},j+\frac{1}{2}} - (uv)^n_{i-\frac{1}{2},j+\frac{1}{2}}}{\Delta x} + \frac{(v^2)^n_{i,j+1} - (v^2)^n_{i,j}}{\Delta y} + \frac{p^n_{i,j+1} - p^n_{i,j}}{\Delta y}$$

$$= \frac{1}{Re}\frac{v^n_{i-1,j+\frac{1}{2}} - 2v^n_{i,j+\frac{1}{2}} + v^n_{i+1,j+\frac{1}{2}}}{(\Delta x)^2} + \frac{1}{Re}\frac{v^n_{i,j-\frac{1}{2}} - 2v^n_{i,j+\frac{1}{2}} + v^n_{i,j+\frac{3}{2}}}{(\Delta y)^2} \tag{8-81}$$

Since the values of the velocity components are known only on the secondary grid (i.e., half points), as provided from Equations (8-80) and (8-81), all the values appearing in terms of the primary grid point must be replaced by their approximate values on the secondary grid. For this purpose, the following approximations are used:

$$(u^2)_{i+1,j} = \frac{1}{4}\left(u_{i+\frac{3}{2},j} + u_{i+\frac{1}{2},j}\right)^2$$

$$(u^2)_{i,j} = \frac{1}{4}\left(u_{i+\frac{1}{2},j} + u_{i-\frac{1}{2},j}\right)^2$$

$$(v^2)_{i,j+1} = \frac{1}{4}\left(v_{i,j+\frac{3}{2}} + v_{i,j+\frac{1}{2}}\right)^2$$

$$(v^2)_{i,j} = \frac{1}{4}\left(v_{i,j+\frac{1}{2}} + v_{i,j-\frac{1}{2}}\right)^2$$

$$(uv)_{i+\frac{1}{2},j+\frac{1}{2}} = \frac{1}{4}\left(u_{i+\frac{1}{2},j} + u_{i+\frac{1}{2},j+1}\right)\left(v_{i,j+\frac{1}{2}} + v_{i+1,j+\frac{1}{2}}\right)$$

$$(uv)_{i+\frac{1}{2},j-\frac{1}{2}} = \frac{1}{4}(u_{i+\frac{1}{2},j} + u_{i+\frac{1}{2},j-1})(v_{i,j-\frac{1}{2}} + v_{i+1,j-\frac{1}{2}})$$

$$(uv)_{i-\frac{1}{2},j+\frac{1}{2}} = \frac{1}{4}(u_{i-\frac{1}{2},j} + u_{i-\frac{1}{2},j+1})(v_{i,j+\frac{1}{2}} + v_{i-1,j+\frac{1}{2}})$$

8.7.2 Implementation of the Boundary Conditions

Due to lack of boundary conditions for pressure in most applications, the use of a staggered grid and the MAC formulation provide an advantage. That is, one may locate the secondary grid along the boundaries of the domain where only specification of velocity boundary conditions is required but not that of pressure. However, this advantage is accompanied by a disadvantage. That is due to the fact that some values of the velocity outside the domain will be required. These values are essentially obtained by extrapolation of the interior points or equivalently by approximation of the derivatives at the boundaries. To illustrate the implementation of the boundary conditions consider the staggered grid at a boundary as shown in Fig. 8-8.

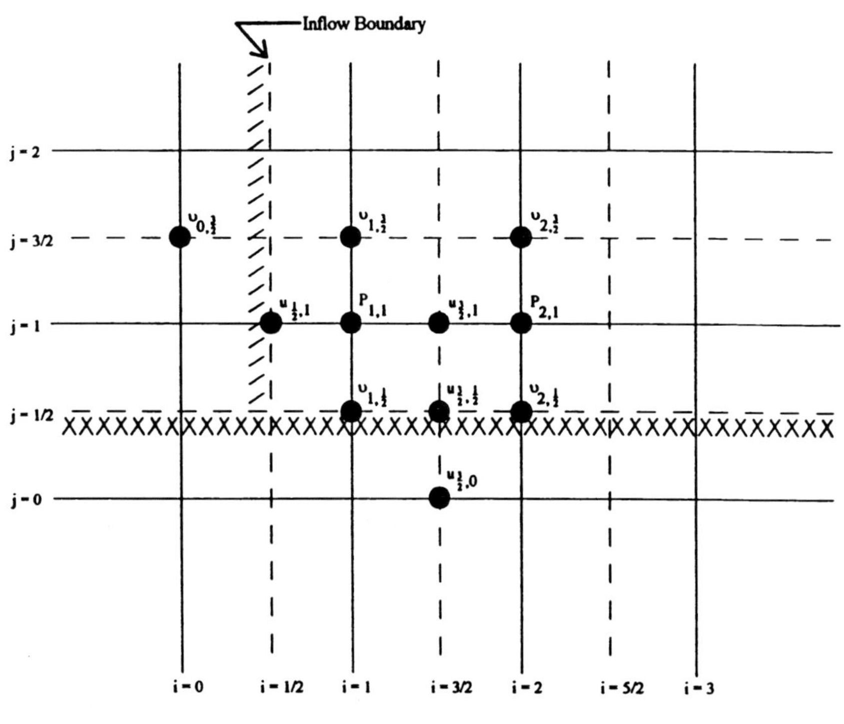

Figure 8-8. Illustration of specification of the boundary conditions on a staggered grid.

Assume a solid surface aligned along $j = \frac{1}{2}$ and an inflow boundary aligned along $i = \frac{1}{2}$. Since the surface is specified as solid, the no slip condition is imposed. Therefore,

$$v_{1,\frac{1}{2}} = v_{2,\frac{1}{2}} = \cdots = 0$$

Similarly, $u_{\frac{3}{2},\frac{1}{2}} = 0$, from which $u_{\frac{3}{2},0}$ is approximated by

$$u_{\frac{3}{2},\frac{1}{2}} = \frac{1}{2}(u_{\frac{3}{2},0} + u_{\frac{3}{2},1}) = 0$$

or

$$u_{\frac{3}{2},0} = -u_{\frac{3}{2},1}$$

At the inflow boundary, the u components are specified directly, i.e., terms like $u_{\frac{1}{2},1}, u_{\frac{1}{2},2}, \ldots$. However, the v components are approximated by extrapolation as

$$v_{0,\frac{3}{2}} = 2v_{\frac{1}{2},\frac{3}{2}} - v_{1,\frac{3}{2}} \tag{8-82}$$

where $v_{\frac{1}{2},\frac{3}{2}}$ is provided by the specified inflow boundary. At the outflow boundary one may use zero velocity gradient, e.g., $\partial u/\partial x = 0$.

8.7.3 Dufort-Frankel Scheme

Recall that Dufort-Frankel algorithm for the linear diffusion equation is unconditionally stable. However, the scheme was shown to be inconsistent and, indeed, the equation to be solved was equivalent to

$$\frac{\partial u}{\partial t} + \alpha k^2 \frac{\partial^2 u}{\partial t^2} = \alpha \frac{\partial^2 u}{\partial x^2}$$

where $k = \dfrac{\Delta t}{\Delta x}$. Extending the scheme to the x-momentum equation for example, one has

$$\frac{\partial u}{\partial t} + \frac{\partial}{\partial x}(u^2 + p) + \frac{\partial}{\partial y}(uv) + \frac{1}{Re}(k_x^2 + k_y^2)\frac{\partial^2 u}{\partial t^2} = \frac{1}{Re}\nabla^2 u$$

where

$$k_x = \frac{\Delta t}{\Delta x} \quad \text{and} \quad k_y = \frac{\Delta t}{\Delta y}$$

To obtain the finite difference equations, second-order approximations for the time derivatives are introduced. The continuity equation is applied at (i, j) to provide

$$\frac{p_{i,j}^{n+1} - p_{i,j}^{n-1}}{2\Delta t} + a^2 \left(\frac{u_{i+\frac{1}{2},j}^n - u_{i-\frac{1}{2},j}^n}{\Delta x} + \frac{v_{i,j+\frac{1}{2}}^n - v_{i,j-\frac{1}{2}}^n}{\Delta y} \right) = 0 \tag{8-83}$$

The x-momentum is applied at $(i+\frac{1}{2},j)$ resulting in the following finite difference equation

$$\frac{u^{n+1}_{i+\frac{1}{2},j}-u^{n-1}_{i+\frac{1}{2},j}}{2\Delta t}+\frac{(u^n_{i+1,j})^2-(u^n_{i,j})^2}{\Delta x}+\frac{p^n_{i+1,j}-p^n_{i,j}}{\Delta x}+\frac{(uv)^n_{i+\frac{1}{2},j+\frac{1}{2}}-(uv)^n_{i+\frac{1}{2},j-\frac{1}{2}}}{\Delta y}$$

$$=\frac{1}{Re}\frac{u^n_{i-\frac{1}{2},j}-2u^n_{i+\frac{1}{2},j}+u^n_{i+\frac{3}{2},j}}{(\Delta x)^2}+\frac{1}{Re}\frac{u^n_{i+\frac{1}{2},j-1}-2u^n_{i+\frac{1}{2},j}+u^n_{i+\frac{1}{2},j+1}}{(\Delta y)^2}$$

$$-\frac{1}{Re}\left[\left(\frac{\Delta t}{\Delta x}\right)^2+\left(\frac{\Delta t}{\Delta y}\right)^2\right]\frac{u^{n+1}_{i+\frac{1}{2},j}-2u^n_{i+\frac{1}{2},j}+u^{n-1}_{i+\frac{1}{2},j}}{(\Delta t)^2} \tag{8-84}$$

Similarly the y-component of the momentum equation is applied at point $(i,j+\frac{1}{2})$ to yield

$$\frac{v^{n+1}_{i,j+\frac{1}{2}}-v^{n-1}_{i,j+\frac{1}{2}}}{2\Delta t}+\frac{(uv)^n_{i+\frac{1}{2},j+\frac{1}{2}}-(uv)^n_{i-\frac{1}{2},j+\frac{1}{2}}}{\Delta x}+\frac{(v^2)^n_{i,j+1}-(v^2)^n_{i,j}}{\Delta y}+\frac{p^n_{i,j+1}-p^n_{i,j}}{\Delta y}$$

$$=\frac{1}{Re}\frac{v^n_{i-1,j+\frac{1}{2}}-2v^n_{i,j+\frac{1}{2}}+v^n_{i+1,j+\frac{1}{2}}}{(\Delta x)^2}+\frac{1}{Re}\frac{v^n_{i,j-\frac{1}{2}}-2v^n_{i,j+\frac{1}{2}}+v^n_{i,j+\frac{3}{2}}}{(\Delta y)^2}$$

$$-\frac{1}{Re}\left[\left(\frac{\Delta t}{\Delta x}\right)^2+\left(\frac{\Delta t}{\Delta y}\right)^2\right]\frac{v^{n+1}_{i,j+\frac{1}{2}}-2v^n_{i,j+\frac{1}{2}}+v^{n-1}_{i,j+\frac{1}{2}}}{(\Delta t)^2} \tag{8-85}$$

The computation proceeds with Equation (8-84) and (8-85) which provide the values of $u_{i+\frac{1}{2},j}$ and $v_{i,j+\frac{1}{2}}$ at the time level $n+1$, respectively. Subsequently, Equation (8-83) is solved for the pressure. The scheme is subject to stability requirement. For example, the following constraints have been suggested

$$\sqrt{2}\left(u^2+v^2\right)^{\frac{1}{2}}\frac{\Delta t}{\Delta x}\leq 1$$

for $\Delta x=\Delta y$. Note that this restriction is a Courant number type requirement.

8.7.4 Use of the Poisson Equation for Pressure

An alternate approach for solving Equation (8-60) directly is the use of the Poisson equation given by Equation (8-46). Thus the system of equations is composed of the x- and y-components of the momentum equation and the Poisson equation expressed as follows:

$$\frac{\partial^2 p}{\partial x^2}+\frac{\partial^2 p}{\partial y^2}=-\frac{\partial D}{\partial t}-\frac{\partial^2}{\partial x^2}(u^2)-2\frac{\partial^2}{\partial x\partial y}(uv)-\frac{\partial^2}{\partial y^2}(v^2)+\frac{1}{Re}\left[\frac{\partial^2}{\partial x^2}(D)+\frac{\partial^2}{\partial y^2}(D)\right] \tag{8-86}$$

$$\frac{\partial u}{\partial t}+\frac{\partial}{\partial x}(u^2+p)+\frac{\partial}{\partial y}(uv)=\frac{1}{Re}\nabla^2 u \tag{8-87}$$

$$\frac{\partial v}{\partial t} + \frac{\partial}{\partial x}(uv) + \frac{\partial}{\partial y}(v^2 + p) = \frac{1}{Re}\nabla^2 v \tag{8-88}$$

The solution approach begins with Equation (8-86), where the value of the pressure at $n+1$ time level is computed. Subsequently, Equations (8-87) and (8-88) are solved for the values of u^{n+1} and v^{n+1}, respectively. A typical numerical scheme on a staggered grid is described in this section. The right-hand side of Equation (8-86) is formulated as follows:

$$\begin{aligned}
(\text{RHS})_{i,j} = {} & \frac{D^n_{i,j}}{\Delta t} - \frac{(u^2)^n_{i-1,j} - 2(u^2)^n_{i,j} + (u^2)^n_{i+1,j}}{(\Delta x)^2} - \frac{(v^2)^n_{i,j-1} - 2(v^2)^n_{i,j} + (v^2)^2_{i,j+1}}{(\Delta y)^2} \\
& - 2\frac{(uv)^n_{i+\frac{1}{2},j+\frac{1}{2}} - (uv)^n_{i-\frac{1}{2},j+\frac{1}{2}} - (uv)^n_{i+\frac{1}{2},j-\frac{1}{2}} + (uv)^n_{i-\frac{1}{2},j-\frac{1}{2}}}{\Delta x \Delta y} \\
& + \frac{1}{Re}\left[\frac{D^n_{i-1,j} - 2D^n_{i,j} + D^n_{i+1,j}}{(\Delta x)^2} + \frac{D^n_{i,j-1} - 2D^n_{i,j} + D^n_{i,j+1}}{(\Delta y)^2}\right]
\end{aligned}$$

Note that $D^{n+1}_{i,j}$ in the first term has been set to zero, i.e., the continuity requirement at $n+1$ time level is enforced. Obviously, all the values appearing on the $(\text{RHS})_{i,j}$ term are known at the time level n. To solve the elliptic Equation (8-86), an iterative scheme is usually used. For example, if one selects the point Gauss-Seidel scheme, the formulation is expressed as

$$p^{k+1}_{i,j} = \frac{1}{2(1+\beta^2)}\left[p^k_{i+1,j} + p^{k+1}_{i-1,j} + \beta^2\left(p^k_{i,j+1} + p^{k+1}_{i,j-1}\right)\right] + (\text{RHS})_{i,j} \tag{8-89}$$

Equation (8-89) is solved at each time level to provide $p_{i,j}$ at the $n+1$ time level. Subsequently, Equations (8-87) and (8-88) are solved for $u_{i,j}$ and $v_{i,j}$ at the time level of $n+1$. A typical finite difference formulation on a staggered grid would be Equations (8-80) and (8-81).

8.7.5 Unsteady Incompressible Navier-Stokes Equations

The solution procedures discussed previously for the steady incompressible Navier-Stokes equations can be easily extended to the unsteady problems. Among the two procedures identified to overcome the difficulty associated with the continuity equation (namely the artificial compressibility and the Poisson equation for pressure), the latter is more commonly used for unsteady problems. As a prototype, the MAC formulation is considered in this section. Revisit the (nondimensional) governing equations provided in Section 8.2 given by

$$\frac{\partial u}{\partial x} + \frac{\partial v}{\partial y} = 0$$

$$\frac{\partial u}{\partial t} + \frac{\partial}{\partial x}(u^2 + p) + \frac{\partial}{\partial y}(uv) = \frac{1}{Re}\nabla^2 u$$

$$\frac{\partial v}{\partial t} + \frac{\partial}{\partial x}(uv) + \frac{\partial}{\partial y}(v^2 + p) + \frac{1}{Re}\nabla^2 v$$

The MAC formulation applied to the momentum equation, from Equations (8-80) and (8-81) provides

$$\begin{aligned} u^{n+1}_{i+\frac{1}{2},j} &= u^n_{i+\frac{1}{2},j} - \frac{\Delta t}{\Delta x}\left[(p^{n+1}_{i+1,j} - p^{n+1}_{i,j}) + (u^2)^n_{i+1,j} - (u^2)^n_{i,j}\right] \\ &- \frac{\Delta t}{\Delta y}\left[(uv)^n_{i+\frac{1}{2},j+\frac{1}{2}} - (uv)^n_{i+\frac{1}{2},j-\frac{1}{2}}\right] + \frac{1}{Re}\frac{\Delta t}{(\Delta x)^2}(u^n_{i-\frac{1}{2},j} - 2u^n_{i+\frac{1}{2},j} + u^n_{i+\frac{3}{2},j}) \\ &+ \frac{1}{Re}\frac{\Delta t}{(\Delta y)^2}(u^n_{i+\frac{1}{2},j-1} - 2u^n_{i+\frac{1}{2},j} + u^n_{i+\frac{1}{2},j+1}) \\ &= -\frac{\Delta t}{\Delta x}(p^{n+1}_{i+1,j} - p^{n+1}_{i,j}) + \mathrm{RHS}U_{i+\frac{1}{2},j} \end{aligned} \tag{8-90}$$

and

$$\begin{aligned} v^{n+1}_{i,j+\frac{1}{2}} &= v^n_{i,j+\frac{1}{2}} - \frac{\Delta t}{\Delta x}\left[(uv)^n_{i+\frac{1}{2},j+\frac{1}{2}} - (uv)^n_{i-\frac{1}{2},j+\frac{1}{2}}\right] \\ &- \frac{\Delta t}{\Delta y}\left[(v^2)^n_{i,j+1} - (v^2)^n_{i,j} + (p^{n+1}_{i,j+1} - p^{n+1}_{i,j})\right] + \frac{1}{Re}\frac{\Delta t}{(\Delta x)^2}(v^n_{i-1,j+\frac{1}{2}} - 2v^n_{i,j+\frac{1}{2}} + v^n_{i+1,j+\frac{1}{2}}) \\ &+ \frac{1}{Re}\frac{\Delta t}{(\Delta y)^2}(v^n_{i,j-\frac{1}{2}} - 2v^n_{i,j+\frac{1}{2}} + v^n_{i,j+\frac{3}{2}}) = -\frac{\Delta t}{\Delta y}(p^{n+1}_{i,j+1} - p^{n+1}_{i,j}) + \mathrm{RHS}V_{i,j+\frac{1}{2}} \end{aligned} \tag{8-91}$$

Now, the continuity equation is applied at $n+1$ to yield

$$\frac{u^{n+1}_{i+\frac{1}{2},j} - u^{n+1}_{i-\frac{1}{2},j}}{\Delta x} + \frac{v^{n+1}_{i,j+\frac{1}{2}} - v^{n+1}_{i,j-\frac{1}{2}}}{\Delta y} = 0 \tag{8-92}$$

Substitution of Equations (8-90), (8-91) and similar terms into Equation (8-92) provides

$$\begin{aligned} &\frac{1}{\Delta x}\left\{\left[-\frac{\Delta t}{\Delta x}(p^{n+1}_{i+1,j} - p^{n+1}_{i,j}) + \mathrm{RHS}U_{i+\frac{1}{2},j}\right] - \left[-\frac{\Delta t}{\Delta x}(p^{n+1}_{i,j} - p^{n+1}_{i-1,j}) + \mathrm{RHS}U_{i-\frac{1}{2},j}\right]\right\} \\ &+\frac{1}{\Delta y}\left\{\left[-\frac{\Delta t}{\Delta y}(p^{n+1}_{i,j+1} - p^{n+1}_{i,j}) + \mathrm{RHS}V_{i,j+\frac{1}{2}}\right] - \left[-\frac{\Delta t}{\Delta y}(p^{n+1}_{i,j} - p^{n+1}_{i,j-1}) + \mathrm{RHS}V_{i,j-\frac{1}{2}}\right]\right\} \\ &= 0 \end{aligned}$$

Rearranging terms yields

$$\begin{aligned} &\frac{p^{n+1}_{i-1,j} - 2p^{n+1}_{i,j} + p^{n+1}_{i+1,j}}{(\Delta x)^2} + \frac{p^{n+1}_{i,j-1} - 2p^{n+1}_{i,j} + p^{n+1}_{i,j+1}}{(\Delta y)^2} \\ &= \frac{1}{\Delta t}\left[\frac{\mathrm{RHS}U_{i+\frac{1}{2},j} - \mathrm{RHS}U_{i-\frac{1}{2},j}}{\Delta x} + \frac{\mathrm{RHS}V_{i,j+\frac{1}{2}} - \mathrm{RHS}V_{i,j-\frac{1}{2}}}{\Delta y}\right] \end{aligned} \tag{8-93}$$

Equation (8-93) may be solved by any iterative scheme for the value of $p_{i,j}^{n+1}$ over the entire domain. Subsequently, Equations (8-90) and (8-91) are used to compute the velocity components u and v.

8.8 Numerical Algorithms: Vorticity-Stream Function Formulation

The governing equation for the incompressible Navier-Stokes equations for two-dimensional applications in vorticity-stream function formulation were reviewed in Sec. 8.2.2. Fundamentally, the system is composed of the vorticity transport equation and the stream function equation. One of the advantages of this formulation is that the pressure term does not appear explicitly in either of the equations. Therefore, the system of equations is solved to provide the velocity field. Now, if the pressure field is required as well, then the Poisson equation for the pressure is subsequently solved. A major disadvantage of the formulation which was previously mentioned is the direct extension to three dimensions. Furthermore, a difficulty associated with the formulation is the specification of the boundary condition on the vorticity, which is due to the lack of physical boundary conditions for vorticity. Therefore, numerical boundary conditions for the vorticity must be derived.

At this point, it is useful to revisit the system of governing equations and review some of its features.

1. The system is composed of:

   a. Vorticity transport equation

$$\frac{\partial \Omega}{\partial t} + u\frac{\partial \Omega}{\partial x} + v\frac{\partial \Omega}{\partial y} = \frac{1}{Re}\left(\frac{\partial^2 \Omega}{\partial x^2} + \frac{\partial^2 \Omega}{\partial y^2}\right) \tag{8-94}$$

   and

   b. Stream function equation

$$\frac{\partial^2 \psi}{\partial x^2} + \frac{\partial^2 \psi}{\partial y^2} = -\Omega \tag{8-95}$$

2. The governing equations given by (8-94) and (8-95) can be used for either steady or unsteady flows. Note that time appears explicitly in the vorticity transport equation which is classified as a parabolic equation. Thus, any scheme previously introduced for the solution of parabolic equations can be utilized to solve Equation (8-94). But, what about Equation (8-95)? Even though time does not appear in that equation, it is still used for unsteady flows.

Therefore, for unsteady flow computations, the stream function equation is solved at each time step by any scheme previously introduced for the solution of elliptic equations.

3. For steady state problems, three options are available.

   a. The unsteady equations are solved until steady state is achieved. If this approach is used, one must pay attention to the total computation time in that it may be too excessive. Obviously, this would be problem dependent and in particular, Reynolds number dependent.

   b. The steady state form of the equations are given by

$$u\frac{\partial \Omega}{\partial x} + v\frac{\partial \Omega}{\partial y} = \frac{1}{Re}\left(\frac{\partial^2 \Omega}{\partial x^2} + \frac{\partial^2 \Omega}{\partial y^2}\right)$$

   and

$$\frac{\partial^2 \psi}{\partial x^2} + \frac{\partial^2 \psi}{\partial y^2} = -\Omega$$

   These equations may be solved directly. Note that the system is now composed of two elliptic equations which may be solved by iterative scheme.

   c. A pseudotransient approach similar to that previously used with regard to the primitive variable formulation may be developed for the vorticity-stream function formulation. The construction of pseudotransient equations yields

$$\frac{\partial \Omega}{\partial t} + u\frac{\partial \Omega}{\partial x} + v\frac{\partial \Omega}{\partial y} = \frac{1}{Re}\left(\frac{\partial^2 \Omega}{\partial x^2} + \frac{\partial^2 \Omega}{\partial y^2}\right)$$

   and

$$\frac{\partial \psi}{\partial t} - \left(\frac{\partial^2 \psi}{\partial x^2} + \frac{\partial^2 \psi}{\partial y^2} + \Omega\right) = 0$$

   Consequently, one now solves two parabolic equations. The equations may be solved sequentially or as a coupled system.

In the following section a limited number of finite difference formulations for the vorticity transport equation and, subsequently, the stream function equation is explored.

8.8.1 Vorticity Transport Equation

Recall that numerous schemes for the solution of parabolic equations were discussed previously in Chapters 3 and 7. These schemes can be adapted for the solution of the vorticity transport equation. The formulation may be explicit or implicit.

Perhaps one of the most simple finite difference formulations applied to a parabolic system is the FTCS scheme. In this method the time derivative is approximated by a first-order forward difference expression whereas second-order central difference relations are used for the spatial derivatives. The resulting finite difference equation of the nonconservative vorticity transport equation is written as

$$\frac{\Omega_{i,j}^{n+1}-\Omega_{i,j}^{n}}{\Delta t}+u_{i,j}^{n}\frac{\Omega_{i+1,j}^{n}-\Omega_{i-1,j}^{n}}{2\Delta x}+v_{i,j}^{n}\frac{\Omega_{i,j+1}^{n}-\Omega_{i,j-1}^{n}}{2\Delta y}$$

$$=\frac{1}{Re}\left[\frac{\Omega_{i+1,j}^{n}-2\Omega_{i,j}^{n}+\Omega_{i-1,j}^{n}}{(\Delta x)^2}+\frac{\Omega_{i,j+1}^{n}-2\Omega_{i,j}^{n}+\Omega_{i,j-1}^{n}}{(\Delta y)^2}\right]$$

If the conservative form of the equation is used, the convective terms are approximated by

$$\frac{u_{i+1,j}^{n}\Omega_{i+1,j}^{n}-u_{i-1,j}^{n}\Omega_{i-1,j}^{n}}{2\Delta x}+\frac{v_{i,j+1}^{n}\Omega_{i,j+1}^{n}-v_{i,j-1}^{n}\Omega_{i,j-1}^{n}}{2\Delta y}$$

Recall that central difference approximation of convective terms does not model the physics of the problem accurately in that it does not correctly represent the directional influence of a disturbance. Therefore, the use of an upwind differencing scheme may be more appropriate in particular if the flowfield of interest is convection dominated. First-order approximation of the convective terms yields

$$\frac{\Omega_{i,j}^{n+1}-\Omega_{i,j}^{n}}{\Delta t}+\frac{1}{2}(1-\epsilon_x)\frac{u_{i+1,j}^{n}\Omega_{i+1,j}^{n}-u_{i,j}^{n}\Omega_{i,j}^{n}}{\Delta x}+\frac{1}{2}(1+\epsilon_x)\frac{u_{i,j}^{n}\Omega_{i,j}^{n}-u_{i-1,j}^{n}\Omega_{i-1,j}^{n}}{\Delta x}$$

$$+\frac{1}{2}(1-\epsilon_y)\frac{v_{i,j+1}^{n}\Omega_{i,j+1}^{n}-v_{i,j}^{n}\Omega_{i,j}^{n}}{\Delta y}+\frac{1}{2}(1+\epsilon_y)\frac{v_{i,j}^{n}\Omega_{i,j}^{n}-v_{i,j-1}^{n}\Omega_{i,j-1}^{n}}{\Delta y}$$

$$=\frac{1}{Re}\frac{\Omega_{i+1,j}^{n}-2\Omega_{i,j}^{n}+\Omega_{i-1,j}^{n}}{(\Delta x)^2}+\frac{1}{Re}\frac{\Omega_{i,j+1}^{n}-2\Omega_{i,j}^{n}+\Omega_{i,j-1}^{n}}{(\Delta y)^2} \tag{8-96}$$

Note the following features of the finite difference equation given by Equation (8-96) above: (a) If u is positive, a backward approximation must be utilized. Thus, ϵ_x is set to one. If u is negative, a forward approximation is used and therefore ϵ_x is set equal to -1. The same analogy is applied to v and the corresponding coefficient ϵ_y. It is emphasized again that the upwind formulation allows the information to be convected only to the points in the flow direction and, therefore, more appropriately models the physics of the problem. The formulation (8-96) with appropriately defined coefficients provides this option. One may increase the accuracy by incorporating a second-order approximation. For example, a forward approximation would yield

$$\frac{\partial}{\partial x}(u\Omega) = \frac{-3u_{i,j}\Omega_{i,j} + 4u_{i+1,j}\Omega_{i+1,j} - u_{i+2,j}\Omega_{i+2,j}}{2\Delta x}$$

However, note that at the boundary one may have to revert to the first-order scheme. (b) If ϵ_x and ϵ_y are both set to zero, second-order central difference approximation of the convective terms are recovered.

Recall that the implicit formulation of multi-dimensional problems requires approximate factorization to improve their efficiency. To gain a flavor of finite difference formulations of both nonconservative and conservative forms of the governing equations, the nonconservative form is used in the first formulation, whereas the conservative form will be employed for the second finite difference equation. The nonconservative form of the vorticity transport equation is

$$\frac{\partial \Omega}{\partial t} + u\frac{\partial \Omega}{\partial x} + v\frac{\partial \Omega}{\partial y} = \frac{1}{Re}\left(\frac{\partial^2 \Omega}{\partial x^2} + \frac{\partial^2 \Omega}{\partial y^2}\right)$$

The ADI formulation results in the following two equations which are solved sequentially:

$$\frac{(\Omega_{i,j}^{n+\frac{1}{2}} - \Omega_{i,j}^{n})}{\Delta t/2} + u_{i,j}^{n}\frac{\Omega_{i+1,j}^{n+\frac{1}{2}} - \Omega_{i-1,j}^{n+\frac{1}{2}}}{2\Delta x} + v_{i,j}^{n}\frac{\Omega_{i,j+1}^{n} - \Omega_{i,j-1}^{n}}{2\Delta y}$$

$$= \frac{1}{Re}\frac{\Omega_{i+1,j}^{n+\frac{1}{2}} - 2\Omega_{i,j}^{n+\frac{1}{2}} + \Omega_{i-1,j}^{n+\frac{1}{2}}}{(\Delta x)^2} + \frac{1}{Re}\frac{\Omega_{i,j+1}^{n} - 2\Omega_{i,j}^{n} + \Omega_{i,j-1}^{n}}{(\Delta y)^2} \tag{8-97}$$

and

$$\frac{(\Omega_{i,j}^{n+1} - \Omega_{i,j}^{n+\frac{1}{2}})}{\Delta t/2} + u_{i,j}^{n+\frac{1}{2}}\frac{\Omega_{i+1,j}^{n+\frac{1}{2}} - \Omega_{i-1,j}^{n+\frac{1}{2}}}{2\Delta x} + v_{i,j}^{n+\frac{1}{2}}\frac{\Omega_{i,j+1}^{n+1} - \Omega_{i,j-1}^{n+1}}{2\Delta y}$$

$$= \frac{1}{Re}\frac{\Omega_{i+1,j}^{n+\frac{1}{2}} - 2\Omega_{i,j}^{n+\frac{1}{2}} + \Omega_{i-1,j}^{n+\frac{1}{2}}}{(\Delta x)^2} + \frac{1}{Re}\frac{\Omega_{i,j+1}^{n+1} - 2\Omega_{i,j}^{n+1} + \Omega_{i,j-1}^{n+1}}{(\Delta y)^2} \tag{8-98}$$

Note that in linearization of (8-98) one may use the values of u and v at a time level of n. In that case there is no need to solve for the stream function at level $n+\frac{1}{2}$, which would be necessary to provide u and v at $n+\frac{1}{2}$ if the values of u and v at $n+\frac{1}{2}$ are used in Equation (8-98). Obviously, the computation time is reduced if linearization is based on the values of u and v at the n time level. With this argument, Equations (8-97) and (8-98) are rearranged to provide

$$-\frac{1}{2}\left(\frac{1}{2}c_x + d_x\right)\Omega_{i-1,j}^{n+\frac{1}{2}} + (1+d_x)\Omega_{i,j}^{n+\frac{1}{2}} + \frac{1}{2}\left(\frac{1}{2}c_x - d_x\right)\Omega_{i+1,j}^{n+\frac{1}{2}} = D_x \tag{8-99}$$

and

$$-\frac{1}{2}\left(\frac{1}{2}c_y + d_y\right)\Omega_{i,j-1}^{n+1} + (1+d_y)\Omega_{i,j}^{n+1} + \frac{1}{2}\left(\frac{1}{2}c_y - d_y\right)\Omega_{i,j+1}^{n+1} = D_y \tag{8-100}$$

where the Courant numbers and the diffusion numbers are defined as

$$c_x = u\frac{\Delta t}{\Delta x} \quad , \quad c_y = v\frac{\Delta t}{\Delta y}$$

and

$$d_x = \frac{1}{Re}\frac{\Delta t}{(\Delta x)^2} \quad , \quad d_y = \frac{1}{Re}\frac{\Delta t}{(\Delta y)^2}$$

The right-hand sides of (8-99) and (8-100) defined by D_x and D_y, respectively, are

$$D_x = \frac{1}{2}\left(\frac{1}{2}c_y + d_y\right)\Omega^n_{i,j-1} + (1-d_y)\Omega^n_{i,j} + \frac{1}{2}\left(-\frac{1}{2}c_y + d_y\right)\Omega^n_{i,j+1}$$

and

$$D_y = \frac{1}{2}\left(\frac{1}{2}c_x + d_x\right)\Omega^{n+\frac{1}{2}}_{i-1,j} + (1-d_x)\Omega^{n+\frac{1}{2}}_{i,j} + \frac{1}{2}\left(-\frac{1}{2}c_x + d_x\right)\Omega^{n+\frac{1}{2}}_{i+1,j}$$

Note that, based on previous arguments, the values of u and v used in c_x and v_y which are incorporated in Equation (8-100) may be at n or at $n+\frac{1}{2}$. If the latter option is used, the stream function equation must be solved for at $n+\frac{1}{2}$ level, thus requiring additional computation time. Further definition of the coefficients would provide

$$A_x\,\Omega^{n+\frac{1}{2}}_{i-1,j} + B_x\,\Omega^{n+\frac{1}{2}}_{i,j} + C_x\,\Omega^{n+\frac{1}{2}}_{i+1,j} = D_x \tag{8-101}$$

and

$$A_y\,\Omega^{n+1}_{i,j-1} + B_y\,\Omega^{n+1}_{i,j} + C_y\,\Omega^{n+1}_{i,j+1} = D_y \tag{8-102}$$

where

$$A_x = -\frac{1}{2}\left(\frac{1}{2}c_x + d_x\right)$$

$$B_x = 1 + d_x$$

$$C_x = \frac{1}{2}\left(\frac{1}{2}c_x - d_x\right)$$

$$A_y = -\frac{1}{2}\left(\frac{1}{2}c_y + d_y\right)$$

$$B_y = 1 + d_y$$

$$C_y = \frac{1}{2}\left(\frac{1}{2}c_y - d_y\right)$$

The tridiagonal systems given by Equations (8-101) and (8-102) are solved sequentially by the procedure described in Appendix B.

Now consider the conservative form of the equation given by

$$\frac{\partial\Omega}{\partial t} + \frac{\partial}{\partial x}(u\Omega) + \frac{\partial}{\partial y}(v\Omega) = \frac{1}{Re}\left(\frac{\partial^2\Omega}{\partial x^2} + \frac{\partial^2\Omega}{\partial y^2}\right)$$

As mentioned previously, upwind differencing will be used to approximate the convective terms. In order to preserve the tridiagonal nature of the system, first-order approximation is used. As shown in previous chapters, if higher order upwind approximations are utilized, the off tridiagonal elements may be treated explicitly in order to preserve the tridiagonal nature of the system. As in the explicit formulation, coefficients ϵ_x and ϵ_y will be incorporated to provide the option of second-order central difference approximation. The sequential steps required for ADI formulation are

$$\frac{(\Omega_{i,j}^{n+\frac{1}{2}} - \Omega_{i,j}^{n})}{\Delta t/2} + \frac{1}{2}(1-\epsilon_x)\frac{u_{i+1,j}^{n}\Omega_{i+1,j}^{n+\frac{1}{2}} - u_{i,j}^{n}\Omega_{i,j}^{n+\frac{1}{2}}}{\Delta x}$$

$$+\frac{1}{2}(1+\epsilon_x)\frac{u_{i,j}^{n}\Omega_{i,j}^{n+\frac{1}{2}} - u_{i-1,j}^{n}\Omega_{i-1,j}^{n+\frac{1}{2}}}{\Delta x} + \frac{1}{2}(1-\epsilon_y)\frac{v_{i,j+1}^{n}\Omega_{i,j+1}^{n} - v_{i,j}^{n}\Omega_{i,j}^{n}}{\Delta y}$$

$$+\frac{1}{2}(1+\epsilon_y)\frac{v_{i,j}^{n}\Omega_{i,j}^{n} - v_{i,j-1}^{n}\Omega_{i,j-1}^{n}}{\Delta y}$$

$$= \frac{1}{Re}\frac{\Omega_{i+1,j}^{n+\frac{1}{2}} - 2\Omega_{i,j}^{n+\frac{1}{2}} + \Omega_{i-1,j}^{n+\frac{1}{2}}}{(\Delta x)^2} + \frac{1}{Re}\frac{\Omega_{i,j+1}^{n} - 2\Omega_{i,j}^{n} + \Omega_{i,j-1}^{n}}{(\Delta y)^2}$$

which is rearranged as

$$-\frac{1}{2}\left[\frac{1}{2}(1+\epsilon_x)c_{x(i-1,j)} + d_x\right]\Omega_{i-1,j}^{n+\frac{1}{2}} + \left(1 + d_x + \frac{1}{2}\epsilon_x c_{x(i,j)}\right)\Omega_{i,j}^{n+\frac{1}{2}}$$

$$+\frac{1}{2}\left[\frac{1}{2}(1-\epsilon_x)c_{x(i+1,j)} - d_x\right]\Omega_{i+1,j}^{n+\frac{1}{2}} = D_x \qquad \text{(8-103)}$$

where

$$D_x = \frac{1}{2}\left[\frac{1}{2}(1+\epsilon_y)c_{y(i,j-1)} + d_y\right]\Omega_{i,j-1}^{n} + (1 - d_y - \frac{1}{2}\epsilon_y c_{y(i,j)})\Omega_{i,j}^{n}$$

$$+\frac{1}{2}\left[-\frac{1}{2}(1-\epsilon_y)c_{y(i,j+1)} + d_y\right]\Omega_{i,j+1}^{n}$$

and

$$\frac{(\Omega_{i,j}^{n+1} - \Omega_{i,j}^{n+\frac{1}{2}})}{\Delta t/2} + \frac{1}{2}(1-\epsilon_x)\frac{u_{i+1,j}^{n}\Omega_{i+1,j}^{n+\frac{1}{2}} - u_{i,j}^{n}\Omega_{i,j}^{n+\frac{1}{2}}}{\Delta x}$$

$$+\frac{1}{2}(1+\epsilon_x)\frac{u_{i,j}^{n}\Omega_{i,j}^{n+\frac{1}{2}} - u_{i-1,j}^{n}\Omega_{i-1,j}^{n+\frac{1}{2}}}{\Delta x} + \frac{1}{2}(1-\epsilon_y)\frac{v_{i,j+1}\Omega_{i,j+1}^{n+1} - v_{i,j}\Omega_{i,j}^{n+1}}{\Delta y}$$

$$+\frac{1}{2}(1+\epsilon_y)\frac{v_{i,j}\Omega_{i,j}^{n+1} - v_{i,j-1}\Omega_{i,j-1}^{n+1}}{\Delta y}$$

$$= \frac{1}{Re}\frac{\Omega_{i+1,j}^{n+\frac{1}{2}} - 2\Omega_{i,j}^{n+\frac{1}{2}} + \Omega_{i-1,j}^{n+\frac{1}{2}}}{(\Delta x)^2} + \frac{1}{Re}\frac{\Omega_{i,j+1}^{n+1} - 2\Omega_{i,j}^{n+1} + \Omega_{i,j-1}^{n+1}}{(\Delta y)^2}$$

which is rearranged as

$$-\frac{1}{2}\left[\frac{1}{2}(1+\epsilon_y)c_{y(i,j-1)}+d_y\right]\Omega_{i,j-1}^{n+1}+\left(1+d_y+\frac{1}{2}\epsilon_y c_{y(i,j)}\right)\Omega_{i,j}^{n+1}$$
$$+\frac{1}{2}\left[\frac{1}{2}(1-\epsilon_y)c_{y(i,j+1)}-d_y\right]\Omega_{i,j+1}^{n+1}=D_y \qquad \text{(8-104)}$$

where

$$D_y=\frac{1}{2}\left[\frac{1}{2}(1+\epsilon_x)c_{x(i-1,j)}+d_x\right]\Omega_{i-1,j}^{n+\frac{1}{2}}+(1-d_x-\frac{1}{2}\epsilon_x c_{x(i,j)})\Omega_{i,j}^{n+\frac{1}{2}}$$
$$+\frac{1}{2}\left[-\frac{1}{2}(1-\epsilon_x)c_{x(i+1,j)}+d_x\right]\Omega_{i+1,j}^{n+\frac{1}{2}}$$

As discussed previously, the Courant numbers in Equation (8-104) may be evaluated at the $n+\frac{1}{2}$ level in which case the stream function equation at $n+\frac{1}{2}$ needs to be solved subsequent to the solution of Equation (8-103).

8.8.2 Stream Function Equation

The stream function equation is given by

$$\frac{\partial^2\psi}{\partial x^2}+\frac{\partial^2\psi}{\partial y^2}=-\Omega \qquad \text{(8-105)}$$

which is classified as an elliptic equation. Any numerical scheme introduced in Chapter 5 can be used for its solution. For example, the point Gauss-Seidel formulation yields

$$\psi_{i,j}^{k+1}=\frac{1}{2(1+\beta^2)}\left[(\Delta x)^2\Omega_{i,j}^{n+1}+\psi_{i+1,j}^{k}+\psi_{i-1,j}^{k+1}+\beta^2(\psi_{i,j+1}^{k}+\psi_{i,j-1}^{k+1})\right] \qquad \text{(8-106)}$$

where $\beta=\dfrac{\Delta x}{\Delta y}$.

The computation begins with the solution of the vorticity equation within the domain. Subsequently, the vorticity in Equation (8-106) is updated and the equation is solved for the stream function ψ. The process is repeated until the desired solution is reached.

8.9 Boundary Conditions

In order to solve the vorticity transport and the stream function equations by the numerical scheme described in the previous section, boundary conditions must be prescribed. Generally speaking, boundary conditions are categorized into the following five groups: body surface, far-field, symmetry line, inflow, and outflow boundaries. The specification of boundary conditions for the primitive variable in each category was discussed in Section 8.6. In this section the boundary conditions for the stream function and vorticity are explored.

8.9.1 Body Surface

A solid surface can be considered as a stream line and, therefore, the stream function is constant and its value may be assigned arbitrarily. As mentioned previously, boundary conditions for the vorticity do not exist. Therefore, a set of boundary conditions must be constructed. The procedure involves the stream function equation along with Taylor series expansion of the stream function. As a result, a different formulation with various orders of approximation can be derived. At this point, the construction of a first-order expression is illustrated. Subsequently, a second-order relation is provided. For illustration purposes, assume non-porous and stationary surfaces and a rectangular domain, as shown in Fig. 8-9. The expression for the vorticity to be applied at boundary A is determined initially and, subsequently, the result is extended to the other boundaries at B, C, and D.

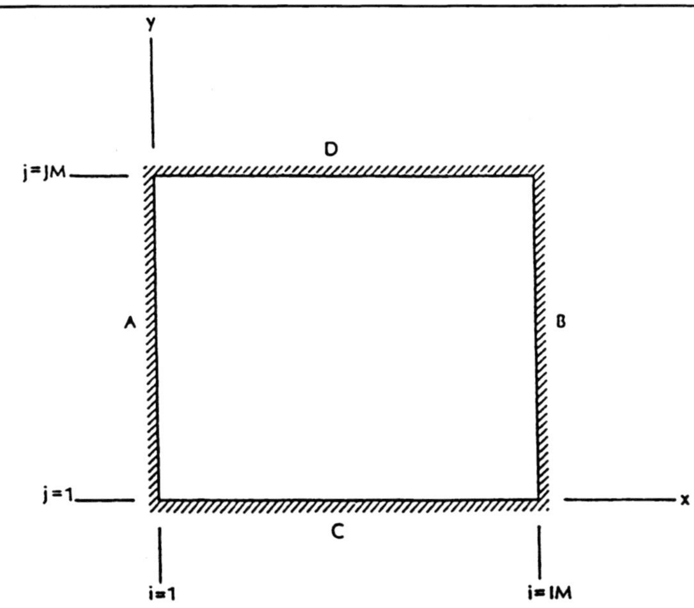

Figure 8-9. The rectangular cavity with fixed surfaces.

Consider Equation (8-105) at point $(1, j)$, i.e.,

$$\left(\frac{\partial^2\psi}{\partial x^2} + \frac{\partial^2\psi}{\partial y^2}\right)_{1,j} = -\Omega_{1,j} \tag{8-107}$$

Along the surface, the stream function is constant, and its value may be specified arbitrarily; for example, $\psi_{1,j} = \psi_1$. Then, along A,

$$\left.\frac{\partial^2\psi}{\partial y^2}\right|_{1,j} = 0$$

and Equation (8-107) is then reduced to

$$\left.\frac{\partial^2\psi}{\partial x^2}\right|_{1,j} = -\Omega_{1,j} \tag{8-108}$$

To obtain an expression for the second-order derivative in the equation above, consider the Taylor series expansion

$$\psi_{2,j} = \psi_{1,j} + \left.\frac{\partial\psi}{\partial x}\right|_{1,j}\Delta x + \left.\frac{\partial^2\psi}{\partial x^2}\right|_{1,j}\frac{(\Delta x)^2}{2} + \cdots$$

Along boundary A

$$v_{1,j} = -\left.\frac{\partial\psi}{\partial x}\right|_{1,j} = 0$$

Therefore,

$$\psi_{2,j} = \psi_{1,j} + \left.\frac{\partial^2\psi}{\partial x^2}\right|_{1,j}\frac{(\Delta x)^2}{2} + O(\Delta x)^3$$

from which

$$\left.\frac{\partial^2\psi}{\partial x^2}\right|_{1,j} = \frac{2(\psi_{2,j} - \psi_{1,j})}{(\Delta x)^2} + O(\Delta x) \tag{8-109}$$

Substitution of (8-109) into (8-108) yields

$$\Omega_{1,j} = \frac{2(\psi_{1,j} - \psi_{2,j})}{(\Delta x)^2} \tag{8-110}$$

A similar procedure is used to derive the boundary conditions at boundaries B, C, and D. The appropriate expressions are, respectively,

$$\Omega_{IM,j} = -\left.\frac{\partial^2\psi}{\partial x^2}\right|_{IM,j} = \frac{2(\psi_{IM,j} - \psi_{IMM1,j})}{(\Delta x)^2} \tag{8-111}$$

$$\Omega_{i,1} = -\left.\frac{\partial^2\psi}{\partial y^2}\right|_{i,1} = \frac{2(\psi_{i,1} - \psi_{i,2})}{(\Delta y)^2} \tag{8-112}$$

$$\Omega_{i,JM} = -\left.\frac{\partial^2\psi}{\partial y^2}\right|_{i,JM} = \frac{2(\psi_{i,JM} - \psi_{i,JMM1})}{(\Delta y)^2} \tag{8-113}$$

Now suppose a boundary is moving with some specified velocity. For example, assume that the upper surface is moving to the right with a constant velocity u_0. Following the procedure described previously, the Taylor series expansion yields

$$\psi_{i,j-1} = \psi_{i,j} - \left.\frac{\partial\psi}{\partial y}\right|_{i,j}\Delta y + \left.\frac{\partial^2\psi}{\partial y^2}\right|_{i,j}\frac{(\Delta y)^2}{2} + \cdots$$

or

$$\psi_{i,JMM1} = \psi_{i,JM} - u_0\Delta y - \Omega_{i,JM}\frac{(\Delta y)^2}{2}$$

from which

$$\Omega_{i,JM} = \frac{2(\psi_{i,JM} - \psi_{i,JMM1})}{(\Delta y)^2} - \frac{2u_0}{\Delta y} \tag{8-114}$$

A second-order equivalent of (8-110) can be expressed as

$$\Omega_{1,j} = \frac{\psi_{3,j} - 8\psi_{2,j} + 7\psi_{1,j}}{2(\Delta x)^2} + O(\Delta x)^2 \tag{8-115}$$

For a moving boundary with a constant velocity of u_0 at $j = JM$, one has

$$\Omega_{i,JM} = \frac{-\psi_{i,JM} + 8\psi_{i,JMM1} - 7\psi_{i,JMM2}}{2(\Delta y)^2} - \frac{3u_0}{\Delta y} \tag{8-116}$$

which is a second-order equivalence of (8-114). Higher order implementation of the boundary conditions in general will increase the accuracy of the solution. However, it has been shown that it may cause instabilities in high Reynolds number flow. A comparative study of the first-order and second-order boundary conditions is reported in Reference [8-7].

8.9.2 Far-Field

For a true far-field boundary which is set parallel to the free stream, the boundary represents a streamline. Therefore, a constant value for the stream function along this boundary can be specified. However, the assignment of a value for the stream function along various boundaries must be consistent with respect to the continuity equation. After all, recall that the difference between the values of stream function represent mass (or volumetric) flow rate. This consideration is important in particular for domains where there are multiple ports for inflow and outflow.

8.9.3 Symmetry

When the symmetry boundary is set along the axis of symmetry and the flow is truly symmetrical, it can be considered a streamline. Therefore, the value of the stream function along this boundary can be specified. Keep in mind the previous discussion on the relation for the values of the stream function and the continuity equation. Obviously, the velocity components normal to the symmetry boundary would be zero, whereas generally the streamwise component is extrapolated from the interior solution.

8.9.4 Inflow

At the inflow the stream function is usually specified whereas the vorticity is determined with the various approximations. First, consider the stream function at the inflow:

(a) The values of ψ along the inflow are specified

(b) Its value is determined from the interior; for example,

$$\psi_{1,j} = \frac{1}{3}(4\psi_{2,j} - \psi_{3,j}) \tag{8-117}$$

The vorticity at the inflow may be determined by either one of the following procedures:

(a) The equation for the stream function is applied at point $(1, j)$ to yield

$$\left(\frac{\partial^2\psi}{\partial x^2} + \frac{\partial^2\psi}{\partial y^2}\right)_{1,j} = -\Omega_{1,j}$$

Utilizing approximation (8-109) and a second-order central difference expression for the y-derivative, one has

$$\Omega_{1,j} = \frac{2(\psi_{1,j} - \psi_{2,j})}{(\Delta x)^2} - \frac{\psi_{1,j+1} - 2\psi_{1,j} + \psi_{1,j-1}}{(\Delta y)^2} \tag{8-118}$$

(b) The specified value of u at the inflow may be used directly by incorporation of the vorticity. Recall that

$$\Omega = \frac{\partial v}{\partial x} - \frac{\partial u}{\partial y}$$

or

$$\Omega = -\frac{\partial^2\psi}{\partial x^2} - \frac{\partial u}{\partial y}$$

This equation is used at $(1, j)$ to evaluate the vorticity at the inflow. A second-order forward difference approximation utilizing the interior values of the stream function is used for $\partial^2\psi/\partial x^2$, whereas a second-order central difference approximation is used for the velocity gradient. Thus, one has

$$\Omega_{1,j} = -\frac{-3\psi_{1,j} + 4\psi_{2,j} - \psi_{3,j}}{(\Delta x)^2} - \frac{u_{1,j+1} - u_{1,j-1}}{2\Delta y} \tag{8-119}$$

8.9.5 Outflow

At the outflow boundary, the value of the stream function is usually extrapolated from the interior solution. Utilizing $\partial\psi/\partial x = 0$, a second-order approximation will provide

$$\psi_{IM,j} = \frac{1}{3}(4\psi_{IM-1,j} - \psi_{IM-2,j}) \tag{8-120}$$

The condition $\partial^2\psi/\partial x^2 = 0$ has also been used at the outflow boundary, as well. A second-order backward approximation yields

$$\psi_{IM,j} = \frac{1}{2}(\psi_{IM-3,j} - 4\psi_{IM-2,j} + 5\psi_{IM-1,j}) \tag{8-121}$$

whereas a first-order approximation yields

$$\psi_{IM,j} = -\psi_{IM-2,j} + 2\psi_{IM-1,j} \tag{8-122}$$

As in the case of the inflow boundary, the vorticity at the outflow may be determined by numerous methods; for example,

$$\Omega_{IM,j} = \frac{2(\psi_{IM,j} - \psi_{IMM1,j})}{(\Delta x)^2} - \frac{(\psi_{IM,j+1} - 2\psi_{IM,j} + \psi_{IM,j-1})}{(\Delta y)^2} \tag{8-123}$$

A simple extrapolation may be used for which one sets

$$\frac{\partial\Omega}{\partial x} = 0$$

to provide

$$\Omega_{IM,j} = \frac{1}{3}(4\Omega_{IM-1,j} - \Omega_{IM-2,j}) \tag{8-124}$$

8.10 Application

Consider the rectangular cavity shown in Figure 8-10 where the upper plate moves to the right with a velocity of u_0. In addition, an inlet on the left boundary and an outlet at the right boundary are specified. Note that, when the inlet and the outlet are shut, the problem is reduced to the classical driven cavity problem.

The boundary conditions are specified as follows (the nomenclature is shown in Figure 8-10).

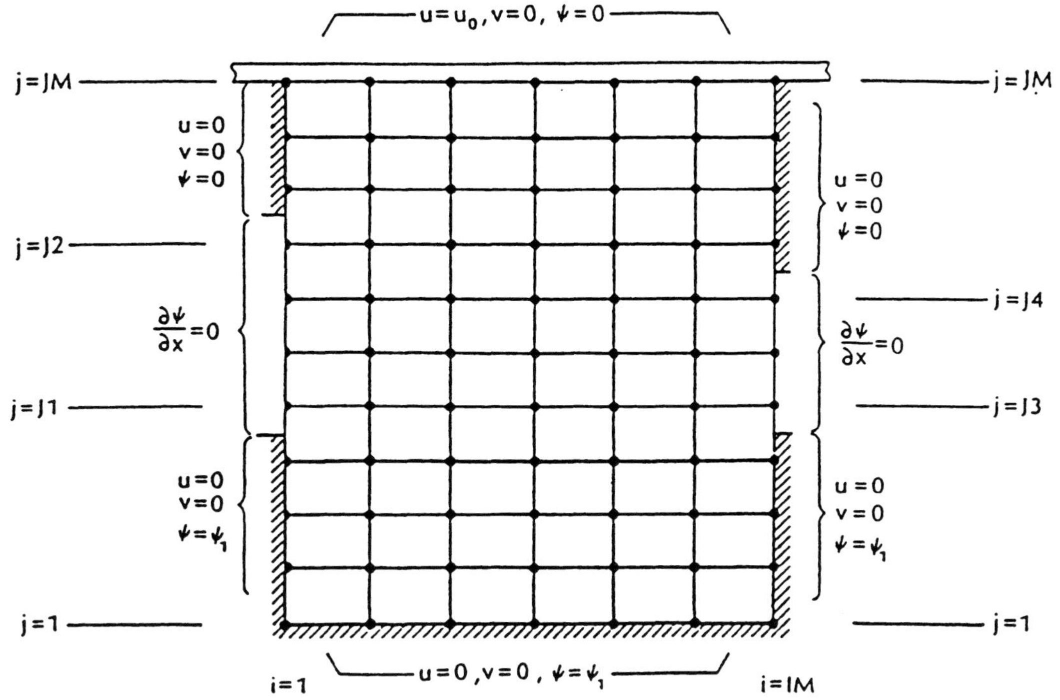

Figure 8-10. The rectangular cavity and the specified boundary conditions.

$$i = 1 \quad : \quad \begin{cases} j = 1 \text{ to } J1 - 1 \quad : & \begin{cases} \psi = \psi_1 \\ u = 0 \\ v = 0 \end{cases} \\ j = J1 \text{ to } J2 \quad : & \left\{ \dfrac{\partial \psi}{\partial x} = 0 \quad (v = 0) \right. \\ j = J2 + 1 \text{ to } JM \quad : & \begin{cases} \psi = 0 \\ u = 0 \\ v = 0 \end{cases} \end{cases}$$

$$i = IM \quad : \quad \begin{cases} j = 1 \text{ to } J3 - 1 \quad : & \begin{cases} \psi = \psi_1 \\ u = 0 \\ v = 0 \end{cases} \\ j = J3 \text{ to } J4 \quad : & \begin{cases} \dfrac{\partial \psi}{\partial x} = 0 \quad (v = 0) \end{cases} \\ j = J4 + 1 \text{ to } JM \quad : & \begin{cases} \psi = 0.0 \\ u = 0 \\ v = 0 \end{cases} \end{cases}$$

$$j = 1 \quad : \quad i = 1 \text{ to } IM \quad : \quad \begin{cases} \psi = \psi_1 \\ u = 0 \\ v = 0 \end{cases}$$

$$j = JM \quad : \quad i = 1 \text{ to } IM \quad : \quad \begin{cases} \psi = 0 \\ u = u_0 \\ v = 0 \end{cases}$$

The kinematic viscosity ν is specified as 0.0025 m^2/sec, the velocity of the upper plate is 5 m/sec, and the cavity is 30 cm by 30 cm. Furthermore,

$$IM = 31 \quad , \quad JM = 31$$
$$J1 = 20 \quad , \quad J2 = 25$$
$$J3 = 5 \quad , \quad J4 = 10$$

Thus, the spatial step sizes are

$$\Delta x = 0.01 \text{ m} \quad \text{and} \quad \Delta y = 0.01 \text{ m}$$

Recall that the FTCS explicit scheme may be used to solve Equation (8-28). Therefore, the numerical scheme is subject to a stability requirement which, for a one-dimensional problem, was provided in Chapter Four. In particular, the stability requirement imposes restrictions on the Courant and Cell Reynolds numbers. A time step of 0.001 sec satisfies the stability requirement and is used in this example.

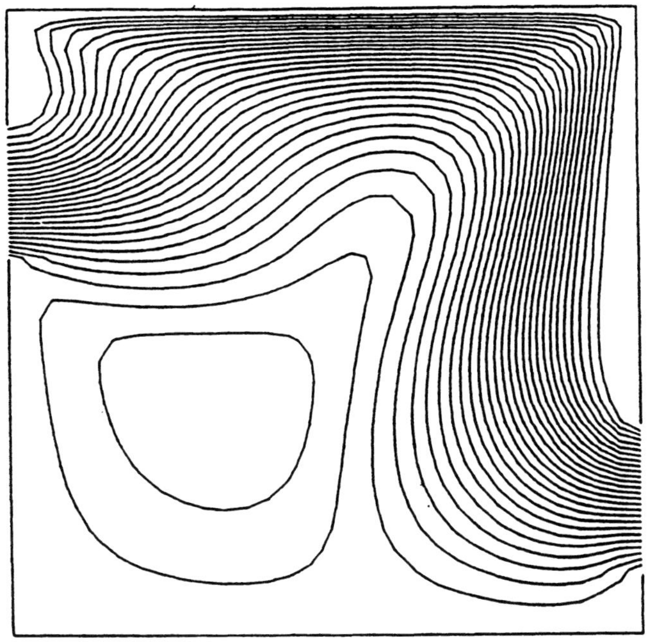

Figure 8-11. The streamline pattern within the rectangular cavity.

The convergence criterion for Equation (8-106) is set by CONGS = 0.001; and, to achieve a steady-state solution, CONSS is set to 0.002.

The streamlines (lines of constant ψ) are shown in Figure 8-11, whereas the velocity vector plot is illustrated in Figure 8-12. A large circulation region within the domain is clearly evident. The solution converged after 123 iterations, where the computation time on a Cyber 175 was 750 seconds.

8.11 Temperature Field

As discussed previously, the energy equation for an incompressible flow with constant properties is decoupled from the continuity and momentum equations. Therefore, the energy equation can be solved subsequent to the computation of the velocity field to provide the temperature distribution if required. In the following sections, the governing equation, numerical schemes, and the boundary conditions are reviewed.

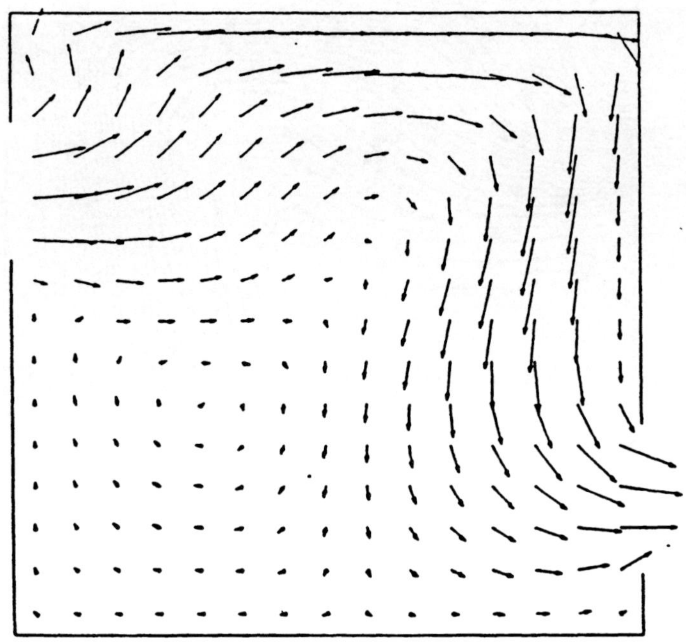

Figure 8-12. The velocity vector plot for the rectangular cavity.

8.11.1 The Energy Equation

The energy equation can be written in various forms depending on the specific form of the energy used. Typically it is written in terms of either one of the following , to name a few: enthalpy, total energy, internal energy, kinetic energy or temperature. Usually the energy equation expressed in terms of the temperature is used for incompressible flow calculations. The energy equation, with no heat generation, and constant flow properties along with the Fourier heat conduction law and Newtonian fluid for a two-dimensional incompressible flow, is expressed as

$$\rho c_p \left(\frac{\partial T}{\partial t} + u \frac{\partial T}{\partial x} + v \frac{\partial T}{\partial y} \right)$$
$$= k \left(\frac{\partial^2 T}{\partial x^2} + \frac{\partial^2 T}{\partial y^2} \right) + \mu \left\{ 2 \left[\left(\frac{\partial u}{\partial x} \right)^2 + \left(\frac{\partial v}{\partial y} \right)^2 \right] + \left(\frac{\partial u}{\partial y} + \frac{\partial v}{\partial x} \right)^2 \right\} \quad (8\text{-}125)$$

The last term is known as the viscous dissipation term and is usually defined by

$$\phi = 2\left[\left(\frac{\partial u}{\partial x}\right)^2 + \left(\frac{\partial v}{\partial y}\right)^2\right] + \left(\frac{\partial u}{\partial y} + \frac{\partial v}{\partial x}\right)^2 \tag{8-126}$$

Thus, one may write

$$\rho c_p \left(\frac{\partial T}{\partial t} + u\frac{\partial T}{\partial x} + v\frac{\partial T}{\partial y}\right) = k\nabla^2 T + \mu\phi \tag{8-127}$$

Observe that Equation (8-127) is written in a nonconservative form. The conservative form is expressed as

$$\rho c_p \left[\frac{\partial T}{\partial t} + \frac{\partial}{\partial x}(uT) + \frac{\partial}{\partial y}(vT)\right] = k\nabla^2 T + \mu\phi \tag{8-128}$$

The energy equation in either form may be written in terms of the thermal diffusivity and kinematic viscosity. For example,

$$\frac{\partial T}{\partial t} + u\frac{\partial T}{\partial x} + v\frac{\partial T}{\partial y} = \alpha\nabla^2 T + \frac{\nu}{c_p}\phi \tag{8-129}$$

Recall that the thermal diffusivity and kinematic viscosity are, respectively, defined as

$$\alpha = \frac{k}{\rho c_p} \quad \text{and} \quad \nu = \frac{\mu}{\rho}$$

The variables in the energy equation are nondimensionalized according to relation (8-21). To nondimensionalize the temperature various reference quantities can be used, among which are the freestream temperature T_∞ or the difference in temperature $(T_w - T_\infty)$. Implementation of $T^* = T/T_\infty$ yields

$$\frac{\partial T^*}{\partial t^*} + u^*\frac{\partial T^*}{\partial x^*} + v^*\frac{\partial T^*}{\partial y^*} = \frac{1}{Pr\,Re}\left(\frac{\partial^2 T^*}{\partial x^{*2}} + \frac{\partial^2 T^*}{\partial y^{*2}}\right) + \frac{(\gamma - 1)M_\infty^2}{Re}\phi^* \tag{8-130}$$

whereas $T^* = \dfrac{T}{T_w - T_\infty}$ yields

$$\frac{\partial T^*}{\partial t^*} + u^*\frac{\partial T^*}{\partial x^*} + v^*\frac{\partial T^*}{\partial y^*} = \frac{1}{Pr\,Re}\left(\frac{\partial^2 T^*}{\partial x^{*2}} + \frac{\partial^2 T^*}{\partial y^{*2}}\right) + \frac{E}{Re}\phi^* \tag{8-131}$$

where $E = \dfrac{u_\infty}{c_p(T_w - T_\infty)}$ is called the Eckert number.

Note that the coefficient of the dissipation term E/Re can be manipulated to provide

$$\frac{E}{Re} = \frac{(\gamma - 1)M_\infty^2}{\left(\frac{T_w}{T_\infty} - 1\right)Re} \tag{8-132}$$

Therefore, Equation (8-131) can also be expressed as

$$\frac{\partial T^*}{\partial t^*} + u^* \frac{\partial T^*}{\partial x^*} + v^* \frac{\partial T^*}{\partial y^*} = \frac{1}{PrRe}\left(\frac{\partial^2 T^*}{\partial x^{*2}} + \frac{\partial^2 T^*}{\partial y^{*2}}\right) + \frac{(\gamma - 1)M_\infty^2}{\left(\frac{T_w}{T_\infty} - 1\right) Re}\phi^* \tag{8-133}$$

Before proceeding to the discussion of numerical schemes, a few comments are in order.

a. The energy equation (in either form, such as (8-130), (8-131), or (8-133)) is linear. Note that the velocity field has been evaluated previously and is fixed for the time level for which the temperature is being computed.

b. Any procedure described in Chapter 7 may be used to solve the energy equation. Furthermore, observe the similarity of the energy equation and the vorticity transport equation discussed in Section 8.2.2. Thus, the numerical procedure reviewed in Section 8.8.1 can be extended to the energy equation.

c. The retention of the viscous dissipation term does not create any difficulty, because it may be computed easily from the velocity field prior to the solution of the energy equation. However, for some applications, the viscous dissipation term is relatively small and may be dropped from the energy equation. A simple order of magnitude analysis of relation (8-132) will provide a guideline on the retention or deletion of the viscous dissipation term. Since M_∞ for an incompressible flow is relatively small, relation (8-132) indicates that E would be small as well if $T_w/T_\infty \gg 1$. On the other hand, it is observed that the value of E (and subsequently of viscous dissipation) would be appreciable if $(T_w/T_\infty - 1)$ is a small quantity. Therefore, the viscous dissipation term is retained. It is strongly recommended that the viscous dissipation be retained if there is any doubt on the magnitude of the term.

8.11.2 Numerical Schemes

Consider the energy equation given by, for example, Equation (8-131). Note that the equation is parabolic in time, linear, and involves only one unknown, namely T^*. Any of the numerical procedures for the solution of parabolic equations described previously may be used to solve the energy equation. Since these procedures were extensively described and revisited in conjunction with the vorticity transport equation, no further discussion is warranted at this point. However, some deliberation of the boundary conditions are appropriate.

8.11.3 Boundary Conditions

The boundary conditions along the inflow and far-field are typically specified as that of the temperature of the freestream. The value at the outflow may be extrapolated from the interior solution similar to that of the velocity. The treatment of the boundary condition at the solid surface, however, warrants some deliberation. Typically two types of boundary conditions are frequently specified. In the first category, the wall temperature (or its distribution along the wall) is provided. Thus, the boundary condition is a Dirichlet type. Implementation of such a boundary condition is straightforward. In the second category, an adiabatic wall is specified requiring zero heat transfer at the surface. At this point a brief review is in order. Recall that the heat transfer rate is frequently expressed in terms of the heat transfer coefficient, h, given by

$$h = \frac{-k\left(\dfrac{\partial T}{\partial n}\right)}{T_w - T_\infty} \tag{8-134}$$

where n designates the direction normal to the surface and k is the thermal conductivity. Furthermore, a popular nondimensional heat transfer parameter is the Nusselt number which is defined as

$$NU = \frac{hL}{k} \tag{8-135}$$

Nondimensionalization of (8-134) along with the use of (8-135) yields

$$NU = -\frac{\partial T^*}{\partial n^*} \tag{8-136}$$

To incorporate the adiabatic boundary condition, instead of directly setting the normal gradient to zero, consider the following Taylor series expansion, where index k represents grid points along the surface normal,

$$T_{k+1} = T_k + \frac{\partial T}{\partial n}\Delta n + \frac{\partial^2 T}{\partial n^2}\frac{(\Delta n)^2}{2} + O(\Delta n)^3 \tag{8-137}$$

where the superscript $*$ has been dropped. Utilizing (8-136) one has

$$T_{k+1} = T_k - (NU)\Delta n + O(\Delta n)^2$$

Now, for an adiabatic wall, the heat transfer is zero, implying that $NU = 0$, and therefore,

$$T_{k+1} = T_k + O(\Delta n)^2$$

Application of the adiabatic wall condition to the surface shown in Fig. 8-13 provides

$$T_{i,1} = T_{i,2} \tag{8-138}$$

$$T_{1,j} = T_{2,j} \tag{8-139}$$

An important point to emphasize here is that the approximations (8-138) and (8-139) are indeed second-order accurate!

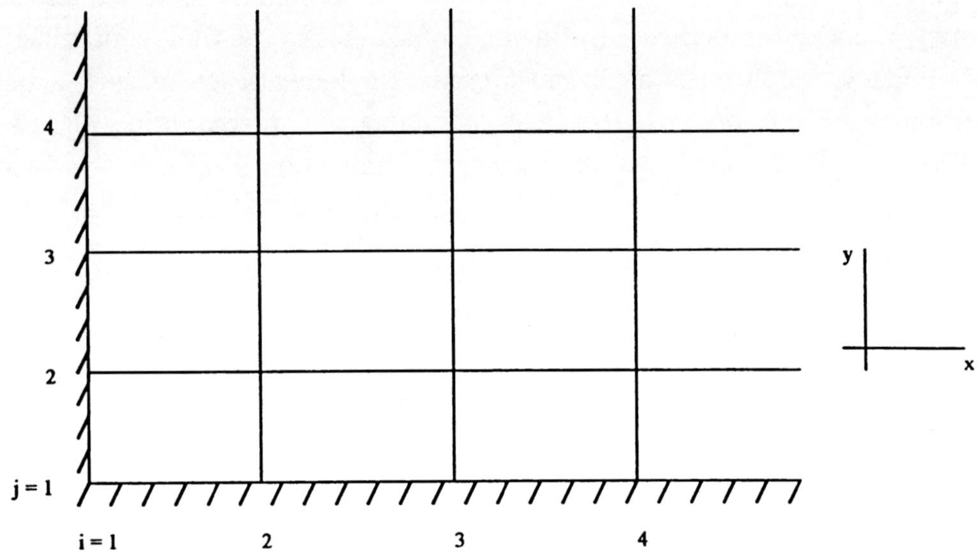

Figure 8-13.

8.12 Problems

8.1 The upper plate of a rectangular cavity shown in Figure P8-1 moves to the right with a velocity of u_o.

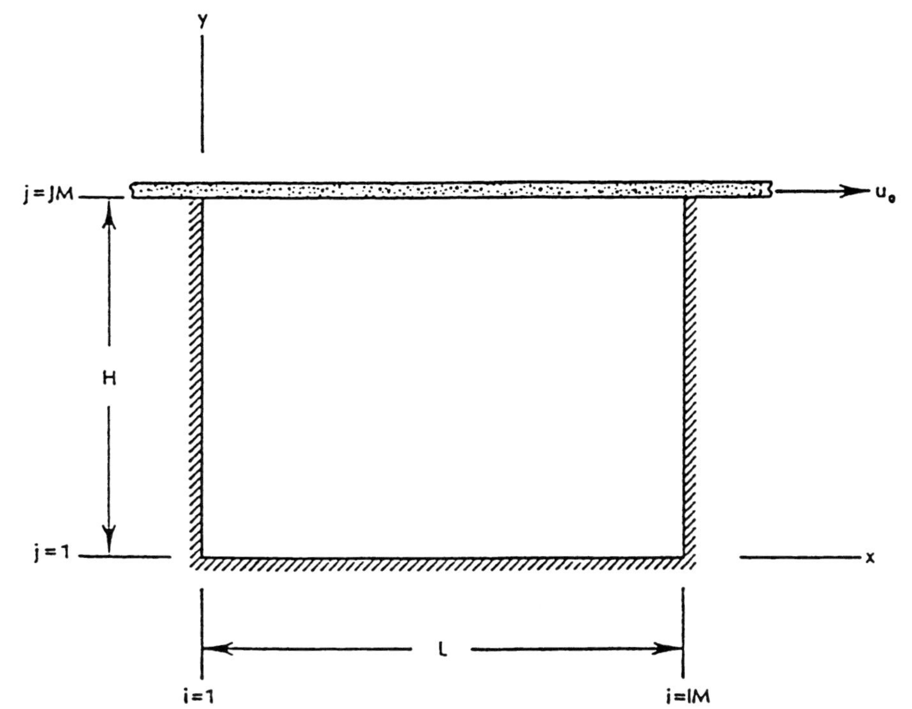

Figure P8-1. The domain of solution for the driven cavity problem.

The rectangular cavity has dimensions of L by H. The fluid within the cavity has a kinematic viscosity of 0.0025 m^2/sec. Use the FTCS explicit scheme and the point Gauss-Seidel formulation to solve for the vorticity and the stream function equations, respectively. The following data are specified

$$L = 40 \text{ cm} , \quad \Delta x = 0.01 \text{ m} , \quad IM = 41$$
$$H = 30 \text{ cm} , \quad \Delta y = 0.01 \text{ m} , \quad JM = 31$$
$$u_o = 5 \text{ m/sec.} , \quad \Delta t = 0.001 \text{ sec.}$$

Use a convergence criterion of 0.001 for the point Gauss-Seidel formulation and set the convergence to steady state at 0.002 for the vorticity equation.

Print the steady-state solution at increments of 0.1 m and 0.05 m in the x- and y-directions, respectively. The values of the vorticity and stream function, and the x-component of the velocity are to be printed. Plot the streamline pattern and the velocity field.

Chapter 9

Grid Generation – Structured Grids

9.1 Introductory Remarks

In order to numerically solve the governing partial differential equations (PDEs) of fluid mechanics, approximations to the partial differentials are introduced. These approximations convert the partial derivatives to finite difference expressions, which are used to rewrite the PDEs as algebraic equations. The approximate algebraic equations, referred to as finite difference equations (FDEs), are subsequently solved at discrete points within the domain of interest. Therefore, a set of grid points within the domain, as well as the boundaries of the domain, must be specified.

Typically, the computational domain is selected to be rectangular in shape where the interior grid points are distributed along grid lines. Therefore, the grid points can be identified easily with reference to the appropriate grid lines. This type grid is known as the *structured grid* and is the focus of this chapter. A second category of grid system may be constructed where the grid points cannot be associated with orderly defined grid lines. This type grid system is known as the *unstructured grid* and will be introduced in Chapter 15. In the remainder of this chapter, *structured grid generation* is implied wherever we refer to grid generation.

The generation of a grid, with uniform spacing, is a simple exercise within a rectangular physical domain. Grid points may be specified as coincident with the boundaries of the physical domain, thus making specification of boundary conditions considerably less complex. In the previous chapters, this restriction on the physical domain was enforced, i.e., all applications were limited to rectangular-type domains.

Unfortunately, the majority of the physical domains of interest are nonrectangular. Therefore, imposing a rectangular computational domain on such a physical domain will require some sort of interpolation for the implementation of the boundary conditions. Since the boundary conditions have a dominant influence on the solution of the equation, such an interpolation causes inaccuracies at the place of greatest sensitivity. In addition, unequal grid spacing near the boundaries creates further complications with the FDEs since approximations with nonequal stepsizes must be used. This form of the FDE changes from node to node, creating cumbersome programming details.

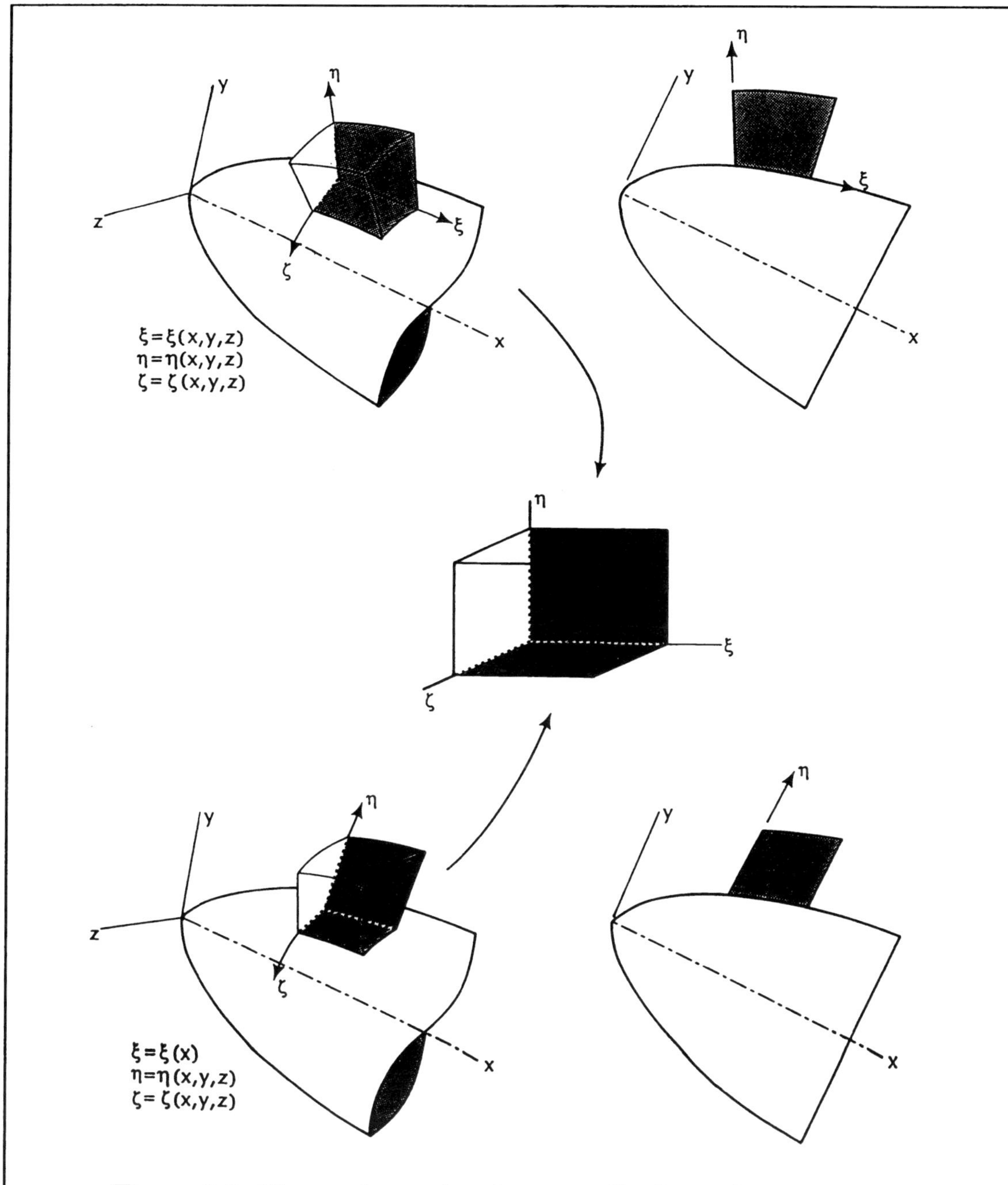

Figure 9-1. Nomenclature for the generalized coordinate system.

To overcome these difficulties, a transformation from physical space to computational space is introduced. This transformation is accomplished by specifying a generalized coordinate system which will map the nonrectangular grid system in the physical space to a rectangular uniform grid spacing in the computational space. The generalized coordinate system may be expressed in many ways. Two examples are illustrated in Figure 9-1. The first example shows the so-called body-fitted coordinate system where two coordinate lines, ξ and ζ, are aligned on the surface along the streamwise and circumferential directions, and where the third coordinate, η, is normal to the surface. In the second example, the ξ coordinate is aligned along the body axis, ζ is in the circumferential direction and η is normal to the body axis.

For illustrative purposes, two-dimensional (2-D) problems will be considered in detail, with a description of the three-dimensional (3-D) problems to follow. A 2-D domain is illustrated in Figure 9-2.

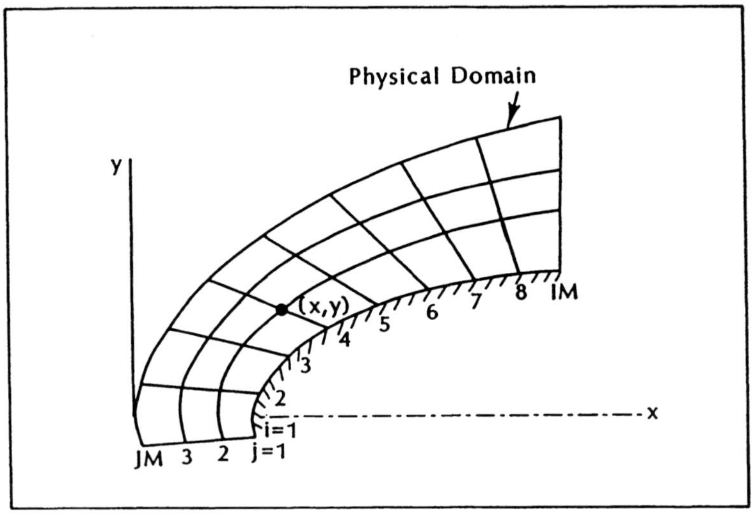

Figure 9-2. A typical 2-D domain for the axisymmetric blunt body problem.

In order to eliminate the difficulty associated with the nonequal stepsizes used in the FDEs, the physical domain is transformed into a rectangular, constant step-size, computational domain. A typical computational domain is shown in Figure 9-3. Note that the computational domain is obtained by deformation of the physical domain, i.e., by twisting and stretching, etc.

The central issue at this point is identifying the location of the grid points in the physical domain. That is, what are the x and y coordinates of a grid point in physical space (Figure 9-2) that correspond to a grid point (i, j) specified in the computational domain (Figure 9-3)? In determining the grid points, a few constraints must be imposed. First, the mapping must be one-to-one; i.e., grid lines of the same family cannot cross each other. Second, from a numerical point of view, a smooth grid point distribution with minimum grid line skewness, orthogonality or near orthogonality, and a concentration of grid points in regions where high flow gradients occur are required. Obviously, all of the desired features may not be met by the use of a particular grid generation technique. Grid generation techniques which emphasize any one or a combination of these features will be presented in the following sections.

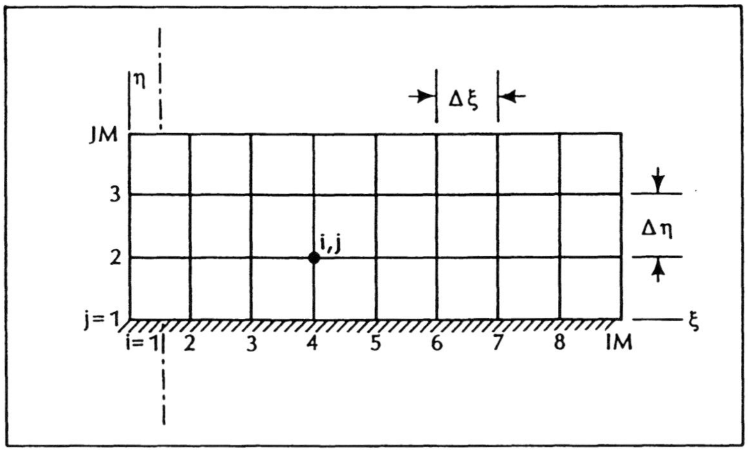

Figure 9-3. Computational domain with constant stepsizes $\Delta\xi$ and $\Delta\eta$.

In general, grid generation techniques may be classified as (1) algebraic methods, (2) partial differential methods, or (3) conformal mappings based on complex variables. In addition, the grid system may be categorized as fixed or adaptive. Obviously, a fixed grid system is generated prior to the solution of the governing equations of fluid motion and remains fixed independent of the solution. On the other hand, an adaptive grid system evolves as a result of the solution of the equations of fluid motion. For example, grid points may move toward regions of high gradients such as in the neighborhood of a shock wave.

Conformal mappings based on complex variables are limited to 2-D problems and require a reasonable knowledge of complex variables. In addition, the determination of the mapping function is sometimes a difficult task. Therefore, this method will not be discussed here. In the following sections, grid generation techniques based on algebraic and PDE methods are presented.

9.2 Transformation of the Governing Partial Differential Equations

The equations of fluid motion include the continuity, momentum and energy equations. For a single phase continuum flow, the transformation of this system of equations will be presented in Chapter 11. In this section, a simple 2-D problem is proposed to familiarize the reader with the processes involved in the transformation of a PDE and the complexity of the resultant equation. It should be mentioned that the form and type of the transformed equation remains the same as the original PDE; i.e., if the original equation is parabolic, then the transformed equation is also parabolic. A mathematical proof is given in Reference [9-1]. Now, define the following relations between the physical and computational spaces:

$$\xi = \xi(x, y) \tag{9-1}$$

$$\eta = \eta(x, y) \tag{9-2}$$

The chain rule for partial differentiation yields the following expression:

$$\frac{\partial}{\partial x} = \frac{\partial \xi}{\partial x}\frac{\partial}{\partial \xi} + \frac{\partial \eta}{\partial x}\frac{\partial}{\partial \eta} \tag{9-3}$$

The partial derivatives will be denoted using the subscripts notation, i.e., $\frac{\partial \xi}{\partial x} = \xi_x$. Hence,

$$\frac{\partial}{\partial x} = \xi_x \frac{\partial}{\partial \xi} + \eta_x \frac{\partial}{\partial \eta} \tag{9-4}$$

and similarly,

$$\frac{\partial}{\partial y} = \xi_y \frac{\partial}{\partial \xi} + \eta_y \frac{\partial}{\partial \eta} \tag{9-5}$$

Now consider a model PDE, such as

$$\frac{\partial u}{\partial x} + a\frac{\partial u}{\partial y} = 0 \tag{9-6}$$

This equation may be transformed from physical space to computational space using Equations (9-4) and (9-5). As a result,

$$\xi_x \frac{\partial u}{\partial \xi} + \eta_x \frac{\partial u}{\partial \eta} + a\left(\xi_y \frac{\partial u}{\partial \xi} + \eta_y \frac{\partial u}{\partial \eta}\right) = 0$$

which may be rearranged as

$$(\xi_x + a\xi_y)\frac{\partial u}{\partial \xi} + (\eta_x + a\eta_y)\frac{\partial u}{\partial \eta} = 0 \tag{9-7}$$

This equation is the one which will be solved in the computational domain. Also note that the transformation derivatives ξ_x, ξ_y, η_x, and η_y must be determined from the functional relations (9-1) and (9-2). The determination of the transformation derivatives will be addressed briefly in the next section and in more detail for a 3-D case in Chapter 11. Comparing the original PDE (9-6) and the transformed equation given in (9-7), it is obvious that the transformed equation is more complicated than the original equation. Generally that is always the case. Thus, a trade-off is introduced whereby advantages gained by using the generalized coordinates are somehow counterbalanced by the resultant complexity of the PDE. However, the advantages by far outweigh the complexity of the transformed PDE.

9.3 Metrics and the Jacobian of Transformation

Recall that in Equations (9-4) and (9-5), terms such as ξ_x, ξ_y, η_x, and η_y appear. These transformation derivatives are defined as the metrics of transformation or simply as the metrics. The interpretation of the metrics is obvious considering the following approximation:

$$\xi_x = \frac{\partial \xi}{\partial x} \cong \frac{\Delta \xi}{\Delta x}$$

This expression indicates that the metrics represent the ratio of arc lengths in the computational space to that of the physical space. The computation of the metrics is considered next.

From Equations (9-1) and (9-2) the following differential expressions are obtained

$$d\xi = \xi_x dx + \xi_y dy \tag{9-8}$$

$$d\eta = \eta_x dx + \eta_y dy \tag{9-9}$$

which are written in a compact form as

$$\begin{bmatrix} d\xi \\ d\eta \end{bmatrix} = \begin{bmatrix} \xi_x & \xi_y \\ \eta_x & \eta_y \end{bmatrix} \begin{bmatrix} dx \\ dy \end{bmatrix} \tag{9-10}$$

Reversing the role of independent variables, i.e.,

$$x = x(\xi, \eta)$$

$$y = y(\xi, \eta)$$

The following may be written

$$dx = x_\xi d\xi + x_\eta d\eta \tag{9-11}$$

$$dy = y_\xi d\xi + y_\eta d\eta \tag{9-12}$$

In a compact form they are written as

$$\begin{bmatrix} dx \\ dy \end{bmatrix} = \begin{bmatrix} x_\xi & x_\eta \\ y_\xi & y_\eta \end{bmatrix} \begin{bmatrix} d\xi \\ d\eta \end{bmatrix} \tag{9-13}$$

Comparing Equations (9-10) and (9-13), it can be concluded that

$$\begin{bmatrix} \xi_x & \xi_y \\ \eta_x & \eta_y \end{bmatrix} = \begin{bmatrix} x_\xi & x_\eta \\ y_\xi & y_\eta \end{bmatrix}^{-1}$$

from which

$$\xi_x = J y_\eta \tag{9-14}$$

$$\xi_y = -J x_\eta \tag{9-15}$$

$$\eta_x = -J y_\xi \tag{9-16}$$

$$\eta_y = J x_\xi \tag{9-17}$$

where

$$J = \frac{1}{x_\xi y_\eta - y_\xi x_\eta} \tag{9-18}$$

and is defined as the Jacobian of transformation.

The Jacobian, J, is interpreted as the ratio of the areas (volumes in 3-D) in the computational space to that of the physical space.

Note that the actual values of the metrics or the Jacobian could be negative. Obviously, the values depend on the specification of physical and computational grid systems. The computed values of the metrics for various grid systems will be presented shortly. For grid generation methods where analytical expressions for the metrics can be written, they may be analytically evaluated or determined numerically by the use of finite difference expressions. This point is illustrated in the example problem given in Section 9.5.

9.4 Grid Generation Techniques

Before proceeding with the investigation of various grid generation techniques, the objectives will be summarized. A grid system with the following features is desired:

(1) A mapping which guarantees one-to-one correspondence ensuring grid lines of the same family do not cross each other;

(2) Smoothness of the grid point distribution;

(3) Orthogonality or near orthogonality of the grid lines;

(4) Options for grid point clustering.

As mentioned previously, some of the features enumerated, as in (2)–(4) above, may not be achievable with a particular grid generation technique. In the next few sections, various grid generation techniques are introduced.

9.5 Algebraic Grid Generation Techniques

The simplest grid generation technique is the algebraic method. The major advantage of this scheme is the speed with which a grid can be generated. An algebraic equation is used to relate the grid points in the computational domain to those of the physical domain. This objective is met by using an interpolation scheme between the specified boundary grid points to generate the interior grid points. Clearly, many algebraic equations (or interpolation schemes, if preferred) can be introduced for this purpose. To illustrate the procedure, consider the simple physical domain depicted in Figure 9-4.

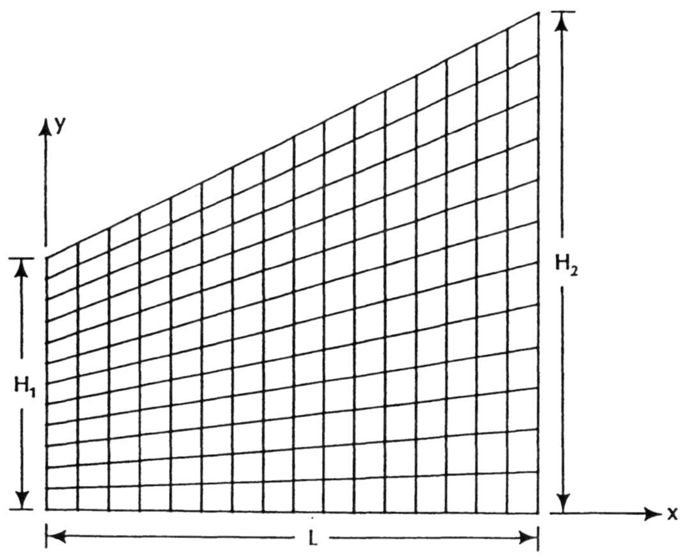

Figure 9-4. The physical space which must be transformed to a uniform rectangular computational space.

Introducing the following algebraic relations will transform this nonrectangular physical domain into a rectangular domain:

$$\xi = x \tag{9-19}$$

$$\eta = \frac{y}{y_t} \tag{9-20}$$

In (9-20), y_t represents the upper boundary which is expressed as

$$y_t = H_1 + \frac{H_2 - H_1}{L} x$$

Thus, it may be written that

$$\xi = x \tag{9-21}$$

$$\eta = \frac{y}{H_1 + \frac{H_2 - H_1}{L} x} \tag{9-22}$$

which can be rearranged as

$$x = \xi \tag{9-23}$$

and

$$y = \left(H_1 + \frac{H_2 - H_1}{L} \xi \right) \eta \tag{9-24}$$

The grid system is generated as follows. The geometry in the physical space is defined. For this particular problem, it is accomplished by specifying values of L, H_1, and H_2. Next, the desired number of grid points defined by IM (the maximum number of grid points in ξ) and JM (the maximum number of grid points in η) is specified. The equal grid spacing in the computational domain is produced as follows:

$$\Delta\xi = \frac{L}{IM - 1} \tag{9-25}$$

$$\Delta\eta = \frac{1.0}{JM - 1} \tag{9-26}$$

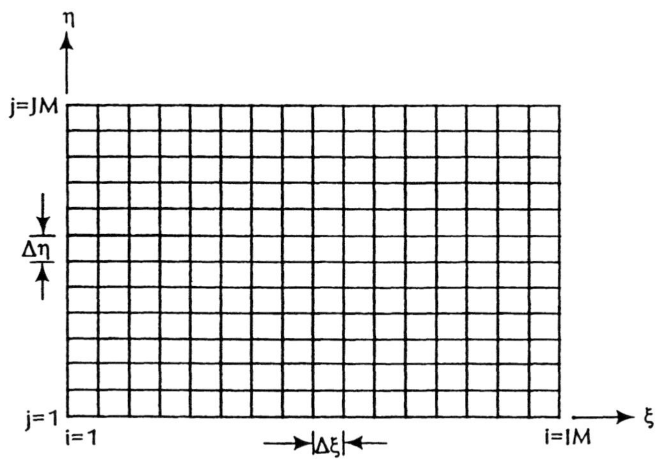

Figure 9-5. The rectangular computational domain with uniform grid spacing.

Note that in Equation (9-20), η has been normalized, i.e., its value varies from zero to one. With the equal step-sizes provided by Equations (9-25) and (9-26), the uniform computational domain is constructed, which is shown in Figure 9-5 for a 17×13 grid system. Therefore, the values of ξ and η are known at each grid point within this domain. Now Equations (9-23) and (9-24) are used to identify the corresponding grid points in the physical space. For illustrative purposes, the grid system generated for a physical domain defined by $L = 4$, $H_1 = 2$, and $H_2 = 4$ is shown in Figure 9-4.

As discussed previously, the metrics and the Jacobian of the transformation must be evaluated before any transformed PDEs can be solved. In many instances, when an algebraic model is used the metrics may be calculated analytically. This aspect is obviously an advantage of the algebraic methods, since numerical computation of the metrics will require additional computation time and may introduce some errors into the system of equations of motion that are to be solved. To illustrate this point, the metrics in the proposed example are computed both analytically and numerically. From Equation (9-21), metrics ξ_x and ξ_y can be determined analytically, resulting in

$$\xi_x = 1 \tag{9-27}$$

$$\xi_y = 0 \tag{9-28}$$

Similarly, the following is obtained from Equation (9-22)

$$\eta_x = -\frac{(H_2 - H_1)y/L}{[H_1 + (H_2 - H_1)x/L]^2} \quad \text{or}$$

$$\eta_x = -\frac{H_2 - H_1}{L}\,\frac{\eta}{[H_1 + (H_2 - H_1)\xi/L]} \tag{9-29}$$

and

$$\eta_y = \frac{1}{H_1 + (H_2 - H_1)x/L} \quad \text{or}$$

$$\eta_y = \frac{1}{H_1 + (H_2 - H_1)\xi/L} \tag{9-30}$$

and the Jacobian of transformation is

$$J = \frac{1}{x_\xi y_\eta - y_\xi x_\eta}$$

To compute the metrics numerically, Equations (9-14) through (9-18) are used. Thus, the terms x_ξ, y_ξ, x_η, and y_η are computed initially, from which the Jacobian may be evaluated. These expressions are computed numerically using finite difference approximations. For example, a second-order central difference approximation may be used to compute the transformation derivative x_η for the interior grid points, i.e.,

$$x_\eta = \frac{x_{i,j+1} - x_{i,j-1}}{2\Delta\eta}$$

The transformation derivatives at the boundaries are evaluated with forward or backward second-order approximations. For example, x_η at the $j = 1$ boundary is computed using the forward difference approximation

$$x_\eta = \frac{-3x_{i,1} + 4x_{i,2} - x_{i,3}}{2\Delta\eta}$$

A comparison of the metric η_x evaluated numerically and analytically by Equation (9-29) and the error introduced in the numerical computations is shown in Table (9-1).

j	Analytic	Numerical	ERROR
4	$-.52632E-01$	$-.52632E-01$	$.44409E-15$
8	$-.12281E+00$	$-.12281E+00$	$-.13323E-14$
12	$-.19298E+00$	$-.19298E+00$	$-.26645E-14$

Table 9-1. Comparison of the metric η_x at $i = 4$ evaluated analytically and numerically.

Notice that for this simple grid the errors in the numerical computation of the metrics are extremely small (practically zero!). The distributions of metrics are shown in Figures 9-6a through 9-6d. Each tic mark on the vertical axis represents a numerical change of 0.25 in Figures 9-6a and 9-6b, −0.1 in Figure 9-6c, and 0.125 in Figure 9-6d. These figures illustrate the smooth variations of the metrics. Erratic metric distributions, especially those with some sort of discontinuity, will certainly invite disaster! It is strongly recommended that the metric distributions be investigated prior to solving the governing PDEs of fluid motion.

The simple algebraic model just investigated does not include an option for clustering. Next, some algebraic expressions which employ clustering techniques are presented.

For flow problems where large gradients are concentrated in a specific region, additional resolution of the flow properties is essential. As an example, flow in the vicinity of a solid surface in a viscous fluid possesses large flow gradients. Accurate computation of flow gradients in this region will require many grid points within the domain. Rather than using a nearly uniform grid distribution in the physical domain, grid points may be clustered in the regions of high flow gradients, which reduces the total number of grid points and thus increases efficiency. Some examples of such algebraic expressions with clustering options are provided below. As a first example, consider the transformation given by

$$\xi = x \tag{9-31}$$

$$\eta = 1 - \frac{ln\{[\beta + 1 - (y/H)]/[\beta - 1 + (y/H)]\}}{ln[(\beta+1)/(\beta-1)]} \tag{9-32}$$

In this equation, β is the clustering parameter within the range of 1 to ∞. As the value of β approaches 1, more grid points are clustered near the surface, where $y = 0$. From Equations (9-31) and (9-32), the inverse of the transformation can be written as

$$x = \xi \tag{9-33}$$

$$y = H\,\frac{(\beta+1) - (\beta-1)\,\{[(\beta+1)/(\beta-1)]^{1-\eta}\}}{[(\beta+1)/(\beta-1)]^{1-\eta} + 1} \tag{9-34}$$

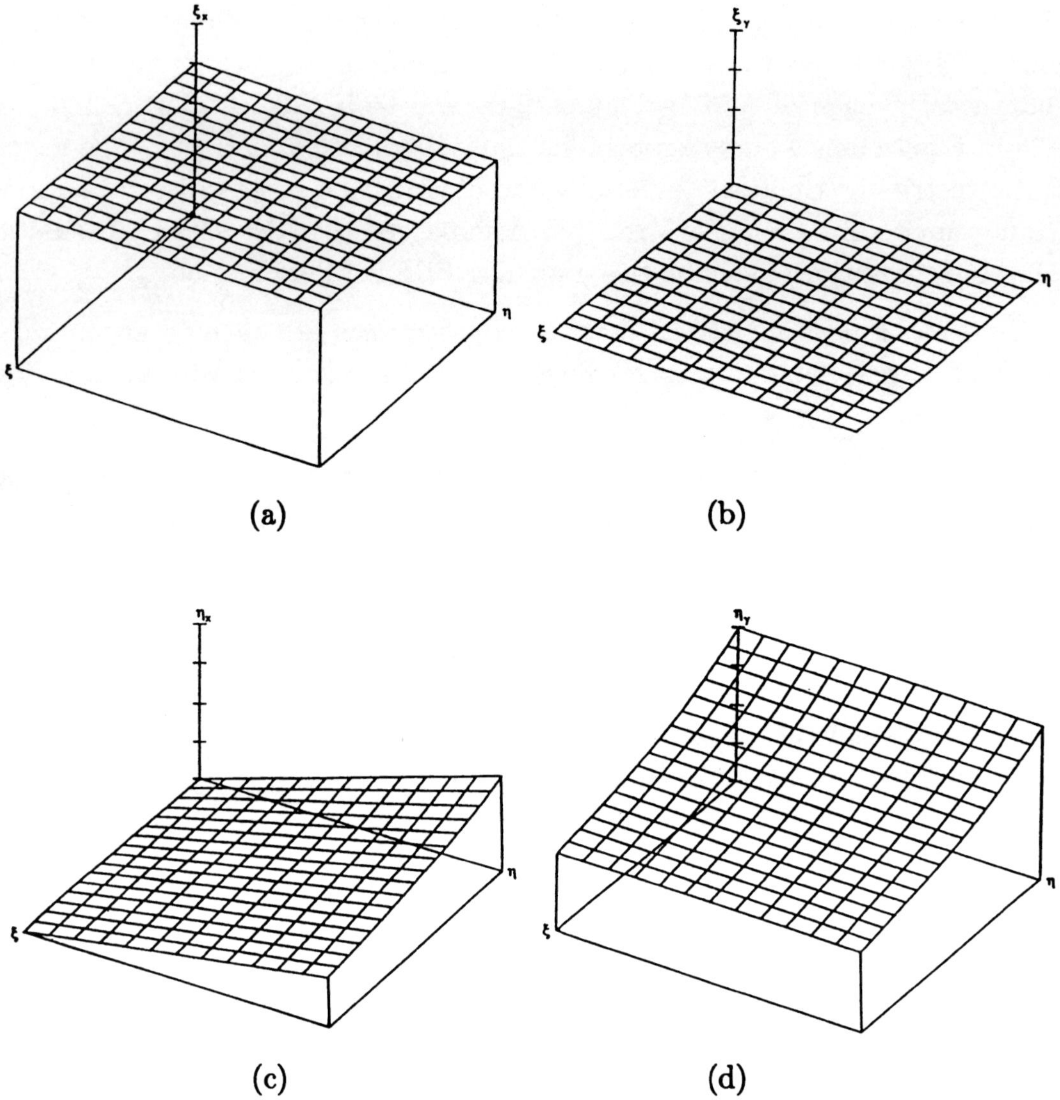

Figures 9-6a, b, c, d. Distribution of metrics for the domain shown in Figure 9-4.

The metrics of transformation may be determined analytically from the algebraic relations (9-31) and (9-32) and are given below:

$$\xi_x = 1 \tag{9-35}$$

$$\eta_x = 0 \tag{9-36}$$

$$\xi_y = 0 \tag{9-37}$$

$$\eta_y = \frac{2\beta}{H\left\{\beta^2 - [1-(y/H)]^2\right\}\left\{ln[(\beta+1)/(\beta-1)]\right\}} \tag{9-38}$$

The physical and computational domains, using the transformations (9-31) and (9-32), are shown in Figures 9-7a, 9-7b, and 9-7c. The grid system is generated for a 21×24 grid, and clustering parameter values of 1.05 and 1.2 are used, respectively. The distributions of metric η_y are shown in Figures 9-8a and 9-8b. Each tic mark on the vertical axis represents a change of 0.5 in Figures 9-8a and 9-8b. Obviously, this type of grid is suitable for boundary-layer type computations, where grid clustering near the surface is required.

For a physical domain enclosed by lower and upper solid surfaces, clustering at both surfaces must be considered. A flow field within a 2-D duct is such an example. The following algebraic equations may be employed for this purpose:

$$\xi = x \tag{9-39}$$

$$\eta = \alpha + (1-\alpha)\,\frac{ln\,\{\{\beta + [(2\alpha+1)y/H] - 2\alpha\}/\{\beta - [(2\alpha+1)y/H] + 2\alpha\}\}}{ln[(\beta+1)/(\beta-1)]} \tag{9-40}$$

where β is the clustering parameter, and α defines where the clustering takes place. When $\alpha = 0$, the clustering is at $y = H$; whereas, when $\alpha = 1/2$, clustering is distributed equally at $y = 0$ and $y = H$. The inverse transformation is given by

$$x = \xi \tag{9-41}$$

$$y = H\,\frac{(2\alpha+\beta)\,[(\beta+1)/(\beta-1)]^{(\eta-\alpha)/(1-\alpha)} + 2\alpha - \beta}{(2\alpha+1)\,\{1 + [(\beta+1)/(\beta-1)]^{(\eta-\alpha)/(1-\alpha)}\}} \tag{9-42}$$

Analytical expressions for the metrics are determined from Equations (9-39) and (9-40) as

$$\xi_x = 1 \tag{9-43}$$

$$\xi_y = 0 \tag{9-44}$$

$$\eta_x = 0 \tag{9-45}$$

$$\eta_y = \frac{2\beta(2\alpha+1)(1-\alpha)}{H\,\{\beta^2 - [(2\alpha+1)y/H - 2\alpha]^2\}\,ln[(\beta+1)/(\beta-1)]} \tag{9-46}$$

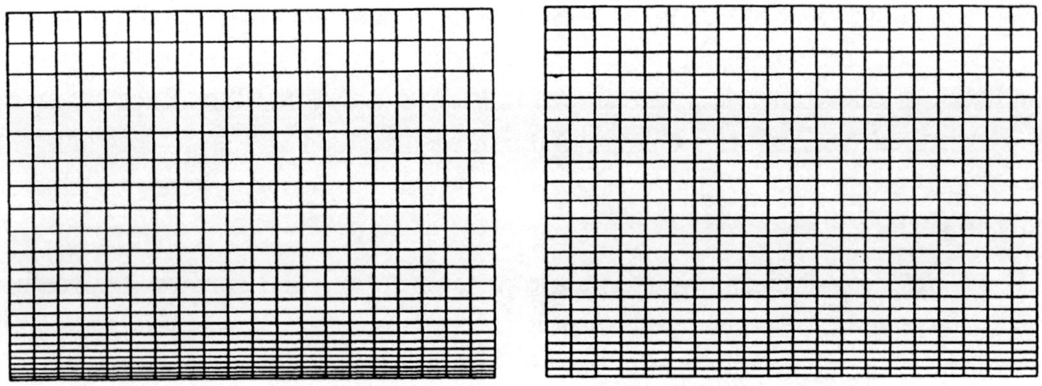

(a) $\beta = 1.05$ (b) $\beta = 1.2$

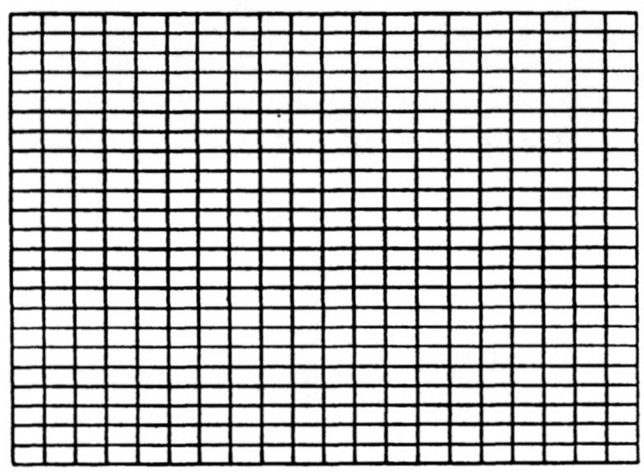

(c) Computational domain.

Figures 9-7a, b, c. The physical and computational domains generated by transformation functions (9-31) and (9-32).

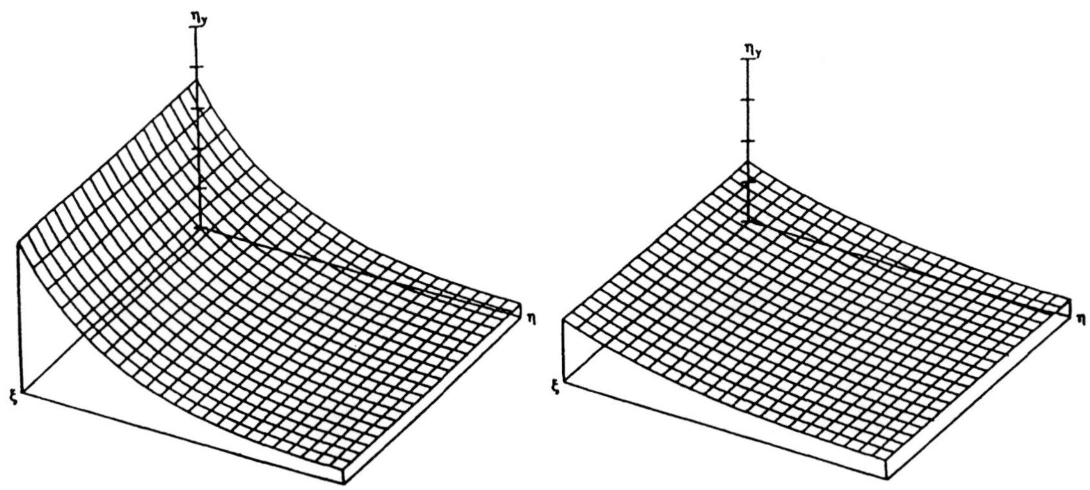

Figures 9-8a, b. Metric distributions for the domains of Figures 9-7a and 9-7b.

Grid systems generated for a domain with 21 × 24 grid points are shown in Figures 9-9a and 9-9b for $\alpha = 1/2$ and clustering parameters of 1.05 and 1.2, respectively.

The corresponding η_y metric distributions are shown in Figures 9-10a and 9-10b. Each tic mark on the vertical axis represents a change of 0.5 in Figures 9-10a and 9-10b.

For problems where clustering in the interior of the domain is required, the following relations may be utilized

$$\xi = x \tag{9-47}$$

$$\eta = A + \frac{1}{\beta}\sinh^{-1}\left[\left(\frac{y}{D} - 1\right)\sinh(\beta A)\right] \tag{9-48}$$

where

$$A = \frac{1}{2\beta}\, ln \left[\frac{1 + (e^{\beta} - 1)(D/H)}{1 + (e^{-\beta} - 1)(D/H)}\right]$$

In Equation (9-48), β is the clustering parameter in the range of $0 < \beta < \infty$, and D is the y coordinate where clustering is desired. The inverse transformation is given by:

$$x = \xi \tag{9-49}$$

$$y = D\left\{1 + \frac{\sinh[\beta(\eta - A)]}{\sinh(\beta A)}\right\} \tag{9-50}$$

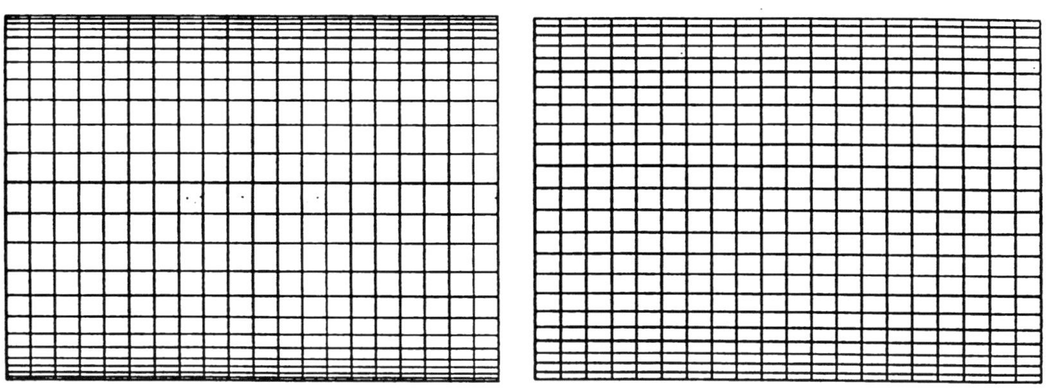

(a) $\beta = 1.05$ (b) $\beta = 1.2$

Figures 9-9a, b. Grid system generated by the transformation function (9-42).

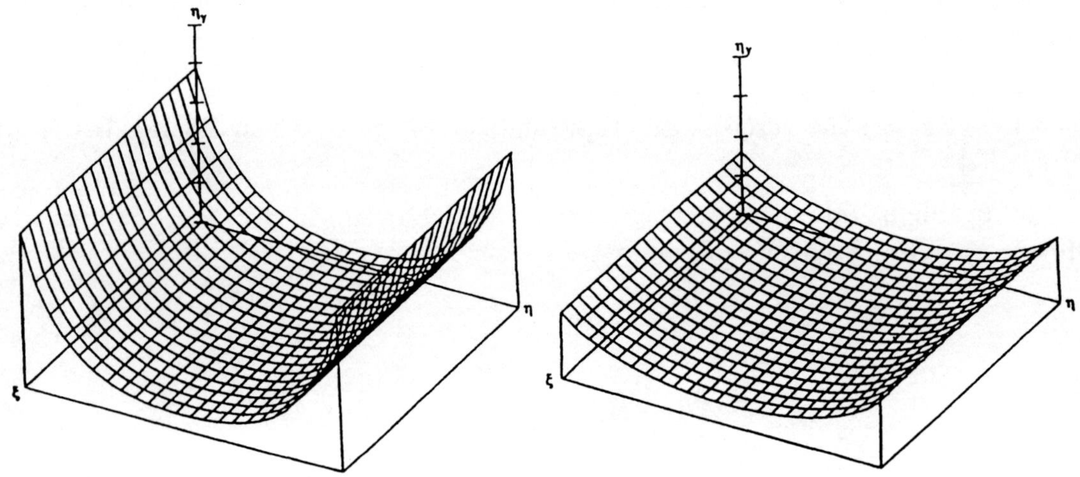

Figures 9-10a, b. Metric distributions for the domains shown in Figure 9-9.

Note that for $\beta = 0$, no clustering is enforced, while a denser clustering of grid points near D is produced for larger values of β. Analytical expressions for the metrics are determined from Equations (9-47) and (9-48) and are given by:

$$\xi_x = 1 \tag{9-51}$$

$$\xi_y = 0 \tag{9-52}$$

$$\eta_x = 0 \tag{9-53}$$

$$\eta_y = \frac{\sinh(\beta A)}{\beta D\left\{1 + [(y/D) - 1]^2 \sinh^2(\beta A)\right\}^{0.5}} \tag{9-54}$$

The algebraic expressions given by (9-49) and (9-50) are used to generate the grid systems shown in Figures 9-11a through 9-11c. The clustering is specified at $D = H/2$ for the domains shown in Figures 9-11a and 9-11b and at $D = H/4$ for the domain shown in Figure 9-11c. The values of β are 5, 10, and 5, respectively. The distributions of metric η_y are illustrated in Figures 9-12a through 9-12c. Each tic mark on the vertical axis represents a change of 0.5 in Figures 9-12a through 9-12c.

Next, an algebraic expression for generating a grid system around an arbitrary shape is investigated. For simplicity, a conical body shape with a circular cross-section will be assumed. The grid system is determined at cross sections where the relevant PDEs are to be solved. An example would be the solution of parabolized Navier-Stokes equations over a conical configuration in a supersonic flow field. The flow field is depicted in Figure 9-13 where cross-sectional planes are normal to the body axis.

An alternate choice would be the selection of a grid system normal to the body surface. For many applications where the flow is symmetrical, only half of the domain needs to be solved. With that in mind, consider the generation of a grid system at a cross section.

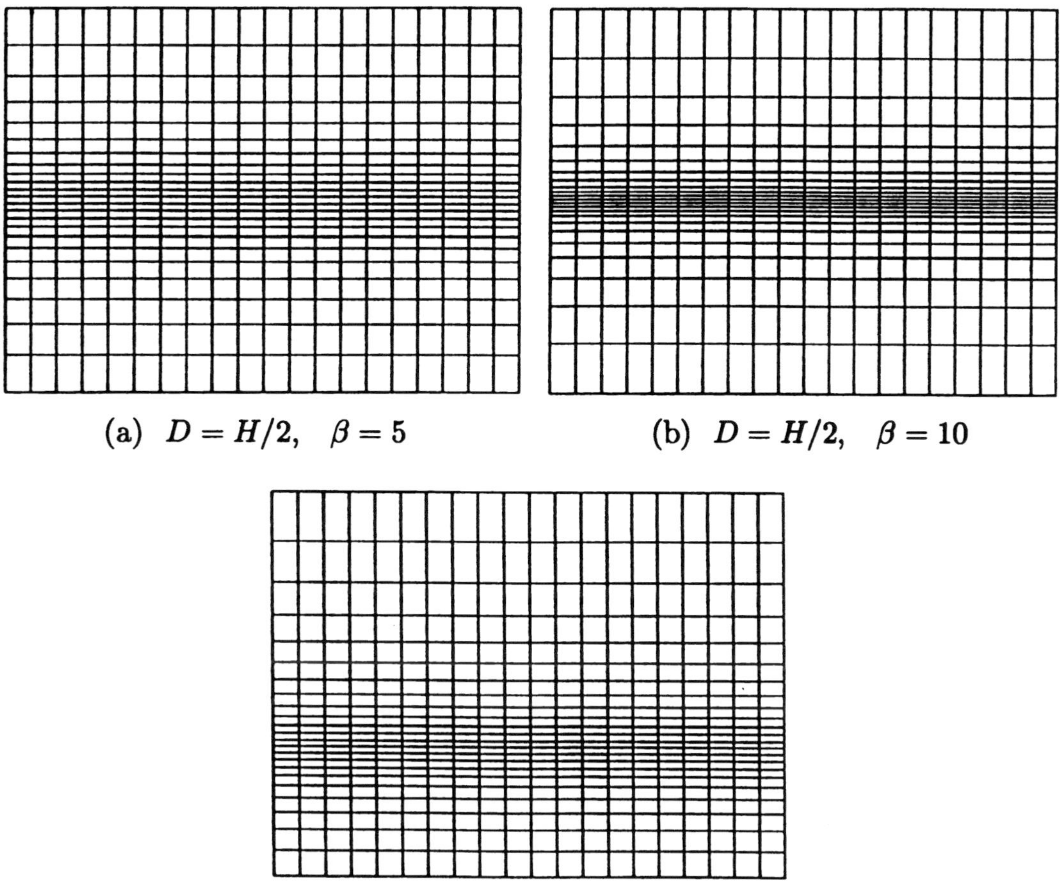

(a) $D = H/2, \quad \beta = 5$

(b) $D = H/2, \quad \beta = 10$

(c) $D = H/4, \quad \beta = 5$

Figures 9-11a, b, c. Grid systems generated by the transformation functions (9-49) and (9-50).

Note that since the physical domain is changing at each streamwise station, a new grid system must be generated at each station. However, no difficulties arise since the grid generating procedure is coded as a subroutine and called at each x-station.

A typical procedure is described next.

First a domain bounded by the body, the plane of symmetry, and the free stream must be defined. The outer boundary in the free stream must be far enough away to include the bow shock generated by the conical configuration. An alternate selection for the outer boundary would be at the bow shock itself. However, the shock location is unknown; therefore, Rankine-Hugoniot relations must be utilized to determine its location. This procedure is known as shock fitting and will be investigated later in Chapter 13.

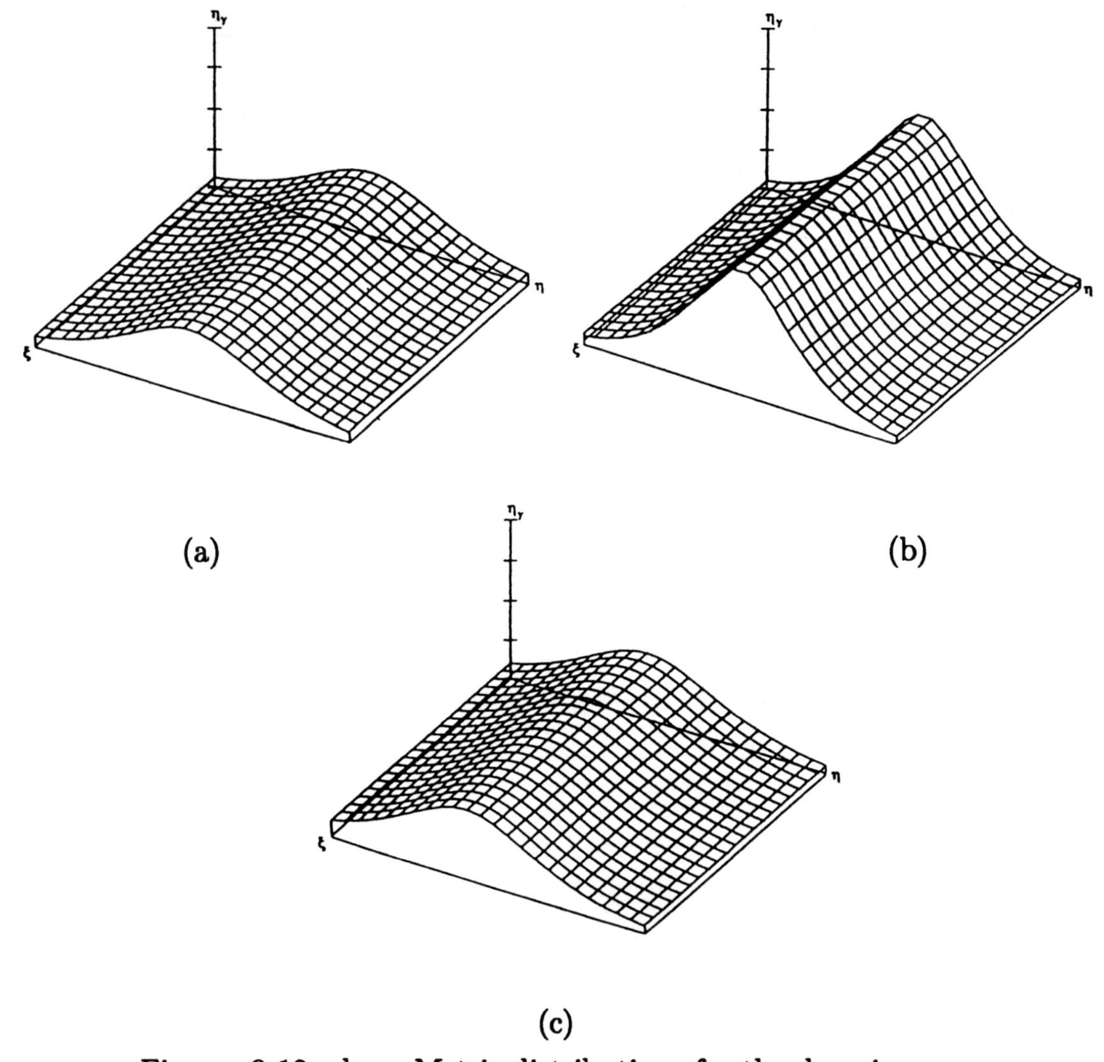

Figures 9-12a, b, c. Metric distributions for the domains shown in Figure 9-11.

For now, the outer boundary is specified as the free stream. This specification is accomplished by defining two elliptical shapes (or any other geometries you may choose) with semi-major and semi-minor axes denoted by a_1, b_1, and b_2. This specification is illustrated in Figure 9-14. The inner boundary represents the circular cone and will be defined by its radius, R. The number of grid points in the circumferential direction is specified by KM, and for this example they are distributed equally around the body. This distribution is accomplished by defining the incremental angle DELTHET as DELTHET $= \pi/(KM - 1)$, and subsequently computing θ at each k. The nomenclature is shown in Figure 9-15. Note that $k = 1$ is chosen on the windward side of the body and $k = KM$ is on the leeward side. The y and z coordinates of the grid points on the body are easily computed from

$$y(1,k) = -R\cos\theta \tag{9-55}$$

$$z(1,k) = R\sin\theta \tag{9-56}$$

The corresponding grid points on the outer boundary are defined by rays emanating from the origin with the appropriate angular positions.

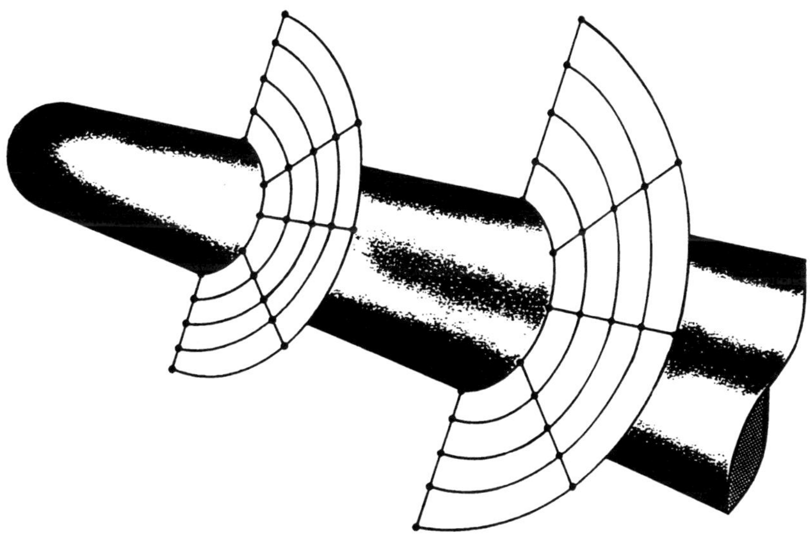

Figure 9-13. The grid system in the physical domain.

For this purpose, the length of the rays are obtained from the equation of an ellipse as

$$r = \frac{1}{\sqrt{\left(\frac{\sin^2\theta}{a^2}\right) + \left(\frac{\cos^2\theta}{b^2}\right)}}$$

where appropriate values of a and b are used. Subsequently, the y and z coordinates of the grid points on the outer boundary are determined by similar expressions given in Equations (9-55) and (9-56).

For viscous flow computations, clustering in the vicinity of the body is desirable. For this purpose the following expression is used:

$$c(k,j) = \delta\left\{1 - \beta\left[\left(\frac{\beta+1}{\beta-1}\right)^{\eta} - 1\right] \Big/ \left[\left(\frac{\beta+1}{\beta-1}\right)^{\eta} + 1\right]\right\} \tag{9-57}$$

In (9-57), δ is the radial distance between the body and the outer boundary, i.e.,

$$\delta(k) = r(k) - R(k) \tag{9-58}$$

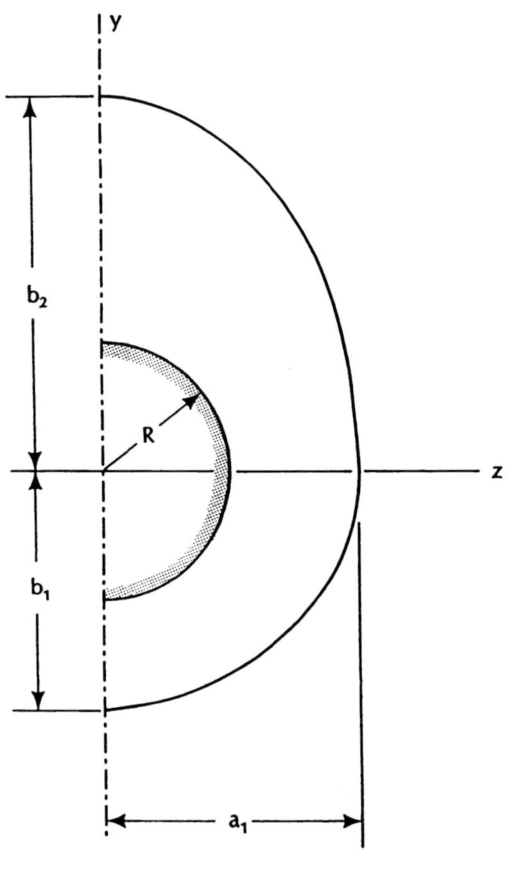

Figure 9-14. Nomenclature required to describe the physical domain.

and β is the clustering parameter. The clustering function given by Equation (9-57) places $\eta = 0.0$ at the outer boundary and $\eta = 1.0$ at the surface. The computational domain is shown in Figure 9-16. A computational domain for the entire problem equivalent to the physical space depicted in Figure 9-13 is shown in Figure 9-17. The y and z coordinates within the physical domain are evaluated from the following equations

$$y(k,j) = y(k,1) - c(k,j)\cos\alpha(k) \tag{9-59}$$

and

$$z(k,j) = z(k,1) + c(k,j)\sin\alpha(k) \tag{9-60}$$

where α is defined as the angle between the normal to the body in the cross-sectional plane and the vertical direction. This relationship is also shown in Figure 9-15.

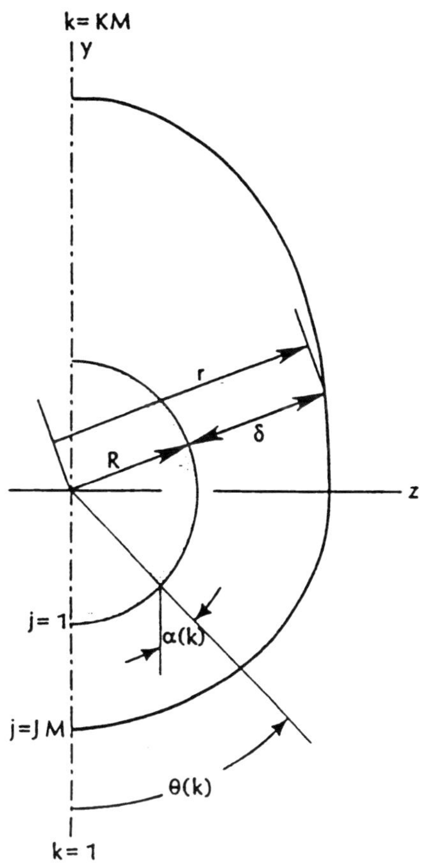

Figure 9-15. Nomenclature required to define the grid system.

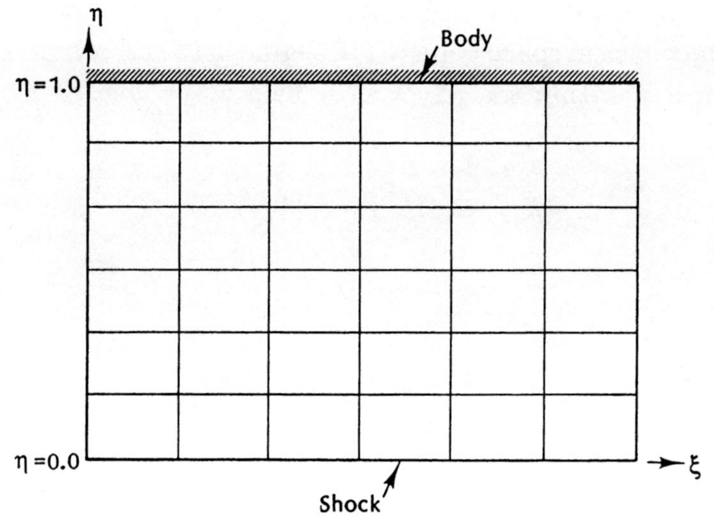

Figure 9-16. Computational grid system at a cross section.

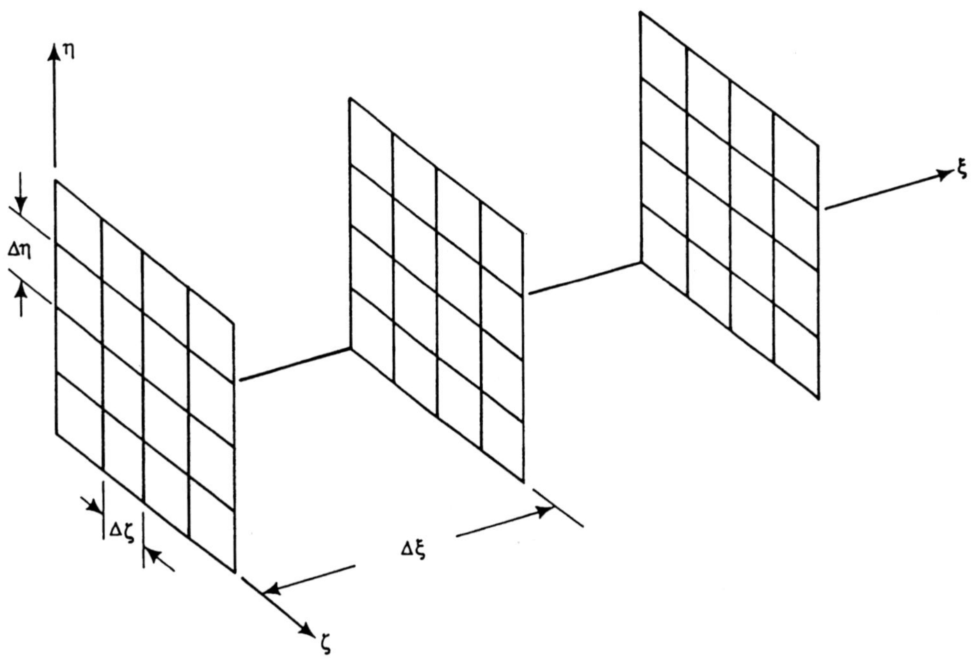

Figure 9-17. Computational domain for the 3-D problem.

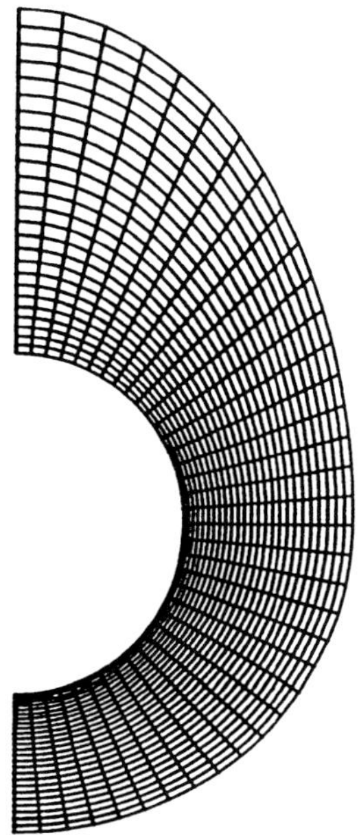

Figure 9-18. Cross-sectional grid system generated for a circular conical configuration.

A grid system generated for a 37 × 26 grid is shown in Figure 9-18 for $\beta = 1.4$. The configuration is defined by $R = 1.0$, $a_1 = 2.0$, $b_1 = 1.8$, and $b_2 = 3.0$. For this problem the metrics are evaluated numerically by using a second-order central difference approximation in the interior of the domain and by second-order forward and backward difference approximations at the boundaries. The metric distributions are illustrated in Figures 9-19a through 9-19d, where the smoothness of the metrics is clearly evident. Note that if the metric distribution had discontinuities, further investigation of the suggested grid system and the solution procedure used to obtain the grid and the metrics would be required.

The algebraic expressions used to generate the grid systems just presented are a few among many appearing in various literature. However, the procedures used to generate grids by algebraic methods are fundamentally similar.

For many applications, algebraic models provide a reasonable grid system with continuous and smooth metric distributions. However, if grid smoothness, skewness, and orthogonality are of concern, grid systems generated by solving PDEs must be used. This option includes elliptic, parabolic, or hyperbolic grid generators, which will be discussed in the next section. For the time being, briefly consider the elliptic grid generators. In this technique, some type of elliptic PDEs is solved to identify the coordinates of the grid points in the physical space. This approach is similar to what was just accomplished with algebraic models, except now a system of PDEs must be solved.

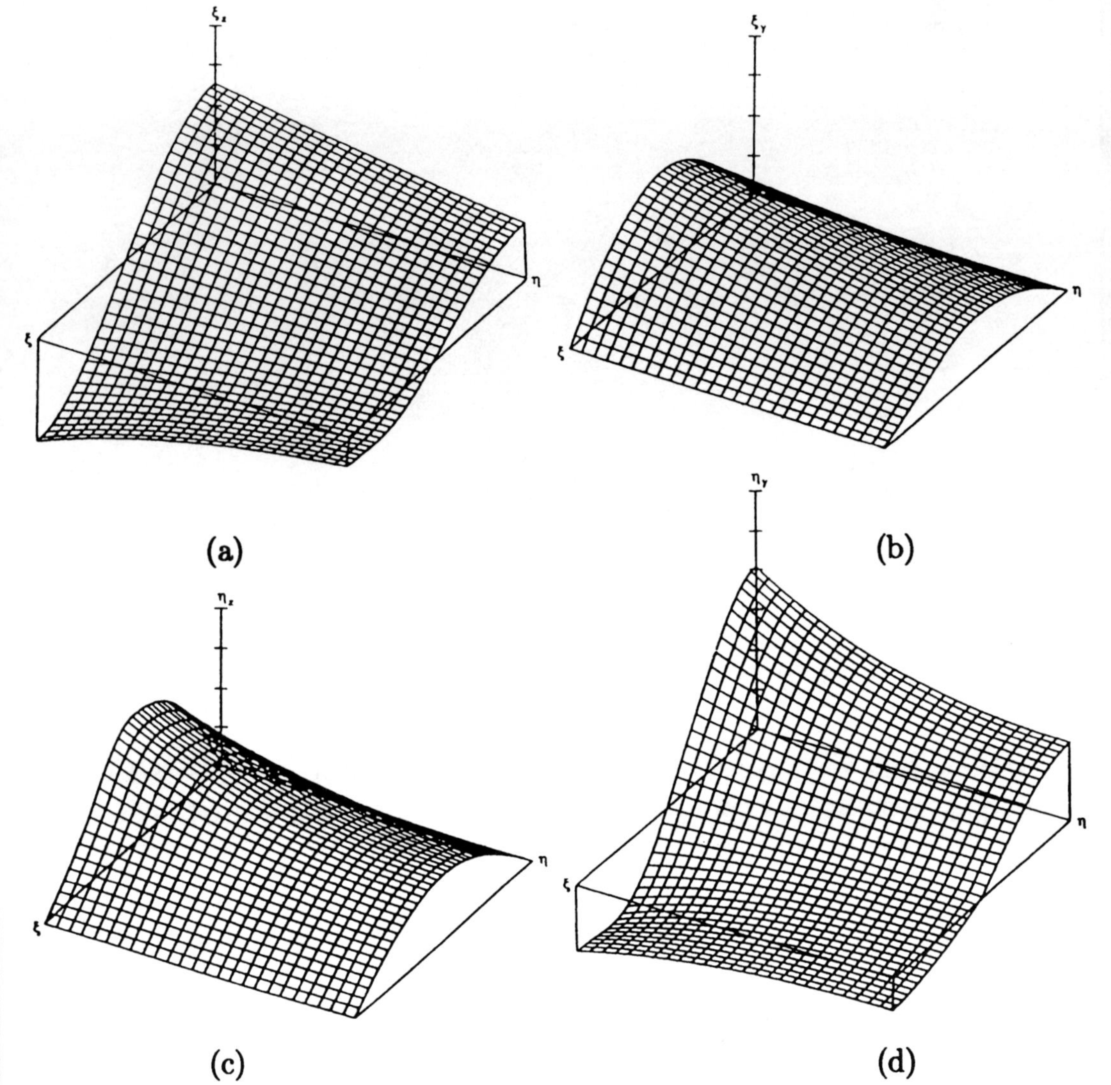

Figures 9-19a, b, c, d. Metric distributions for the domain shown in Figure 9-18.

Recall that iterative procedures were recommended to solve elliptic PDEs. To start such a method, an initial estimate of the grid points must be provided. This initial distribution is easily obtained by use of algebraic methods. Thus, algebraic techniques are used not only to generate grid points for which the equations of fluid motion are solved, but they may also be used as an initial grid point distribution to start an iterative solution procedure for elliptic grid generators. This point will be illustrated in the next section.

Before closing this section, the following conclusions are drawn. The advantages of the algebraic grid generation methods are:

(1) Computationally, they are very fast;

(2) Metrics may be evaluated analytically, thus avoiding numerical errors;

(3) The ability to cluster grid points in different regions can be easily implemented.

The disadvantages are:

(1) Discontinuities at a boundary may propagate into the interior region which could lead to errors due to sudden changes in the metrics;

(2) Control of grid smoothness and skewness is a difficult task.

Some of the disadvantages of the algebraic grid generators are overcome by the use of PDE grid generators which is, of course, accomplished with increased computational time. This procedure will be described in the following section.

9.6 Partial Differential Equation Techniques

A grid generation scheme which is gaining popularity is one in which PDEs are used to create the grid system. In these methods, a system of PDEs is solved for the location of the grid points in the physical space, whereas the computational domain is a rectangular shape with uniform grid spacing. These methods may be categorized as an elliptic, parabolic, or hyperbolic system of PDEs. The elliptic grid generator is the most extensively developed method. It is commonly used for 2-D problems and the procedure has been extended to 3-D problems. Parabolic and hyperbolic grid generators are not as well developed but have some very interesting features. In this section all three methods will be introduced.

The presentation of various schemes will be limited to 2-D problems; however, methods which can be extended to 3-D problems will be identified. By 3-D problems (in grid generation), we refer to situations where all three coordinates are transformed. For some 3-D applications, a coordinate transformation in the streamwise direction may not be required, i.e., $\xi = x$. For such problems, the grid system will be generated only in a 2-D sense at each streamwise location as needed. Thus, the 2-D grid generators are used extensively for 3-D computations. A typical grid system was shown in Figure 9-13.

Due to the advantages previously introduced, the generalized coordinate system will be used in the grid generation techniques which are presented next.

9.7 Elliptic Grid Generators

For domains where all the physical boundaries are specified, elliptic grid generators work very well. A system of elliptic equations in the form of Laplace's equation or Poisson's equation is introduced, which is solved for the coordinates of the grid points in the physical domain. Any iterative scheme such as Gauss-Seidel, point successive over relaxation (PSOR), etc., may be used to solve the elliptic PDEs.

Before proceeding with the mathematical development, consider the fundamental reasoning behind this procedure. Recall that the heat conduction equation for a steady, 2-D problem is reduced to an elliptic PDE. If a rectangular domain with the values of temperature on the boundaries is specified, the temperature distribution within the interior points is easily obtained by any iterative scheme. The solution provides the isothermal lines. Consider two such solutions shown in Figure 9-20.

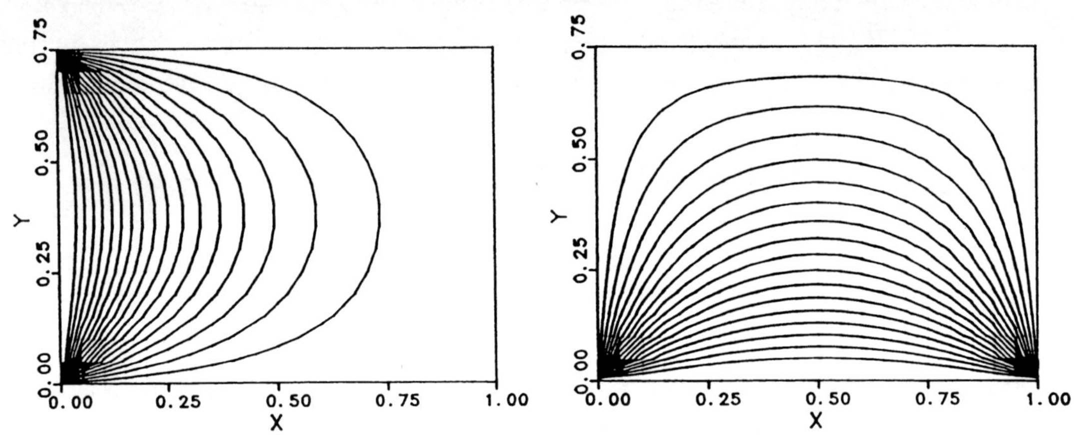

Figure 9-20. Isothermal lines for the two rectangular domains.

Superimposing the two solutions yields a solution shown in Figure 9-21.

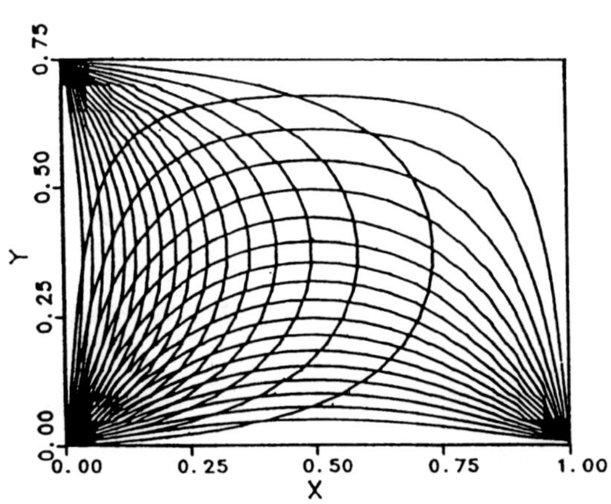

Figure 9-21. The superimposed solution for the rectangular domain.

Note that the governing equation is linear; thus, addition of solutions is allowed. If a heat source within the domain is specified, the isothermal lines will be altered and can be oriented to a particular region. Now, the following question is posed. What if the isothermal lines are to represent grid lines? Indeed, that is precisely the fundamental idea of using elliptic PDEs to generate a grid system. The dependent variables are the x and y coordinates of the grid points in the physical space. Thus, for a closed domain, the distribution of grid points on the boundaries are specified and a set of elliptic PDEs is solved to locate the coordinates of the interior grid points.

Consider a system of elliptic PDEs of the form

$$\xi_{xx} + \xi_{yy} = 0 \tag{9-61}$$

$$\eta_{xx} + \eta_{yy} = 0 \tag{9-62}$$

where ξ and η represent the coordinates in the computational domain. Equations (9-61) and (9-62) may be solved by any of the iterative techniques introduced previously in Chapter 5. However, computation must take place in a rectangular domain with uniform grid spacing as described earlier. To transform the elliptic PDEs, the dependent and independent variables are interchanged. The mathematical expressions employed for this purpose are derived in Appendix F. With the use of Equation (F-10), the elliptic equations (9-61) and (9-62) become:

$$ax_{\xi\xi} - 2bx_{\xi\eta} + cx_{\eta\eta} = 0 \tag{9-63}$$

$$ay_{\xi\xi} - 2by_{\xi\eta} + cy_{\eta\eta} = 0 \tag{9-64}$$

where

$$a = x_\eta^2 + y_\eta^2 \tag{9-65}$$

$$b = x_\xi x_\eta + y_\xi y_\eta \tag{9-66}$$

$$c = x_\xi^2 + y_\xi^2 \tag{9-67}$$

The system of elliptic equations (9-63) and (9-64) is solved in the computational domain (ξ, η) in order to provide the grid point locations in the physical space (x, y). Note that the equations are nonlinear; thus, a linearization procedure must be employed. For simplicity, lagging of the coefficients will be used, i.e., the coefficients a, b, and c are evaluated at the previous iteration level.

Three categories of physical domains will be considered here. They are: (1) a simply-connected domain, (2) a doubly-connected domain, and (3) a multiply-connected domain. The description of each domain, specification of boundary points (conditions), examples, and analyses are given next.

9.7.1 Simply-Connected Domain

By definition a simply-connected region is one which is reducible and can be contracted to a point. Thus, for a simply-connected region, there are no objects within the domain. An example of a simply-connected domain is shown in Figure 9-22a. The corresponding computational domain is shown in Figure 9-22b. Another example is shown in Figure 9-23a, whereas Figure 9-23b represents the corresponding computational domain.

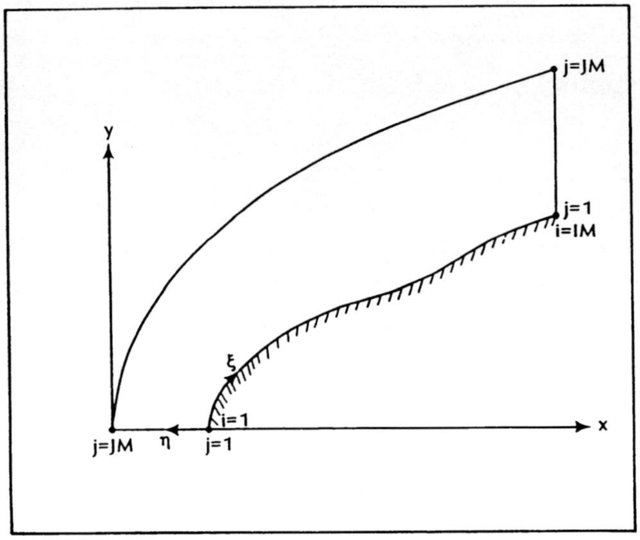

Figure 9-22a. Physical domain for simply-connected region.

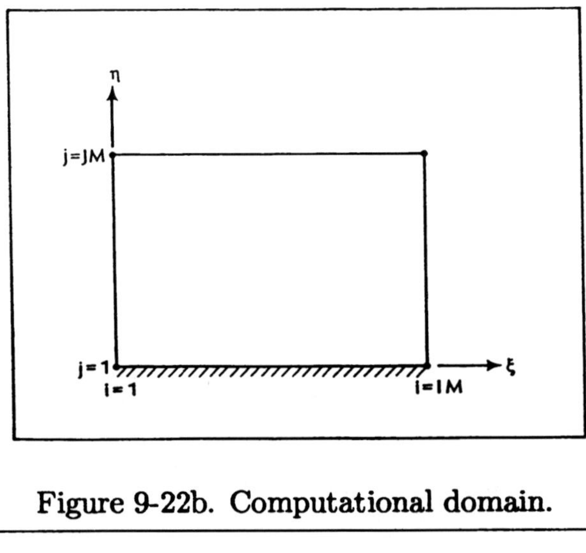

Figure 9-22b. Computational domain.

Now return to the elliptic equations (9-63) and (9-64). In order to investigate a few iterative solution schemes, the FDE is obtained by replacing the partial derivatives with a second-order central difference approximation. From Equation (9-63):

$$a\left[\frac{x_{i+1,j}-2x_{i,j}+x_{i-1,j}}{(\Delta\xi)^2}\right]-2b\left[\frac{x_{i+1,j+1}-x_{i+1,j-1}+x_{i-1,j-1}-x_{i-1,j+1}}{4\Delta\xi\Delta\eta}\right]$$

$$+c\left[\frac{x_{i,j+1}-2x_{i,j}+x_{i,j-1}}{(\Delta\eta)^2}\right]=0$$

If the Gauss-Seidel iterative method is used, the equation is rearranged as:

$$2\left[\frac{a}{(\Delta\xi)^2}+\frac{c}{(\Delta\eta)^2}\right]x_{i,j}=\frac{a}{(\Delta\xi)^2}[x_{i+1,j}+x_{i-1,j}]$$

$$+\frac{c}{(\Delta\eta)^2}[x_{i,j+1}+x_{i,j-1}]-\frac{b}{2\Delta\xi\Delta\eta}[x_{i+1,j+1}-x_{i+1,j-1}+x_{i-1,j-1}-x_{i-1,j+1}]$$

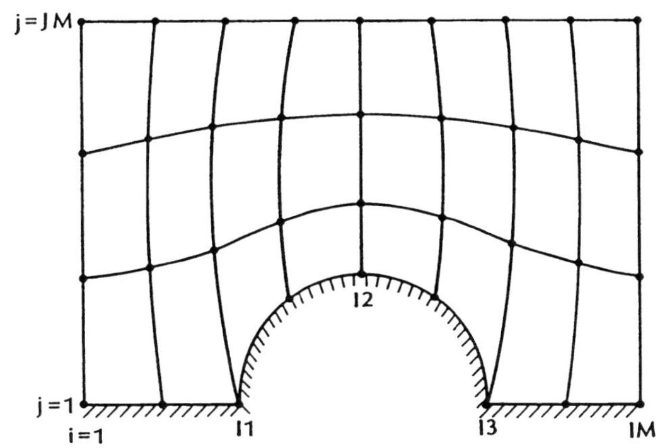

Figure 9-23a. Physical domain.

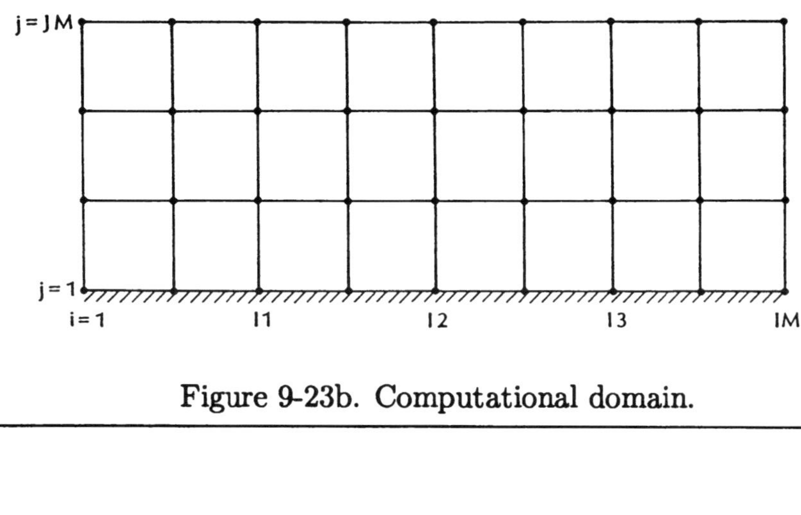

Figure 9-23b. Computational domain.

From which

$$x_{i,j} = \left\{\frac{a}{(\Delta\xi)^2}[x_{i+1,j} + x_{i-1,j}] + \frac{c}{(\Delta\eta)^2}[x_{i,j+1} + x_{i,j-1}] - \frac{b}{2\Delta\xi\Delta\eta}[x_{i+1,j+1} - x_{i+1,j-1} + x_{i-1,j-1} - x_{i-1,j+1}]\right\} /2\left[\frac{a}{(\Delta\xi)^2} + \frac{c}{(\Delta\eta)^2}\right] \quad (9\text{-}68)$$

Similarly, from Equation (9-64)

$$y_{i,j} = \left\{\frac{a}{(\Delta\xi)^2}[y_{i+1,j} + y_{i-1,j}] + \frac{c}{(\Delta\eta)^2}[y_{i,j+1} + y_{i,j-1}] - \frac{b}{2\Delta\xi\Delta\eta}[y_{i+1,j+1} - y_{i+1,j-1} + y_{i-1,j-1} - y_{i-1,j+1}]\right\} /2\left[\frac{a}{(\Delta\xi)^2} + \frac{c}{(\Delta\eta)^2}\right] \quad (9\text{-}69)$$

To start the solution, an initial distribution of x and y coordinates of the grid points within the physical domain must be provided. As discussed previously, this distribution may be obtained by using an algebraic model. The coefficients a, b, and c appearing in Equations (9-68) and (9-69) are determined from Equations (9-65) through (9-67) using finite difference approximations. The x and y values in these expressions are provided by the initial distributions for the first iteration, and subsequently from the previous iteration, i.e., the computation of coefficients lag by one iterative level. The iterative solution continues until a specified convergence criterion is met. For this purpose the total changes in the dependent variables are evaluated as

$$\text{ERRORX} = \sum_{\substack{i=2\\j=2}}^{\substack{j=JMM1\\i=IMM1}} ABS\left[x_{i,j}^{k+1} - x_{i,j}^{k}\right]$$

$$\text{ERRORY} = \sum_{\substack{i=2\\j=2}}^{\substack{j=JMM1\\i=IMM1}} ABS\left[y_{i,j}^{k+1} - y_{i,j}^{k}\right]$$

$$\text{ERRORT} = \text{ERRORX} + \text{ERRORY}$$

where k represents the iterative level. The convergence criterion is set as ERRORT < ERRORMAX where ERRORMAX is a specified input.

Other iterative schemes may be used to solve the system of equations given by (9-63) and (9-64). For example, the line Gauss-Seidel formulation yields:

$$\frac{a}{(\Delta\xi)^2}x^{k+1}_{i+1,j} - 2\left[\frac{a}{(\Delta\xi)^2} + \frac{c}{(\Delta\eta)^2}\right]x^{k+1}_{i,j} + \frac{a}{(\Delta\xi)^2}x^{k+1}_{i-1,j} =$$
$$\frac{b}{2\Delta\xi\Delta\eta}\left[x^k_{i+1,j+1} - x^{k+1}_{i+1,j-1} + x^{k+1}_{i-1,j-1} - x^k_{i-1,j+1}\right] - \frac{c}{(\Delta\eta)^2}\left[x^k_{i,j+1} + x^{k+1}_{i,j-1}\right] \qquad (9\text{-}70)$$

$$\frac{a}{(\Delta\xi)^2}y^{k+1}_{i+1,j} - 2\left[\frac{a}{(\Delta\xi)^2} + \frac{c}{(\Delta\eta)^2}\right]y^{k+1}_{i,j} + \frac{a}{(\Delta\xi)^2}y^{k+1}_{i-1,j} =$$
$$\frac{b}{2\Delta\xi\Delta\eta}\left[y^k_{i+1,j+1} - y^{k+1}_{i+1,j-1} + y^{k+1}_{i-1,j-1} - y^k_{i-1,j+1}\right] - \frac{c}{(\Delta\eta)^2}\left[y^k_{i,j+1} + y^{k+1}_{i,j-1}\right] \qquad (9\text{-}71)$$

and the line SOR results in

$$\omega\frac{a}{(\Delta\xi)^2}x^{k+1}_{i+1,j} - 2\left[\frac{a}{(\Delta\xi)^2} + \frac{c}{(\Delta\eta)^2}\right]x^{k+1}_{i,j} + \omega\frac{a}{(\Delta\xi)^2}x^{k+1}_{i-1,j} =$$
$$-2(1-\omega)\left[\frac{a}{(\Delta\xi)^2} + \frac{c}{(\Delta\eta)^2}\right]x^k_{i,j}$$
$$+\omega\frac{b}{2\Delta\xi\Delta\eta}\left[x^k_{i+1,j+1} - x^{k+1}_{i+1,j-1} + x^{k+1}_{i-1,j-1} - x^k_{i-1,j+1}\right]$$
$$-\omega\frac{c}{(\Delta\eta)^2}\left[x^k_{i,j+1} + x^{k+1}_{i,j-1}\right] \qquad (9\text{-}72)$$

$$\omega\frac{a}{(\Delta\xi)^2}y^{k+1}_{i+1,j} - 2\left[\frac{a}{(\Delta\xi)^2} + \frac{c}{(\Delta\eta)^2}\right]y^{k+1}_{i,j} + \omega\frac{a}{(\Delta\xi)^2}y^{k+1}_{i-1,j} =$$
$$-2(1-\omega)\left[\frac{a}{(\Delta\xi)^2} + \frac{c}{(\Delta\eta)^2}\right]y^k_{i,j}$$
$$+\omega\frac{b}{2\Delta\xi\Delta\eta}\left[y^k_{i+1,j+1} - y^{k+1}_{i+1,j-1} + y^{k+1}_{i-1,j-1} - y^k_{i-1,j+1}\right]$$
$$-\omega\frac{c}{(\Delta\eta)^2}\left[y^k_{i,j+1} + y^{k+1}_{i,j-1}\right] \qquad (9\text{-}73)$$

Next, an example is given where the elliptic grid generator just described is used to create the required grid system. Consider an axisymmetric blunt body at zero degree angle of attack. The configuration is an elliptical shape defined by the semi-major axis a_1 and semi-minor axis b_1. Similarly, an elliptically shaped outer boundary is defined by specifying a_2 and b_2. The nomenclature is shown in Figure 9-24. Since the grid system generated here will be used in the future to solve the Euler equations, the first grid line, i.e., $i = 1$, is specified below the stagnation line (which is aligned along the x-axis). Symmetry of grid lines about the x-axis for grid lines $i = 1$ and $i = 2$ is enforced.

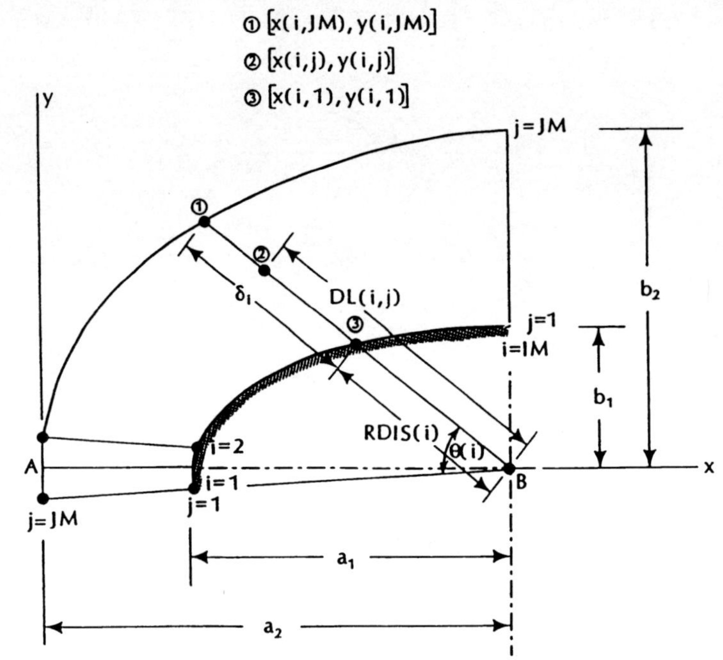

Figure 9-24. Illustration of the nomenclature for the proposed example.

The reason for this selection of grid lines is that some difficulty in solving the Euler equations is observed if grid line $i = 1$ is chosen along the stagnation line. Before the elliptic equations (9-63) and (9-64) can be solved, the grid point distribution along the boundaries and an initial grid point distribution within the domain must be specified. The distribution of grid points along the boundaries may be accomplished by using various procedures. One may select equal spacing of grid points on the body or distribute them by defining an angular position θ. Clustering of points may be specified in different regions. For example, one may choose to cluster grid points in the vicinity of the stagnation point. For the application shown, an equally divided angular position is used. The intersection of a ray originating from point B (Figure 9-24) at an angular position $\theta(i)$ with the body surface will define the x and y coordinate of grid point i at the surface where $j = 1$. Similarly, the grid point distribution on the outer boundary, where $j = JM$, is determined. In the radial direction, an algebraic expression with a clustering option is used to distribute the grid points. Since the step-size in the computational domain denoted by $\Delta\xi$ and $\Delta\eta$ can be selected arbitrarily, it is sometimes set to unity, i.e., $\Delta\xi = \Delta\eta = 1.0$.

That is the case for the algebraic expression given by

$$c(i,j) = \delta(i)\left\{1 + \beta\frac{\left[1 - \left(\frac{\beta+1}{\beta-1}\right)^{1-\gamma}\right]}{\left[1 + \left(\frac{\beta+1}{\beta-1}\right)^{1-\gamma}\right]}\right\} \tag{9-74}$$

where $\gamma = (j-1)/(JM-1)$. The radial location of grid points is determined from

$$DL(i,j) = RDIS(i,1) + c(i,j) \tag{9-75}$$

where the nomenclature is given in Figure (9-24). This procedure is also employed to distribute the grid points on the radial boundaries along $i = 2$ and $i = IM$.

A grid system constructed by the algebraic procedure just described is shown in Figure 9-25a.

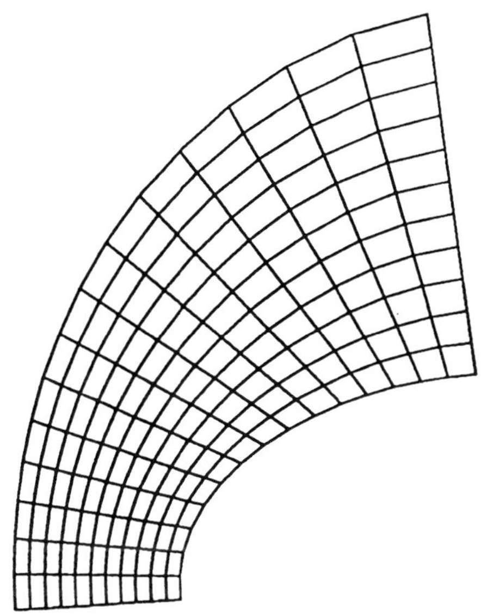

Figure 9-25a. Algebraic grid system for the simply-connected domain.

The relevant data includes $a_1 = 2$, $b_1 = 1.25$, $a_2 = 3.0$, $b_2 = 3.5$, $IM = 16$, $JM = 12$, and the clustering parameter is set to 5. With this high value of β, clustering of grid points is not reinforced. This setting is chosen for the purpose of clarity in the figure. With the initial grid point distribution available, it is now possible to use the elliptic grid generator, i.e., Equations (9-63) and (9-64). The convergence criterion ERRORMAX is set to 0.1. The solution is presented in Figure 9-25b. The metrics ξ_x, ξ_y, η_x, and η_y are computed numerically using a second-order approximate finite difference scheme. A central difference expression is used for the interior points, whereas forward and backward approximation schemes are utilized at the boundaries. The metric distributions are presented in Figures 9-26a through 9-26d. Smooth distribution of the metrics is evident. It must be emphasized again that before an attempt is made to solve the governing PDEs of fluid motion in a grid system, the metrics must be very carefully investigated. This step will prevent many unnecessary problems due to the failure of a code which may have its roots in the grid generation routine.

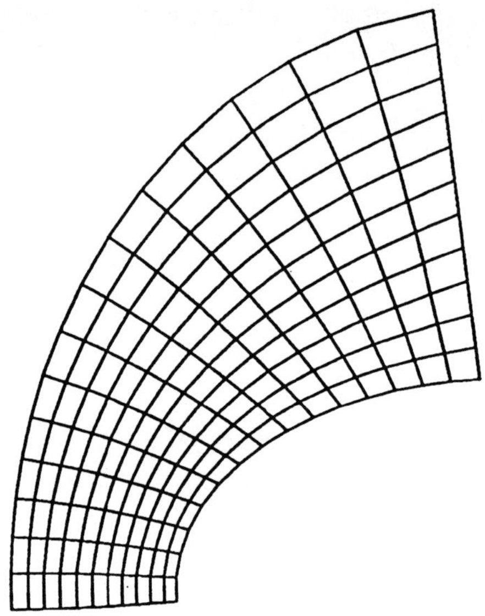

Figure 9-25b. Grid system obtained by solving a system of elliptic partial differential equations.

Hence, it is a good practice to analyze the metric distributions prior to solving the equations of fluid motion. The analysis is facilitated by plotting the metric distributions as shown in Figure 9-26.

Two observations can be made with regard to the grid system shown in Figure 9-25b. First, there is clearly a high degree of skewness in some regions of the domain. This skewness will cause some difficulty and inaccuracy in the computation of the normal gradients of flow properties at the surface. To overcome this problem, the grid lines should be perpendicular at the surface, which will improve the accuracy and simplify the computation of the normal gradients at the surface. For example, heat transfer calculations require the normal gradient of the temperature at the surface. Thus, it seems extremely useful to include an option where orthogonality of the grid lines can be reinforced at the surface. This issue will be addressed in the section entitled "Coordinate System Control".

The second observation is related to grid point clustering. The elliptic equations (9-63) and (9-64) do not include an option for grid point clustering. To include this option, Poisson's equation is used. This equation is also introduced in the section entitled "Coordinate System Control."

Note that some sort of grid point clustering can be used in the initial grid system obtained from an algebraic method. Subsequently, elliptic equations (9-63) and (9-64) may be solved.

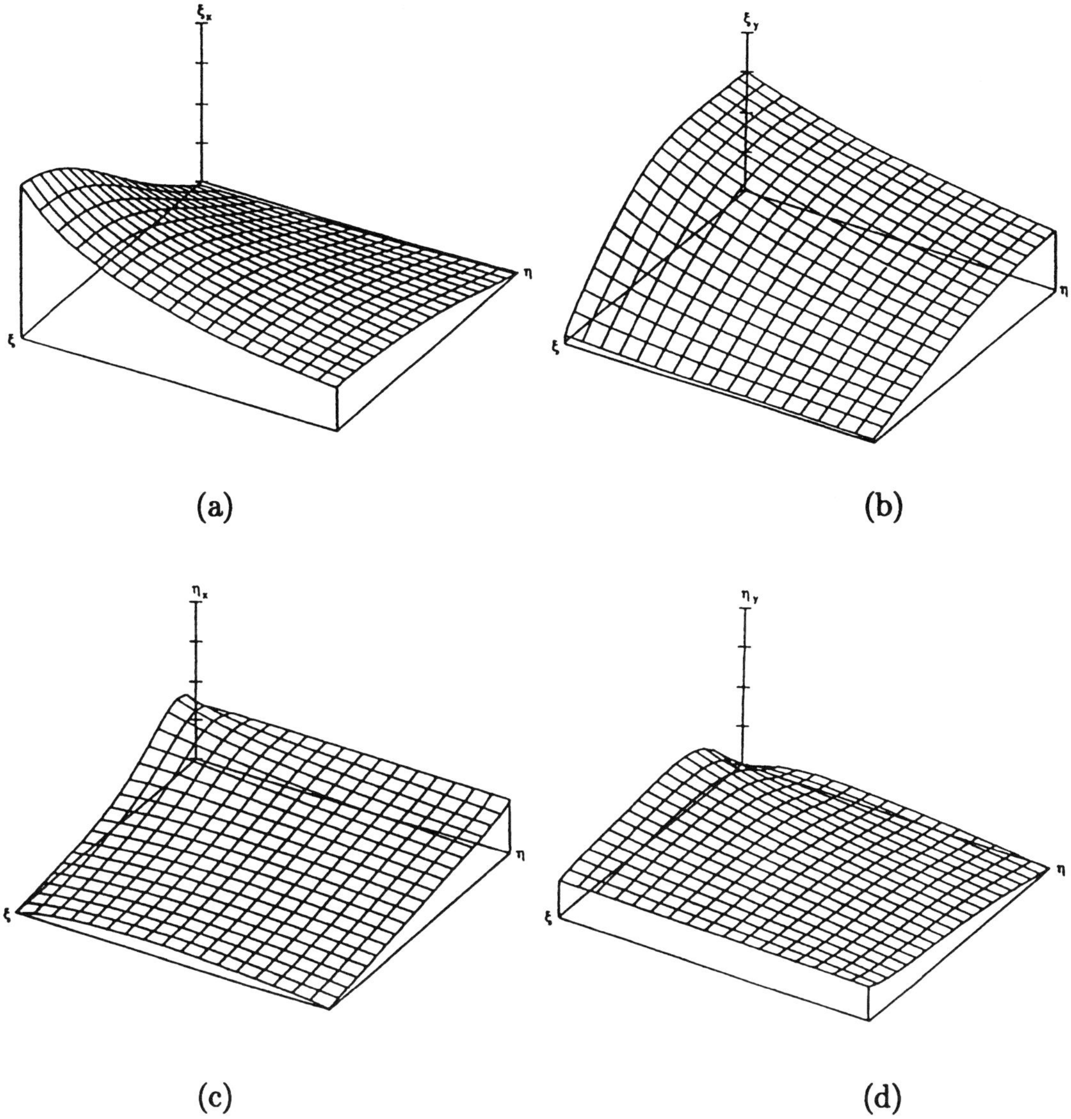

Figures 9-26a, b, c, d. Metric distributions for the grid system shown in Figure 9-25b.

However, the clustering option is not within the elliptic grid generator; and, in a classical sense, grid point control is not being used in the elliptic system. To illustrate this point, the previous domain is used with a grid clustering parameter of $\beta = 1.2$ for the algebraic model. The results are shown in Figures 9-27a

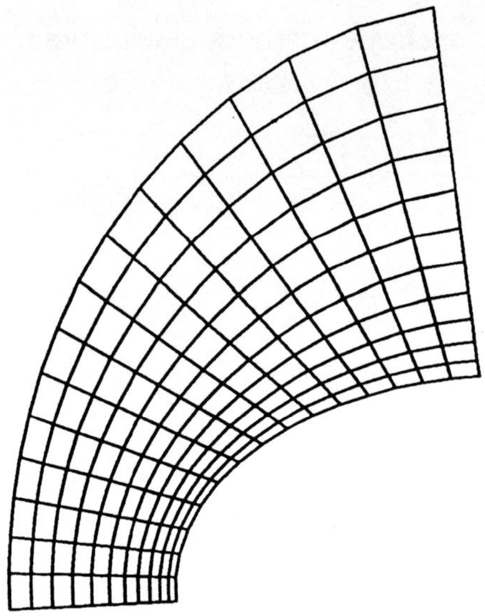

Figure 9-27a. Grid system generated by the algebraic model.

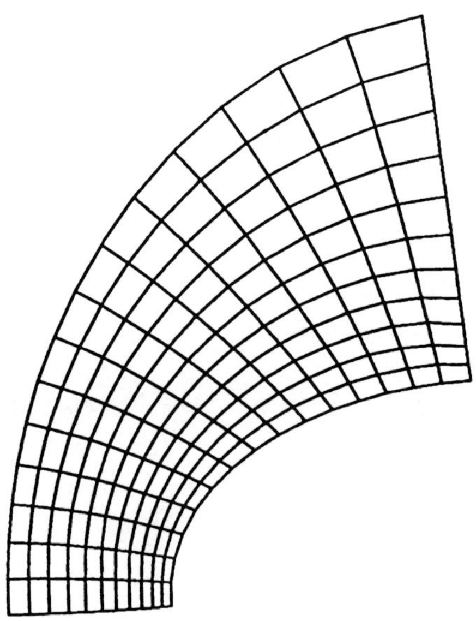

Figure 9-27b. Grid system generated by elliptic system.

and 9-27b. A wide range of grid clustering options are available for elliptic grid generators which will be reviewed shortly.

9.7.2 Doubly-Connected Domain

A doubly-connected domain is defined as a region which is not reducible. A detailed description of doubly- and multiply-connected regions is given in Reference [9-2]. For our purposes, a domain which includes one configuration within the region of interest is classified as a doubly-connected domain. A doubly-connected region may be rendered simply-connected by introducing a suitable branch cut. This procedure is accomplished by inserting a branch cut that extends from a point on the body (interior boundary) to a point on the outer boundary. As an example, consider the airfoil shown in Figure 9-28. Select an outer boundary by specifying some geometrical configuration such as a circle, ellipse, rectangle, etc. In order to unwrap the domain such that a rectangular computational domain can be created, a branch cut, shown as line AC in Figure 9-28, is introduced. An intermediate step is shown in Figure 9-29.

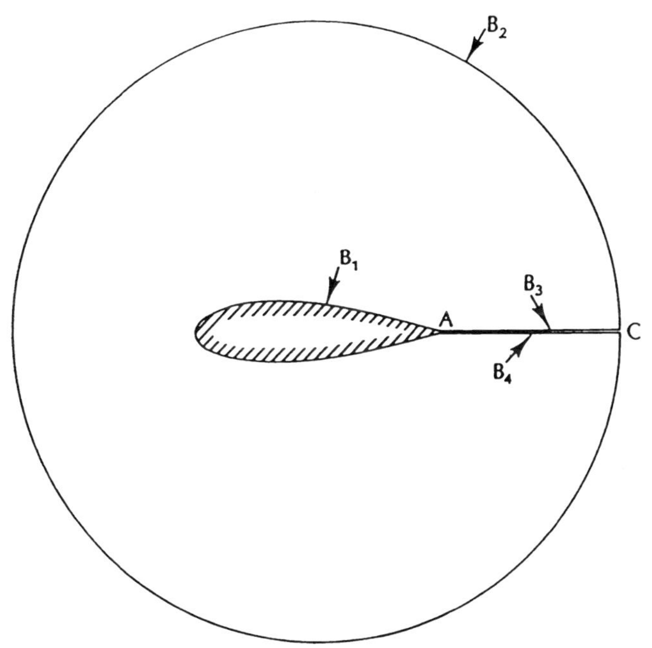

Figure 9-28. Doubly-connected region and the branch cut.

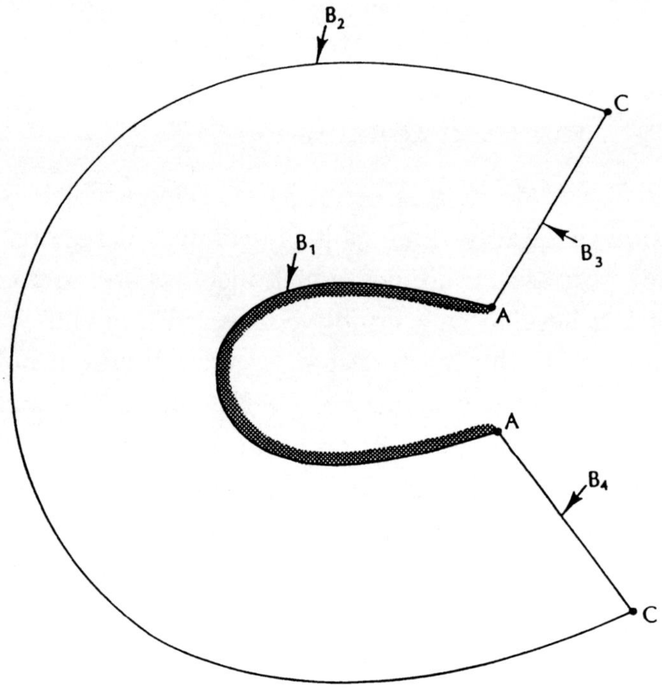

Figure 9-29. Unwrapping of the doubly-connected region.

The domain is stretched and deformed to create a rectangular shape (computational domain) as shown in Figure 9-30.

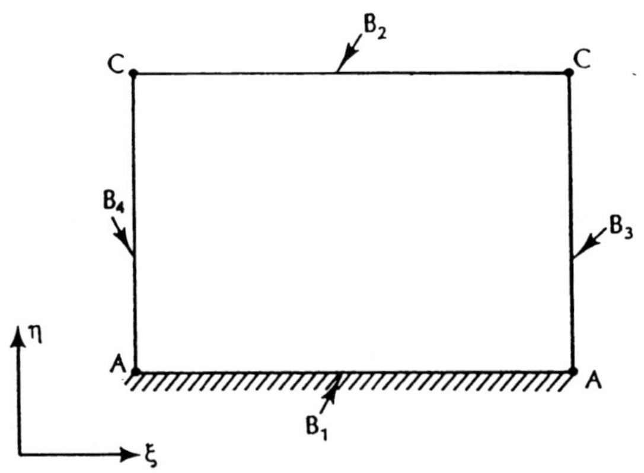

Figure 9-30. Computational domain.

The boundaries of the domain are identified by B_1, B_2, B_3, and B_4. A uniformly distributed grid system is constructed in the computational space; therefore, the location of every grid point in the computational domain, including the boundaries, is known. The object then is to employ the elliptic grid generator to determine the location of grid points in the physical space. Thus, the elliptic equations to be solved are Equations (9-63) and (9-64). The procedure is similar to the one used for simply-connected regions. That is, the grid point distribution on the boundaries of the physical domain must be defined and an initial grid point distribution for the interior region must be provided. One distinct difference is the treatment of grid points on the boundaries B_3 and B_4, i.e., on the branch cut. These points must be free to float, i.e., the location of the grids along line AC must be updated. This update is accomplished by computing new values of $x_{1,j}$ and $y_{1,j}$ after each iteration. It is not necessary to compute the values of $x_{IM,j}$ and $y_{IM,j}$, since grid lines $i = 1$ and $i = IM$ are coincident; therefore, $x_{IM,j} = x_{1,j}$ and $y_{IM,j} = y_{1,j}$. For the Gauss-Seidel formulation, $x_{1,j}$ and $y_{1,j}$ are computed according to

$$\begin{aligned} x_{1,j} = & \left\{ \frac{a}{(\Delta\xi)^2}\left[x_{2,j} + x_{IM1,j}\right] + \frac{c}{(\Delta\eta)^2}\left[x_{1,j+1} + x_{1,j-1}\right] \right. \\ & \left. - \frac{b}{2\Delta\xi\Delta\eta}\left[x_{2,j+1} - x_{2,j-1} + x_{IM1,j-1} - x_{IM1,j+1}\right] \right\} / 2\left[\frac{a}{(\Delta\xi)^2} + \frac{c}{(\Delta\eta)^2}\right] \end{aligned} \tag{9-76}$$

$$\begin{aligned} y_{1,j} = & \left\{ \frac{a}{(\Delta\xi)^2}\left[y_{2,j} + y_{IM1,j}\right] + \frac{c}{(\Delta\eta)^2}\left[y_{1,j+1} + y_{1,j-1}\right] \right. \\ & \left. - \frac{b}{2\Delta\xi\Delta\eta}\left[y_{2,j+1} - y_{2,j-1} + y_{IM1,j-1} - y_{IM1,j+1}\right] \right\} / 2\left[\frac{a}{(\Delta\xi)^2} + \frac{c}{(\Delta\eta)^2}\right] \end{aligned} \tag{9-77}$$

where $IM1 = IM - 1$.

These equations are used after each iteration to determine the new location of the grid points on the branch cut. Note that if the grid points on the branch cut are kept fixed, highly skewed grids at the branch cut are produced, which is undesirable.

To illustrate the solution procedure, the following example is proposed. Consider an airfoil whose geometry is given by

$$y = \frac{t}{0.2}(0.2969x^{\frac{1}{2}} - 0.126x - 0.3516x^2 + 0.2843x^3 - 0.1015x^4) \tag{9-78}$$

In this equation, t is the maximum thickness in percentage of the chord and the origin of the coordinate system is located at the leading edge. It is required that a grid system be generated between the airfoil and an outer boundary defined by a circle of radius, R, with the origin at the mid-chord. The physical domain is shown in Figure 9-31. The grid point distribution on the boundaries may be specified using various methods. Here an increment, Δx, is determined as

$$\Delta x = \frac{c}{(IM + 1)/2 - 1}$$

where IM must be specified as an odd number, forcing grid points at the leading and trailing edges and symmetry of grid point distribution between the upper and lower surfaces. Thus, the x coordinates of the grid points on the body can be defined for each $(i, 1)$ around the airfoil. Once the x coordinates are known, the y coordinates are evaluated by employing Equation (9-78). The grid points on the outer boundary are evaluated as follows. Locate the origin of the circle at the mid-chord and use equally spaced angular positions for each point on the outer boundary, i.e.,

$$\begin{aligned}
\text{DELTAI} &= \frac{2\pi}{IM-1} \\
\text{DELTA}(i) &= i * (\text{DELTAI}) \\
x(i, JM) &= R * \cos(\text{DELTA}(i)) \\
y(i, JM) &= -R * \sin(\text{DELTA}(i))
\end{aligned}$$

where R is the radius of the circle defining the outer boundary. Since all the grid

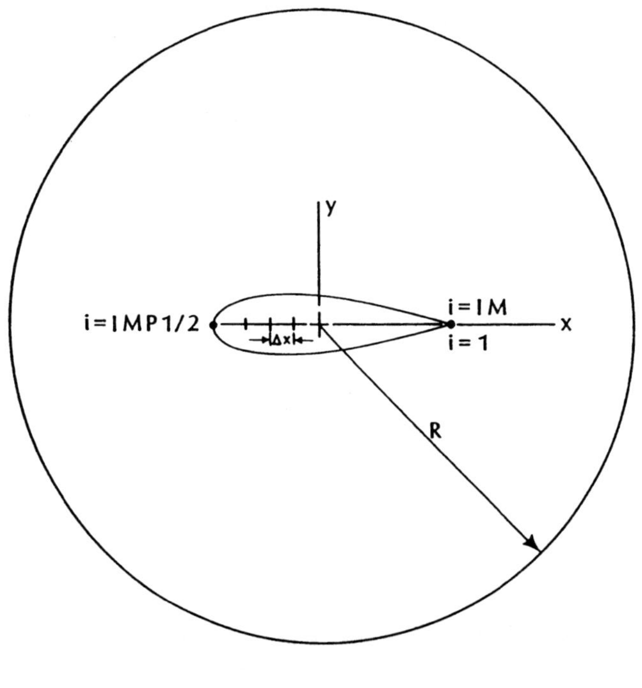

Figure 9-31. The physical domain used to illustrate the application of an elliptic grid generator to a doubly-connected region.

points should be expressed in the same coordinate system, a shift of the coordinates of the inner grid points on the airfoil to mid-chord is imposed. Now, the corresponding grid points for each i on the body (where $j = 1$) and on the outer boundary (where $j = JM$) are connected by straight lines. The grid points in the interior region are distributed along these lines. This distribution may be equally spaced, or some sort of grid clustering scheme employed. Thus, an algebraic grid system is constructed which is shown in Figure 9-32.

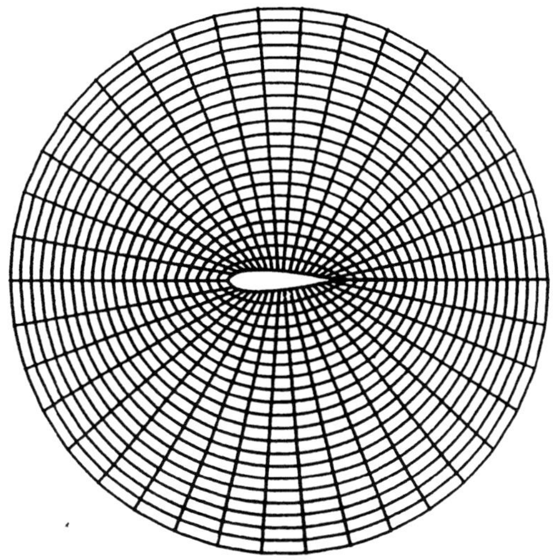

Figure 9-32. Algebraic grid system used as initial grid point distribution required by the elliptic grid generator.

Now the elliptic PDEs given by Equations (9-63) and (9-64) can be solved. The resulting grid system is shown in Figure 9-33.

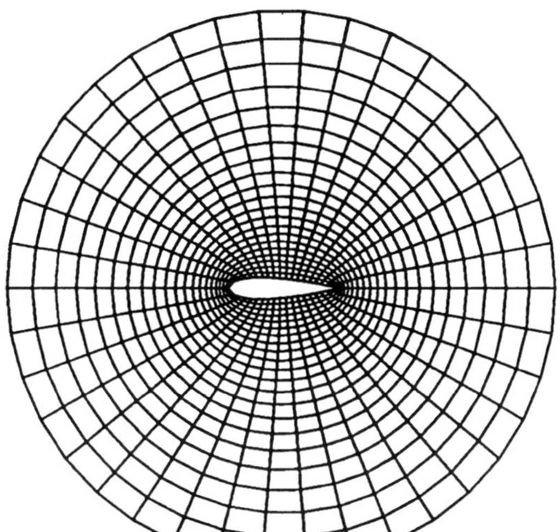

Figure 9-33. Grid system obtained by elliptic grid generator.

Careful examination of this figure clearly indicates the smooth distribution of the grid points within the domain. Also note the adjustment of the grid points on the branch cut, which was used to define the two boundaries (B_3 and B_4 shown in Figure 9-28). This type grid is known as the "O" grid for obvious reasons! If the outer boundary is defined in a C-shape configuration, it is referred to as the "C" grid.

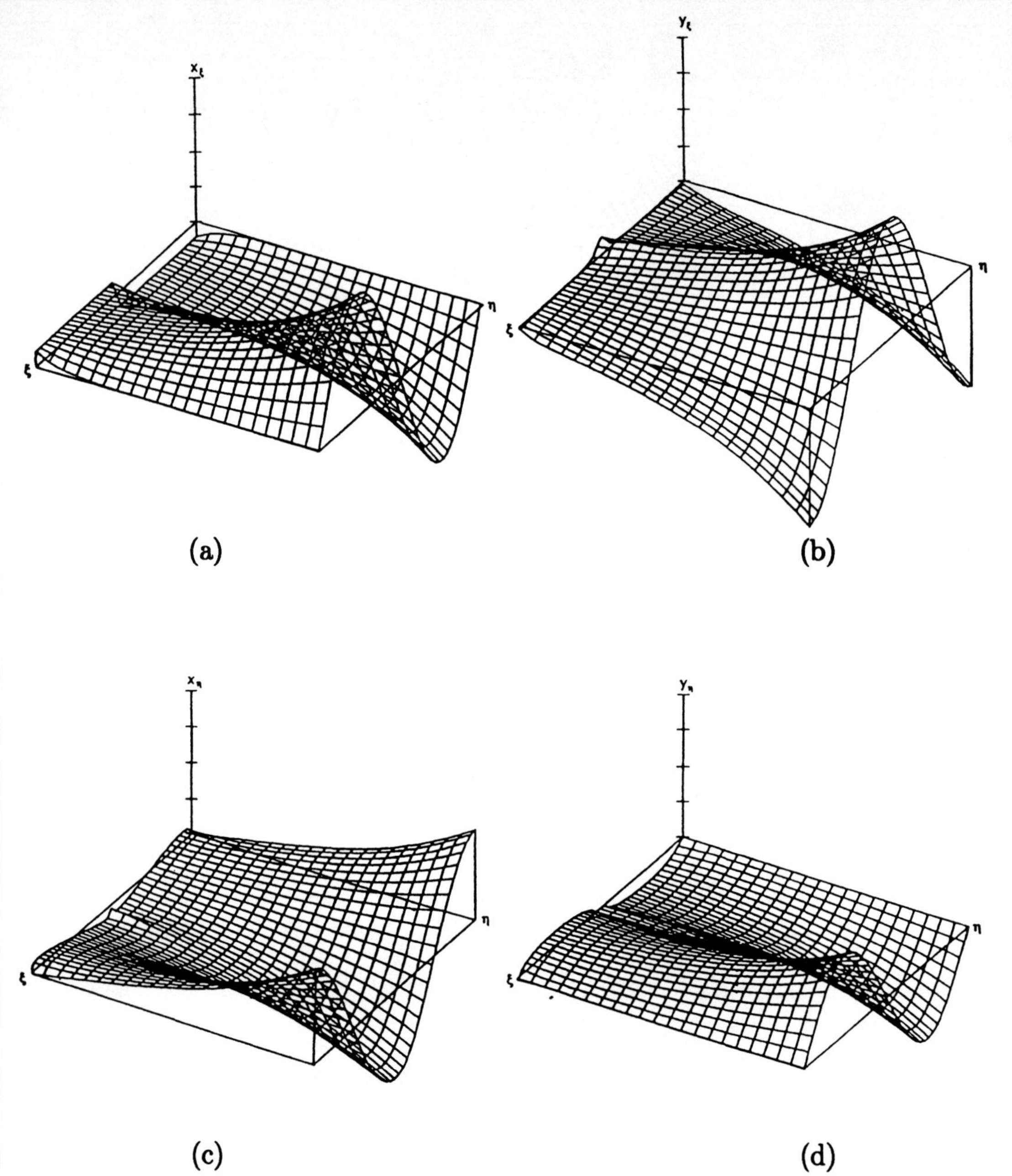

Figures 9-34a, b, c, d. Metric distributions for the domain shown in Figure 9-33.

Clearly the selection of the physical domain is dependent on the particular fluid mechanics problem to be solved. The methods by which the initial grid distribution is created is arbitrary and any procedure may be used for this purpose.

Before solving the differential equations of fluid motion, the grid system must be investigated carefully for grid point smoothness, skewness, orthogonality, and the metric distributions. The distributions of the transformation derivatives x_ξ, y_ξ, x_η, and y_η are shown in Figures 9-34a through 9-34d, which indicate well-behaved distributions.

9.7.3 Multiply-Connected Domain

The same procedure which applied to a doubly-connected region may be extended to a multiply-connected region. For a multiply-connected region, more than one object is located within the domain. A branch cut is introduced to connect one body to the outer boundary. In addition, other cuts are inserted between various objects within the domain. A graphical illustration is shown in Figure 9-35 which represents the physical domain.

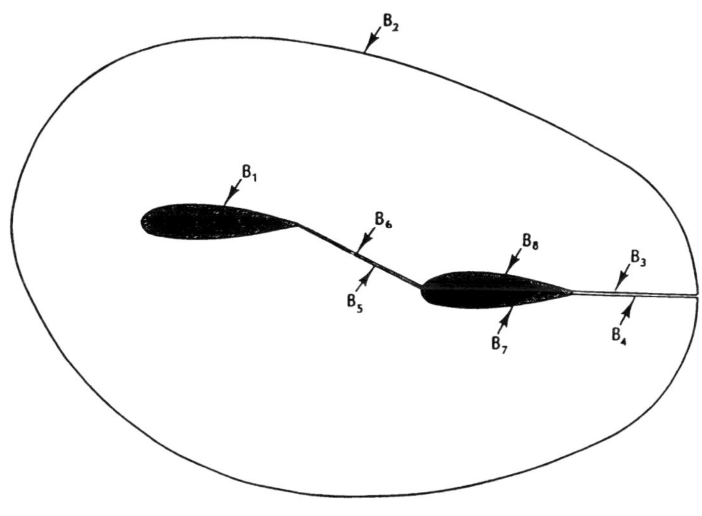

Figure 9-35. Physical domain for a multiply-connected region.

Figure 9-36 shows a typical intermediate step, and Figure 9-37 represents the computational space.

The elliptic equations (9-63) and (9-64) may be solved by any iterative method for this domain. The procedure for specifying the boundary points and initial distribution of the grid points within the domain is similar to that of a doubly-connected region. Again, the grid points on the branch cuts must be computed after each iteration level, i.e., the grid points on the branch cuts and, thus, the shape of the branch cuts, are changing from one iteration to the next. The iterative procedure is continued until a specified convergence criterion is satisfied.

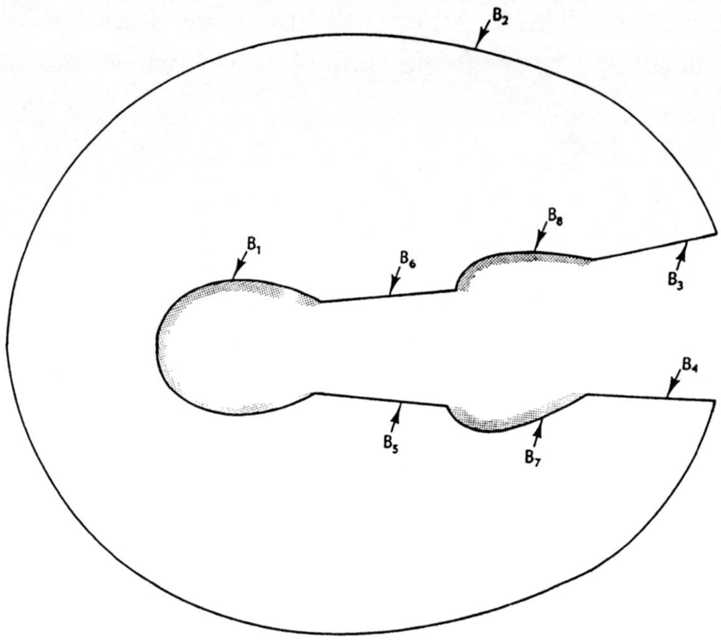

Figure 9-36. Unwrapping of a multiply-connected region.

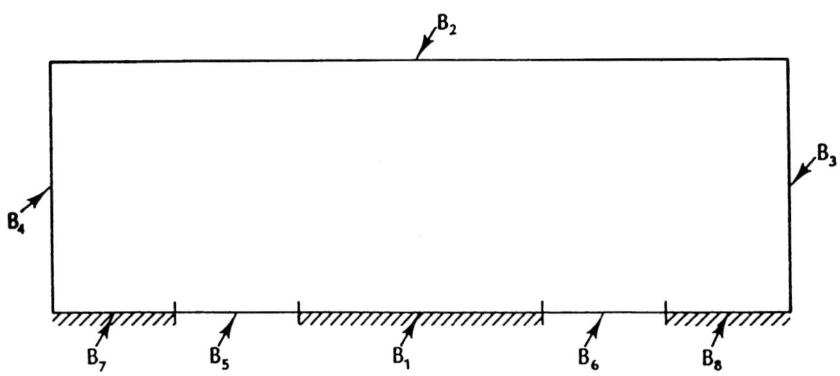

Figure 9-37. Computational domain.

For illustrative purposes, consider the airfoil described in the previous application. Now, insert two such airfoils within the circular domain. Grid points are distributed on the bodies and the outer boundary as well as the branch cuts. Once the grid points on the boundaries are specified, the grid point distribution within the domain can be determined by any algebraic procedure. For example, grid points may be distributed on straight rays with equal spacing. This method was used to generate the grid system shown in Figure 9-38.

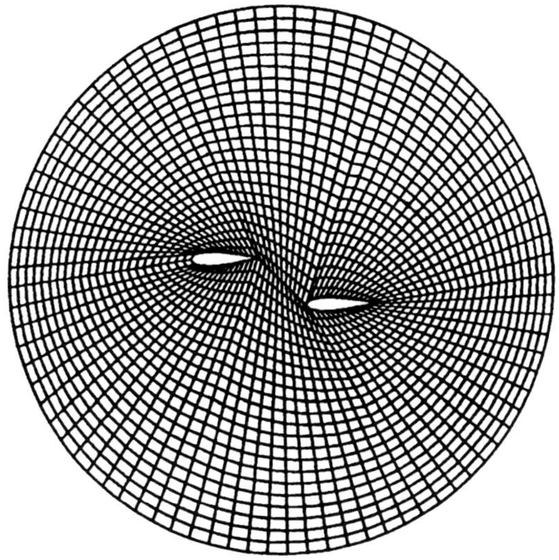

Figure 9-38. Algebraic grid point distribution for the multiply-connected region.

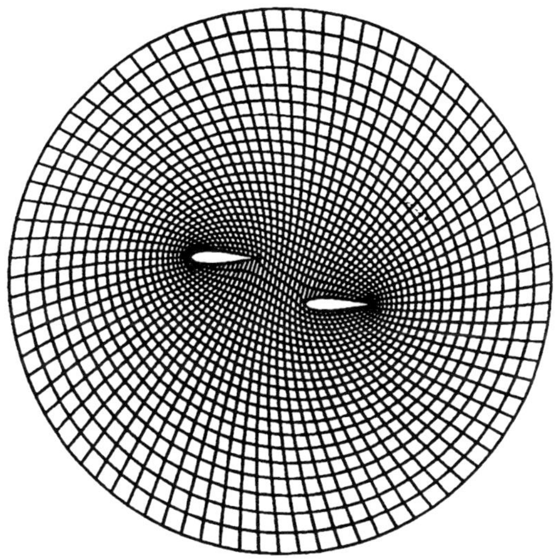

Figure 9-39. Grid system generated by elliptic PDEs.

The crude distribution of grid points within the domain is clearly evident. This distribution is used as an initial condition to start the elliptic grid generator, i.e., Equations (9-63) and (9-64). The solution is shown in Figure 9-39. Two observations can be made. First, the grid points are distributed smoothly within the domain; second, the original branch cuts have been reshaped due to readjustment of the grid points.

9.8 Coordinate System Control

The elliptic PDEs given in Equations (9-61) and (9-62) do not include any option for grid point control within the domain. The location of grid points around the body and outer boundary is input; therefore, control of grid spacing on the boundaries is established by the user. In order to use the grid point control within the domain, Poisson's equation must be introduced. The logic behind the fundamental idea is similar to that found in the previous discussion which is briefly reviewed here. Consider the steady state heat conduction in 2-D with a source term. By varying the location and strength of the heat source, the isothermal lines are modified. This reasoning is adapted to a grid generation technique produced by elliptic PDEs. Thus, a source term is added to the right-hand sides of Equations (9-61) and (9-62). The resulting Poisson equations are:

$$\xi_{xx} + \xi_{yy} = P(\xi, \eta) \tag{9-79}$$

$$\eta_{xx} + \eta_{yy} = Q(\xi, \eta) \tag{9-80}$$

Since the numerical solution of Equations (9-79) and (9-80) is performed in a rectangular domain with equal spacing, a transformation of equations and associated boundary conditions is required. For this purpose, equations derived in Appendix F are used. The resulting transformed equations are:

$$a x_{\xi\xi} - 2b x_{\xi\eta} + c x_{\eta\eta} = -\frac{1}{J^2}(P x_\xi + Q x_\eta) \tag{9-81}$$

$$a y_{\xi\xi} - 2b y_{\xi\eta} + c y_{\eta\eta} = -\frac{1}{J^2}(P y_\xi + Q y_\eta) \tag{9-82}$$

These elliptic equations may be solved by any iterative scheme, provided P and Q are given.

The functions P and Q are selected according to a specific need. The requirement may be a clustering of grid points at some prescribed location or forcing orthogonality at the surface. Among several expressions available, two such functions will be investigated here.

9.8.1 Grid Point Clustering

Grid point clustering is enforced by selection of the functions P and Q. The selection is based on grid point attraction in the vicinity of a defined grid line(s) or grid point(s) or a combination of both. For example, it might be desirable to cluster the grid points in the vicinity of line η_k or near point (ξ_i, η_j). A particular functional relation with such capabilities was introduced in Reference [9-1]. The relations are given in exponential form as

$$P = -\sum_{IS=1}^{NI} a(IS)\frac{I-IAL(IS)}{ABS(I-IAL(IS))}\exp[-c(IS)ABS(I-IAL(IS))] + \tag{9-83}$$

$$-\sum_{JS=1}^{NJ} b(JS)\frac{I-IAL(JS)}{ABS(I-IAL(JS))}\exp\left\{-d(JS)\left[(I-IAL(JS))^2+(J-JAL(JS))^2\right]^{\frac{1}{2}}\right\}$$

$$Q = -\sum_{IS=1}^{NI} a(IS)\frac{J-JAL(IS)}{ABS(J-JAL(IS))}\exp[-c(IS)ABS(J-JAL(IS))] + \tag{9-84}$$

$$-\sum_{JS=1}^{NJ} b(JS)\frac{J-JAL(JS)}{ABS(J-JAL(JS))}\exp\left\{-d(JS)\left[(I-IAL(JS))^2+(J-JAL(JS))^2\right]^{\frac{1}{2}}\right\}$$

Note that these equations are used when $\Delta\eta = \Delta\xi = 1$. Otherwise,

$$\begin{aligned} P &= -\sum_{m=1}^{M} a_m \frac{\xi-\xi_m}{|\xi-\xi_m|}\exp(-c_m|\xi-\xi_m|) \\ &\quad -\sum_{n=1}^{N} b_n \frac{\xi-\xi_n}{|\xi-\xi_n|}\exp\left\{-d_n\left[(\xi-\xi_n)^2+(\eta-\eta_n)^2\right]^{\frac{1}{2}}\right\} \end{aligned} \tag{9-85}$$

$$\begin{aligned} Q &= -\sum_{m=1}^{M} a_m \frac{\eta-\eta_m}{|\eta-\eta_m|}\exp(-c_m|\eta-\eta_m|) \\ &\quad -\sum_{n=1}^{N} b_n \frac{\eta-\eta_n}{|\eta-\eta_n|}\exp\left\{-d_n\left[(\xi-\xi_n)^2+(\eta-\eta_n)^2\right]^{\frac{1}{2}}\right\} \end{aligned} \tag{9-86}$$

where the amplification factors a and b and the decay factors c and d are inputs. NI or M is the number of constant grid lines ξ and/or η around which clustering is enforced, whereas $IAL(IS)$ and $JAL(IS)$ specify the lines. Similarly, NJ or N is the number of grid points around which clustering is enforced with $IAL(JS)$ and $JAL(JS)$ specifying the points.

Each term in these equations has a special role. The first exponential term in Equation (9-83) attracts the lines of constant ξ to the vicinity of lines ξ_i, where ξ_i may be specified as one line or a series of lines. For example, it may be advantageous to attract grid lines toward a particular grid line corresponding to $i = 16$, in which case $NI = 1$ and $IAL(1) = 16$. For grid line attraction toward a set of grid lines (such as $i = 15$, 16, and 17), $NI = 3$ and

$$\begin{aligned} IAL(1) &= 15 \\ IAL(2) &= 16 \\ IAL(3) &= 17 \end{aligned}$$

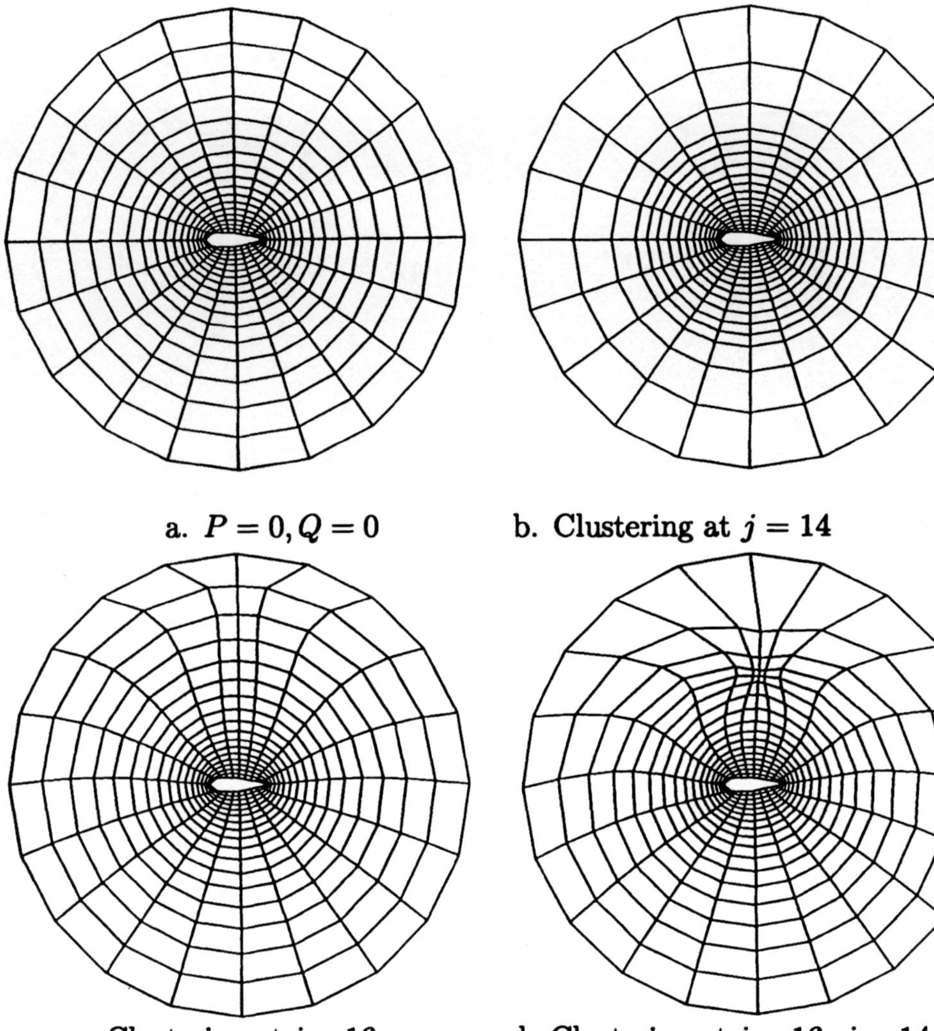

a. $P = 0, Q = 0$ b. Clustering at $j = 14$

c. Clustering at $i = 16$ d. Clustering at $i = 16$, $j = 14$

Figure 9-40. Grid systems generated by elliptic PDEs.

Similar effects are produced by Equation (9-84) for the lines of constant η. The inclusion of the sign function [term $[I - IAL(IS)]/[ABS(I - IAL(IS))]$ in Equation (9-83) and similar term in Equation (9-84)] enables the attraction on both sides of ξ_i and η_i lines. Elimination of this term will cause attraction on one side and repulsion on the other side.

The second term in Equation (9-83) clusters the lines of constant ξ around point(s) ξ_i, η_j. A similar effect is produced by the second term of Equation (9-84), where the lines of constant η are attracted toward point(s) ξ_i, η_j.

To illustrate the effect of each term, they are investigated separately. Consider the symmetrical airfoil described in Equation (9-78), where a value of 0.25 is used for t. When P and Q are set equal to zero, a grid system with no clustering is generated, i.e., Equations (9-63) and (9-64) are solved. The grid system is shown in Figure 9-40a where IM and JM are set to 21 and 18, respectively. When the first term of Equation (9-84) is employed, clustering in the vicinity of constant η lines is enforced. For the example shown in Figure 9-40b, clustering at a grid line corresponding to $j = 14$ was enforced. An amplification factor of $a = 40$ and a decay factor of $c = 2.0$ was used. A similar effect is shown when the first term of Equation (9-83) is used. The result is illustrated in Figure 9-40c, where the grid line attraction is enforced toward the line $i = 16$ with $a = 10$ and $c = 1.2$. Finally, the second terms in P and Q are used to cluster grid points near point $i = 16$, $j = 14$. The amplification and decay factors were specified as 200 and 1.5, respectively. The result is shown in Figure 9-40d.

An obvious question at this point is: How does one decide on the values of the amplification and decay factors required for P and Q? These values depend on the particular problem, i.e., the given physical domain, the number of grid points used, boundary points, etc. However, there is no dependency function which will automatically provide the values of a, b, c, and d. Thus, a trial and error procedure is used. Note that the values used in one problem are not necessarily valid for a different problem. Usually a few sets of values must be investigated in order to obtain the desired grid point distribution.

9.8.2 Orthogonality at the Surface

In the previous section the option of grid point clustering was investigated. The grid system generated by these methods may not be satisfactory for some applications due to large skewness of the grid lines, especially when they occur at the surface. The difficulty is encountered when normal gradients of flow properties are required. To overcome this deficiency, a forcing function is used which will enforce orthogonality of grid lines at the surface. The resulting grid system simplifies the computation of the normal gradients and increases their accuracy. A forcing function to meet this criterion was introduced in Ref. [9-3]. The proposed functions are

$$P = P_1 \exp[-a(\eta - \eta_1)] \tag{9-87}$$

and

$$Q = Q_1 \exp[-b(\eta - \eta_1)] \tag{9-88}$$

where a and b are specified as positive constants and P_1 and Q_1 are given by

$$P_1 = J(y_\eta R_1 - x_\eta R_2)|_{\eta=\eta_1} \tag{9-89}$$

$$Q_1 = J(-y_\xi R_1 + x_\xi R_2)|_{\eta=\eta_1} \tag{9-90}$$

whereas

$$R_1 = -J^2(ax_{\xi\xi} - 2bx_{\xi\eta} + cx_{\eta\eta})|_{\eta=\eta_1} \tag{9-91}$$

$$R_2 = -J^2(ay_{\xi\xi} - 2by_{\xi\eta} + cy_{\eta\eta})|_{\eta=\eta_1} \tag{9-92}$$

The governing elliptic PDEs with the forcing functions are

$$ax_{\xi\xi} - 2bx_{\xi\eta} + cx_{\eta\eta} = -\frac{1}{J^2}\Big\{P_1\Big[\exp\{-a(\eta-\eta_1)\}\Big]x_\xi + Q_1\Big[\exp\{-b(\eta-\eta_1)\}\Big]x_\eta\Big\} \tag{9-93}$$

and

$$ay_{\xi\xi} - 2by_{\xi\eta} + cy_{\eta\eta} = -\frac{1}{J^2}\Big\{P_1\Big[\exp\{-a(\eta-\eta_1)\}\Big]y_\xi + Q_1\Big[\exp\{-b(\eta-\eta_1)\}\Big]y_\eta\Big\} \tag{9-94}$$

At this point the central issue is how the constraints must be imposed on orthogonality and clustering. The orthogonality of the grid lines, ξ, to the body surface defined by $\eta = \eta_1$ is considered first. This condition implies that

$$\nabla\xi\cdot\nabla\eta = |\nabla\xi||\nabla\eta|\cos\theta \tag{9-95}$$

where θ of 90°provides orthogonality. This equation is expanded as

$$\xi_x\eta_x + \xi_y\eta_y = \left[\left(\xi_x^2+\xi_y^2\right)\left(\eta_x^2+\eta_y^2\right)\right]^{\frac{1}{2}}\cos\theta \tag{9-96}$$

The following equation results after substitution of the transformation derivatives given by Equations (9-14) through (9-17):

$$x_\eta x_\xi + y_\xi y_\eta = -\left[\left(x_\eta^2+y_\eta^2\right)\left(x_\xi^2+y_\xi^2\right)\right]^{\frac{1}{2}}\cos\theta \tag{9-97}$$

A second condition is introduced to define the grid line spacing between η_1 (at the surface) and η_2. This specification is accomplished by considering $ds = [(dx)^2 + (dy)^2]^{\frac{1}{2}}$, where s is the distance along the constant ξ lines. Utilizing Equations (9-11) and (9-12), ds can be expressed as

$$ds = [(x_\xi d\xi + x_\eta d\eta)^2 + (y_\xi d\xi + y_\eta d\eta)^2]^{\frac{1}{2}}$$

which may be reduced to

$$ds = [(x_\eta)^2 + (y_\eta)^2]^{\frac{1}{2}}d\eta \tag{9-98}$$

when applied along the lines of constant ξ, i.e., when $d\xi = 0$.

In order to solve the elliptic equations given by (9-93) and (9-94), the forcing terms on the right-hand sides must be computed. This procedure will require computation of R_1, R_2, P_1, and Q_1, which are all evaluated at $\eta = \eta_1$. Since the grid line η_1 is the body surface, grid point distribution along this boundary is specified. Therefore, the ξ derivatives along the surface may be computed according to

$$x_\xi(i,1) = \frac{x_{i+1,1} - x_{i-1,1}}{2\Delta\xi} \tag{9-99}$$

$$y_\xi(i,1) = \frac{y_{i+1,1} - y_{i-1,1}}{2\Delta\xi} \tag{9-100}$$

$$x_{\xi\xi}(i,1) = \frac{x_{i+1,1} - 2x_{i,1} + x_{i-1,1}}{(\Delta\xi)^2} \tag{9-101}$$

$$y_{\xi\xi}(i,1) = \frac{y_{i+1,1} - 2y_{i,1} + y_{i-1,1}}{(\Delta\xi)^2} \tag{9-102}$$

To compute the η derivatives, Equations (9-97) and (9-98) are solved for x_η and y_η, yielding

$$x_\eta = s_\eta(-x_\xi \cos\theta - y_\xi \sin\theta)/(x_\xi^2 + y_\xi^2)^{\frac{1}{2}}$$

and

$$y_\eta = s_\eta(-y_\xi \cos\theta + x_\xi \sin\theta)/(x_\xi^2 + y_\xi^2)^{\frac{1}{2}}$$

When θ is equal to 90°, the equations above are reduced to

$$x_\eta = s_\eta(-y_\xi)/(x_\xi^2 + y_\xi^2)^{\frac{1}{2}} \tag{9-103}$$

and

$$y_\eta = s_\eta(x_\xi)/(x_\xi^2 + y_\xi^2)^{\frac{1}{2}} \tag{9-104}$$

where

$$s_\eta = \left.\frac{ds}{d\eta}\right|_{\xi=\text{constant}} = \frac{\Delta s}{\Delta\eta} \tag{9-105}$$

The second-order derivatives with respect to η are computed according to

$$x_{\eta\eta}(i,1) = \frac{-7x_{i,1} + 8x_{i,2} - x_{i,3}}{2(\Delta\eta)^2} - 3x_\eta(i,1)/\Delta\eta \tag{9-106}$$

and

$$y_{\eta\eta}(i,1) = \frac{-7y_{i,1} + 8y_{i,2} - y_{i,3}}{2(\Delta\eta)^2} - 3y_\eta(i,1)/\Delta\eta \tag{9-107}$$

This completes the equations necessary to evaluate all the derivatives at $\eta = \eta_1$ used in the computation of P_1 and Q_1.

The second-order derivatives on the left-hand sides of Equations (9-93) and (9-94) are approximated by central difference expressions. The first order derivatives appearing on the right-hand side are approximated by forward or backward formulas. The reason for this approximation is to prevent any instabilities associated with central differencing, which could occur for large values of P_1 and Q_1. For example, if P_1 is positive, x_ξ and y_ξ are approximated by forward differencing; and if P_1 is negative, a backward differencing is used. Similarly, if Q_1 is positive, x_η and y_η are approximated by forward differencing—otherwise a backward differencing is used.

To solve the elliptic system given by Equations (9-93) and (9-94), any iterative scheme may be used. A typical solution is shown in Figure 9-41.

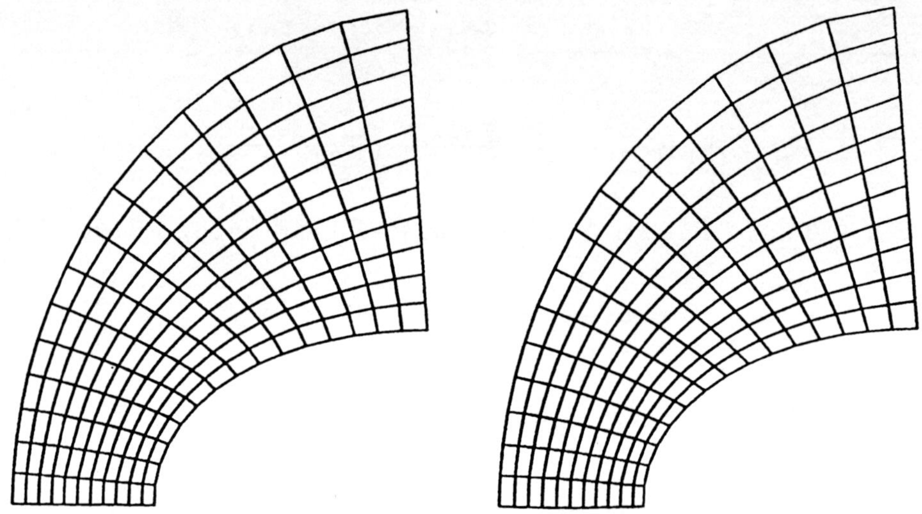

Figure 9-41. Grid systems generated with and without orthogonality condition.

In this example, a grid system is generated for a blunt body configuration. For comparative purposes a similar grid system with no orthogonality at the surface is also shown. The values of constants a and b were specified as 1.0, whereas Δs was given as a function of i locations, i.e., Δs increased along the surface.

Before closing this section, consider the following observations on elliptic grid generators. The advantages of this class of grid generators are:

(1) Will provide smooth grid point distribution, i.e., if a boundary discontinuity point exists, it will be smoothed out in the interior domain;
(2) Numerous options for grid clustering and surface orthogonality are available;
(3) Method can be extended to 3-D problems.

Among disadvantages of the method are:

(1) Computation time is large (compared to algebraic methods or hyperbolic grid generators);
(2) Specification of the forcing functions P and Q (or the constants used in these functions) is not easy;
(3) Metrics must be computed numerically.

9.9 Hyperbolic Grid Generation Techniques

It has been shown how elliptic PDEs are used to generate grid systems for closed domains. For open domains, where the outer boundary cannot be prescribed, hyperbolic PDEs may be developed to provide the required grid generators. Since a marching procedure is used to solve such a system, computationally they are faster than elliptic grid generators.

The mathematical development is based on two constraints. The first constraint imposes orthogonality of grid lines at the surface as well as the interior domain. Mathematically the orthogonality is described by the slopes being negative reciprocals. The slope of constant ξ and η lines are determined as follows. Along a line of constant ξ, the differential $d\xi$ is zero. Considering a 2-D problem, it may be written that

$$d\xi = \xi_x dx + \xi_y dy$$

For $d\xi = 0$, the slope is

$$\left.\frac{dy}{dx}\right|_{\xi=c} = -\frac{\xi_x}{\xi_y} \tag{9-108}$$

Similarly, for lines of constant η,

$$d\eta = \eta_x dx + \eta_y dy$$

from which is obtained

$$\left.\frac{dy}{dx}\right|_{\eta=c} = -\frac{\eta_x}{\eta_y} \tag{9-109}$$

Now, imposing the orthogonality condition,

$$\left.\frac{dy}{dx}\right|_{\xi=c} \left.\frac{dy}{dx}\right|_{\eta=c} = -1$$

or using Equations (9-108) and (9-109),

$$\left(-\frac{\xi_x}{\xi_y}\right)\left(-\frac{\eta_x}{\eta_y}\right) = -1$$

or

$$\xi_x \eta_x + \xi_y \eta_y = 0 \tag{9-110}$$

Equation (9-110) is rearranged using Equations (9-14) through (9-17) to give

$$x_\xi x_\eta + y_\xi y_\eta = 0 \tag{9-111}$$

The second constraint must include some geometrical consideration. To achieve this objective two approaches are investigated. In the first approach, the Jacobian of transformation is specified. Various schemes can be used for this purpose which will be discussed shortly. This method will be referred to as the *cell area approach*. In the second approach, the arc length from one grid point to the next is prescribed. This procedure will be referred to as the *arc-length approach*.

In the first approach, the definition for the Jacobian of transformation is used as the second hyperbolic equation, i.e.,

$$x_\xi y_\eta - x_\eta y_\xi = F(\xi, \eta) \tag{9-112}$$

Thus, a system of hyperbolic equations which is solved for the grid point distribution in the physical space is

$$x_\xi x_\eta + y_\xi y_\eta = 0 \tag{9-113a}$$

$$x_\xi y_\eta - x_\eta y_\xi = F \tag{9-113b}$$

In order to solve this hyperbolic system, F must be provided. Three procedures described in References [9-4], [9-5], and [9-6] are considered here. Reference [9-4] suggests employing concentric circles to determine $F(\xi, \eta)$ distribution. To do so, a circle is defined, whose perimeter is that of the inner boundary of the physical domain. Then a set of concentric circles at various radii is specified. This specification could be accomplished by using an algebraic function (similar to those used for grid point clustering). Subsequently, grid points on the inner circle (whose perimeter is that of the inner boundary) are distributed and grid points on all concentric circles are determined by rays emanating from the origin, as shown in Figure 9-42. Now, the Jacobian F is computed for this grid point distribution. Note the manner in which concentric circles are specified is used for grid line clustering.

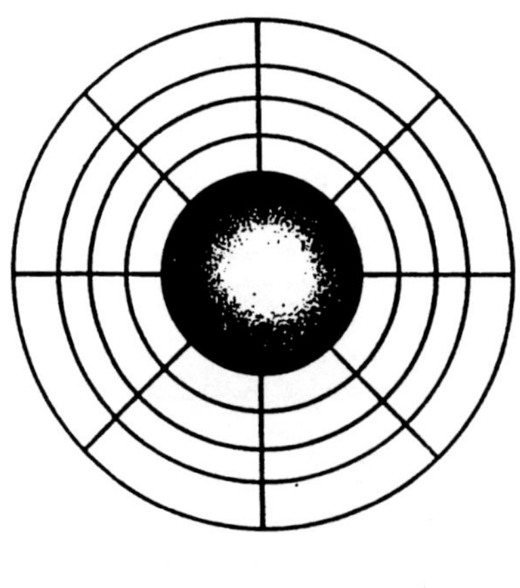

Figure 9-42. Concentric circles used to determine F.

A second approach to determine F is discussed in Reference [9-5]. In this procedure, the length of the inner boundary is drawn as a straight line with the same grid point distribution. Subsequently, parallel grid lines (constant η) are created to produce a nonuniform grid spacing in a rectangular domain. Again clustering of grid lines may be used in this procedure. Now, the Jacobian is calculated to provide the cell area function, F. This procedure is illustrated graphically in Figure 9-43.

A third scheme is given in Reference [9-6]. In this procedure, an algebraic grid system is used to determine the cell area function. This scheme proved to be more robust and clustering of grid points can be implemented with ease.

Now return to the system of PDEs given by Equations (9-113a, b). Before considering a solution procedure, a few observations are made with regard to this system. First, the system is nonlinear and, therefore, a linearization procedure must be utilized. Second, since the system is hyperbolic, a marching procedure will be used. For this formulation, marching will be in the η direction. Third, for a hyperbolic system, an initial condition and a set of boundary conditions must be specified. Finally, to prevent any oscillation in the solution, a fourth-order damping term may be added to the right-hand sides of the equations.

To linearize the equations, Newton's iterative scheme is used. Recall that a nonlinear term is approximated according to

$$AB = A^{k+1}B^k + B^{k+1}A^k - A^kB^k \tag{9-114}$$

where the level k is the known state. For this formulation, the $k+1$ notation will

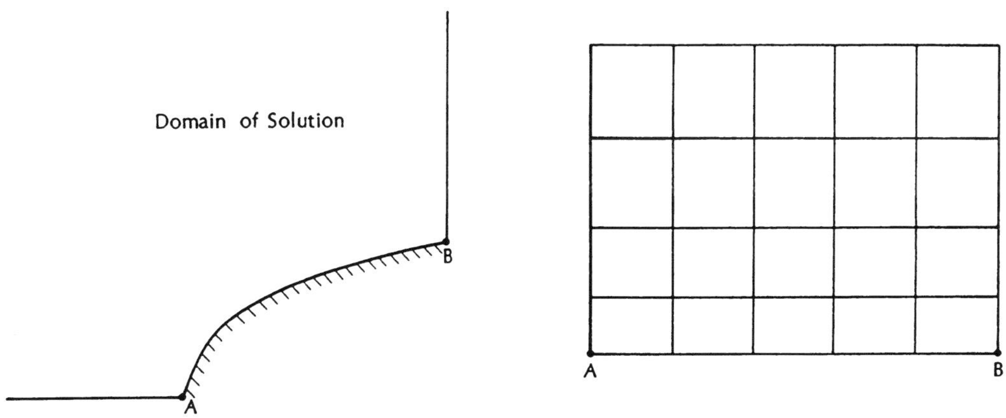

Figure 9-43. Rectangular grid point distribution used to determine cell area function, F.

be dropped; thus, any variable without a superscript denotes the unknowns at the $k+1$ level. For example, a nonlinear term such as $x_\xi y_\eta$ is linearized to provide

$$x_\xi y_\eta = x_\xi y_\eta^k + x_\xi^k y_\eta - x_\xi^k y_\eta^k$$

Hence, the system of hyperbolic equations is expressed in linear form as

$$x_\xi x_\eta^k + x_\xi^k x_\eta - x_\xi^k x_\eta^k + y_\xi y_\eta^k + y_\xi^k y_\eta - y_\xi^k y_\eta^k = 0 \qquad \text{(9-115a)}$$

$$x_\xi y_\eta^k + x_\xi^k y_\eta - x_\xi^k y_\eta^k - x_\eta y_\xi^k - x_\eta^k y_\xi + x_\eta^k y_\xi^k = F \qquad \text{(9-115b)}$$

These equations are simplified by observing that

$$x_\xi^k x_\eta^k + y_\xi^k y_\eta^k = 0$$

obtained from Equation (9-113a), and

$$x_\eta^k y_\xi^k - x_\xi^k y_\eta^k = -F^k$$

according to Equation (9-113b). Thus, the reduced form of the equations is

$$x_\xi x_\eta^k + x_\xi^k x_\eta + y_\xi y_\eta^k + y_\xi^k y_\eta = 0$$

$$x_\xi y_\eta^k + x_\xi^k y_\eta - x_\eta y_\xi^k - x_\eta^k y_\xi = F + F^k$$

This system of equations is written in a compact form as

$$[A]R_\xi + [B]R_\eta = H \qquad \text{(9-116)}$$

where

$$R = \begin{bmatrix} x \\ y \end{bmatrix} \quad , \quad A = \begin{bmatrix} x_\eta^k & y_\eta^k \\ y_\eta^k & -x_\eta^k \end{bmatrix} \quad , \quad B = \begin{bmatrix} x_\xi^k & y_\xi^k \\ -y_\xi^k & x_\xi^k \end{bmatrix}$$

and

$$H = \begin{bmatrix} 0 \\ F + F^k \end{bmatrix}$$

By definition, the system of equations given by (9-116) is hyperbolic if the eigenvalues of $[B]^{-1}[A]$ are real. Note that,

$$[B]^{-1} = \frac{1}{DN} \begin{bmatrix} x_\xi^k & -y_\xi^k \\ y_\xi^k & x_\xi^k \end{bmatrix}$$

and

$$[C] = [B]^{-1}[A] = \frac{1}{DN}\begin{bmatrix} x_\xi^k x_\eta^k - y_\xi^k y_\eta^k & x_\xi^k y_\eta^k + x_\eta^k y_\xi^k \\ x_\xi^k y_\eta^k + x_\eta^k y_\xi^k & -(x_\xi^k x_\eta^k - y_\xi^k y_\eta^k) \end{bmatrix}$$

where

$$DN = (x_\xi^k)^2 + (y_\xi^k)^2$$

The eigenvalues of $[C]$ are

$$\lambda_{1,2} = \pm\left[\frac{(x_\eta^k)^2 + (y_\eta^k)^2}{DN}\right]^{\frac{1}{2}}$$

It is recognized that the eigenvalues are always real; however, it is required that

$$(x_\xi^k)^2 + (y_\xi^k)^2 \neq 0$$

To obtain the FDE of (9-116), a second-order central difference approximation for ξ derivatives and a first-order backward difference approximation for the η derivatives are employed. The resulting FDE is

$$\frac{R_{i,j} - R_{i,j-1}}{\Delta\eta} + [B]^{-1}[A]\frac{R_{i+1,j} - R_{i-1,j}}{2\Delta\xi} = [B]^{-1}H_{i,j} \tag{9-117}$$

where $[A]$ and $[B]^{-1}$ are evaluated at $(j-1)$ grid line. Equation (9-117) is rearranged for $\Delta\xi = \Delta\eta = 1$ as

$$-\frac{1}{2}[C]_{i,j-1}R_{i-1,j} + [I]R_{i,j} + \frac{1}{2}[C]_{i,j-1}R_{i+1,j} = [B]_{i,j-1}^{-1}H_{i,j} + R_{i,j-1}$$

where $[I]$ is the identity matrix. Further simplification is introduced by the following definitions:

$$\begin{aligned} [AA] &= -\frac{1}{2}[C]_{i,j-1} \\ [BB] &= [I] \\ [CC] &= \frac{1}{2}[C]_{i,j-1} \end{aligned}$$

and

$$[DD] = [B]_{i,j-1}^{-1}H_{i,j} + R_{i,j-1}$$

Hence,

$$[AA]R_{i-1,j} + [BB]R_{i,j} + [CC]R_{i+1,j} = [DD]_{i,j} \tag{9-118}$$

Once Equation (9-118) is written for all i's at a j level, a block-tridiagonal system is obtained

$$\begin{bmatrix} [BB]_2 & [CC]_2 & & & \\ [AA]_3 & [BB]_3 & [CC]_3 & & \\ & & & & \\ & [AA]_{IM2} & [BB]_{IM2} & [CC]_{IM2} & \\ & & [AA]_{IM1} & [BB]_{IM1} \end{bmatrix} \begin{bmatrix} R_2 \\ R_3 \\ | \\ R_{IM2} \\ R_{IM1} \end{bmatrix} = \begin{bmatrix} [DD]_2 \\ [DD]_3 \\ | \\ [DD]_{IM2} \\ [DD]_{IM1} \end{bmatrix}$$

Note that the boundary conditions affect $[DD]_2$ and $[DD]_{IM1}$ according to

$$[DD]_2 = [DD]_2 - [AA]_2 R_1$$

$$[DD]_{IM1} = [DD]_{IM1} - [CC]_{IM1} R_{IM}$$

The block-tridiagonal system (9-118) is solved by marching in the η direction, provided an initial distribution of grid points on the surface and boundary lines are given. With the known grid point distribution at the surface, the derivatives x_ξ^k and y_ξ^k are computed by central difference approximation. At the boundaries, forward or backward approximations are used. To evaluate x_η^k and y_η^k, Equations (9-113a) and (9-113b) are employed. These equations at level k are

$$x_\xi^k x_\eta^k + y_\xi^k y_\eta^k = 0$$

$$x_\xi^k y_\eta^k - x_\eta^k y_\xi^k = F^k$$

which, when solved simultaneously, give

$$x_\eta^k = -\frac{y_\xi^k F^k}{(x_\xi^k)^2 + (y_\xi^k)^2}$$

$$y_\eta^k = \frac{x_\xi^k F^k}{(x_\xi^k)^2 + (y_\xi^k)^2}$$

With all the metrics at level k computed, the block matrices $[AA]$, $[BB]$, and $[CC]$ are determined. Subsequently, the block-tridiagonal system (9-118) is solved by any standard technique (a procedure is described in Appendix E). Note that the boundary points which are specified *a priori* to the solution must be free to float during the solution procedure, i.e., their values along the two boundaries are updated according to the interior solution. Some difficulty is observed if this updating procedure is not used. The reason for this problem is that the boundary points cannot be specified and kept fixed since that would define an outer domain for the hyperbolic system along these boundaries, which is not allowed. To achieve this goal, the orthogonality condition at the boundaries $i = 1$ and $i = IM$ is used to update the boundary points.

The result obtained for a simply-connected region is shown in Figure 9-44.

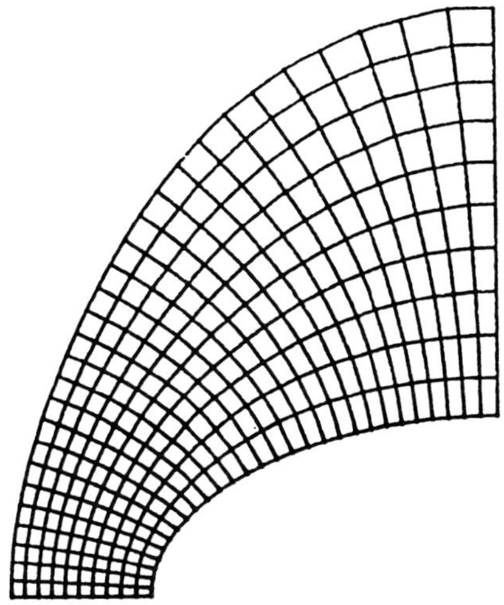

Figure 9-44. Grid system generated for the axisymmetric blunt body by hyperbolic PDEs.

As mentioned previously, a second set of hyperbolic equations may be developed. This second set is referred to as the arc-length approach. In this system, the first equation is the same as the cell-area approach, i.e., the equation is obtained by imposing orthogonality of the grid lines. Thus,

$$x_\xi x_\eta + y_\xi y_\eta = 0 \tag{9-119}$$

The geometric constraint is provided by considering

$$(ds)^2 = (dx)^2 + (dy)^2$$

which may be rearranged by using Equations (9-11) and (9-12), i.e.,

$$(ds)^2 = (x_\xi d\xi + x_\eta d\eta)^2 + (y_\xi d\xi + y_\eta d\eta)^2 \tag{9-120}$$

or

$$(\Delta s)^2 = x_\xi^2 + y_\xi^2 + x_\eta^2 + y_\eta^2 \tag{9-121}$$

when $\Delta\xi = \Delta\eta = 1.0$. The value of Δs is specified by the user. The system of hyperbolic equations composed of (9-119) and (9-121) may be solved to generate the grid system. Some difficulty was observed using this form of the equation. To overcome this problem, Δs along lines of constant ξ was specified. Therefore

Equation (9-120) was applied along the lines of constant ξ, which will then reduce Equation (9-120) to

$$(\Delta s)^2 = x_\eta^2 + y_\eta^2 \tag{9-122}$$

The system composed of Equations (9-119) and (9-122) works very well. Note that Δs distribution may be used for clustering purposes, since these values are specified by the user. The same procedure discussed previously for linearization and finite difference approximations is used for this hyperbolic system.

This section is closed by summarizing the following. The advantages of hyperbolic grid generators are:

(1) The grid system is orthogonal in two-dimensions;

(2) Since a marching scheme is used for the solution of the system, computationally they are much faster compared to elliptic systems;

(3) Grid line spacing may be controlled by the cell area or arc-length functions.

The disadvantages are:

(1) Extension to three-dimensions where complete orthogonality exists is not possible;

(2) They cannot be used for domains where the outer boundary is specified;

(3) Boundary discontinuity may be propagated into the interior domain;

(4) Specifying the cell-area or arc-length functions must be handled carefully. A bad selection of these functions easily leads to undesirable grid systems.

9.10 Parabolic Grid Generators

At this point, two sets of PDEs have been investigated in order to generate grid systems: elliptic and hyperbolic grid generators. Some advantages and deficiencies of each technique have been identified. One of the benefits of the elliptic system is that it smoothed out the grid point distribution, i.e., a boundary discontinuity does not propagate into the domain. This smoothness is due to the diffusive nature of the elliptic equations. On the other hand, a boundary discontinuity may propagate into the domain when a hyperbolic system is used because the hyperbolic system lacks a diffusive mechanism. Another distinct difference between the two systems was identified by the relative computation time. A hyperbolic system is much faster than an elliptic system, because a marching scheme is used for the solution of the system compared to an iterative solution of an elliptic system. Specification of the boundary conditions was also different for each method.

To combine the benefits of elliptic and hyperbolic systems, a parabolic system provides a compromise. A parabolic system includes a second-order derivative which introduces natural diffusion; as a consequence, propagation of boundary discontinuity is prevented. On the other hand, marching schemes are used to solve parabolic PDEs, thus reducing computation time compared to an elliptic system.

Development of parabolic systems is relatively new; therefore, much work remains. As a result, they will be reviewed here briefly for conceptual purposes only.

A parabolic system proposed in Reference [9-7] for investigative purposes is

$$\frac{\partial x}{\partial \eta} - A\frac{\partial^2 x}{\partial \xi^2} = S_x \tag{9-123}$$

$$\frac{\partial y}{\partial \eta} - A\frac{\partial^2 y}{\partial \xi^2} = S_y \tag{9-124}$$

where S_x and S_y are source terms and A is a specified constant. The solution procedure is similar to that of hyperbolic systems. The boundary grid points on the body surface are specified and a marching scheme is used to advance in the outward direction. Grid point clustering similar to that used in elliptic systems and local orthogonality may be enforced. Detailed discussion is given in Reference [9-7]. The method is extended to 3-D in Reference [9-8].

Much work needs to be done for perfection and robustness of this technique, especially in relation to the specification of control functions. Therefore, only a brief introduction has been presented to familiarize the reader with the concepts involved.

9.11 Problems

9.1 A grid system for a blunt conical configuration at zero degree angle of attack is required. The configuration is defined by its radius R, cone angle δ, and body length L as shown in Figure P9.1.

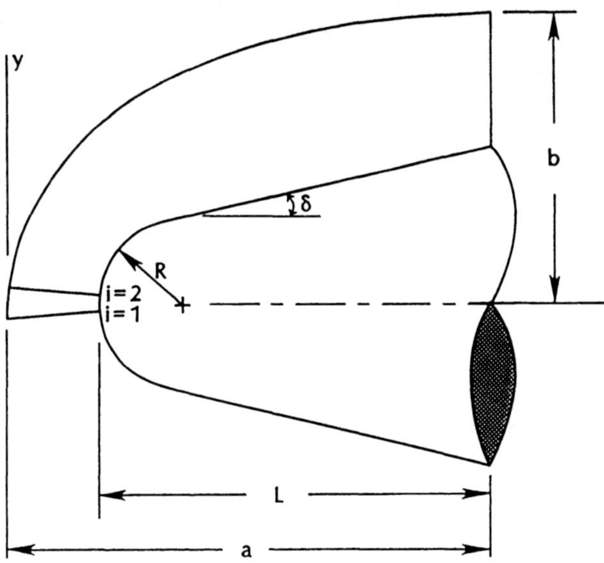

Figure P9.1. Nomenclature used for problem 9.1.

An outer boundary is defined by an ellipse with semi-major and semi-minor axes a and b. The following criteria are set for the grid system: (1) Grid line $i = 2$ must be above the body axis, whereas grid line $i = 1$ is determined by symmetry about the body axis; (2) Include an option for grid point clustering in the vicinity of the blunt nose. This option may be accomplished either by angular clustering or length clustering; (3) The grid system is generated (a) by an algebraic scheme and (b) by an elliptic method; (4) The following set of data are specified: $R = 1.0$, $\delta = 5.0°$, $L = 3.0$, $a = 4.0$, $b = 3.5$, $IM = 30$, $JM = 22$; and (5) The following results are required: (a) Plot of the grid system generated by both schemes (i.e., algebraic and elliptic), and (b) Plot of the metrics η_x, η_y, ξ_x, and ξ_y for both schemes.

9.2 A symmetrical airfoil is to be tested in a wind tunnel. Being a CFD expert, your help in determination of the flow field is desperately needed. Your computed results will be used to locate instruments in critical areas and, in addition, to verify experimental work! Thus, a set of equations of fluid motion is to be solved within the specified domain. The objective of this assignment is to generate the required grid system. The airfoil is described by the following equation:

$$y = \frac{t}{0.2}\left(0.2969x^{\frac{1}{2}} - 0.126x - 0.3516x^2 + 0.2843x^3 - 0.1015x^4\right)$$

The domain of interest is shown in Figure P9.2, where $c = 1$ is the chord length.

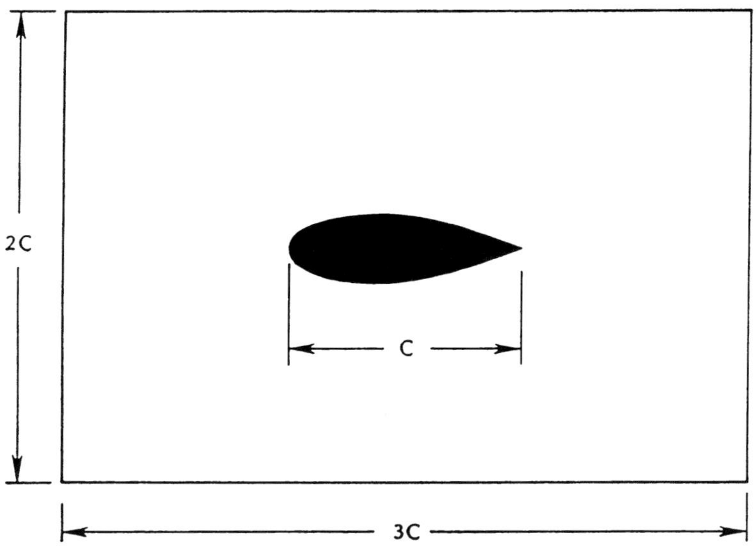

Figure P9.2. Physical domain for problem 9.2.

Grid point distribution on the body surface is determined by specifying the x-coordinates by two different methods: (a) the x-spacing of grid points are equal, and (b) clustering near the leading and trailing edges are used. Set $c = 1.0$, $t = 0.2$, $IM = 35$, $JM = 21$, then plot the grid system obtained by (a) an algebraic scheme and (b) by an elliptic scheme. Also, compute and plot the metrics η_x, η_y, ξ_x, and ξ_y.

9.3 Use Taylor series expansion to derive Equation (9-106).

9.4 Consider a circular cylinder of radius R located in a rectangular domain with dimensions of L and H as shown in Figure F9.4. Distribute grid points on the cylinder and the rectangular boundary with equal spacing, though grid spacing along L is not equal to grid spacing along H. (a) Generate an algebraic grid where grid points in the radial direction are equally spaced; (b) use an elliptic grid generator; and (c) use an elliptic grid generator with clustering option.

The following data are specified:

$$R = 1.0 \quad , \quad H = 4.0 \quad , \quad L = 6.0$$

$IM = 53$ (Maximum grid points along the cylinder)

$JM = 14$ (Maximum grid points along the radial)

For part (c), enforce clustering of the first five grid lines near the cylinder, with clustering coefficients of $a = -75.0$ and $c = 2.0$.

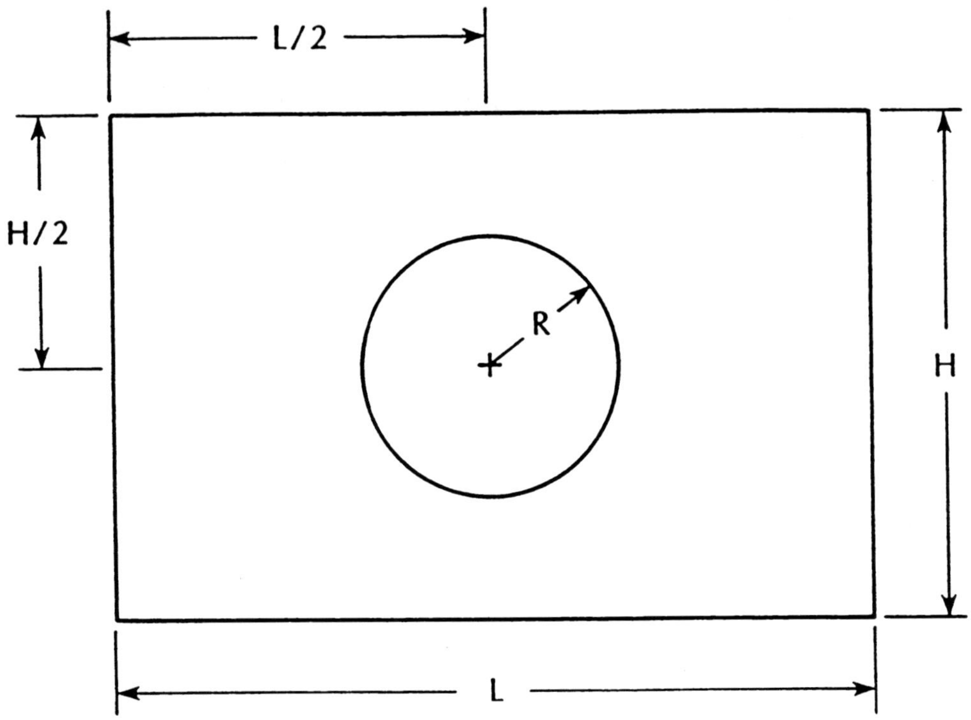

Figure P9.4. The nomenclature for problem 9.4.

9.5 The domain shown in Figure P9.5 is defined by the following data: $A1 = 3.0$, $A2 = 2.0$, $D = 2.0$, $H = 4.0$. Generate a grid by (a) an algebraic scheme with an option for grid clustering near the cylindrical surface, and (b) an elliptic scheme. Use the following data: $I1 = 10$, $IMID = 14$, $I2 = 18$, $IM = 28$, and $JM = 30$.

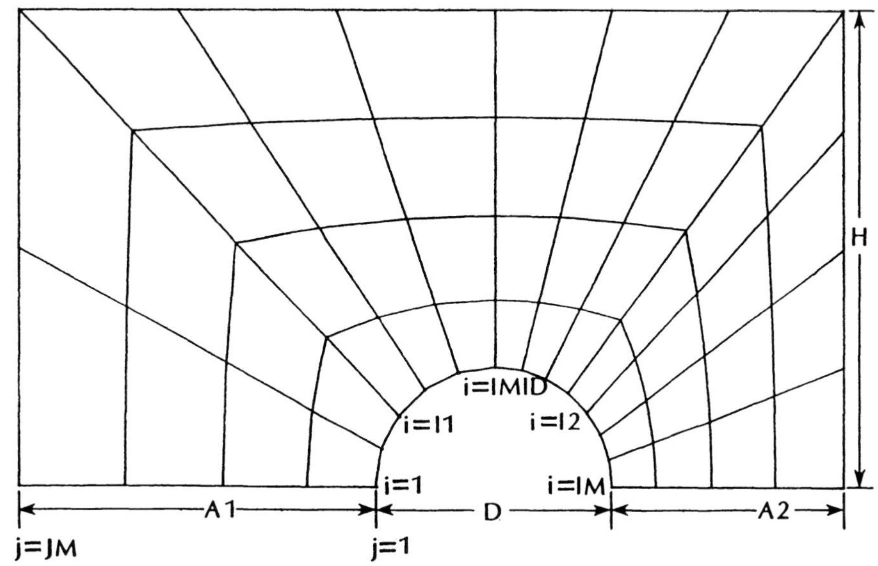

Figure P9.5. The nomenclature for the domain of problem 9.5.

9.6 Consider an airfoil whose geometry is described by Equation (9-78). An outer C-shaped domain is specified by a half circle and two parallel lines, as shown in Figure P9.6a. Grid point distribution on the airfoil surface is to be clustered near the leading and trailing edges.

(a) Generate a grid by an algebraic method with grid point clustering near the airfoil surface and the wake region, similar to the one shown in Figure P9.6b.

(b) Generate a grid by the elliptic scheme with no grid clustering.

(c) Generate a grid by the elliptic scheme with grid clustering. Specify clustering of the first 20 grid lines near the surface. The clustering coefficient a is specified as -200 for the first 10 grid lines and -150 for the next 10 grid lines. The clustering coefficient c is set to 5.0 for all lines.

The following set of data is to be used:

$$t = 0.2 \quad , \quad c = 1.0 \quad , \quad L = 1.5$$
$$IT1 = 20 \quad , \quad IL = 38 \quad , \quad JM = 30$$

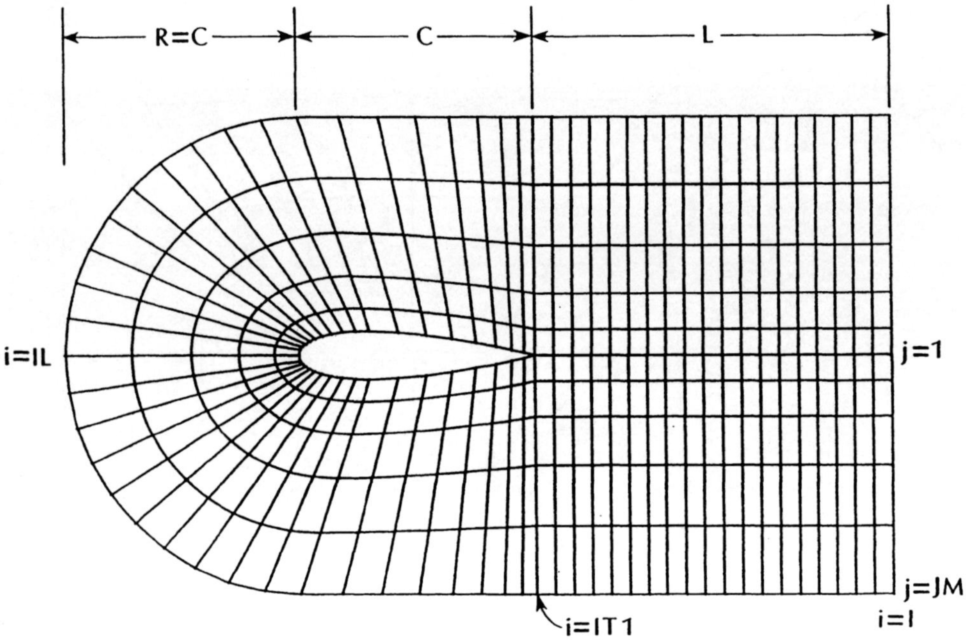

Figure P9.6a. The nomenclature for problem 9.6.

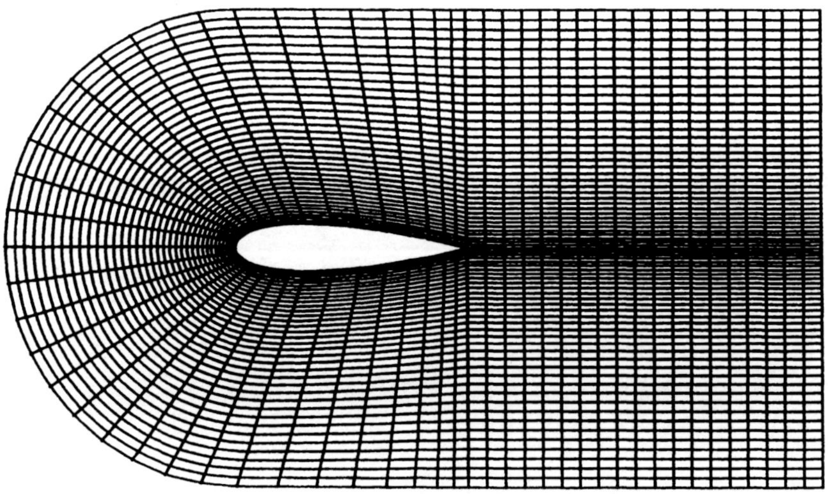

Figure P9.6b. A typical algebraic grid system to be generated.

9.7 Consider the flow described in Problem 7.3. The governing equation in nondimensional form is expressed by

$$\frac{\partial T^*}{\partial t^*} + u^* \frac{\partial T^*}{\partial x^*} = \alpha^* \frac{\partial^2 T^*}{\partial x^*}$$

where

$$t^* = \frac{tU_o}{L} \qquad x^* = \frac{x}{L} \quad , \quad u^* = \frac{u}{U_o} \quad T^* = \frac{T - T_c}{T_h - T_c}$$

and

$$\alpha^* = \frac{\alpha}{LU_o}$$

The initial and boundary conditions are specified as

$$\begin{array}{lll} t^* = 0 & T^*(x^*, 0) = 0.0 & \\ t^* \geq 0 & x^* = 0.0 & T^* = 0.0 \\ & x^* = 1.0 & T^* = 1.0 \end{array}$$

(a) Transform the governing equation from the physical space (x) to a computational space (ξ).

(b) Write an explicit finite difference equation which is first order in time and second order in space.

(c) Implement the grid clustering scheme proposed in Problem 3.9 to obtain the numerical solution for the following set of data:

$$\begin{array}{l} T_c = 20°\text{C} \quad , \quad T_h = 100°\text{C} \quad , \quad U_o = 0.2 \text{ m/sec} \\ \alpha = 0.04 \text{ m}^2/\text{sec.} \quad , \quad L = 2 \text{ m} \quad , \quad \Delta t = 0.01 \text{ sec.} \\ IM = 21 \quad , \quad \beta = 1.1 \end{array}$$

Print the solutions at time levels of 1, 2, and 3 seconds. Plot the temperature and error distributions at a time level of 3 seconds where the error distribution is determined, as defined in Problem 7.2, Part (g).
Hint: Use the following relation in Part (a)

$$\frac{\partial}{\partial x} = \xi_x \frac{\partial}{\partial \xi}$$

APPENDIX A:

An Introduction to Theory of Characteristics: Wave Propagation

A.1 Introductory Remarks

A technique which has been proven to be a powerful scheme for the solution of two-dimensional hyperbolic equations is known as the *method of characteristics.* Generally speaking, the application of the scheme in fluid mechanics is limited to two-dimensional, steady, isentropic, adiabatic, irrotational flow of a perfect gas. However, the scheme has been extended to inviscid rotational flows as well as three-dimensional problems. Furthermore, special procedures to cross discontinuities such as shock waves have been developed. Due to the introductory level of this appendix, the extension to three dimensions and special considerations will not be explored. Instead, the objectives are to review the theory of characteristics and to introduce the concepts of compatibility equations, Reimann invariants, and specification of boundary conditions.

These concepts are to be explored in one-dimensional space at first, and subsequently they will be extended to two-dimensional problems in Appendix G. In the process, the mathematics will be related to the physical aspect of the flowfield.

A.2 The Wave Equation

Perhaps the simplest hyperbolic equation to employ as a model equation, for the purpose of exploring the various concepts of characteristics, is the simple wave equation in one-space dimension.

Consider a disturbance which is gradually being transmitted into an undisturbed domain. For example, the disturbance can be created by applying an impulse to a piston in a tube. The small disturbance, which travels with the speed of sound, creates changes in the flow properties. Designate the properties in the undisturbed region by "∞" and the change in properties by "1," which may be referred to as *perturbation properties.* It is assumed that the changes are small. Now, the pressure, density, and velocity of the fluid are expressed as

$$p = p_\infty + p_1 \tag{A-1a}$$

$$\rho = \rho_\infty + \rho_1 \tag{A-1b}$$

$$u = u_1 \tag{A-1c}$$

Keep in mind the assumption of small disturbance, i.e., $p_1 \ll p_\infty$, $\rho_1 \ll \rho_\infty$, and $u_1 \ll a_\infty$. Furthermore, $u_\infty = 0$. Assuming a one-dimensional flow, then in general the disturbed properties are functions of x and t. Schematically, the wave propagation of sinusoidal shape is shown in Figure A-1. In order to describe the

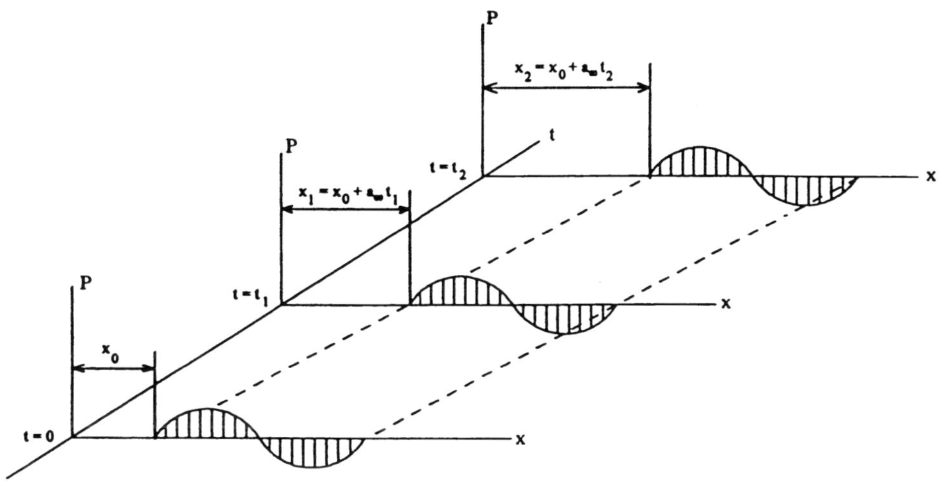

Figure A-1. Propagation of sinusoidal wave.

changes in the flowfield due to wave propagation, the conservation laws must be utilized. Re-emphasizing the assumptions of inviscid, adiabatic flow, the governing equations are:

Conservation of mass

$$\frac{\partial \rho}{\partial t} + \nabla \cdot \left(\rho \vec{V}\right) = 0 \tag{A-2}$$

and *Conservation of momentum*

$$\rho\left(\frac{\partial \vec{V}}{\partial t}+\vec{V}\cdot\nabla\vec{V}\right)=-\nabla p \tag{A-3}$$

For a one-dimensional flow in the Cartesian coordinate system, the equations are expressed as

$$\frac{\partial \rho}{\partial t}+\frac{\partial}{\partial x}(\rho u)=0 \tag{A-4}$$

$$\frac{\partial u}{\partial t}+u\frac{\partial u}{\partial x}=-\frac{1}{\rho}\frac{\partial p}{\partial x} \tag{A-5}$$

Relations (A-1) may be substituted into (A-4) to provide

$$\frac{\partial}{\partial t}(\rho_\infty+\rho_1)+\frac{\partial}{\partial x}\left[u_1\left(\rho_\infty+\rho_1\right)\right]=0$$

or

$$\frac{\partial \rho_1}{\partial t}+\rho_\infty\frac{\partial u_1}{\partial x}+\rho_1\frac{\partial u_1}{\partial x}+u_1\frac{\partial \rho_1}{\partial x}=0 \tag{A-6}$$

At this point, the assumption of small disturbance is imposed. Thus, all the second-order terms which are assumed to be much smaller compared to the first-order terms are dropped, resulting in

$$\frac{\partial \rho_1}{\partial t}+\rho_\infty\frac{\partial u_1}{\partial x}=0 \tag{A-7}$$

Now, recall from Thermodynamics that any thermodynamic property can be expressed in terms of any two-state variables; for example, one may write

$$p=p(\rho,s)$$

from which

$$dp=\left(\frac{\partial p}{\partial \rho}\right)_s d\rho+\left(\frac{\partial p}{\partial s}\right)_\rho ds \tag{A-8}$$

For an isentropic flow $ds=0$, and relation (A-8) is reduced to

$$dp=\left(\frac{\partial p}{\partial \rho}\right)_s d\rho$$

or

$$\frac{\partial p}{\partial x}=\left(\frac{\partial p}{\partial \rho}\right)_s\frac{\partial \rho}{\partial x}$$

Furthermore, recall that the speed of sound is given by

$$a^2=\left(\frac{\partial p}{\partial \rho}\right)_s \tag{A-9}$$

Therefore,

$$\frac{\partial p}{\partial x} = a^2 \frac{\partial \rho}{\partial x} \tag{A-10}$$

Now Equation (A-5) becomes

$$\frac{\partial u}{\partial t} + u\frac{\partial u}{\partial x} = -\frac{1}{\rho} a^2 \frac{\partial \rho}{\partial x}$$

Upon substitution of relations (A-1), one has

$$\frac{\partial u_1}{\partial t} + u_1 \frac{\partial u_1}{\partial x} = -\frac{1}{(\rho_\infty + \rho_1)} a^2 \frac{\partial}{\partial x}(\rho_\infty + \rho_1)$$

or

$$\rho_\infty \frac{\partial u_1}{\partial t} + \rho_1 \frac{\partial u_1}{\partial t} + \rho_\infty u_1 \frac{\partial u_1}{\partial x} + \rho_1 u_1 \frac{\partial u_1}{\partial x} = -a^2 \frac{\partial \rho_1}{\partial x} \tag{A-11}$$

Imposing the assumption of small disturbance, Equation (A-11) is reduced to

$$\rho_\infty \frac{\partial u_1}{\partial t} = -a^2 \frac{\partial \rho_1}{\partial x} \tag{A-12}$$

Recall that, from basic fluid mechanics, one may write

$$a^2 = a_\infty^2 - \frac{1}{2}(\gamma - 1)u_1^2$$

Based on the assumption of small disturbance, since $u_1 \ll a_\infty$, the relation above is reduced to

$$a^2 = a_\infty^2$$

Thus, Equation (A-12) can be written as

$$\rho_\infty \frac{\partial u}{\partial t} = -a_\infty^2 \frac{\partial \rho_1}{\partial x}$$

Finally, the governing equations of wave propagation are summarized by

$$\frac{\partial \rho_1}{\partial t} + \rho_\infty \frac{\partial u_1}{\partial x} = 0 \tag{A-13}$$

and

$$\rho_\infty \frac{\partial u_1}{\partial t} + a_\infty^2 \frac{\partial \rho_1}{\partial x} = 0 \tag{A-14}$$

Equations (A-13) and (A-14) are referred to as the *acoustic equations* which describe the changes in the domain due to passage of a sound wave. An important aspect of these equations is that they are linear. The acoustic equations given by (A-13) and (A-14) may be combined by the following manipulations. Taking the derivative, with respect to t of Equation (A-13) and with respect to x of Equation (A-14), one has

$$\frac{\partial^2 \rho_1}{\partial t^2} + \rho_\infty \frac{\partial^2 u_1}{\partial x \partial t} = 0$$

and

$$\rho_\infty \frac{\partial^2 u_1}{\partial x \partial t} + a_\infty^2 \frac{\partial^2 \rho_1}{\partial x^2} = 0$$

From which

$$\frac{\partial^2 \rho_1}{\partial t^2} - a_\infty^2 \frac{\partial^2 \rho_1}{\partial x^2} = 0 \tag{A-15}$$

Similarly, eliminating ρ_1, one has

$$\frac{\partial^2 u_1}{\partial t^2} - a_\infty^2 \frac{\partial^2 u_1}{\partial x^2} = 0 \tag{A-16}$$

and, in terms of the pressure,

$$\frac{\partial^2 p_1}{\partial t^2} - a_\infty^2 \frac{\partial^2 p_1}{\partial x^2} = 0 \tag{A-17}$$

The second-order, linear partial differential equation, given by any one of Equations (A-15) through (A-17), is called the *wave equation*, which can be expressed in a general form as

$$\frac{\partial^2 \phi}{\partial t^2} = a_\infty^2 \frac{\partial^2 \phi}{\partial x^2}$$

Now, consider Equation (A-17), where its general solution is given by arbitrary functions f and g, by the following

$$\frac{p_1}{p_\infty} = f(x - a_\infty t) + g(x + a_\infty t) \tag{A-18}$$

Since functions f and g are arbitrary, select a family of solutions such that $g = 0$. Therefore, one may write

$$\frac{p_1}{p_\infty} = f(x - a_\infty t) \tag{A-19}$$

Referring to Figure (A-1), solution (A-19) at various time levels can be written as

$$t = 0 \;:\; \frac{p_1}{p_\infty} = f(x_0) \tag{A-20a}$$

$$t = t_1 \;:\; \frac{p_1}{p_\infty} = f(x_1 - a_\infty t_1) \tag{A-20b}$$

$$t = t_2 \;:\; \frac{p_1}{p_\infty} = f(x_2 - a_\infty t_2) \tag{A-20c}$$

Substitution of $x_1 = x_0 + a_\infty t_1$ and $x_2 = x_0 + a_\infty t_2$ into Equations (A-20b) and (A-20c) show that both of these equations are represented by

$$\frac{p_1}{p_\infty} = f(x_0)$$

which is valid for the subsequent time levels as well. Therefore, one may conclude that

$$x - a_\infty t = \text{constant}$$

which represents the propagation of the pressure change to the right with the propagation speed of a_∞. It is important to note that the shape of the disturbance does not change as it propagates within the domain. Of course, that is due to the linear nature of the governing equations. Similarly, if one considers the solution g, it may be concluded that

$$x + a_\infty t = \text{constant}$$

That is, the disturbance (in this example, the pressure change) propagates to the left with the speed of a_∞. If one now plots the lines of constant pressures on an x-t plane, a family of straight lines with slopes of $+a_\infty$ or $-a_\infty$ is obtained, as shown in Figure A-2.

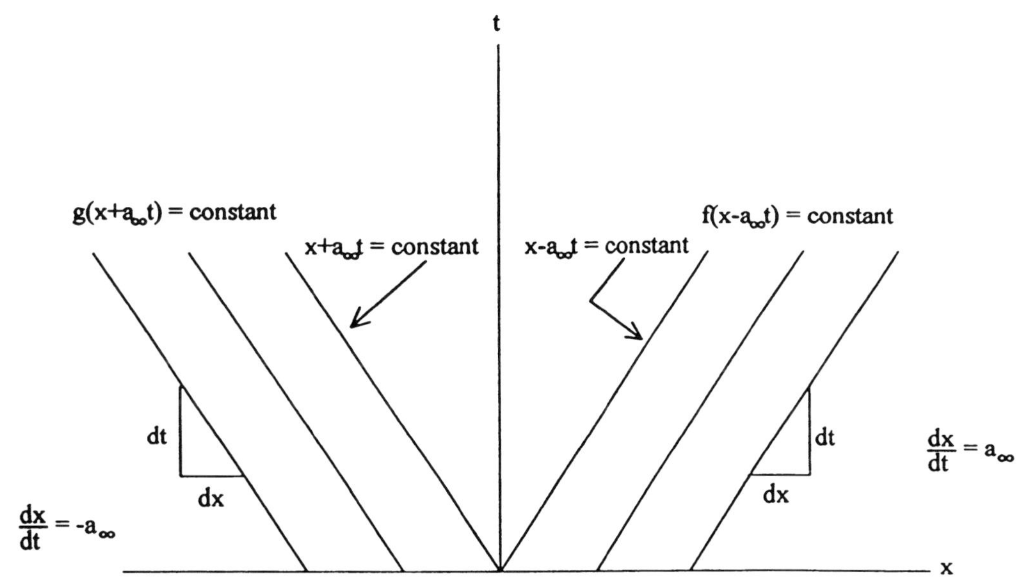

Figure A–2. Illustration of characteristics of the one-dimensional wave equation.

The lines of constants f and g are known as the *characteristic lines* or, simply, the *characteristics*. Those with positive slopes are called *right-running characteristics*, and those with negative slopes are called *left-running characteristics*.

Before proceeding to the next section, the following observations are made:

1. The wave equation is linear;
2. A small disturbance propagates along lines called characteristics with the speed of sound;
3. The shape of the disturbance remains the same;
4. The characteristics are straight lines with slopes of $+a_\infty$ or $-a_\infty$.

At this point, the assumption of small disturbance is removed, and one allows large changes to occur as the wave propagates within the domain.

The description of the flowfield is governed by Equations (A-2) and (A-3) where, for a one-dimensional flow, they are expressed as

$$\frac{\partial \rho}{\partial t} + \frac{\partial}{\partial x}(\rho u) = 0 \tag{A-21}$$

and

$$\frac{\partial u}{\partial t} + u\frac{\partial u}{\partial x} = -\frac{1}{\rho}\frac{\partial p}{\partial x} \tag{A-22}$$

The development of equations follows that of Reference [A-1].

Recall from previous discussion that, for an isentropic flow, one may write

$$d\rho = \frac{1}{a_\infty^2} dp \tag{A-23}$$

Now, rewrite Equation (A-21) as

$$\frac{\partial \rho}{\partial t} + \rho\frac{\partial u}{\partial x} + u\frac{\partial \rho}{\partial x} = 0 \tag{A-24}$$

Differentiate Equation (A-23) with respect to time and with respect to x to obtain

$$\frac{\partial \rho}{\partial t} = \frac{1}{a_\infty^2}\frac{\partial p}{\partial t} \tag{A-25a}$$

and

$$\frac{\partial \rho}{\partial x} = \frac{1}{a_\infty^2}\frac{\partial p}{\partial x} \tag{A-25b}$$

Substitution of relations (A-25a) and (A-25b) into Equation (A-24) yields

$$\frac{1}{a_\infty^2}\left(\frac{\partial p}{\partial t} + u\frac{\partial p}{\partial x}\right) + \rho\frac{\partial u}{\partial x} = 0 \tag{A-26}$$

Multiplication of the modified continuity equation (A-26) by a_∞/ρ and subsequent addition to the momentum equation given by (A-22) provides

$$\frac{\partial u}{\partial t} + (u+a)\frac{\partial u}{\partial x} + \frac{1}{\rho a}\left[\frac{\partial p}{\partial t} + (u+a)\frac{\partial p}{\partial x}\right] = 0 \qquad \text{(A-27)}$$

Similarly, subtraction of the two equations provides

$$\frac{\partial u}{\partial t} + (u-a)\frac{\partial u}{\partial x} - \frac{1}{\rho a}\left[\frac{\partial p}{\partial t} + (u-a)\frac{\partial p}{\partial x}\right] = 0 \qquad \text{(A-28)}$$

Now, since $u = u(x,t)$, a change in u, namely du, corresponding to dx and dt is given by

$$du = \frac{\partial u}{\partial t}\,dt + \frac{\partial u}{\partial x}\,dx \qquad \text{(A-29)}$$

At this point, consider a specific path upon which the changes take place; in particular, consider the change given by

$$dx = (u+a)dt \qquad \text{(A-30a)}$$

Substitution into (A-29) yields

$$du = \left[\frac{\partial u}{\partial t} + (u+a)\frac{\partial u}{\partial x}\right] dt \qquad \text{(A-31a)}$$

Similarly,

$$dp = \left[\frac{\partial p}{\partial t} + (u+a)\frac{\partial p}{\partial x}\right] dt \qquad \text{(A-31b)}$$

Thus, from Equation (A-27), one obtains

$$du + \frac{1}{\rho a}\,dp = 0 \qquad \text{(A-32a)}$$

which represents changes in u and p along the path specified by Equation (A-30). Similarly, along the path defined by

$$dx = (u-a)\,dt \qquad \text{(A-30b)}$$

one has

$$du - \frac{1}{\rho a}\,dp = 0 \qquad \text{(A-32b)}$$

At this point, let's pause a moment and review the mathematical work. A specific path given by Equation (A-30a) was selected, and it was shown that the governing partial differential equation given by (A-27) is reduced to an ordinary differential equation given by (A-32a) along that path. This path is the characteristic line along which a disturbance propagates with the speed of $(u + a)$. The ordinary differential equation (A-32) is called *the compatibility equation.* Since two characteristics are identified, they will be designated as C^+ and C^- characteristics. Schematically, they are shown in Figure A-3. It is then obvious that the primary advantage of the method of characteristics is the reduction of the governing partial differential equation to an ordinary differential equation along the characteristics. Solution of an ordinary differential equation is much simpler than that of partial differential equations.

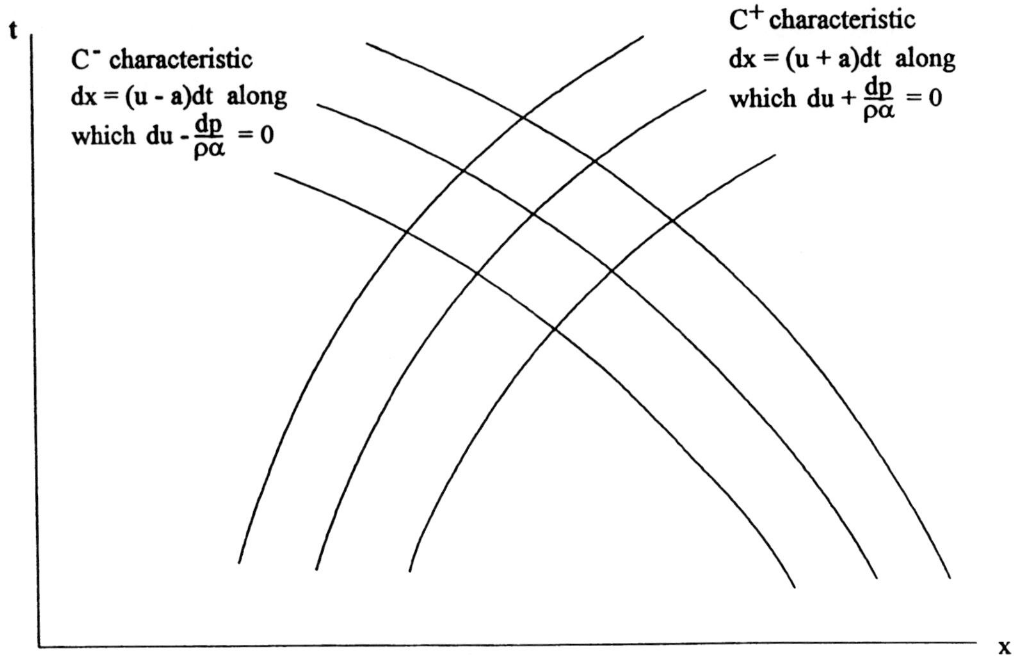

Figure A-3. Illustration of right- and left-running characteristics.

Proceeding with the solution of the compatibility equations (A-32a) and (A-32b), one integrates the equations along the corresponding characteristics to obtain

$$R^+ = u + \int \frac{dp}{\rho a} = \text{constant} \tag{A-33a}$$

and

$$R^- = u - \int \frac{dp}{\rho a} = \text{constant} \tag{A-33b}$$

A relation between the density and pressure can be established by an equation of state. If one imposes the assumption of perfect gas, then

$$p = \rho RT$$

and the speed of sound can be expressed as

$$a^2 = \gamma \frac{p}{\rho} = \gamma RT$$

Furthermore, recall that for an isentropic process

$$\frac{p_2}{p_1} = \left(\frac{T_2}{T_1}\right)^{\frac{\gamma}{\gamma-1}} = \left(\frac{\rho_2}{\rho_1}\right)^{\gamma} = \left(\frac{a_2}{a_1}\right)^{\frac{2\gamma}{\gamma-1}}$$

or

$$p = c a^{\frac{2\gamma}{\gamma-1}}$$

from which

$$dp = c\left(\frac{2\gamma}{\gamma-1}\right) a^{\frac{\gamma+1}{\gamma-1}} da \tag{A-34a}$$

and

$$\rho = c\gamma a^{\frac{2}{\gamma-1}} \tag{A-34b}$$

Substitution of (A-34a) and (A-34b) into (A-33a) and (A-33b) yields

$$R^+ = u + \frac{2a}{\gamma-1} \tag{A-35a}$$

and

$$R^- = u - \frac{2a}{\gamma-1} \tag{A-35b}$$

where R^+ and R^- are defined as the Riemann invariants.

Now, the following conclusions with regard to the finite wave are stated.

1. The governing equations are nonlinear.
2. Property changes, which can be large, propagate along the characteristics with the speed of $u \pm a$.
3. The shape of disturbance may change with time.
4. The characteristics are curved lines.

It is helpful to relate the preceding mathematical analyses to the corresponding physical analog. Recall from basic fluid mechanics that Equations (A-21) and (A-22) may be used to describe propagation of an expansion wave in a tube, as shown in Figure (A-4a). Flow properties continuously change across the expansion wave, and a typical distribution is illustrated in Figure (A-4b).

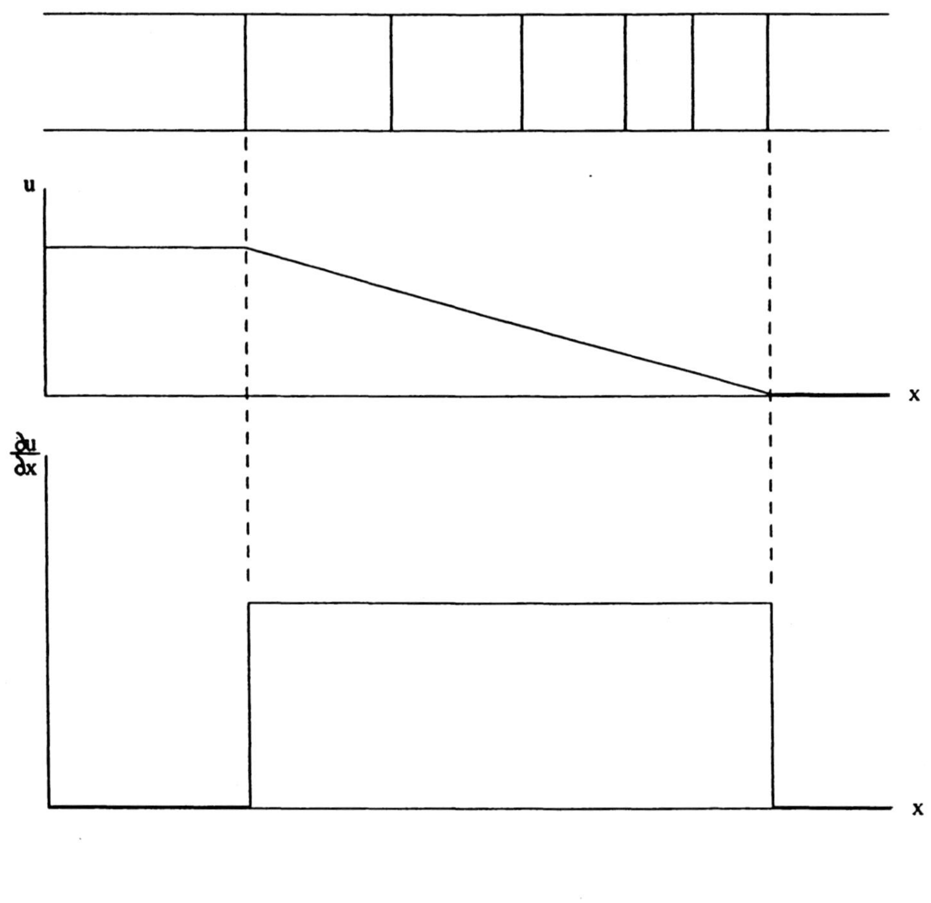

Figure A-4. Propagation of an expansion wave.

However, if one considers the gradient of the property, such as $\partial u/\partial x$, discontinuities at the leading and trailing edges are encountered, as shown in Figure (A-4c). From this simple example, one can generalize an important aspect of characteristics as follows:

1. Characteristics are paths of propagation of physical disturbance.
2. Along the characteristics, the governing partial differential equations are reduced by one space dimension, which are called *compatibility equations.* For example, the governing partial equations in two-space dimensions are reduced to one-space dimension and, hence, to an ordinary differential equation.
3. The physical properties are continuous across the characteristics, whereas there may be discontinuities in their derivatives.
4. With regard to physics of fluid motion, the following observations are made:
   (a) For a steady, two-dimensional, isentropic, irrotational flow, real characteristics exist only for supersonic flow.
   (b) Mach lines are characteristics of the flow.

APPENDIX B: Tridiagonal System of Equations

The formulation of many implicit methods for a scalar PDE results in the following equation:

$$a_i^n u_{i-1}^{n+1} + b_i^n u_i^{n+1} + c_i^n u_{i+1}^{n+1} = D_i^n \tag{B-1}$$

Once this equation is applied to all the nodes at the advanced level, a system of linear algebraic equations is obtained. When these equations are represented in a matrix form, the coefficient matrix is tridiagonal. We will take advantage of the tridiagonal nature of the coefficient matrix and review a very efficient solution procedure.

To see the matrix formulation of the equations, consider the Laasonen implicit formulation of our diffusion model equation, i.e., Equation (3-12), presented here as

$$(d)u_{i-1}^{n+1} - (2d+1)u_i^{n+1} + (d)u_{i+1}^{n+1} = -u_i^n \tag{B-2}$$

where $d = \alpha\Delta t/(\Delta x)^2$ is the diffusion number. Define the following coefficients of (B-2) according to the formulation of (B-1):

$$\begin{aligned} a &= d \\ b &= -(2d+1) \\ c &= d \\ D &= \text{RHS} \end{aligned}$$

Applying Equation (B-2) to all the grid points will result in the following set of linear algebraic equations:

$$i = 2 \qquad a_2 u_1 + b_2 u_2 + c_2 u_3 = D_2 \tag{B-3}$$

$$i = 3 \qquad a_3u_2 + b_3u_3 + c_3u_4 = D_3$$

$$i = 4 \qquad a_4u_3 + b_4u_4 + c_4u_5 = D_4$$

$$\vdots$$

$$i = IM2 \qquad a_{IM2}u_{IM3} + b_{IM2}u_{IM2} + c_{IM2}u_{IM1} = D_{IM2}$$

$$i = IM1 \qquad a_{IM1}u_{IM2} + b_{IM1}u_{IM1} + c_{IM1}u_{IM} = D_{IM1} \tag{B-4}$$

where $IM1 = IM - 1$, $IM2 = IM - 2$, and so on. In addition, note that the superscript $n+1$ has been dropped from the equations. Assume that Dirichlet boundary conditions are imposed and, therefore, the values of the dependent variable u at $i = 1$ (the lower boundary) and at $i = IM$ (the upper boundary) are given. Then the first equation, (B-3), and the last equation, (B-4), can be written as:

$$b_2u_2 + c_2u_3 = D_2 - a_2u_1$$

and

$$a_{IM1}u_{IM2} + b_{IM1}u_{IM1} = D_{IM1} - c_{IM1}u_{IM}$$

Now, the set of equations in matrix formulation is

$$\begin{bmatrix} b_2 & c_2 & & & & & \\ a_3 & b_3 & c_3 & & & & \\ & a_4 & b_4 & c_4 & & & \\ & & & & & & \\ & & & & a_{IM2} & b_{IM2} & c_{IM2} \\ & & & & & a_{IM1} & b_{IM1} \end{bmatrix} \begin{bmatrix} u_2 \\ u_3 \\ u_4 \\ | \\ u_{IM2} \\ u_{IM1} \end{bmatrix} = \begin{bmatrix} D_2 - a_2u_1 \\ D_3 \\ D_4 \\ | \\ D_{IM2} \\ D_{IM1} - c_{IM1}u_{IM} \end{bmatrix}$$

Applying other types of boundary conditions does not change the tridiagonal form of the coefficient matrix, and can be easily implemented. An example is given below, where the Neumann boundary condition at $i = IM$ is applied. Impose the boundary condition $\partial u/\partial x = 0$ (at $i = IM$). Then

$$\frac{\partial u}{\partial x} = \frac{u_{IMP1} - u_{IM}}{\Delta x}$$

or

$$u_{IM} = u_{IMP1}$$

Note that we have introduced a fictitious boundary at $IMP1 = IM + 1$. The grid points are shown in Figure B-1. Now, at $i = IM$ (the upper boundary), we have

$$a_{IM}u_{IM1} + b_{IM}u_{IM} + c_{IM}u_{IMP1} = D_{IM}$$

or

$$a_{IM}u_{IM1} + (b_{IM} + c_{IM})u_{IM} = D_{IM}$$

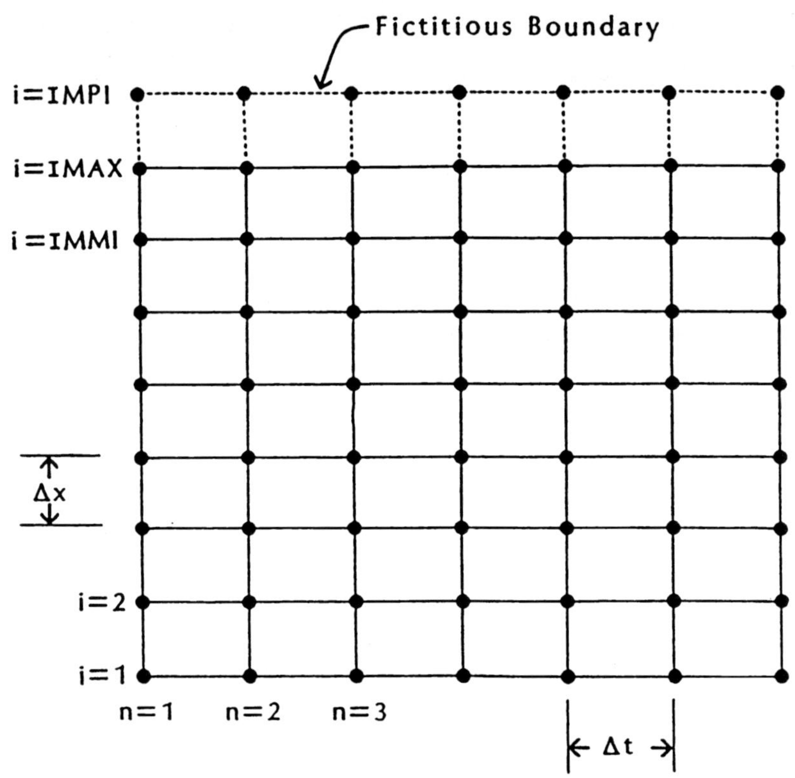

Figure B-1. Illustration of grid points with fictitious boundary.

and the matrix representation takes the following form:

$$\begin{bmatrix} b_2 & c_2 & & & & \\ a_3 & b_3 & c_3 & & & \\ & a_4 & b_4 & c_4 & & \\ & & & & & \\ & & & a_{IM1} & b_{IM1} & c_{IM1} \\ & & & & a_{IM} & (b+c)_{IM} \end{bmatrix} \begin{bmatrix} u_2 \\ u_3 \\ u_4 \\ | \\ u_{IM1} \\ u_{IM} \end{bmatrix} = \begin{bmatrix} D_2 - a_2 u_1 \\ D_3 \\ D_4 \\ | \\ D_{IM1} \\ D_{IM} \end{bmatrix}$$

In the example above, a first-order approximation for the gradient was used. If a second-order approximation is utilized, then

$$\frac{\partial u}{\partial x} = \frac{u_{IMP1} - u_{IM1}}{2\Delta x}$$

or

$$u_{IMP1} = u_{IM1}$$

Thus, the equation at $i = IM$ is modified according to

$$a_{IM} u_{IM1} + b_{IM} u_{IM} + c_{IM} u_{IM1} = D_{IM}$$

or

$$(a_{IM} + c_{IM}) u_{IM1} + b_{IM} u_{IM} = D_{IM}$$

To proceed with the solution technique, assume a solution of the form

$$u_i = -H_i u_{i+1} + G_i \tag{B-5}$$

where u_i is unknown, u_{i+1} is known from the imposed boundary condition, and H_i and G_i are yet to be determined. Apply Equation (B-5) at node $i-1$, then

$$u_{i-1} = -H_{i-1} u_i + G_{i-1} \tag{B-6}$$

Upon substitution of (B-6) into Equation (B-1), one has

$$a_i(-H_{i-1} u_i + G_{i-1}) + b_i u_i + c_i u_{i+1} = D_i$$

or

$$(b_i - a_i H_{i-1}) u_i + c_i u_{i+1} = D_i - a_i G_{i-1}$$

from which

$$u_i = -\frac{c_i}{b_i - a_i H_{i-1}} u_{i+1} + \frac{D_i - a_i G_{i-1}}{b_i - a_i H_{i-1}} \tag{B-7}$$

Comparing Equations (B-7) with (B-5), one concludes that

$$H_i = \frac{c_i}{b_i - a_i H_{i-1}} \tag{B-8}$$

and

$$G_i = \frac{D_i - a_i G_{i-1}}{b_i - a_i H_{i-1}} \tag{B-9}$$

Now that H_i and G_i have been determined, the recursion equation, (B-5), can be used to solve for all the unknowns.

To see how this procedure is applied, consider the parabolic equation investigated in Chapter 3, i.e., the suddenly accelerated plane. At the lower boundary $i = 1$, $u_1 = UWALL$ is specified for all times; therefore, (B-5) at $i = 1$ becomes

$$u_1 = -H_1 u_2 + G_1$$

Since this equation must hold for all u_2, $H_1 = 0$ and $G_1 = u_1 = UWALL$. With the values of H_1 and G_1 provided from the boundary condition, Equations (B-8) and (B-9) can be solved for the values of H_i and G_i at the second node. Subsequently, (B-8) and (B-9) are sequentially applied to all grid points to obtain the values of H_i and G_i. Note that the computation of H_i and G_i starts from the lower boundary and proceeds upward. Now, Equation (B-5) is used for the computation of u_i. This calculation is performed inward from the upper boundary. At $i = IM1$, Equation (B-5) provides

$$u_{IM1} = -H_{IM1} u_{IM} + G_{IM1}$$

In this equation, u_{IM} is specified from the upper boundary condition, with H and G at $IM1$ previously determined. Once u_{IM1} is computed, Equation (B-5) is applied to compute u_{IM2} and so on. The solution procedure may be coded in the program or as a subroutine.

APPENDIX C:

Derivation of Partial Derivatives for the Modified Equation

To determine $\partial^2 u/\partial t^2$, take the derivative of Equation (4-81) with respect to x to obtain

$$\begin{aligned}\frac{\partial^2 u}{\partial x \partial t} &= -a\frac{\partial^2 u}{\partial x^2} - \frac{\Delta t}{2}\frac{\partial^3 u}{\partial x \partial t^2} + \frac{a\Delta x}{2}\frac{\partial^3 u}{\partial x^3} \\ &\quad - \frac{(\Delta t)^2}{6}\frac{\partial^4 u}{\partial x \partial t^3} - \frac{a(\Delta x)^2}{6}\frac{\partial^4 u}{\partial x^4} + O\left[(\Delta t)^3, (\Delta x)^3\right] \qquad \text{(C-1)}\end{aligned}$$

Similarly, the derivative with respect to t of Equation (4-81) is

$$\begin{aligned}\frac{\partial^2 u}{\partial t^2} &= -a\frac{\partial^2 u}{\partial t \partial x} - \frac{\Delta t}{2}\frac{\partial^3 u}{\partial t^3} + \frac{a\Delta x}{2}\frac{\partial^3 u}{\partial t \partial x^2} - \frac{(\Delta t)^2}{6}\frac{\partial^4 u}{\partial t^4} \\ &\quad - \frac{a(\Delta x)^2}{6}\frac{\partial^4 u}{\partial t \partial x^3} + O\left[(\Delta t)^3, (\Delta x)^3\right] \qquad \text{(C-2)}\end{aligned}$$

Multiply Equation (C-1) by $-a$ and add it to Equation (C-2) to obtain

$$\begin{aligned}\frac{\partial^2 u}{\partial t^2} &= a^2\frac{\partial^2 u}{\partial x^2} + \frac{\Delta t}{2}\left[a\frac{\partial^3 u}{\partial x \partial t^2} - \frac{\partial^3 u}{\partial t^3}\right] \\ &\quad + \frac{\Delta x}{2}\left[a\frac{\partial^3 u}{\partial t \partial x^2} - a^2\frac{\partial^3 u}{\partial x^3}\right] + O\left[(\Delta t)^2, (\Delta x)^2\right] \qquad \text{(C-3)}\end{aligned}$$

This equation requires that we determine $\partial^3 u/\partial t^3$, $(\partial^3 u)/(\partial t \partial x^2)$, and $(\partial^3 u)/(\partial t^2 \partial x)$, which are first-order accurate. One such calculation is as follows. Taking the second derivative of Equation (4-81) with respect to time, we obtain

$$\frac{\partial^3 u}{\partial t^3} = -a\frac{\partial^3 u}{\partial t^2 \partial x} - \frac{\partial^4 u}{\partial t^4}\frac{\Delta t}{2} + a\frac{\partial^4 u}{\partial t^2 \partial x^2}\frac{\Delta x}{2} + O\left[(\Delta t)^2, (\Delta x)^2\right] \qquad \text{(C-4)}$$

and the derivative with respect to x of (C-3) is

$$\begin{aligned}\frac{\partial^3 u}{\partial x \partial t^2} &= a^2\frac{\partial^3 u}{\partial x^3} + \frac{\Delta t}{2}\left[a\frac{\partial^4 u}{\partial x^2 \partial t^2} - \frac{\partial^4 u}{\partial x \partial t^3}\right] \\ &+ \frac{\Delta x}{2}\left[a\frac{\partial^4 u}{\partial t \partial x^3} - a^2\frac{\partial^4 u}{\partial x^4}\right] + O\left[(\Delta t)^2, (\Delta x)^2\right] \qquad \text{(C-5)}\end{aligned}$$

Multiplying Equation (C-5) by $-a$ and adding it to Equation (C-4), we obtain

$$\begin{aligned}\frac{\partial^3 u}{\partial t^3} &= -a^3\frac{\partial^3 u}{\partial x^3} + \frac{\Delta t}{2}\left[a\frac{\partial^4 u}{\partial x \partial t^3} - a^2\frac{\partial^4 u}{\partial x^2 \partial t^2} - \frac{\partial^4 u}{\partial t^4}\right] \\ &+ \frac{\Delta x}{2}\left[a\frac{\partial^4 u}{\partial t^2 \partial x^2} - a^2\frac{\partial^4 u}{\partial t \partial x^3} + a^3\frac{\partial^4 u}{\partial x^4}\right] + O\left[(\Delta t)^2, (\Delta x)^2\right]\end{aligned}$$

Since we are interested in the first-order accurate relation, we may write

$$\frac{\partial^3 u}{\partial t^3} = -a^3\frac{\partial^3 u}{\partial x^3} + O[(\Delta t), (\Delta x)] \qquad \text{(C-6)}$$

Similarly, we may derive

$$\frac{\partial^3 u}{\partial t \partial x^2} = -a\frac{\partial^3 u}{\partial x^3} + O[(\Delta t), (\Delta x)] \qquad \text{(C-7)}$$

and

$$\frac{\partial^3 u}{\partial t^2 \partial x} = a^2\frac{\partial^3 u}{\partial x^3} + O[(\Delta t), (\Delta x)] \qquad \text{(C-8)}$$

Substituting (C-6) through (C-8) into (C-3) yields:

$$\begin{aligned}\frac{\partial^2 u}{\partial t^2} &= a^2\frac{\partial^2 u}{\partial x^2} + \frac{\Delta t}{2}a\left\{a^2\frac{\partial^3 u}{\partial x^3} + O[(\Delta t), (\Delta x)]\right\} \\ &- \frac{\Delta t}{2}\left\{-a^3\frac{\partial^3 u}{\partial x^3} + O[(\Delta t), (\Delta x)]\right\} + \frac{a\Delta x}{2}\left\{-a\frac{\partial^3 u}{\partial x^3} + O[(\Delta t), (\Delta x)]\right\} \\ &- a^2\frac{\Delta x}{2}\frac{\partial^3 u}{\partial x^3} + O[(\Delta t)^2, (\Delta x)^2]\end{aligned}$$

Hence,

$$\begin{aligned}\frac{\partial^2 u}{\partial t^2} &= a^2\frac{\partial^2 u}{\partial x^2} + (a^3\Delta t - a^2\Delta x)\frac{\partial^3 u}{\partial x^3} \\ &+ O[(\Delta t)^2, (\Delta x)(\Delta t), (\Delta x)^2] \qquad \text{(C-9)}\end{aligned}$$

APPENDIX D:

Basic Equations of Fluid Mechanics

D.1 Introductory Remarks

The fundamental equations of fluid motion are based on three conservation laws. Additional conservation equations will also be required if, for example, a fluid is composed of various chemical species with mass diffusion and/or chemical reactions. Since, for most engineering applications, the average measurable values of the flow properties are desired, the assumption of continuous distribution of matter is imposed. This assumption is known as *continuum* and is valid as long as the characteristic length in a physical domain is much larger than the mean free path of molecules. The assumption of continuum is imposed on the equations of fluid motion presented throughout the text.

The basic equations of fluid motion are derived in either integral form or differential form from the

1. Conservation of mass (continuity),
2. Conservation of linear momentum (Newton's second law),
3. Conservation of energy (first law of thermodynamics).

The conservation of linear momentum is a vector equation and, therefore, provides three scalar equations for a three dimensional problem. The conservation of linear momentum in differential form was derived originally by Stokes and independently by Navier and, therefore, is known as the Navier-Stokes equation. It is common to refer to the entire system of equations in differential form composed of conservations of mass, momentum, and energy as *the Navier-Stokes equations.*

The system of equations may contain nine unknowns which include ρ (density), u, v, w (components of the velocity vector), e_t (total energy [or h (enthalpy)]), p

(thermodynamic pressure), T (temperature), μ (dynamic viscosity), and k (thermal conductivity). The system of equations is closed by introducing thermodynamic relations and auxiliary relations for the coefficient of viscosity and thermal conductivity. Functional relations for thermodynamic properties may be expressed as

$$\rho = \rho(p, T)$$

and

$$h = h(p, T)$$

which may be in the form of equations or tables (or charts). The transport properties μ and k are expressed as

$$\mu = \mu(p, T)$$

and

$$k = k(p, T)$$

Hence, a system of nine equations is available which must be solved simultaneously for the nine unknowns. Other variables may also appear in the governing equations, depending on the choice of dependent variables. For example, the energy equation may include the specific heats c_p and c_v. But in either case, there are expressions by which c_p and c_v are related to thermodynamic properties. In general, specific heats are functions of temperature and are slightly pressure dependent. The complexity of the system is further increased by the introduction of turbulence and chemistry.

The analytical solution of such a system does not exist. Therefore, numerical techniques are employed to obtain a solution for a specified problem. Since, even with the advancement of computer technology, most solutions are time-consuming and difficult, various assumptions are imposed on the system of equations in order to simplify the solution procedure and to reduce computation time.

In this appendix the governing equations of fluid motion are reviewed. The derivation of equations is not included; however, references are provided for those interested in the derivation of equations. Furthermore, the differential form of the equations is expressed in Cartesian coordinate system. Ref. [D.1] may be consulted for the equations expressed in other coordinate systems.

D.2 Integral Formulations

Integral forms of the equations are used if an average value of the fluid properties at a cross-section is desired. This approach does not provide a detailed analysis of the flow field; however, the application is simple and is used extensively. In general, the integral form of the equation is derived for an extensive property and then the conservation laws are applied.

If N represents an extensive property, then a relation exists between the rate of change of extensive property for a system and the time rate of change of the property within the control volume plus net efflux of the property across the control surfaces. Defining η as the extensive property per unit mass, then

$$\frac{dN}{dt} = \frac{\partial}{\partial t}\int_{C.V.} \eta\rho \, d(\text{Vol}) + \int_{C.S.} \eta(\rho\vec{V} \cdot \vec{n})dS \tag{D-1}$$

where t represents time, ρ is the density of the fluid, $\vec{V}$ is the velocity vector and $\vec{n}$ is the unit vector normal to the control surface in the outward direction as shown in Figure D.1. The derivation of Equation (D.1) may be found in any standard fluid mechanics text such as (D.2–D.4). d/dt is used to represent the total or substantial derivative and is composed of a local derivative $\partial/\partial t$ and a convective derivative $\vec{V} \cdot \nabla$. Thus,

$$\frac{d}{dt} = \frac{\partial}{\partial t} + \vec{V} \cdot \nabla \tag{D-2}$$

and in Cartesian coordinates it is expressed as

$$\frac{d}{dt} = \frac{\partial}{\partial t} + u\frac{\partial}{\partial x} + v\frac{\partial}{\partial y} + w\frac{\partial}{\partial z} \tag{D-3}$$

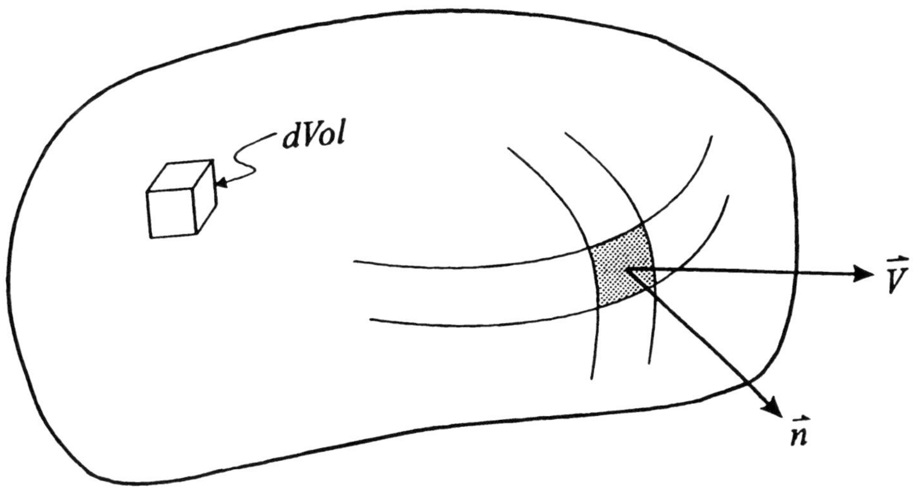

Figure D.1 Illustration of control volume.

D.2.1. Conservation of Mass: This conservation law requires that mass is neither created nor destroyed; mathematically this is expressed as $dM/dt = 0$. Using Equation (D-1), where $N = M$ and $\eta = 1$, the integral form of the conservation of mass is obtained as

$$\frac{\partial}{\partial t}\int_{C.V.} \rho \, d(\text{Vol}) + \int_{C.S.} \rho\vec{V} \cdot \vec{n} dS = 0 \tag{D-4}$$

The physical interpretation of Equation (D-4) is as follows: The sum of the rate of change of mass within the control volume and net efflux of mass across the control surfaces is zero.

D.2.2. Conservation of Linear Momentum: Newton's second law applied to a nonaccelerating control volume which is referenced to a fixed coordinate system, will result in the integral form of the momentum equation once Equation (D-1) is utilized. In this case the linear momentum $\vec{G} = m\vec{V}$ is taken to be the extensive property and therefore $\eta = \vec{V}$. Newton's second law in an inertial reference is expressed as $\Sigma\vec{F} = d\vec{G}/dt$. Thus, use of Equation (D-1) yields

$$\Sigma\vec{F} = \frac{\partial}{\partial t}\int_{C.V.} \rho\vec{V} \, d(\text{Vol}) + \int_{C.S.} \vec{V}(\rho\vec{V} \cdot \vec{n}) dS \tag{D-5}$$

This equation states that the sum of the forces acting on a control volume is equal to the sum of the rate of change of linear momentum within the control volume and net efflux of the linear momentum across the control surfaces. The forces acting on a control volume usually represent a combination of body force (such as gravity) and surface forces (such as viscous force). Note that Equation (D-5) is a vector equation from which scalar components may be obtained. For example, the x-component is expressed as

$$\Sigma F_x = \frac{\partial}{\partial t}\int_{C.V.} \rho u \, d(\text{Vol}) + \int_{C.S.} u(\rho\vec{V} \cdot \vec{n}) dS \tag{D-6}$$

D.2.3. Conservation of Energy: This conservation law is based on the first law of thermodynamics, which may be expressed as

$$\frac{d(\rho e_t)}{dt} = \frac{\partial Q}{\partial t} + \frac{\partial W}{\partial t} \tag{D-7}$$

In this relation, e_t represents the total energy of the system per unit mass, while $\partial Q/\partial t$ and $\partial W/\partial t$ represent the rate of heat transfer to the system and the rate of work done on the system, respectively. Usually heat added to the system and work done on the system are defined positive. In order to use Equation (D-1), ρe_t is selected to represent the extensive property N, and $\eta = e_t$, the total energy per unit mass. Hence,

$$\frac{\partial}{\partial t}\int_{C.V.} \rho e_t \, d(\text{Vol}) + \int_{C.S.} e_t(\rho\vec{V}\cdot\vec{n})dS = \dot{Q} + \dot{W} \qquad \text{(D-8)}$$

In general, the total energy is the sum of the internal energy, kinetic energy and potential energy, or in terms of energy per unit mass,

$$e_t = e + \tfrac{1}{2}V^2 + gz \qquad \text{(D-9)}$$

It is convenient to categorize the work as either flow work or shaft work. Flow work W_f is due to stresses producing work on the control surfaces where there is fluid flow.

The work transferred through the control surfaces is called the shaft work, $\dot{W}_S$. With such classification of work, the energy equation may be expressed as

$$\frac{\partial}{\partial t}\int_{C.V.} \rho e_t \, d(\text{Vol}) + \int_{C.S.} [h + \tfrac{1}{2}V^2 + gz](\rho\vec{V}\cdot\vec{n})dS = \dot{Q} + \dot{W}_S \qquad \text{(D-10)}$$

where the rate of flow work given by $\int (p/\rho)\,(\rho\vec{V}\cdot\vec{n})ds$ is moved to the left-hand side and added to the internal energy. Recall that the combination of $e + p/\rho = h$ is called enthalpy.

D.3 Differential Formulations

The differential forms of the equations of motion are utilized for situations where a detailed solution of the flow field is required. These equations are obtained by the application of conservation laws to an infinitesimal fixed control volume. A typical differential element for a Cartesian coordinate system is shown in Figure D.2. The flow properties at the control surfaces are expanded in Taylor series and integral forms of the equations are used. After a limiting process for which the differential element approaches zero, all the higher order terms are dropped and the differential form of the equations are produced. The differential equations which are derived based on fixed coordinate system and control volume are known as *Eulerian* approach. On the other hand, if we were to move with the fluid element, the approach is called *Lagrangian*. Clearly the Eulerian approach is much desirable in fluid mechanics, and in this text methods based on Eulerian formulation are used.

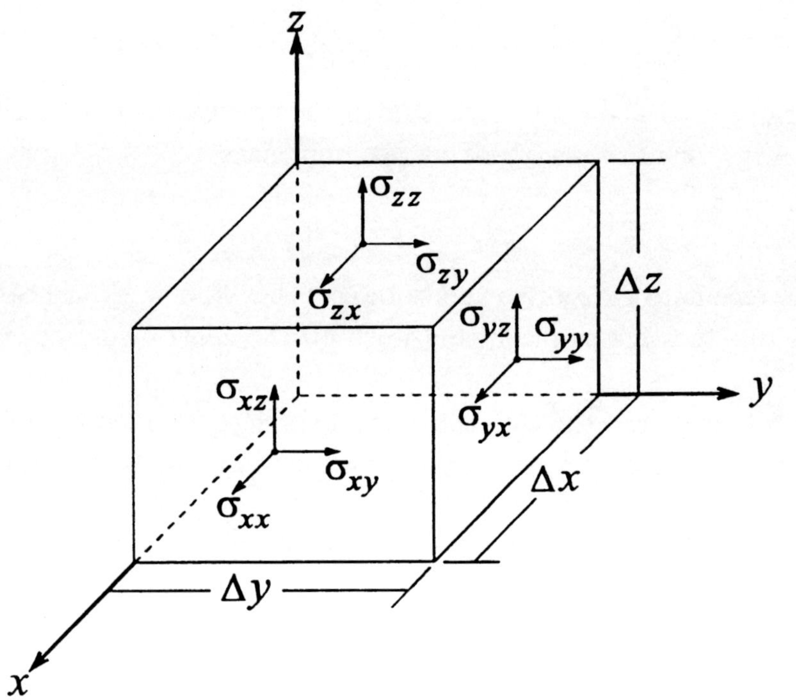

Figure D.2 Differential element for Cartesian coordinate system.

D.3.1 Conservation of Mass

The differential form of the conservation of mass is known as *the continuity equation*, and in a vector form it is expressed as

$$\frac{\partial \rho}{\partial t} + \nabla \cdot (\rho \vec{V}) = 0 \tag{D-11}$$

The derivation of this equation may be found in any standard fluid mechanics text such as (D.2–D.4). This equation may be written in terms of the total derivative as

$$\frac{d\rho}{dt} + \rho \nabla \cdot \vec{V} = 0 \tag{D-12}$$

For a steady state problem $\partial/\partial t \equiv 0$, and, therefore,

$$\nabla \cdot (\rho \vec{V}) = 0$$

For an incompressible flow, where the density variations are considered negligible, continuity is reduced to

$$\nabla \cdot \vec{V} = 0 \tag{D-13}$$

Note that this equation is valid for unsteady problems as well. In Cartesian coordinate system, where the velocity vector is $\vec{V} = u\vec{i} + v\vec{j} + w\vec{k}$, Equation (D-11) becomes

$$\frac{\partial \rho}{\partial t} + \frac{\partial}{\partial x}(\rho u) + \frac{\partial}{\partial y}(\rho v) + \frac{\partial}{\partial z}(\rho w) = 0 \tag{D-14}$$

and, for an incompressible flow,

$$\frac{\partial u}{\partial x} + \frac{\partial v}{\partial y} + \frac{\partial w}{\partial z} = 0 \tag{D-15}$$

D.3.2 Conservation of Linear Momentum

The linear momentum equation, also known as the Navier-Stokes equation, is obtained by the application of Newton's second law to a differential element in an inertial coordinate system. If σ is used to represent stresses acting on the differential element, the components of the Navier-Stokes equations in a Cartesian coordinate system takes the following form:

$$\rho\frac{du}{dt} = \rho f_x + \frac{\partial}{\partial x}(\sigma_{xx}) + \frac{\partial}{\partial y}(\sigma_{yx}) + \frac{\partial}{\partial z}(\sigma_{zx}) \tag{D-16}$$

$$\rho\frac{dv}{dt} = \rho f_y + \frac{\partial}{\partial x}(\sigma_{xy}) + \frac{\partial}{\partial y}(\sigma_{yy}) + \frac{\partial}{\partial z}(\sigma_{zy}) \tag{D-17}$$

$$\rho\frac{dw}{dt} = \rho f_z + \frac{\partial}{\partial x}(\sigma_{xz}) + \frac{\partial}{\partial y}(\sigma_{yz}) + \frac{\partial}{\partial z}(\sigma_{zz}) \tag{D-18}$$

Shear stress σ usually is written in terms of pressure p and viscous stress τ. In tensor notation this is expressed as

$$\sigma_{ij} = -p\delta_{ij} + \tau_{ij} \tag{D-19}$$

where δ_{ij} is the Kronecker delta,

$$\delta_{ij} = \begin{cases} 1 & \text{for } i = j \\ 0 & \text{for } i \neq j \end{cases}$$

The subscripts assigned to the components of stress are as follows. The first subscript denotes the direction of the normal to the surface on which stress acts, and the second denotes the direction in which the stress acts. This is illustrated in Figure D.2.

The components of the linear momentum equation given by Equations (D-16 through D-18) can be written in terms of viscous stresses as

$$\rho\frac{du}{dt} = \rho f_x - \frac{\partial p}{\partial x} + \frac{\partial \tau_{xx}}{\partial x} + \frac{\partial \tau_{yx}}{\partial y} + \frac{\partial \tau_{zx}}{\partial z} \tag{D-20}$$

$$\rho\frac{dv}{dt} = \rho f_y - \frac{\partial p}{\partial y} + \frac{\partial \tau_{xy}}{\partial x} + \frac{\partial \tau_{yy}}{\partial y} + \frac{\partial \tau_{zy}}{\partial z} \tag{D-21}$$

$$\rho\frac{dw}{dt} = \rho f_z - \frac{\partial p}{\partial z} + \frac{\partial \tau_{xz}}{\partial x} + \frac{\partial \tau_{yz}}{\partial y} + \frac{\partial \tau_{zz}}{\partial z} \tag{D-22}$$

Viscous stresses are related to the rates of strain by a physical law. For most fluids, this relation is linear and is known as *Newtonian fluid.* For a Newtonian fluid, viscous stresses in a Cartesian coordinate system are

$$\tau_{xx} = 2\mu\frac{\partial u}{\partial x} + \lambda\nabla\cdot\vec{V} \tag{D-23}$$

$$\tau_{yy} = 2\mu\frac{\partial v}{\partial y} + \lambda\nabla\cdot\vec{V} \tag{D-24}$$

$$\tau_{zz} = 2\mu\frac{\partial w}{\partial z} + \lambda\nabla\cdot\vec{V} \tag{D-25}$$

$$\tau_{xy} = \tau_{yx} = \mu\left(\frac{\partial u}{\partial y} + \frac{\partial v}{\partial x}\right) \tag{D-26}$$

$$\tau_{xz} = \tau_{zx} = \mu\left(\frac{\partial w}{\partial x} + \frac{\partial u}{\partial z}\right) \tag{D-27}$$

$$\tau_{yz} = \tau_{zy} = \mu\left(\frac{\partial v}{\partial z} + \frac{\partial w}{\partial y}\right) \tag{D-28}$$

where μ is known as *the coefficient of viscosity* or *dynamic viscosity* and λ is defined as *the second coefficient of viscosity.* The combination of μ and λ in the following form is known as *the bulk viscosity* k, i.e.,

$$k = \lambda + \tfrac{2}{3}\mu \tag{D-29}$$

If bulk viscosity of a fluid is assumed negligible, then

$$\lambda = -\tfrac{2}{3}\mu \tag{D-30}$$

This is known as *Stokes hypothesis.*

After substitution of viscous stress terms for a Newtonian fluid given by Equations (D-23) through (D-28) into the Navier-Stokes Equations (D-20) through (D-

22), one obtains

$$\rho\frac{du}{dt} = \rho f_x + \frac{\partial}{\partial x}\left[-p + 2\mu\frac{\partial u}{\partial x} + \lambda\nabla\cdot\vec{V}\right] + \frac{\partial}{\partial y}\left[\mu\left(\frac{\partial u}{\partial y} + \frac{\partial v}{\partial x}\right)\right] + \frac{\partial}{\partial z}\left[\mu\left(\frac{\partial w}{\partial x} + \frac{\partial u}{\partial z}\right)\right] \quad \text{(D-31)}$$

$$\rho\frac{dv}{dt} = \rho f_y + \frac{\partial}{\partial x}\left[\mu\left(\frac{\partial u}{\partial y} + \frac{\partial v}{\partial x}\right)\right] + \frac{\partial}{\partial y}\left[-p + 2\mu\frac{\partial v}{\partial y} + \lambda\nabla\cdot\vec{V}\right] + \frac{\partial}{\partial z}\left[\mu\left(\frac{\partial v}{\partial z} + \frac{\partial w}{\partial y}\right)\right] \quad \text{(D-32)}$$

$$\rho\frac{dw}{dt} = \rho f_z + \frac{\partial}{\partial x}\left[\mu\left(\frac{\partial w}{\partial x} + \frac{\partial u}{\partial z}\right)\right] + \frac{\partial}{\partial y}\left[\mu\left(\frac{\partial v}{\partial z} + \frac{\partial w}{\partial y}\right)\right] + \frac{\partial}{\partial z}\left[-p + 2\mu\frac{\partial w}{\partial z} + \lambda\nabla\cdot\vec{V}\right] \quad \text{(D-33)}$$

In applications where the temperature variations are small, viscosity may be assumed constant and in addition to negligible changes in density, the flow is assumed incompressible. For such a flow, the Navier-Stokes equations are reduced to

$$\rho\frac{du}{dt} = \rho f_x - \frac{\partial p}{\partial x} + \mu\left(\frac{\partial^2 u}{\partial x^2} + \frac{\partial^2 u}{\partial y^2} + \frac{\partial^2 u}{\partial z^2}\right) \quad \text{(D-34)}$$

$$\rho\frac{dv}{dt} = \rho f_y - \frac{\partial p}{\partial y} + \mu\left(\frac{\partial^2 v}{\partial x^2} + \frac{\partial^2 v}{\partial y^2} + \frac{\partial^2 v}{\partial z^2}\right) \quad \text{(D-35)}$$

$$\rho\frac{dw}{dt} = \rho f_z - \frac{\partial p}{\partial z} + \mu\left(\frac{\partial^2 w}{\partial x^2} + \frac{\partial^2 w}{\partial y^2} + \frac{\partial^2 w}{\partial z^2}\right) \quad \text{(D-36)}$$

The linear momentum equation can be written in a conservative form by addition of the continuity equation to the left hand side of the equation. For the x-component of the equation given by Equation (D-20) one may add the continuity equation multiplied by the u-component of the velocity to obtain

$$\rho\frac{\partial u}{\partial x} + \rho u\frac{\partial u}{\partial x} + \rho v\frac{\partial u}{\partial y} + \rho w\frac{\partial u}{\partial z} + u\left[\frac{\partial \rho}{\partial t} + \frac{\partial}{\partial x}(\rho u) + \frac{\partial}{\partial y}(\rho v) + \frac{\partial}{\partial z}(\rho w)\right] = \rho f_x - \frac{\partial p}{\partial x} + \frac{\partial}{\partial x}(\tau_{xx}) + \frac{\partial}{\partial y}(\tau_{yx}) + \frac{\partial}{\partial z}(\tau_{zx}) \quad \text{(D-37)}$$

The terms on the left hand of Equation (D-37) may be combined to yield

$$\frac{\partial}{\partial t}(\rho u)+\frac{\partial}{\partial x}(\rho u^2)+\frac{\partial}{\partial y}(\rho uv)+\frac{\partial}{\partial z}(\rho uw)=\rho f_x-\frac{\partial p}{\partial x}+\frac{\partial}{\partial x}(\tau_{xx})+\frac{\partial}{\partial y}(\tau_{yx})+\frac{\partial}{\partial z}(\tau_{zx}) \tag{D-38}$$

Finally, one may write the scalar components of the linear momentum equation in conservation law form as

x-component:

$$\frac{\partial}{\partial t}(\rho u)+\frac{\partial}{\partial x}(\rho u^2+p)+\frac{\partial}{\partial y}(\rho uv)+\frac{\partial}{\partial z}(\rho uw)=\rho f_x+\frac{\partial}{\partial x}(\tau_{xx})+\frac{\partial}{\partial y}(\tau_{xy})+\frac{\partial}{\partial z}(\tau_{xz}) \tag{D-39}$$

y-component:

$$\frac{\partial}{\partial t}(\rho v)+\frac{\partial}{\partial x}(\rho uv)+\frac{\partial}{\partial y}(\rho v^2+p)+\frac{\partial}{\partial z}(\rho vw)=\rho f_y+\frac{\partial}{\partial x}(\tau_{xy})+\frac{\partial}{\partial y}(\tau_{yy})+\frac{\partial}{\partial z}(\tau_{yz}) \tag{D-40}$$

z-component:

$$\frac{\partial}{\partial t}(\rho w)+\frac{\partial}{\partial x}(\rho uw)+\frac{\partial}{\partial y}(\rho vw)+\frac{\partial}{\partial z}(\rho w^2+p)=\rho f_z+\frac{\partial}{\partial x}(\tau_{xz})+\frac{\partial}{\partial y}(\tau_{yz})+\frac{\partial}{\partial z}(\tau_{zz}) \tag{D-41}$$

D.3.3 Energy Equation

The energy equation derived from the first law of thermodynamics may be written in various forms. One such formulation written in terms of the total energy e_t is

$$\rho\frac{de_t}{dt}=\rho(uf_x+vf_y+wf_z)+\frac{\partial}{\partial x}[-pu+u\tau_{xx}+v\tau_{xy}+w\tau_{xz}-q_x]$$
$$+\frac{\partial}{\partial y}[-pv+u\tau_{yx}+v\tau_{yy}+w\tau_{yz}-q_y]+\frac{\partial}{\partial z}[-pw+u\tau_{zx}+v\tau_{zy}+w\tau_{zz}-q_z] \tag{D-42}$$

where q represents heat flux.

Equation (D-42) may be expressed in a conservative form by the addition of the continuity equation as

$$\frac{\partial}{\partial t}(\rho e_t)+\frac{\partial}{\partial x}(\rho ue_t+pu)+\frac{\partial}{\partial y}(\rho ve_t+pv)+\frac{\partial}{\partial z}(\rho we_t+pw)=$$
$$\frac{\partial}{\partial x}\left[u\tau_{xx}+v\tau_{xy}+w\tau_{xz}-q_x\right]+\frac{\partial}{\partial y}\left[u\tau_{yx}+v\tau_{yy}+w\tau_{yz}-q_y\right]$$
$$+\frac{\partial}{\partial z}\left[u\tau_{zx}+v\tau_{zy}+w\tau_{zz}-q_z\right] \tag{D-43}$$

D.3.4 Flux Vector Formulation

The conservative form of the equations of motion in a Cartesian coordinate system assuming negligible body forces is:
Continuity, Equation (D-14)

$$\frac{\partial \rho}{\partial t} + \frac{\partial}{\partial x}(\rho u) + \frac{\partial}{\partial y}(\rho v) + \frac{\partial}{\partial z}(\rho w) = 0$$

x-component of the momentum, Equation (D-20)

$$\frac{\partial}{\partial t}(\rho u) + \frac{\partial}{\partial x}(\rho u^2 + p) + \frac{\partial}{\partial y}(\rho uv) + \frac{\partial}{\partial z}(\rho uw) = \frac{\partial}{\partial x}(\tau_{xx}) + \frac{\partial}{\partial y}(\tau_{xy}) + \frac{\partial}{\partial z}(\tau_{xz})$$

y-component of the momentum, Equation (D-21)

$$\frac{\partial}{\partial t}(\rho v) + \frac{\partial}{\partial x}(\rho uv) + \frac{\partial}{\partial y}(\rho v^2 + p) + \frac{\partial}{\partial z}(\rho vw) = \frac{\partial}{\partial x}(\tau_{xy}) + \frac{\partial}{\partial y}(\tau_{yy}) + \frac{\partial}{\partial z}(\tau_{yz})$$

z-component of the momentum, Equation (D-22)

$$\frac{\partial}{\partial t}(\rho w) + \frac{\partial}{\partial x}(\rho uw) + \frac{\partial}{\partial y}(\rho vw) + \frac{\partial}{\partial z}(\rho w^2 + p) = \frac{\partial}{\partial x}(\tau_{xz}) + \frac{\partial}{\partial y}(\tau_{yz}) + \frac{\partial}{\partial z}(\tau_{zz})$$

Energy, Equation (D-43)

$$\frac{\partial}{\partial t}(\rho e_t) + \frac{\partial}{\partial x}(\rho u e_t + pu) + \frac{\partial}{\partial y}(\rho v e_t + pv) + \frac{\partial}{\partial z}(\rho w e_t + pw) =$$

$$+\frac{\partial}{\partial x}\left[u\tau_{xx} + v\tau_{xy} + w\tau_{xz} - q_x\right] + \frac{\partial}{\partial y}\left[u\tau_{yx} + v\tau_{yy} + w\tau_{yz} - q_y\right]$$

$$+\frac{\partial}{\partial z}\left[u\tau_{zx} + v\tau_{zy} + w\tau_{zz} - q_z\right]$$

It is convenient to write the Cartesian form of the equations of motion in a flux vector form. This formulation may be expressed in a Cartesian coordinate system as

$$\frac{\partial Q}{\partial t} + \frac{\partial E}{\partial x} + \frac{\partial F}{\partial y} + \frac{\partial G}{\partial z} = \frac{\partial E_v}{\partial x} + \frac{\partial F_v}{\partial y} + \frac{\partial G_v}{\partial z} \tag{D-44}$$

where

$$Q = \begin{bmatrix} \rho \\ \rho u \\ \rho v \\ \rho w \\ \rho e_t \end{bmatrix} \tag{D-45}$$

$$E = \begin{bmatrix} \rho u \\ \rho u^2 + p \\ \rho uv \\ \rho uw \\ (\rho e_t + p)u \end{bmatrix} \quad \text{(D-46)} \qquad E_v = \begin{bmatrix} 0 \\ \tau_{xx} \\ \tau_{xy} \\ \tau_{xz} \\ u\tau_{xx} + v\tau_{xy} + w\tau_{xz} - q_x \end{bmatrix} \quad \text{(D-47)}$$

$$F = \begin{bmatrix} \rho v \\ \rho vu \\ \rho v^2 + p \\ \rho vw \\ (\rho e_t + p)v \end{bmatrix} \quad \text{(D-48)} \qquad F_v = \begin{bmatrix} 0 \\ \tau_{yx} \\ \tau_{yy} \\ \tau_{yz} \\ u\tau_{yx} + v\tau_{yy} + w\tau_{yz} - q_y \end{bmatrix} \quad \text{(D-49)}$$

$$G = \begin{bmatrix} \rho w \\ \rho wu \\ \rho wv \\ \rho w^2 + p \\ (\rho e_t + p)w \end{bmatrix} \quad \text{(D-50)} \qquad G_v = \begin{bmatrix} 0 \\ \tau_{zx} \\ \tau_{zy} \\ \tau_{zz} \\ u\tau_{zx} + v\tau_{zy} + w\tau_{zz} - q_z \end{bmatrix} \quad \text{(D-51)}$$

D.3.5 Two-Dimensional Planar and Axisymmetric Formulation

The equations of fluid motion may be expressed in a combined form for a two-dimensional planar flow and an axisymmetric flow. This form of the equation is particularly useful when numerical schemes are utilized for solution. Therefore one computer code may be developed for the solution of two different types of flows.

The equations of motion in a combined form are expressed as

$$\frac{\partial Q}{\partial t} + \frac{\partial E}{\partial x} + \frac{\partial F}{\partial y} + \alpha H = \frac{\partial E_v}{\partial x} + \frac{\partial F_v}{\partial y} + \alpha H_v \quad \text{(D-52)}$$

where

$$\alpha = \begin{cases} 0 \text{ for 2-D planar flow} \\ 1 \text{ for 2-D axisymmetric flow} \end{cases}$$

Note that for a 2-D planar flow, the formulation is based on a Cartesian coordinate system whereas for a 2-D axisymmetric flow, the formulation is based on a cylindrical coordinate system. The flux vectors in Equation (D-52) are defined as

$$Q = \begin{bmatrix} \rho \\ \rho u \\ \rho v \\ \rho e_t \end{bmatrix} \qquad E = \begin{bmatrix} \rho u \\ \rho u^2 + p \\ \rho uv \\ (\rho e_t + p)u \end{bmatrix} \qquad F = \begin{bmatrix} \rho v \\ \rho vu \\ \rho v^2 + p \\ (\rho e_t + p)v \end{bmatrix}$$

$$H = \frac{1}{y}\begin{bmatrix} \rho v \\ \rho uv \\ \rho v^2 \\ (\rho e_t + p)v \end{bmatrix} \qquad E_v = \begin{bmatrix} 0 \\ \tau_{xxp} \\ \tau_{xy} \\ u\tau_{xxp} + v\tau_{xy} - q_x \end{bmatrix} \qquad F_v = \begin{bmatrix} 0 \\ \tau_{xy} \\ \tau_{yyp} \\ u\tau_{xy} + v\tau_{yyp} - q_y \end{bmatrix}$$

$$H_v = \frac{1}{y}\begin{bmatrix} 0 \\ \tau_{xy} - \frac{2}{3}y\frac{\partial}{\partial x}\left(\mu\frac{v}{y}\right) \\ \tau_{yyp} - \tau_{\theta\theta} - \frac{2}{3}\mu\frac{v}{y} - y\frac{2}{3}\frac{\partial}{\partial y}\left(\mu\frac{v}{y}\right) \\ u\tau_{xy} + v\tau_{yyp} - q_y - \frac{2}{3}\mu\frac{v^2}{y} - y\frac{\partial}{\partial y}\left(\frac{2}{3}\mu\frac{v^2}{y}\right) - y\frac{\partial}{\partial x}\left(\frac{2}{3}\mu\frac{uv}{y}\right) \end{bmatrix}$$

where for a Newtonian fluid with Stokes hypothesis,

$$\tau_{xxp} = \frac{4}{3}\mu\frac{\partial u}{\partial x} - \frac{2}{3}\mu\frac{\partial v}{\partial y} \tag{D-53}$$

$$\tau_{xx} = \tau_{xxp} - \frac{2}{3}\mu\frac{v}{y} \tag{D-54}$$

$$\tau_{yyp} = \frac{4}{3}\mu\frac{\partial v}{\partial y} - \frac{2}{3}\mu\frac{\partial u}{\partial x} \tag{D-55}$$

$$\tau_{yy} = \tau_{yyp} - \frac{2}{3}\mu\frac{v}{y} \tag{D-56}$$

$$\tau_{xy} = \mu\left(\frac{\partial u}{\partial y} + \frac{\partial v}{\partial x}\right) \tag{D-57}$$

$$\tau_{\theta\theta} = -\frac{2}{3}\mu\left(\frac{\partial u}{\partial x} + \frac{\partial v}{\partial y}\right) + \frac{4}{3}\mu\frac{v}{y} \tag{D-58}$$

$$q_x = -k\frac{\partial T}{\partial x} \tag{D-59}$$

$$q_y = -k\frac{\partial T}{\partial y} \tag{D-60}$$

Note: Subscript p represents 2-D planar flow.

D.4 Modification of the Navier-Stokes Equation

The Navier-Stokes equation given by (D-44) may be reduced or amended, depending on a particular application. Typically the Navier-Stokes equation is reduced to the thin-layer Navier-Stokes equation by retaining only the normal gradient of the viscous stresses and neglecting the gradients of viscous stresses parallel to the surface. The resulting equation is given by Equation (11-155). For applications where the flow is steady and streamwise flow separation is not an issue, the steady Navier-Stokes equation is modified by neglecting the streamwise gradient of viscous stress and modification of streamwise pressure gradient within the subsonic portion of the viscous region near the surface. The resulting equation is the parabolized Navier-Stokes equation and is given by Equation (11-157).

The reduction of the Navier-Stokes equation to either thin-layer Navier-Stokes equation or parabolized Navier-Stokes equation is primarily introduced from numerical point of view. These reductions reduce the computation time required for a solution.

The Navier-Stokes equation can also be reduced based on the physics of the problem. If the flow can be assumed to be inviscid, then the viscous shear stresses are neglected. The resulting equation is known as the Euler equation and is given by Equation (12-2). For problems where the density variation is negligible, the flow is considered as incompressible, and the Navier-Stokes equation is reduced to the incompressible Navier-Stokes equation given by Equation (11-231).

The Navier-Stokes equation given by (D-44) may be modified and/or amended to account for physical phenomena such as turbulence and chemistry. The consideration for turbulence is addressed in Chapter 21, and chemistry effect is addressed in Chapter 16. A summary of the various forms of the Navier-Stokes equation is provided in Figure D.3.

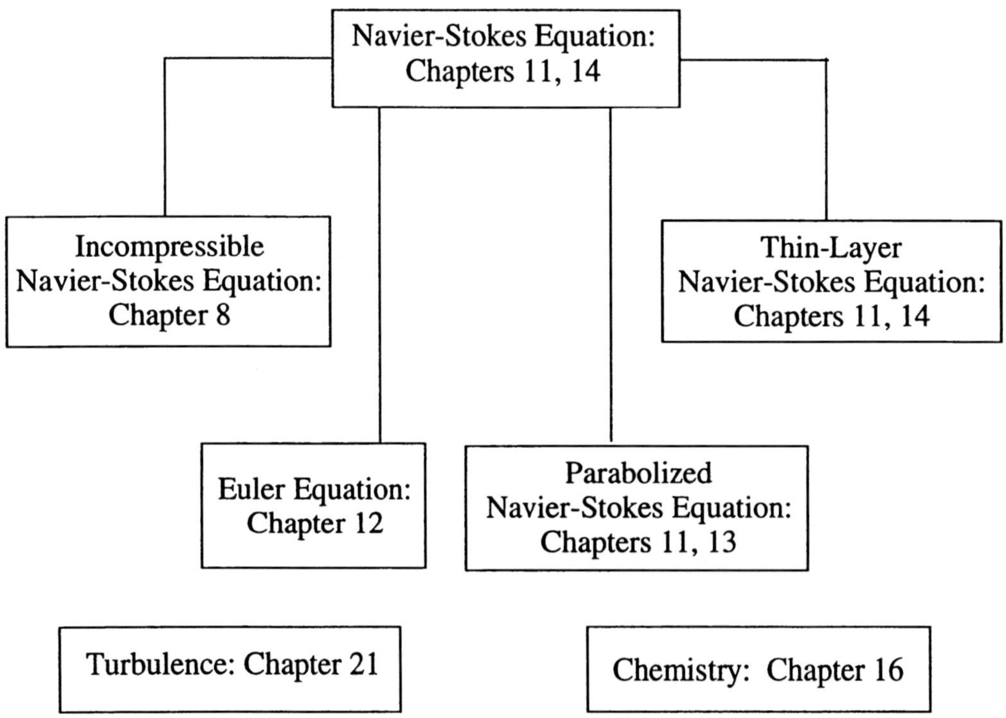

Figure D.3 Summary of the Navier-Stokes Equations.

D.5 Auxiliary Relations

The governing equations of fluid motion typically involve more unknowns than available conservation equations which include conservations of mass, momentum, and energy. Therefore, in order to close the system, additional relations such as the equation of state and relations for the transport properties such as viscosity and thermal conductivity must be introduced. This section will review the appropriate relations which may be required for the solution of the system of equations of fluid motion.

D.5.1 Viscosity

Viscosity of a fluid is a property used to describe the response of a fluid to the imposed shearing forces. In a fluid flow, the rate of deformation is proportional to the shear stress. The constant of proportionality is called the *coefficient of dynamic viscosity.* Typically, it is referred to as the *coefficient of viscosity* or just as *viscosity.* If the relation between the shear stress and the rate of deformation is linear, then the fluid is called a *Newtonian fluid.* These relations are provided as Equations (D-23) through (D-28).

In general, the coefficient of viscosity is a function of composition of the fluid, its temperature, and pressure. In most cases, the pressure dependency is negligible and the coefficient of viscosity is expressed as a function of temperature only. Pressure dependency, however, should be included at very high or at very low pressures.

There are several expressions relating the coefficient of viscosity to the temperature. Perhaps the most commonly used relation for dilute gases is the Sutherland's law, which can be expressed as

$$\mu = c_1 \frac{T^{3/2}}{T + c_2} \tag{D-61}$$

where the constants c_1 and c_2 are given in Table D.1 for selected gases.

Table D.1: Sutherland's constants for various gases.

Gas		SI system $c_1 \times 10^6$ $\left(\frac{\text{kg}}{\text{sec m K}^{1/2}}\right)$	SI system c_2 (K)	British system $c_1 \times 10^8$ $\left(\frac{\text{lbf sec}}{\text{ft}^2\ {}^\circ\text{R}^{1/2}}\right)$	British system c_2 (°R)
Air		1.458	110.4	2.27	198.6
Carbon dioxide	CO_2	1.550	233.0	2.42	420.0
Carbon monoxide	CO	1.400	109.0	2.18	196.2
Hydrogen	H_2	0.649	70.6	1.01	127.0
Nitrogen	N_2	1.390	102.0	2.16	183.6
Oxygen	O_2	1.650	110.0	2.57	198.0

The temperature in Equation (D.61) must be in °R or K, providing the values of viscosity in the units of $\frac{\text{lbf sec}}{\text{ft}^2} = \frac{\text{slug}}{\text{ft sec}}$ or $\frac{\text{kg}}{\text{sec m}} = \frac{\text{N sec}}{\text{m}^2}$, respectively.

D.5.2 Thermal Conductivity

The Fourier's heat conduction law states that the heat conduction per unit area is proportional to the normal gradient of temperature. Mathematically,

$$\frac{q}{A} \propto -\frac{\partial T}{\partial n} \tag{D-62}$$

Expression (D-62) can be expressed as an equation by introducing a constant of proportionality defined as *thermal conductivity* k. Therefore,

$$\frac{q}{A} = -k\frac{\partial T}{\partial n} \tag{D-63}$$

It should be observed that the equation above involving thermal conductivity has similar mathematical form as viscosity. In fact, thermal conductivity also depends on the composition of material, pressure, and temperature. Just as viscosity, the dependency of thermal conductivity at moderate pressures is secondary and, in those applications, it can be determined based on temperature alone.

The thermal conductivity of air below 2000 K at atmospheric pressure may be calculated from the following relation.

$$k = 4.76 \times 10^{-6} \frac{T^{3/2}}{T + 112.0} \tag{D-64}$$

where T is in (K) and k is in [cal/(cm sec K)].

For most applications the thermal conductivity is determined from Prandtl number, as discussed in Section D.5.5.

D.5.3 Specific Heats

Specific heat represents the amount of heat δQ required to produce a variation dT in the temperature of a substance. Specific heat may be defined under a constant pressure process or a constant volume process. Specific heat at constant pressure is defined as

$$c_p = \left(\frac{\partial h}{\partial T}\right)_p \tag{D-65}$$

and specific heat at constant volume is defined as

$$c_v = \left(\frac{\partial e}{\partial T}\right)_v \tag{D-66}$$

The ratio of specific heats is defined as

$$\gamma = \frac{c_p}{c_v} \tag{D-67}$$

In general, the enthalpy h and the internal energy e are functions of two thermodynamic properties such as pressure and temperature. If one assumes that h and e are only functions of temperature, then the gas is called *thermally perfect gas*. For a thermally perfect gas, relations (D-65) and (D-66) can be expressed as

$$c_p = \frac{dh}{dT} \qquad \text{or} \qquad dh = c_p\, dT \tag{D-68}$$

and

$$c_v = \frac{de}{dT} \qquad \text{or} \qquad de = c_v\, dT \tag{D-69}$$

In general, specific heats are a function of both pressure and temperature. Typically, the influence of pressure is less significant, and, therefore, it may be ignored. That is particularly the case for liquids. The effect of pressure on specific heats of liquids becomes important only at extremely high pressures. Even the temperature dependency of specific heats is slight for relatively moderate changes in temperature. However, the temperature effect on the specific heats of gases is more appreciable, and some pressure dependency at extreme values of pressure is present. Typically, the influence of pressure is less at higher temperatures.

For a calorically perfect gas, the specific heats c_p and c_v are constants. Thus, relations (D-68) and (D-69) can be written as

$$h = c_p T \tag{D-70}$$

and

$$e = c_v T \tag{D-71}$$

For a thermally perfect gas, the following relations hold

$$c_p - c_v = R \tag{D-72}$$

$$c_p = \frac{\gamma R}{\gamma - 1} \tag{D-73}$$

$$c_v = \frac{R}{\gamma - 1} \tag{D-74}$$

where R is the gas constant, defined in Sec. D.5.4. According to Kinetic Theory, the ratio of specific heats is

$$\gamma = \frac{5}{3} = 1.67 \qquad \text{for monatomic gases}$$

$$\gamma = \frac{7}{5} = 1.40 \qquad \text{for diatomic gases}$$

$$\gamma = \frac{8}{6} = 1.33 \qquad \text{for polyatomic gases}$$

D.5.4 Equation of State

A relation between the density, pressure, and temperature of a fluid is known as the *equation of state.* This relation may be presented in the form of tables, charts, or an equation. For a thermally perfect gas, the equation of state is

$$p = \rho R T \tag{D-75}$$

where R is the gas constant defined by the universal gas constant $\mathcal{R}$ divided by the molecular weight, i.e.,

$$R = \frac{\mathcal{R}}{MW}$$

The universal gas constant is

$$\mathcal{R} = \begin{cases} 8314.34 \dfrac{\text{N} \cdot \text{m}}{\text{kg mole K}} = 1.987 \dfrac{\text{cal}}{\text{gm mole K}} \\ 1545.33 \dfrac{\text{ft lbf}}{\text{lbm mole} \,^\circ\text{R}} = 4.9723 \times 10^4 \dfrac{\text{ft lbf}}{\text{Slug mole} \,^\circ\text{R}} \end{cases}$$

The gas constant for air is

$$R = \begin{cases} 287.05 \dfrac{\text{N} \cdot \text{m}}{\text{kg K}} \\ 53.34 \dfrac{\text{ft lbf}}{\text{lbm} \,^\circ\text{R}} = 1716.16 \dfrac{\text{ft} \cdot \text{lbf}}{\text{Slug} \,^\circ\text{R}} \end{cases}$$

D.5.5 Prandtl Number

Prandtl number is defined as *the ratio of kinematic viscosity to thermal diffusivity*,

$$Pr = \frac{\nu}{\alpha} = \frac{\mu c_p}{k} \tag{D-76}$$

The Prandtl number represents the ratio of diffusion of momentum by viscosity to the diffusion of heat by conduction.

It should be noted that the Prandtl number is also pressure- and temperature-dependent. In fact, that is easily recognized by considering relation (D-76), because μ, k, and c_p are all pressure- and temperature-dependent. Since expressions or tables are available to estimate the values of μ, k, and c_p, they can be used to determine the Prandtl number.

The Prandtl number for air at standard conditions is 0.72. Since, for most gases the ratio of $\frac{c_p}{Pr}$ is approximately constant, once viscosity is determined the thermal conductivity is computed from

$$k = \frac{\mu c_p}{Pr}$$

APPENDIX E:

Block-Tridiagonal System of Equations

When a system of PDEs is approximated by implicit formulation involving three grid points at each level, a block-tridiagonal system is produced. The resulting block-tridiagonal system may be expressed in a general form as

$$S\Delta Q = R \tag{E-1}$$

where ΔQ and R are m component vectors. The coefficient S represents the block tridiagonal coefficient expressed by

$$S = \begin{bmatrix} B_2 & C_2 & & & & & \\ A_3 & B_3 & C_3 & & & & \\ & A_4 & B_4 & C_4 & & & \\ & & & & & & \\ & & & & A_{IM2} & B_{IM2} & C_{IM2} \\ & & & & & A_{IM1} & B_{IM1} \end{bmatrix}$$

where A_i, B_i, and C_i are matrices of order m.

To investigate a solution scheme, consider the following factorization

$$S = LU =$$

$$\begin{bmatrix} \alpha_2 & & & & & \\ A_3 & \alpha_3 & & & & \\ & A_4 & \alpha_4 & & & \\ & & & & & \\ & & & A_{IM2} & \alpha_{IM2} & \\ & & & & A_{IM1} & \alpha_{IM1} \end{bmatrix} \begin{bmatrix} I & \beta_2 & & & & \\ & I & \beta_3 & & & \\ & & I & \beta_4 & & \\ & & & & & \\ & & & & I & \beta_{IM2} \\ & & & & & I \end{bmatrix} \tag{E-2}$$

where I is the identity matrix of order m. The square matrices α_i and β_i are determined as follows:

$$\alpha_2 = B_2 \quad \text{and} \quad \beta_2 = B_2^{-1}C_2 \tag{E-3}$$

$$\alpha_i = B_i - A_i\beta_{i-1} \quad \text{for } i = 3, 4, \ldots, IM1 \tag{E-4}$$

and

$$\beta_i = \alpha_i^{-1}C_i \quad \text{for } i = 3, 4, \ldots, IM2 \tag{E-5}$$

The system of equations given by (E-1) is now equivalent to

$$LY = R \tag{E-6}$$

where

$$Y = U\Delta Q \tag{E-7}$$

Rewriting (E-6), one has

$$\begin{bmatrix} \alpha_2 & & & & \\ A_3 & \alpha_3 & & & \\ & A_4 & \alpha_4 & & \\ & & & & \\ & & & A_{IM1} & \alpha_{IM1} \end{bmatrix} \begin{bmatrix} Y_2 \\ Y_3 \\ Y_4 \\ | \\ Y_{IM1} \end{bmatrix} = \begin{bmatrix} R_2 \\ R_3 \\ R_4 \\ | \\ R_{IM1} \end{bmatrix}$$

from which

$$Y_2 = \alpha_2^{-1}R_2 \tag{E-8}$$

and

$$Y_i = \alpha_i^{-1}(R_i - A_iY_{i-1}) \quad \text{for} \quad i = 3, 4, \ldots, IM1 \tag{E-9}$$

Equation (E-7) is expressed as

$$\begin{bmatrix} I & \beta_2 & & & & \\ & I & \beta_3 & & & \\ & & I & \beta_4 & & \\ & & & & & \\ & & & & I & \beta_{IM2} \\ & & & & & I \end{bmatrix} \begin{bmatrix} \Delta Q_2 \\ \Delta Q_3 \\ \Delta Q_4 \\ | \\ \Delta Q_{IM2} \\ \Delta Q_{IM1} \end{bmatrix} = \begin{bmatrix} Y_2 \\ Y_3 \\ Y_4 \\ | \\ Y_{IM2} \\ Y_{IM1} \end{bmatrix}$$

from which

$$\Delta Q_{IM1} = Y_{IM1} \tag{E-10}$$

and

$$\Delta Q_i = Y_i - \beta_i \Delta Q_{i+1} \quad \text{for} \quad i = IM1, IM2, \ldots, 3, 2 \tag{E-11}$$

A subroutine which employs the procedure outlined above is listed. This subroutine or other available subroutines may be used to solve the block-tridiagonal system obtained in previous chapters.

To validate this subroutine (or others which may be accessed), a simple problem is proposed. Obviously it is important to verify any external subroutines which are utilized in a program. In this problem the vector X, composed of two components, is the unknown. For simplicity, only three unknown vectors are considered. The problem is formulated as follows:

$$\begin{bmatrix} \begin{bmatrix} 1 & 2 \\ 3 & 4 \end{bmatrix} & \begin{bmatrix} 1 & 3 \\ 2 & 5 \end{bmatrix} & 0 \\ \begin{bmatrix} 9 & 8 \\ 7 & 6 \end{bmatrix} & \begin{bmatrix} 5 & 4 \\ 6 & 8 \end{bmatrix} & \begin{bmatrix} 9 & 11 \\ 13 & 15 \end{bmatrix} \\ 0 & \begin{bmatrix} 7 & 5 \\ 8 & 3 \end{bmatrix} & \begin{bmatrix} 3 & 2 \\ 4 & 5 \end{bmatrix} \end{bmatrix} \begin{bmatrix} \begin{bmatrix} X_1 \\ X_2 \end{bmatrix} \\ \begin{bmatrix} X_3 \\ X_4 \end{bmatrix} \\ \begin{bmatrix} X_5 \\ X_6 \end{bmatrix} \end{bmatrix} = \begin{bmatrix} \begin{bmatrix} 7 \\ 14 \end{bmatrix} \\ \begin{bmatrix} 46 \\ 55 \end{bmatrix} \\ \begin{bmatrix} 17 \\ 20 \end{bmatrix} \end{bmatrix}$$

The solution of the stated problem is

$$\begin{bmatrix} X_1 \\ X_2 \end{bmatrix} = \begin{bmatrix} X_3 \\ X_4 \end{bmatrix} = \begin{bmatrix} X_5 \\ X_6 \end{bmatrix} = \begin{bmatrix} 1 \\ 1 \end{bmatrix}$$

A test program and the required subroutines are provided below.

```
c-----------------------------------------------------------------
c--     TEST is the main program of an example which shows how a
c--     block-tridiagonal system of equations can be solved by
c--     calling the subroutine trisol.
c--
c--**   Notice: parameter (nn=60)  defines the maximum size of the
c--**           system.  nn should be greater than nb, where nb is the
c--**           upper index value of the system.
c--**           make sure to change the value of nn in subroutine
c--**           (invb) and (trisol) if your nb is greater than 60
c--**           Also, the maximum order of the matrices is limited
c--**           to be 5 for the time being, i.e., m should be less
c--**           than 5.  However, you may modify this by changing
c--**           the 5s in the common block /rlh/ to the value
c--**           you need.
c--**
c--**            output will be saved in file "tri.out"
c--
c-------------------------------------------------------------------
      PROGRAM TEST
      parameter (nn=60)
      common/rlh/ ca(nn,5,5),cb(nn,5,5),cc(nn,5,5),beta(nn,5,5),
     #   binv(5,5),cy(nn,5),cr(nn,5)
      open(unit=6,file='tri.out',status='unknown',form='formatted')
c
cc---read the lower (na) and upper (nb) index value of the system
        read(5,*)na,nb
c
cc---read the order of the matrices
        read(5,*)m
c
cc---read the elements of the matrix a (ca), i.e., the
cc---sub-diagonal matrix
        do 100 k=na+1,nb
        do 100 i=1,m
          read(5,*)(ca(k,i,j),j=1,m)
 100    continue
c
cc---read the elements of the matrix b (cb), i.e., the
cc---diagonal matrix
        do 110 k=na,nb
        do 110 i=1,m
          read(5,*)(cb(k,i,j),j=1,m)
 110    continue
c
cc---read the elements of the matrix c (cc), i.e., the
cc---super-diagonal matrix
        do 120 k=na,nb-1
        do 120 i=1,m
          read(5,*)(cc(k,i,j),j=1,m)
 120    continue
c
cc---read the elements of the r.h.s. r vectors (cr)
        do 130 k=na,nb
        do 130 i=1,m
          read(5,*)cr(k,i)
 130    continue
c
cc---call trisol to solve the system, solution of delta q is
cc---saved in cr
      call trisol(na,nb,m)
c
cc---print the solution
      do 140 k=na,nb
        do 140 i=1,m
        write(6,600)k,i,cr(k,i)
```

```
 140    continue
c-------------------------------------------------------------
 600    format(' cr(',i3,',',i3,')= ',f10.5)
        stop
        end
ccccccccccccccccccccccccccccccccccccccccccccccccccccccc
c      invb: is the supporting subroutine of trisol,
c      invb calculates the inverse matrix of cb(i,m,m)
c
c      input: cb(i,m,m),i,m
c      output: binv(m,m)
ccccccccccccccccccccccccccccccccccccccccccccccccccccccc
       subroutine invb(i,m)
       parameter (nn=60)
       common/rlh/  ca(nn,5,5),cb(nn,5,5),cc(nn,5,5),beta(nn,5,5),
     #  binv(5,5),cy(nn,5),cr(nn,5)
       dimension a(5,5),b(5)
       do 200 ii=1,m
            do 100 ia=1,m
            do 100 jb=1,m
100         a(ia,jb)=cb(i,ia,jb)
c
       do 150 jj=1,m
       if(ii.eq.jj)then
            b(jj)=1.0
        else
            b(jj)=0
       endif
150    continue
       call slnpd(a,b,d,m,5)
       do 170 jj=1,m
170    binv(jj,ii)=b(jj)
200    continue
       return
       end
ccccccccccccccccccccccccccccccccccccccccccccccccccccccccccccc
c      slnpd : gauss elimination method to solve linear equations
c              a*x=b, output of x is saved in b (i.e. the original
c              b will be eliminated)
c      slnpd is used to support subroutine invb
c
c      input: a : the l.h.s. n by n matrix
c             b : the r.h.s. n by 1 vector
c             n : the actual dimension used in a and b
c             nx: the dimension of a and b claimed in the calling
c                 subroutine
c      output:
c             b : the solved x vector
c             d : the determinant of diagonal matrix of a;
c                 has value only when a is singular
c
ccccccccccccccccccccccccccccccccccccccccccccccccccccccccccccc
       subroutine slnpd(a,b,d,n,nx)
       dimension a(nx,nx),b(nx)
       n1=n-1
       do 100 k=1,n1
            k1=k+1
            c=a(k,k)
            if(abs(c)-0.000001)1,1,3
1           do 7 j=k1,n
            if(abs(a(j,k))-0.000001)7,7,5
5           do 6 l=k,n
             c=a(k,l)
             a(k,l)=a(j,l)
6            a(j,l)=c
            c=b(k)
```

```
            b(k)=b(j)
            b(j)=c
            c=a(k,k)
            go to 3
 7          continue
            d=0.
            go to 300
 3          c=a(k,k)
            do 4 j=k1,n
 4          a(k,j)=a(k,j)/c
            b(k)=b(k)/c
            do 10 i=k1,n
            c=a(i,k)
                 do 9 j=k1,n
 9               a(i,j)=a(i,j)-c*a(k,j)
 10         b(i)=b(i)-c*b(k)
 100   continue
       if(abs(a(n,n))-0.000001)8,8,101
 8     write(6,600)
 600   format(2x,'... singularity in row')
 101   b(n)=b(n)/a(n,n)
       do 200 l=1,n1
        k=n-l
        k1=k+1
            do 200 j=k1,n
 200        b(k)=b(k)-a(k,j)*b(j)
       d=1.
       do 250 i=1,n
 250   d=d*a(i,i)
 300   return
       end
cccccccccccccccccccccccccccccccccccccccccccccccccccccccccccccc
cc
cc     trisol: block tri-diagonal matrix solver, solving for
cc             S*(Delta Q)=cr and solution of (Delta Q) is saved
cc             back to cr; i.e. original values of cr will be removed
cc
cc     Reference:     "Computational Fluid Dynamics - Volume I "
cc                     K.A. Hoffmann and S.T. Chiang, EES, 1998
cc
cc     the subdiagonal, diagonal, and superdiagonal of marix S
cc     are named as ca, cb, and cc.  the lower index of S
cc     is defined as na and the upper index of S is nb.  m
cc     is the dimension of the element matrix.  therefore S can be
cc     expressed as (the non-zero tri-diagonal is written vertically)
cc
cc  |                                                          |
cc  |       0.           cb(na,m,m)      cc(na,m,m)            |
cc  |   ca(na+1,m,m)     cb(na+1,m,m)    cc(na+1,m,m)          |
cc  |   ca(na+2,m,m)     cb(na+2,m,m)    cc(na+2,m,m)          |
cc  |       .                .               .                 |
cc  |       .                .               .                 |
cc  |       .                .               .                 |
cc  |   ca(nb-1,m,m)     cb(nb-1,m,m)    cc(nb-1,m,m)          |
cc  |   ca(nb,m,m)       cb(nb,m,m)       0.                   |
cc  |                                                          |
cc
cc     input: na,nb,m,ca,cb,cc,cr
cc
cc     output: cr
cc
cccccccccccccccccccccccccccccccccccccccccccccccccccccccccccccc
      subroutine trisol(na,nb,m)
      parameter (nn=60)
      common/rlh/ ca(nn,5,5),cb(nn,5,5),cc(nn,5,5),beta(nn,5,5),
     #  binv(5,5),cy(nn,5),cr(nn,5)
```

```
      do 200 i=na,nb
      if(i.eq.na)then
c
cc----determine the inverse matrix of  B(2) in eq. E-3
cc
           call invb(i,m)
c
cc----determine  beta(2) by using eq.  E-3
cc
           do 50 j=1,m
           do 45 k=1,m
           d=0.
                 do 40 l=1,m
40               d=d+binv(j,l)*cc(i,l,k)
45         beta(i,j,k)=d
50         continue
c
cc----determine y(2) from eq. E-8
cc
           do 55 j=1,m
           d=0.
                do 53 k=1,m
53              d=d+binv(j,k)*cr(i,k)
55         cy(i,j)=d
           go to 200
      endif
c
cc----determine the alpha(i)=B(i)-A(i)*beta(i-1) from eq. E-4
cc----note that the value of alpha(i) is saved into cb(i,j,k).
cc
      do 85 j=1,m
           do 75 k=1,m
           d=0.
                do 70 l=1,m
70              d=d+ca(i,j,l)*beta(i-1,l,k)
75         cb(i,j,k)=cb(i,j,k)-d
85    continue
c
cc----determine the inverse value of alpha(i) (for i=3,4,..iml)
cc
      call invb(i,m)
cc
cc----determine the beta(i) from eq. E-5
cc
      if(i.ne.nb)then
           do 100 j=1,m
                 do 95 k=1,m
               d=0.
                    do 90 l=1,m
90                  d=d+binv(j,l)*cc(i,l,k)
95             beta(i,j,k)=d
100        continue
      endif
cc
cc----determine the value of R(i)-A(i)*Y(i-1) in eq. E-9
cc
      do 115 k=1,m
      d=0.
           do 110 l=1,m
110        d=ca(i,k,l)*cy(i-1,l)+d
115   cr(i,k)=cr(i,k)-d
cc
cc---determine Y(i) from eq. E-9
cc
      do 120 j=1,m
      d=0.
```

```
          do 118 l=1,m
118       d=d+binv(j,l)*cr(i,l)
120        cy(i,j)=d
200   continue
cc
cc----determine the delta Q(iml) from eq. E-10
cc
      do 400 i=na,nb
      ix=nb-i+na
      if(ix.eq.nb)then
          do 210 k=1,m
210       cr(ix,k)=cy(ix,k)
          go to 400
      endif
cc
cc----determine the delta Q(i) from eq. E-11
cc
          do 300 j=1,m
          d=0.
              do 290 k=1,m
290           d=beta(ix,j,k)*cr(ix+1,k)+d
300       cr(ix,j)=cy(ix,j)-d
400   continue
      return
      end
```

Input Data File

```
2,4             < na,nb: lower and upper index of the system
2               < m:     order of matrices Ai, Bi, and Ci
9. 8.           < A3(1,1), A3(1,2)
7. 6.           < A3(2,1), A3(2,2)
7. 5.           <   A4(1,1), A4(1,2)
8. 3.           <   A4(2,1), A4(2,2)
1. 2.           < B2(1,1), B2(1,2)
3. 4.           < B2(2,1), B2(2,2)
5. 4.           <   B3(1,1), B3(1,2)
6. 8.           <   B3(2,1), B3(2,2)
3. 2.           <     B4(1,1), B4(1,2)
4. 5.           <     B4(2,1), B4(2,2)
1. 3.           < C2(1,1), C2(1,2)
2. 5.           < C2(2,1), C2(2,2)
9. 11.          <   C3(1,1), C3(1,2)
13. 15.         <   C3(2,1), C3(2,2)
7.              < R2(1)
14.             < R2(2)
46.             < R3(1)
55.             < R3(2)
17.             < R4(1)
20.             < R4(2)
```

Output

```
cr(  2,   1)=     1.00000
cr(  2,   2)=     1.00000
cr(  3,   1)=     1.00002
cr(  3,   2)=     1.00000
cr(  4,   1)=     0.99999
cr(  4,   2)=     1.00000
```

APPENDIX F:

Derivatives in the Computational Domain

Consider a function f, where it is required to determine its first- and second-order derivatives in the computational domain. The first-order derivatives are evaluated by using Equations (9-4) and (9-5). Recall that,

$$\frac{\partial}{\partial x} = \xi_x \frac{\partial}{\partial \xi} + \eta_x \frac{\partial}{\partial \eta} \tag{F-1}$$

$$\frac{\partial}{\partial y} = \xi_y \frac{\partial}{\partial \xi} + \eta_y \frac{\partial}{\partial \eta} \tag{F-2}$$

Therefore,

$$\frac{\partial f}{\partial x} = f_x = \xi_x f_\xi + \eta_x f_\eta$$

$$\frac{\partial f}{\partial y} = f_y = \xi_y f_\xi + \eta_y f_\eta$$

These equations may be rearranged by utilizing Equations (9-14) through (9-17). Hence,

$$f_x = J y_\eta f_\xi - J y_\xi f_\eta = J(y_\eta f_\xi - y_\xi f_\eta)$$

and

$$f_y = -J x_\eta f_\xi + J x_\xi f_\eta = J(x_\xi f_\eta - x_\eta f_\xi)$$

To determine the second-order derivatives, f_{xx} and f_{yy}, the following mathematical manipulations are performed:

$$\frac{\partial^2 f}{\partial x^2} = \frac{\partial}{\partial x}\left(\frac{\partial f}{\partial x}\right) = \frac{\partial}{\partial x}(\xi_x f_\xi + \eta_x f_\eta)$$

$$= \left(\xi_x \frac{\partial}{\partial \xi} + \eta_x \frac{\partial}{\partial \eta}\right)(\xi_x f_\xi + \eta_x f_\eta)$$

$$
\begin{aligned}
&= \xi_x \frac{\partial}{\partial \xi}(\xi_x f_\xi + \eta_x f_\eta) + \eta_x \frac{\partial}{\partial \eta}(\xi_x f_\xi + \eta_x f_\eta) \\
&= \xi_x^2 f_{\xi\xi} + \xi_x f_\xi \frac{\partial}{\partial \xi}(\xi_x) + \xi_x \eta_x f_{\xi\eta} + \xi_x f_\eta \frac{\partial}{\partial \xi}(\eta_x) \\
&\quad + \eta_x \xi_x f_{\xi\eta} + \eta_x f_\xi \frac{\partial}{\partial \eta}(\xi_x) + \eta_x^2 f_{\eta\eta} + \eta_x f_\eta \frac{\partial}{\partial \eta}(\eta_x)
\end{aligned}
$$

This equation is reduced to the following if relations (9-14) through (9-17) are used, i.e.,

$$
\begin{aligned}
\frac{\partial^2 f}{\partial x^2} &= J^2(y_\eta^2 f_{\xi\xi} - 2y_\xi y_\eta f_{\xi\eta} + y_\xi^2 f_{\eta\eta}) \\
&\quad + J y_\eta \left[f_\xi \frac{\partial}{\partial \xi}(\xi_x) + f_\eta \frac{\partial}{\partial \xi}(\eta_x) \right] \\
&\quad + (-J y_\xi) \left[f_\xi \frac{\partial}{\partial \eta}(\xi_x) + f_\eta \frac{\partial}{\partial \eta}(\eta_x) \right]
\end{aligned}
\tag{F-3}
$$

At this point, the derivatives of the metrics are determined as follows:

$$
\begin{aligned}
\frac{\partial}{\partial \xi}(\xi_x) &= \frac{\partial}{\partial \xi}(J y_\eta) = \frac{\partial}{\partial \xi}\left(\frac{y_\eta}{x_\xi y_\eta - x_\eta y_\xi}\right) \\
&= J^2[y_{\xi\eta}(x_\xi y_\eta - x_\eta y_\xi) - y_\eta(y_\eta x_{\xi\xi} + x_\xi y_{\xi\eta} - x_\eta y_{\xi\xi} - y_\xi x_{\xi\eta})]
\end{aligned}
$$

or

$$
\begin{aligned}
\frac{\partial}{\partial \xi}(\xi_x) &= J^2(x_\xi y_\eta y_{\xi\eta} - x_\eta y_\xi y_{\xi\eta} - y_\eta^2 x_{\xi\xi} - x_\xi y_\eta y_{\xi\eta} \\
&\quad + x_\eta y_\eta y_{\xi\xi} + y_\xi y_\eta x_{\xi\eta})
\end{aligned}
\tag{F-4}
$$

Similarly,

$$
\begin{aligned}
\frac{\partial}{\partial \xi}(\eta_x) &= -J^2(x_\xi y_\eta y_{\xi\xi} - x_\eta y_\xi y_{\xi\xi} - y_\xi y_\eta x_{\xi\xi} - x_\xi y_\xi y_{\xi\eta} \\
&\quad + x_\eta y_\xi y_{\xi\xi} + y_\xi^2 x_{\xi\eta})
\end{aligned}
\tag{F-5}
$$

$$
\begin{aligned}
\frac{\partial}{\partial \eta}(\xi_x) &= J^2(x_\xi y_\eta y_{\eta\eta} - x_\eta y_\xi y_{\eta\eta} - x_\xi y_\eta y_{\eta\eta} - y_\eta^2 x_{\xi\eta} \\
&\quad + y_\xi y_\eta x_{\eta\eta} + x_\eta y_\eta y_{\xi\eta})
\end{aligned}
\tag{F-6}
$$

$$
\begin{aligned}
\frac{\partial}{\partial \eta}(\eta_x) &= -J^2(x_\xi y_\eta y_{\xi\eta} - x_\eta y_\xi y_{\xi\eta} - y_\xi x_\xi y_{\eta\eta} - y_\xi y_\eta x_{\xi\eta} \\
&\quad + y_\xi^2 x_{\eta\eta} + x_\eta y_\xi y_{\xi\eta})
\end{aligned}
\tag{F-7}
$$

Substitution of Equations (F-4) through (F-7) into Equation (F-3) and rearranging terms yields:

$$\begin{aligned}\frac{\partial^2 f}{\partial x^2} &= J^2(y_\eta^2 f_{\xi\xi} - 2y_\xi y_\eta f_{\xi\eta} + y_\xi^2 f_{\eta\eta}) + \\ & J^3\left\{(y_\eta^2 y_{\xi\xi} - 2y_\eta y_\xi y_{\xi\eta} + y_\xi^2 y_{\eta\eta})(x_\eta f_\xi - x_\xi f_\eta) + \right. \\ & \left.(y_\eta^2 x_{\xi\xi} - 2y_\eta y_\xi x_{\xi\eta} + y_\xi^2 x_{\eta\eta})(y_\xi f_\eta - y_\eta f_\xi)\right\} \end{aligned} \tag{F-8}$$

Similarly,

$$\begin{aligned}\frac{\partial^2 f}{\partial y^2} &= J^2(x_\eta^2 f_{\xi\xi} - 2x_\xi x_\eta f_{\xi\eta} + x_\xi^2 f_{\eta\eta}) + \\ & J^3\left\{(x_\eta^2 y_{\xi\xi} - 2x_\xi x_\eta y_{\xi\eta} + x_\xi^2 y_{\eta\eta})(x_\eta f_\xi - x_\xi f_\eta) \right. \\ & \left.(x_\eta^2 x_{\xi\xi} - 2x_\xi x_\eta x_{\xi\eta} + x_\xi^2 x_{\eta\eta})(y_\xi f_\eta - y_\eta f_\xi)\right\} \end{aligned} \tag{F-9}$$

Now, consider the Laplacian,

$$\nabla^2 f = \frac{\partial^2 f}{\partial x^2} + \frac{\partial^2 f}{\partial y^2}$$

and substitute Equations (F-8) and (F-9). After simplification and collection of terms, we obtain

$$\begin{aligned}\nabla^2 f &= J^2\left[(x_\eta^2 + y_\eta^2) f_{\xi\xi} - 2(x_\xi x_\eta + y_\xi y_\eta) f_{\xi\eta}\right. \\ & \left. + (x_\xi^2 + y_\xi^2) f_{\eta\eta}\right] + J^3\left\{\left[(x_\eta^2 + y_\eta^2) y_{\xi\xi} - 2(x_\xi x_\eta + y_\eta y_\xi) y_{\xi\eta}\right.\right. \\ & \left. + (x_\xi^2 + y_\xi^2) y_{\eta\eta}\right](x_\eta f_\xi - x_\xi f_\eta) + \left[(x_\eta^2 + y_\eta^2) x_{\xi\xi}\right. \\ & \left.\left. - 2(x_\xi x_\eta + y_\eta y_\xi) x_{\xi\eta} + (x_\xi^2 + y_\xi^2) x_{\eta\eta}\right](y_\xi f_\eta - y_\eta f_\xi)\right\} \end{aligned}$$

Define the following:

$$\begin{aligned} x_\eta^2 + y_\eta^2 &= a \\ x_\xi x_\eta + y_\xi y_\eta &= b \\ x_\xi^2 + y_\xi^2 &= c \end{aligned}$$

Then

$$\begin{aligned}\nabla^2 f &= J^2(a f_{\xi\xi} - 2b f_{\xi\eta} + c f_{\eta\eta}) \\ & + J^3\left\{(a y_{\xi\xi} - 2b y_{\xi\eta} + c y_{\eta\eta})(x_\eta f_\xi - x_\xi f_\eta)\right. \\ & \left. + (a x_{\xi\xi} - 2b x_{\xi\eta} + c x_{\eta\eta})(y_\xi f_\eta - y_\eta f_\xi)\right\} \end{aligned}$$

and finally,

$$\nabla^2 f = J^2(af_{\xi\xi} - 2bf_{\xi\eta} + cf_{\eta\eta} + df_\eta + ef_\xi) \tag{F-10}$$

where

$$\begin{aligned} d &= J(y_\xi \alpha - x_\xi \beta) \\ e &= J(x_\eta \beta - y_\eta \alpha) \end{aligned}$$

and

$$\begin{aligned} \alpha &= ax_{\xi\xi} - 2bx_{\xi\eta} + cx_{\eta\eta} \\ \beta &= ay_{\xi\xi} - 2by_{\xi\eta} + cy_{\eta\eta} \end{aligned}$$

For illustration purposes, consider the elliptic system

$$\nabla^2 \xi = 0 \tag{F-11}$$

and

$$\nabla^2 \eta = 0 \tag{F-12}$$

We wish to transfer this system to a computational domain. To do so, Equation (F-10) is used—thus, $f = \xi$ for Equation (F-11). The required gradients in Equation (F-10) are

$$\begin{aligned} \xi_\xi &= \frac{\partial \xi}{\partial \xi} = 1 \\ \xi_\eta &= 0 \\ \xi_{\xi\xi} &= \frac{\partial}{\partial \xi}\left(\frac{\partial \xi}{\partial \xi}\right) = 0 \\ \xi_{\eta\eta} &= 0 \\ \xi_{\xi\eta} &= 0 \end{aligned}$$

Therefore, Equation (F-10) yields:

$$J^2 e = 0$$

or

$$J^3(x_\eta \beta - y_\eta \alpha) = 0$$

Similarly, $\nabla^2 \eta = 0$ yields

$$J^2 d = 0$$

or

$$J^3(y_\xi \alpha - x_\xi \beta) = 0$$

Since $J \neq 0$, then

$$x_\eta \beta - y_\eta \alpha \;=\; 0 \tag{F-13}$$

$$y_\xi \alpha - x_\xi \beta \;=\; 0 \tag{F-14}$$

Eliminating α from Equations (F-13) and (F-14) yields:

$$\beta(x_\xi y_\eta - x_\eta y_\xi) = 0$$

But

$$x_\xi y_\eta - x_\eta y_\xi = \frac{1}{J}$$

Thus,

$$\frac{1}{J}\beta = 0$$

Since, $J \neq 0$, then

$$\beta = 0$$

or

$$a y_{\xi\xi} - 2b y_{\xi\eta} + c y_{\eta\eta} = 0$$

We showed that $\beta = 0$ and, therefore, α must also be zero, which results in

$$a x_{\xi\xi} - 2b x_{\xi\eta} + c x_{\eta\eta} = 0$$

REFERENCES

[1-1] Shapiro, A. H., "The Dynamics and Thermodynamics of Compressible Fluid Flow," Vol. 1, The Ronald Press Company, New York, 1953.

[1-2] Anderson, J. D., "Modern Compressible Flow With Historical Perspective," McGraw-Hill Book Company, 1982.

[1-3] Hellwig, G., "Partial Differential Equations: An Introduction," B. G. Teubner, Stuttgart, 1977.

[1-4] Chester, C. R., "Techniques in Partial Differential Equations," McGraw-Hill Book Company.

[6-1] Boris, J. P., and Book, D. L., "Flux Corrected Transport I, SHASTA, a Fluid Transport Algorithm that Works," *Journal of Computational Physics,* 11, pp. 38–69, 1973.

[6-2] Boris, J. P., and Book, D. L., "Solution of the Continuity Equation by the Method of Flux Corrected Transport," *Journal of Computational Physics,* 16, pp. 85–129, 1976.

[6-3] Yee, H. C., "A Class of High-Resolution Explicit and Implicit Shock-Capturing Methods," NASA TM-101088, February 1989.

[6-4] LeVeque, R. J., "Numerical Methods for Conservation Laws," Lectures in Mathematics, Birkhäuser Verlag, 1990.

[6-5] Yang, H. Q., and Przekwas, A. J., "A Comparative Study of Advanced Shock-Capturing Schemes Applied to Burgers' Equation," *Journal of Computational Physics,* 102, pp. 139–159, 1992.

[6-6] Reddy, S., and Papadakis, M., "TVD Schemes and Their Relation to Artificial Dissipation" AIAA Paper 93-0070.

[6-7] Harten, A., "High Resolution Schemes for Hyperbolic Conservation Laws," *Journal of Computational Physics,* Vol. 9, pp. 357–393, 1983.

[6-8] Surely, P. K., "High Resolution Schemes Using Flux Limiters in Hyperbolic Conservation Laws," *SIAM Journal of Numerical Analysis,* Vol. 21, pp. 995–1011, 1984.

[6-9] Roe, P. L., "A Survey of Upwind Differencing Techniques," *Proceedings of the 11th International Conference on Numerical Methods in Fluid Dynamics,* June 1988.

[6-10] Roe, P. L., "Approximate Riemann Solvers, Parameter Vectors and Difference Schemes," *Journal of Computational Physics,* Vol. 43, pp. 357–372, 1981.

[6-11] Harten, A., Hyman, J. M., and Lax, P. D., "On Finite-Difference Approximations and Entropy Conditions for Shocks," *Comm. Pure and Appl. Mathematics,* Vol. 29, pp. 297–322, 1976.

[6-12] Grundall, M. G., and Majda, A., "Monotone Difference Approximations for Scalar Conservation Laws," *Math. Comp.,* Vol. 34, No. 149, pp. 1–21, 1980.

[7-1] Benton, E. R., and Platzman, G. W., "A Table of Solutions of the One-Dimensional Burgers Equation," *Quarterly of Applied Mathematics*, Vol. 30, July 1972.

[8-1] Aziz, K., and Hellums, J. D., "Numerical Solution of the Three-Dimensional Equations of Motion for Laminar Natural Convection," *Phys. Fluids,* Vol. 10, pp. 314–324.

[8-2] Mallinson, G. D., and deVahl, Davis , "Three-Dimensional Natural Convection in a Box: A Numerical Study," *J. of Fluid Mechanics,* Vol. 83, pp. 1–31, 1977.

[8-3] Aregbesola, Y. A. S., and Burley, D. M., "The Vector and Scalar Potential Method for the Numerical Solution of Two- and Three-Dimensional Navier-Stokes Equations," *J. Comput. Physics,* Vol. 24, p. 398, 1977.

[8-4] Wong, A. K., and Reizer, J. A., "An Effective Vorticity – Vector Potential Formulation for Numerical Solution of Three-Dimensional Duct Flow Problems," *J. of Comput. Physics,* Vol. 55, pp. 98–114, 1984.

[8-5] Chorin, A. J., "A Numerical Method for Solving Incompressible Viscous Flow Problems," *Journal Computational Physics,* Vol. 2, pp. 12–26, 1967.

[8-6] Harlow, F. H., and Welch, J. E., "Numerical Calculation of Time-Dependent Viscous Incompressible Flow of Fluid with Free Surface," Vol. 8, pp. 2182–2189, 1965.

[8-7] Gupta, M. M., and Manohar, R. P., "Boundary Approximation and Accuracy in Viscous Flow Computations," *J. of Comput. Physics,* Vol. 31, pp. 265–288, 1979.

[9-1] Thompson, J. F., Thames, F. C., and Mastin, C. W., "Boundary-Fitted Curvilinear Coordinate Systems for the Solution of Partial Differential Equations on Fields Containing Any Number of Arbitrary Two-Dimensional Bodies," NASA CR-2729, July 1977.

[9-2] Karamcheti, K., "Principles of Ideal-Fluid Aerodynamics," John Wiley, 1966.

[9-3] Steger, J. L., and Sorenson, R. L., "Automatic Mesh-Point Clustering Near a Boundary in Grid Generation with Elliptic Partial Differential Equations," *Journal of Computational Physics,* Vol. 33, 1979.

[9-4] Thompson, J. F., Warsi, Z. U. A., and Mastin, C. W., "Numerical Grid Generation Foundation and Applications," North-Holland, 1985.

[9-5] Steger, J. L., and Sorenson, R. L., "Use of Hyperbolic Differential Equations to Generate Body Fitted Coordinates," Numerical Grid Generation Techniques, NASA CP-2166, 1980.

[9-6] Hoffmann, K. A., Rutledge, W. H., and Rodi, P. E., "Hyperbolic Grid Generation Techniques for Blunt Body," Numerical Grid Generation in Computational Fluid Dynamics, Pineridge Press, 1988.

[9-7] Nakamura, S., "Marching Grid Generation Using Parabolic Partial Differential Equations," Numerical Grid Generation, North-Holland, 1982.

[9-8] Nakamura, S., "Noniterative Grid Generation Using Parabolic Difference Equations for Fuselage-Wing Flow Calculations," Proceedings of the Eighth International Conference on Numerical Methods in Fluid Dynamics, Springer-Verlag, 1982.

[A-1] Anderson, J. D., "Modern Compressible Flow with Historical Perspective," McGraw-Hill Book Company, 1982.

[D-1] Hoffmann, K. A., Chiang, S. T., Siddiqui, M. S., and Papadakis, M., "Fundamental Equations of Fluid Mechanics," EES, 1996.

[D-2] Fox, R. W., and McDonald, A. T., "Introduction to Fuid Mechanics," John Wiley & Sons, 1985.

[D-3] Kreider, J. F., "Principles of Fluid Mechanics," Allyn and Bacon, 1985.

[D-4] Shames, I. H., "Mechanics of Fluids," McGraw-Hill, 1962.

INDEX

The numbers 1, 2 or 3 preceding the page numbers refer to Volumes 1, 2 and 3, respectively.